COP 1

A11701 567328

OCT 29 '64

Planning for Man and Motor

L'homme et la Motorisation, Problèmes d'urbanisme

Planung für Mensch und Motor

To ROBERT MAXWELL M.C.
for his help and encouragement

Planning for Man and Motor

by

PAUL RITTER

M.C.D., B. Arch., A.R.I.B.A., A.M.T.P.I.

Kern County Free Library
Bakersfield, California

A PERGAMON PRESS BOOK

R 711

THE MACMILLAN COMPANY
NEW YORK
1964

COP 1

The Macmillan Company
60 FIFTH AVENUE
NEW YORK 11, N.Y.

This book is distributed by
THE MACMILLAN COMPANY
pursuant to a special arrangement with
PERGAMON PRESS LIMITED
Oxford, England

COPYRIGHT © 1964
PERGAMON PRESS LTD.

Library of Congress Card Number 63-22671

Set in 11 on 13pt Baskerville by
SANTYPE OF SALISBURY

Printed in Great Britain by
CRIPPLEGATE PRINTING CO. LTD., LONDON

Contents

I. Acknowledgements

I.1. Inspiration 2

I.2. Personal Help 2

I.3. Information and Permission 2

I.4. Official Recognition 2

I.5. Presentation 2

II. Context and Introduction

II.1. Function of the Book 4

II.1.1. Integration of Knowledge 4
II.1.2. Man and Motor 4
II.1.3. Prejudice and Conflict 4

II.2. Environment and Scientific Method 4

II.2.1. Limitation of Science at Present 4
II.2.2. Criteria 5
II.2.3. Creative Experiment 6
II.2.4. Responsibility 6

II.3. Power and Pattern of Plans 6

II.3.1. The Power of Plans 6
II.3.2. Profusion of Alternatives 7
II.3.3. The Opportunist Idealist 7
II.3.4. Basis for Rational Criticism 7

III. The Man–Vehicle Relationship

III.1. Summary of Man–Vehicle Characteristics 10

III.2. Physical Characteristics of Man and Vehicles 11
- III.2.1. Scale 11
- III.2.2. Relative and Cumulative Size 13
- III.2.3. Tactile Associations 13

III.3. Movement 14
- III.3.1. Speed and Range 14
- III.3.2. Route Capacity 14
- III.3.3. Momentum and Danger 16
  - III.3.3.1. Accidents in Traffic-Segregated areas 19
- III.3.4. Curvature of Path 21
- III.3.5. Mutual Frustration 22

III.4. Ecological Considerations 23
- III.4.1. Fumes and Poison 23
- III.4.2. Noise 24
- III.4.3. Vibrations 26

III.5. Sociological Considerations 27
- III.5.1. Influence of Private Cars on Social Patterns 27
  - III.5.1.1. Constructive Effects 27
  - III.5.1.2. Destructive Effects 27
- III.5.2. Research Project 27
  - III.5.2.1. Design of Research Project 27
  - III.5.2.2. Presentation of Results 27
  - III.5.2.3. Summary of Whole Survey 31
  - III.5.2.4. Survey at Stevenage 32
- III.5.3. Social Aspects of Public Transport 32
- III.5.4. Moral and Legal Considerations 32
- III.5.5. Crowds of People and Crowds of Cars 33
- III.5.6. Psychosocial Factors 34

III.6. Planning Solutions 34
- III.6.1. Functional Traffic Arrangements 34
- III.6.2. Character of Roads and Paths 35
- III.6.3. Graphic Analysis of Needs for Pedestrian Movement and Rest 36

IV. Needs of Man as an Organism

IV.1. Biological Function of Walking 38

IV.2. Qualities Required in the Pedestrian's Environment 39
- IV.2.1. Man Sized Spaces 41
- IV.2.2. Detail Matters 43
- IV.2.3. Endless Variety 44

IV.3. Study of an Urban Pedestrian Piazza 46
- IV.3.1. Minute-by-Minute Survey Recording People Entering and Leaving 46
- IV.3.2. Visual Analysis 47
- IV.3.3. Attempt to Explain why Piazza Works so Well 50

IV.4. Convenience 51

IV.5. Mechanical Aids 52
- IV.5.1. Pedal Cycles 52
- IV.5.2. Pedestrian Conveyors 53
- IV.5.3. Escalators and Lifts 54
- IV.5.4. Trolleys 54
- IV.5.5. "Pedestrians" 54
- IV.5.6. "Carveyors" 55

IV.6. Information 55

IV.7. Comparison of Path Systems 55

V. Needs and Nature of Vehicles

V.1. Man as Driver **58**
V.1.1. Perception 58
V.1.2. Road Furniture 59
V.1.3. Training 59
V.1.4. Emotional State 59

V.2. Motorization Forecasts, General **60**
V.2.1. Transportation Studies and Surveys 62
V.2.2. Vehicles and Town Space 65
V.2.3. General Economic Considerations 67

V.3. Routes of Vehicles **68**
V.3.1. Administration of Road Planning 68
V.3.2. Road Space, Shape, Noise and Dangers 68
V.3.3. Capacity 70
V.3.4. Speed 73
V.3.5. Types of Roads 75
V.3.5.1. General 75
V.3.5.2. Motorways and Freeways 75
V.3.5.3 Main Roads 79
V.3.5.4. Service Roads 79
V.3.5.5. Summary 79
V.3.6. Junctions, General 79
V.3.6.1. Junction Flow 81
V.3.6.2. Junctions of Three Routes 81
V.3.6.3. Junctions of Four Routes 82

V.4. Parking **84**
V.4.1. General and Basic Requirements 84
V.4.2. Parking Lots, Ground Level only 84
V.4.3. Multilevel Parking Garages 85
V.4.4. Mechanical Parking 88

V.5. Functions of Private Vehicles **88**
V.5.1. Journey to Work 89
V.5.2. Schools and Universities 89
V.5.3. Shopping 89
V.5.4. Visiting 90
V.5.5. Sport and Cultural Events 90
V.5.6. Week-ends and Half Day Holidays 90
V.5.7. Long Holidays 90
V.5.8. Interpretation of Statistics 91

V.6. The Forms of Private Vehicles **91**
V.6.1. Factors Affecting Form 91

V.7. Commercial Vehicles, Vans and Lorries **93**
V.7.1. Numbers and Growth 93
V.7.2. Characteristics of Size, Speed and Flow 93
V.7.3. Effect on Flow 93

V.8. Public Transport **94**
V.8.1. General Attitude 94
V.8.2. Principles of Public Transport Systems 95
V.8.3. Types of Rapid Transit 95
V.8.3.1. Buses 95
V.8.3.2. Duo Rail Systems 96
V.8.3.3. Monorail 96
V.8.3.4. Underground Railways 98
V.8.3.5. Self-Guided Bus Trains 99
V.8.4. Comparison of Typical Features 100
V.8.5. Examples 101
V.8.5.1. Philadelphia 101
V.8.5.2. St. Louis 102
V.8.5.3. Boston 103
V.8.5.4. San Francisco 103

V.9. Glossary of Terms **105**

VI. New Towns with Traffic Segregation

VI.1. Basic Requirements **108**

VI.2. Motor Traffic in New Towns, General **108**
- VI.2.1. Parking in New Town Centres 109
- VI.2.2. Cumbernauld, Research into C.A./Retailing and Parking 110

VI.3. Fully Motorized Town Centre for 250,000 **111**

VI.4. Town Design, Based on Walking Distances **111**
- VI.4.1. Density and Town Size 111
  - VI.4.1.1. Town Size, Population and Density Study 111
  - VI.4.1.2. Town Density Diagrams 112
- VI.4.2. Comparison of Towns and their Walking Distances 113
- VI.4.3. Town Shape 114

VI.5. Land Use **114**

VI.6. Open Space **114**

VI.7. The Neighbourhood Concept **115**

VI.8. The Principles of Growth **115**

VI.9. Examples **117**
- VI.9.1. Cumbernauld, Scotland 117
- VI.9.2. Hook, England 121
- VI.9.3. Erith, England 123
- VI.9.4. Toulouse-le-Mirail, France 126
- VI.9.5. Vällingby, Sweden 128
- VI.9.6. Farsta, Sweden 130
- VI.9.7. Cologne, Germany 133
- VI.9.8. Sabende, Guinea 134
- VI.9.9. Sennestadt, Germany 136
- VI.9.10. Neu-Winsen, Germany 138
- VI.9.11. Alkmaar, Holland 139
- VI.9.12. Chandigarh, India 140
- VI.9.13. Kitimat, Canada 143
- VI.9.14. Tapiola, Finland 144
- VI.9.15. Sputnik, Russia 146
- VI.9.16. Cambridge Village, Cambs., England 147
- VI.9.17. Reston, nr. Washington, U.S.A. 148

VII. Urban Renewal

VII.1. General Policy **152**
- VII.1.1. The Case of New York 153
- VII.1.2. The Case of San Francisco 154
- VII.1.3. The Case of Detroit 154

VII.2. Design for Urban Renewal **155**
- VII.2.1. Basic Principles 155
- VII.2.2. Horizontal Traffic Segregation 157
  - VII.2.2.1. Segregation by Time 158
- VII.2.3. Vertical Segregation, Pedestrian Above 158
- VII.2.4. Vertical Segregation, Vehicles Above Ground Level 159
- VII.2.5. Vertical Segregation, Vehicles at Ground Level, Pedestrians Below 159

VII.3. Urban Roads **159**
- VII.3.1. Improving Existing Road Systems 159
- VII.3.2. Introduction of Motorways or Freeways 160
- VII.3.3. Motorways in the Townscape 161

VII.4. Parking **162**
- VII.4.1. Parking Estimates 162
- VII.4.2. Other Factors Affecting Parking 163
- VII.4.3. Kerbside Parking 164
- VII.4.4. Parking Meters 164

VII.5. Economic Considerations, General **165**
- VII.5.1. Cost of Full Motorization 165
- VII.5.2. Comparative Cost of Commuter Journeys 166
- VII.5.3. Cost of Traffic Improvements 166
- VII.5.4. Cost of Congestion 167
- VII.5.5. Economics of Pedestrian Shopping Streets 167
- VII.5.6. Comparative Values 168

VII.6. Examples **170**
- VII.6.1. Coventry, Development and Traffic Plan 170
  - VII.6.1.1. Coventry, Shopping Precinct 172
- VII.6.2. Stockholm, Development and Traffic Plan 173
  - VII.6.2.1. Stockholm, Sergelgaten 174
- VII.6.3. Philadelphia, Development and Traffic Plan 176
  - VII.6.3.1. Market East Section 177
- VII.6.4. Liverpool, Development and Traffic Plan 180
  - VII.6.4.1. Ravenseft Area 182
  - VII.6.4.2. Civic and Social Centre 184
  - VII.6.4.3. Paradise Street—Strand Area 184
- VII.6.5. Newcastle, Development and Traffic Plan 186

VII.6.5.1. Central Area, Overall Development 187
VII.6.6. City of Victoria, Hong-Kong, Central Area 188
VII.6.7. London, Development and Traffic Plan 189
VII.6.7.1. Trafalgar Square Area 191
VII.6.8. Fort Worth, Texas, U.S.A., Central Area Reconstruction 192
VII.6.9. Nottingham, Central Area and Traffic Plan 194
VII.6.10. Banbury, Central Area Redevelopment 197
VII.6.11. Swindon, Central Area Redevelopment 198
VII.6.12. Västerås, Sweden, Development and Traffic Plan 199
VII.6.13. Bishop's Stortford, Central Area Redevelopment 201
VII.6.14. Sutton Coldfield, Central Area Redevelopment 202
VII.6.15. High Wycombe, Central Area Redevelopment 205
VII.6.16. Beeston, Central Area Redevelopment 206
VII.6.17. Horsholm, Denmark 208
VII.6.18. Chester City, Central Development 208
VII.6.19. Andover, Town Development 210
VII.6.20. Lijnbaan, Rotterdam, Holland 212
VII.6.21. South Bank Development, London 213
VII.6.22. Stevenage, Town Centre 214
VII.6.23. Stevenage, Shephall Centre 215
VII.6.24. Treppenstrasse, Kassel, Germany 215
VII.6.25. Pedestrian Shopping Street, Essen, Germany 215
VII.6.26. Greyfriars Redevelopment, Ipswich 216
VII.6.27. Sea-Front Site, Brighton 216
VII.6.28. Strøget, Shopping Street, Copenhagen, Denmark 216
VII.6.29. Ellor Street, Comprehensive Redevelopment Area, Salford 218
VII.6.30. Hammersmith Central Development, London 219
VII.6.31. Piccadilly Circus, London 220
VII.6.32. Knightsbridge Green, London 220
VII.6.33. Basildon New Town, Main Centre 221
VII.6.34. Elizabeth New Town, Australia, Main Centre 222

VIII. Traffic Segregation in Residential Areas

VIII.1. Scale of Livability Applied to Radburn Principles **224**
VIII.1.1. Psychosomatic Criteria, Radburn Possibilities 224
VIII.1.2. Psychosocial Criteria 225
VIII.1.3. Ecological Criteria 227
VIII.1.4. Economic and Administrative Criteria 227

VIII.2. Costs **228**
VIII.2.1. Comparison of Costs of Streets and Sewer Works 228
VIII.2.2. Economic Analysis 229

VIII.3. Dwelling Plans **232**
VIII.3.1. Analysis of Relationship of Dwelling Plans to Access 232
VIII.3.2. Car Storage 235
VIII.3.3. Cost and Density 235

VIII.4. Paths, System of Routes **236**
VIII.4.1. Analysis 236

VIII.5. Plot Shapes **238**
VIII.5.1. Houses 238
VIII.5.2. Flats and Deck Housing 239

VIII.6. Residential Feeder Roads **240**
VIII.6.1. Size and Frequency of Junctions 240
VIII.6.2. Feeder Road Design, General 240
VIII.6.2.1. Culs-de-sac 240
VIII.6.2.2. Loops 240
VIII.6.3. Basic Alternatives 241
VIII.6.3.1. Culs-de-sac 241
VIII.6.3.2. Loops 242
VIII.6.4. Examples 243
VIII.6.4.1. Pioneer Schemes of the Early Fifties 243
VIII.6.4.2. Projects of the Mid Fifties 244
VIII.6.4.3. Projects of the Late Fifties and Early Sixties 246
VIII.6.4.4. Dimensions of Service Culs-de-sac 248
VIII.6.4.5. Links between Culs-de-sac and Paths 249
VIII.6.5. Mixing Pedestrians and Vehicles in Culs-de-sac 250

VIII.7. Superblock Design **252**
VIII.7.1. Basic Factors 252
VIII.7.2. Application in Redevelopment 253

VIII.8. Underpasses and Bridges **253**
VIII.8.1. Position and Importance 253
VIII.8.2. Design and Cost Alternatives 255

VIII.9. Open Space 256

VIII.10 Contours 256

VIII.11. Street Names and Numbers 257

VIII.12. Examples of Traffic-Segregated Residential Areas 258
VIII.12.1. Carbrain, Cumbernauld, nr. Glasgow 258
VIII.12.2. Baldwin Hills Village, Los Angeles 260
VIII.12.3. Caversham, Berks. 262
VIII.12.4. Bron-y-Mor, Wales 263
VIII.12.5. Huntingdon, L.C.C. 265
VIII.12.6. Jackson Estate, Letchworth 266
VIII.12.7. Ilkeston Road, Beeston, Notts. 266
VIII.12.8. Albertslund, Copenhagen 267
VIII.12.9. Eastwick, Philadelphia 268
VIII.12.10. Clements Area, Haverhill, Suffolk 269
VIII.12.11. Cité Jardin, Montreal 271
VIII.12.12. Marly-le-Roi, Paris 273
VIII.12.13. Baronbackarna, Orebro, Sweden 274
VIII.12.14. Lafayette Park, Title I, Detroit 276
VIII.12.15. Biskopsgaden, Göteborg 277
VIII.12.16. Willenhall Wood I, Coventry 278
VIII.12.17. Kildrum, Cumbernauld 280
VIII.12.18. Elm Green I and II, Stevenage 281
VIII.12.19. Bedmont, Hertfordshire 282
VIII.12.20. Inchview, Prestonpans, Edinburgh 283
VIII.12.21. Parkleys, Ham Common, London 285
VIII.12.22. Park Hill, Sheffield 286
VIII.12.23. Barbican, London 288
VIII.12.24. Flemingdon Park, East York Toronto 289
VIII.12.25. South Carbrain, Cumbernauld 290
VIII.12.26. Oakdale Manor, North York, Toronto 294
VIII.12.27. Kentucky Road, Toronto 295
VIII.12.28. Solna, Sweden 296
VIII.12.29. Spon End, Coventry 297
VIII.12.30. Married Quarters, British War Office, (A) 297
VIII.12.31. Married Quarters, British War Office, (B) 298
VIII.12.32. Rowlatts Hill, Leicester 298
VIII.12.33. Brandon Estate, London 299
VIII.12.34. Royal Victoria Yard, Deptford, London 299
VIII.12.35. Brunswick Redevelopment Area, Manchester 300
VIII.12.36. Primrose Hill, Birmingham 300
VIII.12.37. Almhög, Nydala, Hermodsdal, Malmö 300
VIII.12.38. Mellanhaden, Malmö 300
VIII.12.39. Washington, Durham 301
VIII.12.40. Les Buffets, Fontenay-aux-Roses, Paris 304
VIII.12.41. Hillfields, Coventry 304
VIII.12.42. Highfields, Leicester 304a

IX. Residential Renewal

IX.1. General Principles 306
IX.1.1. Research into Grid Iron Plan Accidents 306

IX.1. Examples 308
IX.2.1. Street Improvement Scheme, Pershing Field Area, Fulton Neighbourhood, Minneapolis, U.S.A. 308
IX.2.2. Mill Creek Redevelopment Area, Philadelphia City Planning Commission, 1954 308
IX.2.3. British Bye-Law Housing, Nottingham 309
IX.2.4. Clearing Space at the Back of Flats 310
IX.2.5. Interwar Housing, Nottingham 310
IX.2.6. Transformation of a City, Chicago 312

X. History of Traffic Segregation

X.1. Summary 314

X.2. History of Traffic Segregation, Illustrated 314
X.2.1. Examples pre-dating Petrol Engines 314
X.2.2. Planning for "Life in Spite of the Motor Car" 319
X.2.3. Planning for Traffic Segregation in Europe 322

X.3. Chart Summary of Dates in the History of Planning for Man and Motor 328

XI. Bibliography

XI.1. Books 331

XI.2. Reports, Articles, Theses 333

XII. Appendices

XII.1. Conversion Tables 347

XII.2. Abbreviations 347

XII.3. Schema of Transportation Study 348

XII.4. Index 350

XII.5. French Translation 357
XII.5.1. Contents List 357
XII.5.2. Context and Introduction 366

XII.6. German Translation 371
XII.6.1. Contents List 371
XII.6.2. Context and Introduction 380

I. Acknowledgements

I. 1. Inspiration. 2

I. 2. Personal Help. 2

I. 3. Information and Permission. 2

I. 4. Official Recognition. 2

I. 5. Presentation. 2

I. Acknowledgements

When a project has taken twelve years to come to fruition it is very difficult to acknowledge adequately all the inspiration, help, encouragement and knowledge received during its development. To name all is impossible. To mention some is not intended to belittle others.

I. 1. Inspiration

I am grateful to Professor Gordon Stephenson, J. L. Womersley, R. E. M. McCaughan, Lewis Mumford, Professor Henrik Infield, Clarence Stein, and J. B. Bakema, for their inspiration, and, for any capacity for integration shown in this book, to the work of the late Dr. Wilhelm Reich.

The very sources of my strength have been my wife and six children. To them I owe thanks for tolerance of my preoccupation and for support when things went wrong.

I. 2. Personal Help

My special thanks are due to Derek Lyddon for reading the MS, giving me valuable advice and stimulating contact over many years, and to John Middleton-Harwood, for information on transportation engineering and for reading the relevant sections of the MS.

For personal help with aspects of the MS (beyond the credited examples given in the book), I am indebted to the following: Bruno Alm, J. B. Bakema, Professor Bendsten, H. Blachnicki, Colin Buchanan, Wilfred Burns, G. R. Chadwick, Dr. Dennis Chapman, Geoffrey Copcutt, Oliver Cox, Michael Dower, R. Falk, Maxwell Fry, Sir Donald Gibson, James A. Grey, Walter Gropius, Werner Grossman, Sir William Holford, Lennart Holm, Professor John Howard, Paul Kirby, Arthur Ling, Professor Sune Lindström, Jack Lynn, Hugh Morris, Christopher Millard, Lewis Mumford, Erik Petersen, Professor Steen Eiler Rasmussen, Professor H. B. Reichow, Professor Lloyd Rodwin, Sven Thiberg, Paulette Thivierge, Robert Schoenauer, H. Skaarup, Mieczyslaw Skrzypczak-Spak, Dr. R. J. Smeed, W. K. Smigielski, Ivor Smith, Mrs. Solomon, Professor W. H. Sprott, Professor Stephenson, H. R. Wedgewood, Hugh Wilson, J. L. Womersley, and Professor Wortmann.

I acknowledge the aid given by Robin Birch, David Brindle, Sylvia Cooper, Tony Gwilliam, Garry Gunn, Ian Hunt, Eric Lee, John Morton, Erica Ritter and John York, and many of my present and past students.

Thanks are due for the reliable photographic work by Mr. Middleton, and Midland Engraving Ltd., and to my assistant and secretary, Sheila Smith for typing and retyping the MS with intelligence, patience and good temper.

I. 3. Information and Permission

I am obliged to the numerous authorities, societies and people who have sent me material and to all those photographers who have given permission to reprint, and who have supplied the pictures not taken by myself.

I acknowledge the permission given to quote from the following: The *Traffic Quarterly*; *Architectural Design*; Publications of the Road Research Laboratory; *Architects' Journal*; California Highways and Public Works; *Journal of the Town Planning Institute*; *Traffic Engineering and Control*, and all other sources of excerpts referred to in the text. The indulgence shown by many libraries, who have kindly lent me books for lengthy periods, is appreciated.

I. 4. Official Recognition

I am indebted to the Civic Trust and its Director, Colonel Kenneth Post, for the generous financial help with the preparation of the manuscript; to the Royal Institute of British Architects for awarding me the Henry L. Florence Book Prize; to the Town Planning Institute for the President's Prize.

I. 5. Presentation

Standardization and cross indexing are designed to facilitate the use of this book. Most plans, however, have not been redrawn. Interest springs from the manifold character as imparted by different offices.

Throughout the book red indicates the pedestrian's realm. The variety of plan used demands varied use of the colour: either it shows the actual areas the pedestrian occupies; or the buildings bordering these; or the outline of them; or paths with public buildings. All these and other alternatives are used to make clear realistically or symbolically the extent of the environment for people on foot.

Roads are indicated by black, dwellings usually by grey, car storage by dark grey, unless a key shows otherwise. A large number of the plans are adaptions of the multi-coloured panels of the International Traffic Separation Travelling Research Exhibition designed by the author.

Six standard scales make comparison of dimensions easy. Metric equivalents are given in brackets after all measurements, and French and German translations of the extended Contents List and the Introduction appended. At the end of each section, blank pages are provided to allow the reader to make notes and comments where he will find them again in context.

II. Context and Introduction

II. 1. Function of the Book. 4
II. 1.1. Integration of Knowledge. 4
II. 1.2. Man and Motor. 4
II. 1.3. Prejudice and Conflict. 4

II. 2. Environment and Scientific Method. 4
II. 2.1. Limitations of Science at Present. 4
II. 2.2. Criteria. 5
II. 2.3. Creative Experiment. 6
II. 2.4. Responsibility. 6

II. 3. Power and Pattern of Plans. 6
II. 3.1. The Power of Plans. 6
II. 3.2. The Profusion of Alternatives. 7
II. 3.3. The Opportunist Idealist. 7
II. 3.4. Basis for Rational Criticism. 7

II. Context and Introduction

II. 1. Function of the Book

II. 1.1. *Integration of Knowledge*

Architects, planners, engineers, developers, administrators and legislators all plan. Education and textbooks do not describe, stress, recommend or even define fundamental, common functional principles. A proper basis for decisions and judgements, the starting point, matrix and meeting ground for interprofessional activity, is missing. No wonder there is, in practice, little joint concerted action.

It is the intention of this book not only to dispense, but to integrate knowledge and methods of approach and to illustrate the fundamental elements common to the various professions. Taking as the primary motive for all of them the creation of a good environment, the student can see what is directly relevant to him and, in the context of the book as a whole, how this forms part of the larger pattern.

II. 1.2. *Man and Motor*

The motor car is the most versatile and useful vehicle invented by man. In its present primitive state engines and fuels still produce poisonous fumes, but developed versions will appear and only a positive attitude, planning for the optimum use of vehicles, is rational.

No development is likely to supersede the wheeled vehicle except the hovercraft. Although quality of surface would become unimportant, this would require similar space and route provision as a car, as it is connected to the ground by its air cushion. Air transport will not be important until electronic devices can guide, along individual routes, at varying heights, some kind of aerodynamic invention which is practicable in the ways the helicopter of today is not.

The following chapters analyse the different nature of man and vehicle, so that their respective needs can be judged sensitively. The aim is an ecologically harmonious environment for man in which an efficient use of the vehicle plays a crucial part.

This publication is timely not only because the need for such information is pressing but also because at this point in time, 1963, all the basic solutions which seem theoretically possible have been tried in practice and can therefore be discussed and illustrated. Whether regarded as a "domestic animal" (Smigielski), or a "two toned tyrannosaurus" (Chermayeff), the universal ownership, and/or use, of cars demands a fresh approach to almost every environmental problem, from the plan for a house to the zoning of regions.

II. 1.3. *Prejudice and Conflict*

Without an organized field of knowledge, lacking a frame of reference, the picture is further confused by the emotions stirred up through continual friction in areas crowded with men and vehicles but not designed for such mixed use.

The resentment of pedestrian and cyclist, unable to breathe fresh air and forced off the street, is rational. The town lover's horror at the fitted carpet of cars covering every corner of the townscape is rational. The driver's annoyance is likewise justified, although perhaps the same person as before, but now at the wheel. Infuriated this time by congestion, when he wishes to make use of the great speed his conveyance offers, and later, arriving at his destination, at having to keep on driving merely because there is neither room nor permission to stop.

It does not help to create a clear picture by threatening one faction with a Los Angeles type townscape and the other with forecasts of long, tiring walking distances. Fear and hate do not encourage reason. Only the recognition of the pattern underlying the dilemma of our present position and an understanding of the needs and proper functions of man and his superbly useful aid, the car, will lead to the bold, properly balanced and original working solutions required. Those who are unable to see that there is a growing and genuinely frightening problem, or who become used to slaughter and mediocrity in human life, or regard adopted solutions as mere fashions to copy: such people need shocking out of their complacency.

II. 2. Environment and Scientific Method

II. 2.1. *Limitations of Science at Present*

What is a good environment? What are the fundamental values which should guide us in planning? We don't know, because the tools for finding out are not developed.

It is a widespread fallacy that sociology should provide the answers to these questions. This is a dangerous mistake. Faith is pinned onto a discipline which specifically excludes value judgements and therefore also recommendations for design. Even if the more fruitful approach of anthropology is included, as by the Institute of Community Studies, it is still a matter of studying only what exists, what we often know to be ill conceived beforehand through experience. Original ideas are excluded. Sociology in its present form is not meant to give information to guide planners in the subtle, qualitative problems on which information and inspiration is urgently required.

The position of the science is frequently criticized from within. A thorough work, John Madge's book *The Tools of Social Science*,(i) says, " ... in no science is the pursuit of objective knowledge more futile than in social science". And later on suggests that, " ... The social survey is a movement as well as a fact finding expedition". Yet he knows his views find no echo in British academic sociology. Such work has " ... a rather depressed status", and has used " ... obsolete techniques".

The transportation engineer is in a similar plight. Although he lacks a reliable or tested framework of knowledge, he is called upon to make decisions involving an immense number of variables, the effects of which are at the moment entirely incalculable, often out of his control and beyond the costs his community can afford. Rational transportation of all things required in a given area is a complex task, requiring much original thought. We must conclude that sciences and technologies, with their present ineffective methods for studying man and his environment, cannot help adequately.

II. 2.2. *Criteria*

Doctors know that good food, fresh air and exercise are biologically desirable. There is no counterpart to guide us in the question of what is healthy sociologically, even if one is a conformist. The planner, starting from common sense, and using the knowledge available outside those fields blighted by academic sterility, must form his own picture, acting as philosopher, prophet and social therapist combined.

The need for opportunities to co-operate, for all age groups, in work and ceremony, play, friendship or love has to be deduced from common awareness and the works of people like Professor H. Infield. (See Bibliography.)

The need for privacy is stressed *ad nauseam*. It is a basic need, to be sure. Clever solutions are possible but, essentially, a curtain and a thick wall are all that are needed. It is often a matter of finding privacy within a house and a family, more difficult and often best solved by getting out. This leads to provision for other social groups than the family.

To plan for co-operation is more important by far. It is complex, but the possibilities are immense, the surface of the great mass of opportunity is hardly scratched. This, rather than any other field, will influence the next development of man from the sociological point of view of organization into groups, of citizen participation and "work democracy" (Reich).

(i) MADGE J. *The Tools of Social Science*. London 1953.

Chermayeff, in a penetrating speech, puts movement and mobility into perspective in his vision of the future:

"*All this presupposes a very careful scrutiny of the whole spectrum of experience as related to mobility, speed, and dimension of the space inhabited. It is only then that we shall be able to reclaim, at one end of the spectrum, the pedestrian world: more than the pedestrian world—the world of the man at rest in the fullest sense of the word. When we lie on a lawn, close to a crocus, we have a worm's-eye view of the world and it is a very beautiful world. When we go up in a jet plane, in another scale, we don't wish to deny anyone the pleasure of this astonishing miracle, this entirely new view of the universe and the joys of the landscape. All the way from the crocus to the skyscape I wish to have it, and I would rather have all these pleasures than merely two cars in every garage.*" ...

... "*I am absolutely convinced that it is not the 'city' that is here to stay but conglomerations of people who get together for very good reasons. I hope that we will cultivate the getting together for old and new purposes but not on the horrid 'togetherness' level where everybody really begins to hate the others. We don't get together any longer necessarily because we are neighbors; we get together because we have something in common or a task to perform. I am convinced that we are going to regroup, largely because of the compulsions of the rise in population, the rise in technology, and the infinite complexity of our systems. And I don't wish to accommodate any transportation system at the expense of other values.*"(i)

The picture which emerges for the future includes the selective, practicable and rational use of the private vehicle and also greater citizen participation in the planning process and in the use of the environment planned to extend the social groups of today. Society displays a growing mobility which requires accommodation that facilitates the feeling of belonging and quickens the process of striking roots.(ii)

Yet, perhaps the full circle is nearly turned: Reston, the first(iii) balanced New Town for the U.S.A., planned in 1963, is based on the conviction that "Americans want the stability of belonging to one community for a life time. They are tired of rootlessness.". The great mobility Westerners show may be a symptom of their desire to gain rapid material advantages. It is not an end in itself. To stay put and prosperous may be the real luxury of life in an age of plenty, given highly developed means of travel and communications.

(i) CHERMAYEFF S. *The New Nomads*. T. Q. Saugatuck. April 1960.

(ii) WHYTE W. H. *The Organisation Man*. London 1957.

(iii) See Reston, p.148.

II. 2.3. *Creative Experiment*

It takes the whole man, the feeling philosopher, not only the objective scientist in each one of us, to take properly into consideration functional and ethical implications, imagination, aspirations. The planner must be a man who can imagine, visualize and subscribe to the vision of another, and then work for its realization.

As Madge has said, the scientific method limits itself unnecessarily to the isolation of factors and their analysis. To exclude the subjective element of the experimenter is an unvarying aim. What is particularly required in planning is an equally effective but quite opposite approach, still valid as a scientific method of hypothesis and experiment. Normally the hypothesis is tested numerically, but this again is not valid in planning, where original advantages may occur as a result of experiment.

Let us regard the statement, "I can throw this stone across the river," as a hypothesis. The experiment then includes the ingenuity to find the narrowest spot, use the wind, find the right stone etc., etc. In other words, the various factors are not isolated. As many as can be conducive to the correctness of the hypothesis at any one time, and to the success of the experiment in throwing the stone across, are eagerly sought.

Such is the basic concept of the Creative Experiment which I have formulated as a means of fruitfully introducing and testing new ideas in planning. The results of such an experiment will be of substance and meaning. Its process involves participants dynamically, improving and developing methods and environment all the time. The relevance and interaction of all factors are of major importance, and are observed and modified whereever it may help to do so. The theory, sometimes adopted, that people ought not to know that they are part of an experiment, is pathetically ineffective and dangerous, and it has no place in the creative experiment, where this very awareness is a powerful aid to success.(i) Success in such experiments means the opportunity for a good life for all age groups, with progressively less and less frustration and misery. One cannot feel objective about this. The planner must desire it strongly and strive for it by all means— just like getting the stone across the river.

Those who plan should analyse needs and resources, create imaginatively according to the analysis, organize the execution of the design opportunely, and instigate communication to initiate its proper use. Having come to a conclusion regarding the underlying principles governing a project, the emphasis shifts to the best ways of implementing them. The principles are the hypothesis: the creative experiment lies in their successful realization.

It is important that as many schemes as possible are taken into account in such experiments and the results carefully collected to swell the small volume of useful knowledge.

II. 2.4. *Responsibility*

This is the century of the common man. The luxury of planning is to be for all. The motor vehicle is to serve every family. All this is a challenge of immense proportions. It is not a matter of mere accommodation, not a matter simply of good looks or smooth traffic flow. The task is to provide an environment conducive to better satisfaction of primary needs. The positive attributes of a wonderful mobility given efficient vehicular service, and the planning of communities taking into account the advantages of co-operation, by those who wish to co-operate, give an increase in the choices available to each individual, extending his capacity and his opportunities. It is reasonable to anticipate larger expenditure on our environment, per person, than we have hitherto seen. To have as a sole aim the reduction of rates is as yet as wide-spread as it is short-sighted.

The subject brings with it a heavy responsibility. In the U.S.A. a number of factors have resulted in solutions as a whole not relevant or acceptable to Europe nor indeed to many other, under-developed, countries. We must assess anew the needs and functions of man and motor in the context of each environment. "Under developed countries" [and new towns!] "can achieve traffic order without the pioneer's pains."(i) One country after another becomes richer and more thronged with people and cars. Each day the number of the earth's inhabitants grows by over 100,000: a city a day. And yet, research on all aspects of environment, the world over, is abysmally small and ineffective.

II. 3. Power and Pattern of Plans

II. 3.1. *The Power of Plans*

The plan is a powerful instrument. Convenience can reduce the need and use of private vehicles in the centres of towns. An environment, sheltered, inviting, can materially increase walking and cycling or preclude them by unfavourable conditions. A plan

(i) Mrs. Demers's report of Park Hill, Sheffield, (1962) and the authors work, show that fears of annoying tenants by involving them in research were unfounded. People relish the idea of taking part in an important experiment.

(i) Chermayeff S. *The New Nomads*. T. Q. Saugatuck. April 1960.

can kill a whole area of a town by making it inaccessible to private cars without providing public transport. The potential of planning is excitingly great, our shortcomings monstrous. The enormous volume of ill effects continues. We sacrifice pleasures, we sacrifice life and limb. The iniquity of the (city) fathers is truly visited upon the third and fourth generation.

II. 3.2. *The Profusion of Alternatives*

The interrelated way in which actions affect many fields complicates the work of the planner. At the present time this is particularly serious, as the interrelatedness of factors is not even recognized or included in calculations and forecasts, trends and surveys. Railways and roads, economic development and overspill problems, car manufacture and road provision, and a host of closely connected factors, are dealt with separately with disastrously nonsensical results.

The interrelated nature of factors also allows any problem to be tackled from a number of angles. Let us take, for example, a highly congested road:

It may be widened;
a motorway built;
obstructions may be taken out of the road, such as pedestrians, traffic lights and parking;
traffic may be diverted;
paths or public transport may be made more attractive;
the goals towards which the traffic streams may have their position changed;
the working hours which create the congestion could be staggered;
parking costs at the destination could be raised, discouraging travel;
the use of the road could be charged for with a similar effect;
smaller vehicles could be used;
more people could travel per car;[(i)]
any combination of the above could be tried.[(ii)]

The obvious first solution, to widen the road, is, ironically, likely to be the least effective.

The point is that a thorough knowledge of a particular problem, and of the possible solutions and the chain effects of any actions, make a satisfactory answer more likely. Particularly if it is part of a larger vision or carefully considered master plan.

II. 3.3. *The Opportunist Idealist*

The planner must beware of idealism. Unless the emotional limp of mankind is clearly recognized,[(i)] the many varied ulterior motives understood, allowed for and counteracted in every effective way possible, cynicism is inevitable. There are things we know to be responsible for injury, things which are easy to remove, and yet no effective action is taken by anyone. The inertia of society, which is so exasperating to all those who are mobile and young in mind, which springs from the emotional limp, must be regarded as a disease. Then diagnosed in its various forms, it can be treated in whatever way is available: from cutting out the offending part in surgical action, to loving a thing better, the whole gamut of social therapy is relevant. But some considerable aspect of the inertia, of the "sitting characteristic" of society, will remain. Compromise should be wielded intelligently to change the central theme of a vision as little as possible.

Opportunism is a natural principle of growth—the bean climbs the available pole feeling around for something better all the while—not merely a matter of borrowing ten shillings from someone to depart in a slow boat for the Far East the next day. In two senses planning should be opportunist.

The planner's first opportunities arise out of the latent possibilities of any situation and he should make the very best possible use of these. The planner's opportunism in the second sense applies to the realization of any scheme. To get done what is possible when it is possible and where it is possible and with whom, should advance the progress of every scheme. And, like the bean, a keen lookout for possibilities in every sense is part of the art. The more comprehensive, flexible, bold and all-inclusive the master plan is, the more of the possibilities fulfilled will make sense according to the dynamic pattern of the plan. Ideals are best regarded as dynamic aims with which we work opportunely.

II. 3.4. *Basis for Rational Criticism*

The examples illustrated in this book cover a vast variety of ideas and the author obviously does not like some as much as others. Each, however, is judged to be worth while trying out and assessing as it emerges.

First, however, the critera of each designer must be understood. They may be regarded as invalid. But whether they are or not, the design itself must be judged by the criteria which it was meant to fulfil, and not by those held by the critic. No more important advice on learning from the work of others can be given. It is vital to know: do we reject the designer's or society's criteria? or do we fault the plan's inability to satisfy them?

(i) Hoyt H. The Effect of the Automobile on Patterns of Urban Growth. *Traffic Quarterly*. April 1963.

(ii) The total number of possible combinations is 1024.

(i) Reich, W. *The Emotional Plague of Mankind*, Vol. 1. New York, 1951.

III. The Man–Vehicle Relationship

III. 1. Summary of Man–Vehicle Characteristics. **10**

III. 2. Physical Characteristics of Man and Vehicles. **11**

III. 2.1. Scale. 11
III. 2.2. Relative and Cumulative Size. 13
III. 2.3. Tactile Associations. 13

III. 3. Movement. **14**
III. 3.1. Speed and Range. 14
III. 3.2. Route Capacity. 14
III. 3.3. Momentum and Danger. 16
III. 3.3.1. Accidents in traffic segregated areas. 19
III. 3.4. Curvature of Path. 21
III. 3.5. Mutual Frustration. 22

III. 4. Ecological Considerations. **23**
III. 4.1. Fumes and Poison. 23
III. 4.2. Noise. 24
III. 4.3. Vibrations. 26

III. 5. Sociological Considerations. **27**
III. 5.1. Influence of Private Cars on Social Patterns. 27
III. 5.1.1. Constructive effects. 27
III. 5.1.2. Destructive effects. 27
III. 5.2. Research Project. 27
III. 5.2.1. Design of research project. 27
III. 5.2.2. Presentation of results. 27
III. 5.2.3. Summary of whole survey. 31
III. 5.2.4. Survey at Stevenage. 32
III. 5.3. Social Aspects of Public Transport. 32
III. 5.4. Moral and Legal Considerations. 33
III. 5.5. Crowds of People and Crowds of Cars. 33
III. 5.6. Psychosocial Factors. 34

III. 6. Planning Solutions. **34**
III. 6.1. Functional Traffic Arrangements. 34
III. 6.2. Character of Roads and Paths. 35
III. 6.3. Graphic Analysis of Needs for Pedestrian Movement and Rest. 36

III. The Man–Vehicle Relationship

III. 1. Summary of Man–Vehicle Characteristics

	MAN	VEHICLE
SIZE:	Small (toddler to adult variation).	Big (motor scooter to double decker bus variation).
TACTILITY:	Soft.	Hard.
SPEED and RANGE:	Slow and small.	Fast (potentially) and great.
MOMENTUM:	Slight, safe.	Great, dangerous.
MOVEMENT:	Organic.	Organic tendencies through driver only.
RHYTHM:	Organic patterns, spontaneous.	Mechanical patterns, predetermined lines.
ROUTES:	No site lines, surprise, sudden changes.	Site lines and curvature and junctions according to speed and formulae.
ECOLOGICAL:	Harmonious, basically in smell, sound, feel and waste products.	Petrol fuel disruptive to life. Poisonous, (carbon monoxide) carcinogenic agents, sulphur tri-oxide, ozone, eye, throat and nose irritation serious, destructive of plant life and many crops (Smog).
SOCIOLOGICAL:	Needs security conducive to friendship and co-operation within narrow field and as a general characteristic.	Allows meetings of distant friends but where present is conducive to antisocial behaviour and disruptive of co-operative tendencies, particularly while driven.
DAMAGE:	Care increases with damage. Injury and death irrevocable and therefore tragic. Average life, long.	Care decreases with damage. "Injury" and "death" means insurance, scrap heap and a new car. Average life, short.

III. 2. Physical Characteristics of Men and Vehicles

III. 2.1. *Scale*

The Child's Eye View Exhibition(i) made vivid differences in scale. One section, a domestic interior, designed so that its size and detail were to the adult as the normal house is to the toddler, led to the town planning section where this lesson was applied: the problems of scale in townscape and landscape were seen in relation to the differences between man and his vehicles.

Fig. 3.1. Dining room.

Fig. 3.2. Living room, bathroom at the back. Note washing on the line and bathroom taps.

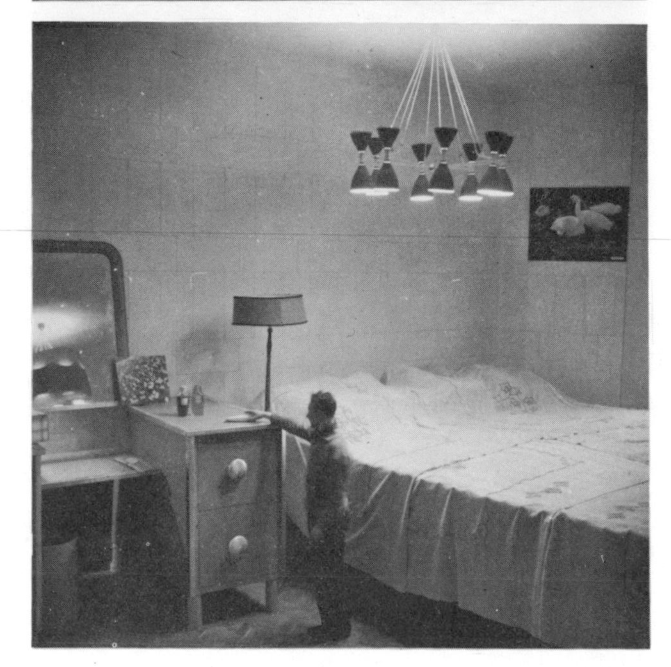

Fig. 3.3. Bedroom.

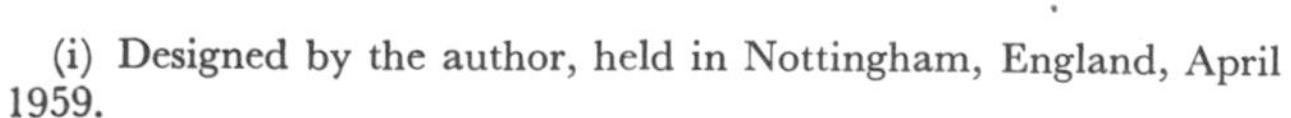

(i) Designed by the author, held in Nottingham, England, April 1959.

Fig. 3.4. Urban path, human scale, Gothenburg, Sweden. Study of the past, a fertile imagination, and full use of present techniques are keys for successful design. Planned to-day the steps shown might well be, in part, escalators.

Fig. 3.5. Urban Motorway Los Angeles. Vehicular scale. Like a great river it may flow with majesty if designed as part of the whole environment. But, as yet, motorways are hardly ever conceived or realised like that.

III. 2.2. *Relative and Cumulative Size*

Fig. 3.6. To stand comfortably a man needs:
2 ft^2, and 10 ft^3 (0·2 m^2, and 0·28 m^3).
An average car:
150 ft^2, and 750 ft^3 (14 m^2, and 21·2 m^3).

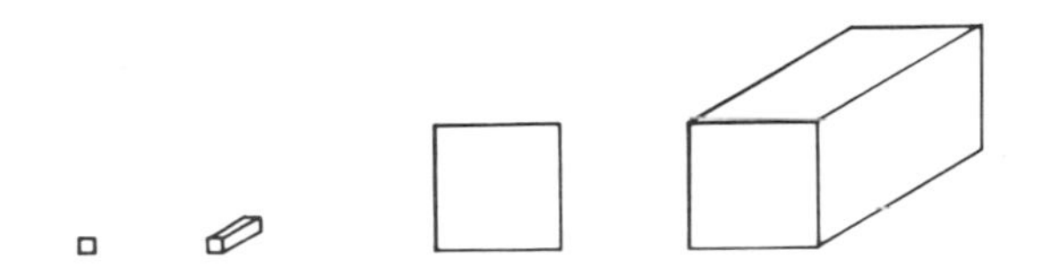

Fig. 3.7. Office work takes per person:
32 ft^2, 275 ft^3 (3·0 m^2, 7·8 m^3).
A parked car of a worker:
250 ft^2, 1500 ft^3 (23·3 m^2, 42·45 m^3).

Fig. 3.8. Car storage provision of one car per family, on one level with 5-storey flats.
(1·8 car per dwelling is the minimum now permitted in Stockholm.)

Fig. 3.9. A bus at peak hour, takes, say 66 persons, and covers 240 ft^2 (22·3 m^2) + 120 ft^2 (11·15 m^2) breaking space, i.e. 360 ft^2 (33·4 m^2) area in private cars 1·2 occupied 12,500 ft^2 (1161 m^2).(i)

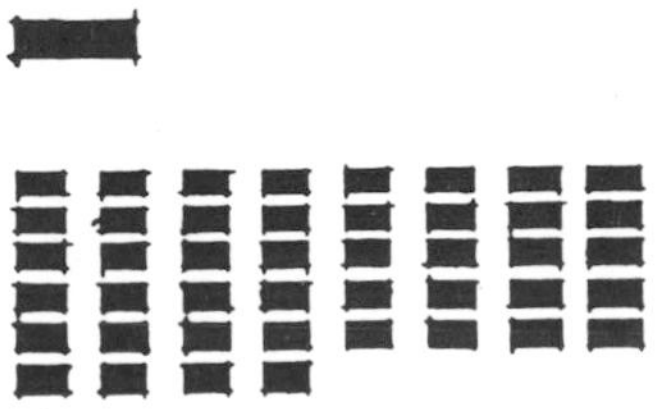

(i) This is a London peak hour figure. A study of twelve U.S. cities gave 1·5 for all trips. From 1·67 per car for those earning less than $2000 to 1·13 for those earning more than $9000, occupancy rate decreases evenly as income increases. (Chicago, Detroit, Washington, Pittsburgh, St. Louis, Houston, Kansas City, Phoenix, Nashville, Fort Landerdale, Charlotte, Reno.) From *Future Highway and Urban Growth.* Smith & Associates, New Haven, 1961.

Fig. 3.10. 50 standing people take: 100 ft^2, 600 ft^3 (9·2 m^2, 17 m^3).
50 parked cars: 12,500 ft^2, 72,500 ft^3 (1161 m^2, 2050 m^3).
50 parked bicycles: 300 ft^2, 1800 ft^3 (28 m^2, 51 m^3).

Fig. 3.11. 100,000 standing people: 200,000 ft^2 (18,581 m^3).
Cars to serve 100,000 people:
25,000 cars, 8,750,000 ft^2, over 200 acres (812,900 m^2, over 81 hectares).
Approximately 0·33 of a mile^2.
San Francisco Giants' Candlestick capacity crowd.

III. 2.3. *Tactile Associations*

Vehicles splash mud and sludge onto people, but keep those inside snug and dry and clean. Their hardness makes them vulnerable to damage through children's play, i.e. scratching. This incompatibility, between the parked car and children's play, is important to bear in mind in the designing of residential areas.

III. 3. Movement

III. 3.1. *Speed and Range*

Ten Minute Journeys:

MAN: Journey to work, 3 mph (5 km/hr); $\frac{1}{2}$ mile (0·8 km); 880 yd (800 m).
Average walk, $2\frac{1}{2}$ mph (4 km/hr); 733 yd (670 m).
Old and very young, $1\frac{1}{2}$ mph (2·5 km/hr); 110 yd (100 km).
Assisted by Mechanical conveyor, 5 mph (8 km/hr); 1 mile (1·6 km).

VEHICLE: 18 mph (29 km/hr); 3 miles (4·83 km).
24 mph (39 km/hr); 4 miles (6·44 km).
45 mph (72·5 km/hr); 7·5 miles (12 km).
72 mph (116 km/hr); 12 miles (19·3 km).

Travelling at high speed has the following effects:

a) As the eye of the driver is focused many hundred yards ahead and leaps great distances at a time, there is a reluctance to stop. (This is borne out by the universal experience of passengers suggesting a good stopping place, e.g. for a picnic, which the driver passes although there was ample time to stop in safety.)

b) The space–time scale, in a fast moving vehicle, makes appreciation of detail impossible and the rhythm of any route, the punctuation and events placed along it, must be, in size and frequence, in scale with the speed. Similarly, the appreciation of detail while walking makes the consideration of this particularly necessary. The executive director of the Philadelphia Planning Commission put it this way:[i]

"*The actual nature of design of each movement system relates to the tempo of movement, its purpose and characteristics. Expressway movements require free flowing forms and curves and articulation widely spaced to accord with the rhythm of fast vehicular movement. On the other extreme, pedestrian movement systems require interest and variety in spaces producing impressions of rapid change under slow foot movement. The system requires frequent punctuation by focal points and symbolic objectives, usually consisting of a series of short sections at different angles with definite visual termini to produce the conditions desired.*" (See also p. 58.)

(i) BACON, E. N. *The Space between Buildings*. Reprint of lecture Harvard 1962.

III. 3.2. *Route Capacity*

Leibrand[i] measured the performance of a nine foot wide (3 metres) lane in town centres during the peak hours and arrived at the following figures:

TABLE III. 1.

Mode of movement	Maximum capacity per hour	Average speed	Area required moving for each item
Pedestrian	16,000 pers/hr	2·5 mph 4·0 km/hr	8·1 ft^2/pers 0·75 m^2/p
Mechanical conveyor(ii)	25,000 pers/hr	1·98 mph 3·2 km/hr	1·0 ft^2/p 0·11 m^2/p
Cyclist	5400 pers/hr	7·5 mph 15·0 km/hr	72·1 ft^2/p 6·7 m^2/p
Motorbike	2400 pers/hr	9·3 mph 15·0 km/hr	193·8 ft^2/p 18·0 m^2/p
Private car	1200 pers/hr	7·5 mph 12·0 km/hr	322·9 ft^2/p 30·0 m^2/p
Bus:			
32 places	5600 pers/hr	6·2 mph 10·0 km/hr	
55 places	8300 pers/hr		
80 places	11,000 pers/hr		
150 places	18,000 pers/hr		
Tram (streetcar): 200 places	24,000 pers/hr	6·2 mph 10·0 km/hr	16·1–53·8 ft^2/p (average 21·5 ft^2/p) 1·5–5·0 m^2/p (average 2·0 m^2/p)
Rapid Track Transit	40,000 pers/hr	15·5 mph 25·0 km/hr	

From these figures Leibrand gives utilization figures to each form of transport relating to the respective areas taken up in transit by a person at the above speeds:

(Typical calculation of peak hour traffic)

Private car (one person)	1
Motorbike	2·5
Cyclist	4·5
Pedestrian	4·5
Conveyor[ii]	5·7
Tramways	16·5
Rapid Track Transit	70·0

(i) LEIBRAND K. *Verkehrsingenieurwesen*. Basel–Stuttgart 1957.

(ii) Added by author.

Area required per person during the peak period for a one mile journey, as assessed by Dr. Smeed.(i)

TABLE III. 2.

	Speed		Area	
	mph	km/hr	ft²	m²
Urban street, 24 ft wide, mixed traffic				
Cars with driver only	15	24	105	9.7
	10	16	64	5·9
Cars with 1·5 persons	15	24	69	6·4
	10	16	42	3·9
Cars with 4 persons	15	24	26	2·4
	10	16	16	1·5
Buses with 32 persons	8·6	14	10	0·9
	6·7	11	6	0·6
Urban street, 44 ft wide, mixed traffic				
Cars with driver only	15	24	59	5·5
	10	16	42	3·9
Cars with 1·5 persons	15	24	39	3·6
	10	16	28	2·6
Cars with 4 persons	15	24	15	1·4
	10	16	10	0·9
Buses with 32 persons	8·6	14	6	0·6
	6·7	11	4	0·4
Urban motorways				
Cars with driver only	40	64	21	2·0
Cars with 1·5 persons	40	64	14	1·3
Cars with 4 persons	40	64	5	0·5
Footway	2·5	4	3	0·3
Urban railway line	18	29	0·9	0·08
Suburban railway line	30	48	1·2	0·11

However, assessing the private car at low speeds such as 10 mph on urban crowded roads leads to very varied results, as comparison of Leibrand and the two Smeed figures show: i.e.

Leibrand: 322 ft² at 7·5 mph
(29·9 m² at 12 km/hr)

Smeed: 105 ft² at 15 mph
(9·8 m² at 24 km/hr)
64 ft² at 10 mph
(5·9 m² at 16·1 km/hr)

Using the graphs prepared by Middleton-Harwood(ii) we see clearly that results at speeds from 5 to 15 mph do not yield useful figures for accurate comparison and that a distinction between 10 mph and 15 mph as observed and used by Smeed may be misleading, without full explanation of the meaning and origin of figures including a description of the prevailing traffic conditions. (See also p. 70.)

(i) SMEED R. J. *The Traffic Problem in Towns*. Manchester. Statistical Society 1961.

(ii) MIDDLETON-HARWOOD J. Planning Department City of Leicester. City Planning Officer, Smigielski W. K.

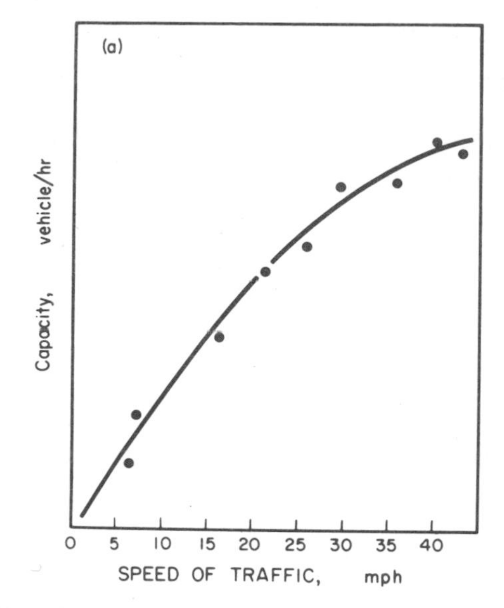

Fig. 3.12. a) Working at maximum capacity (Liverpool), by Middleton-Harwood.

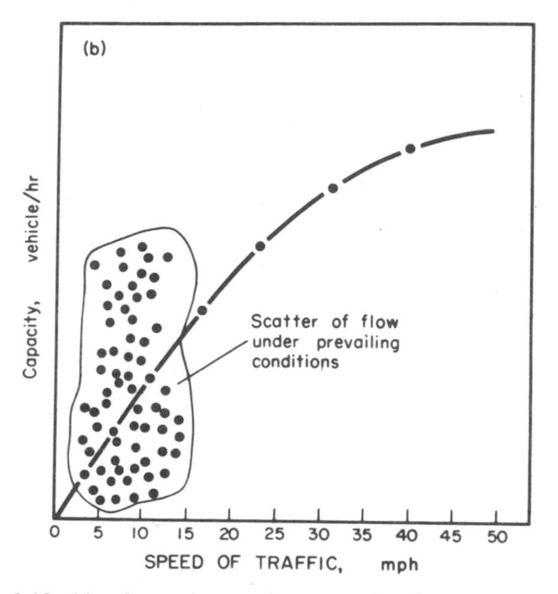

Fig. 3.13. b) Operating under normal urban street conditions (Liverpool), by Middleton-Harwood.

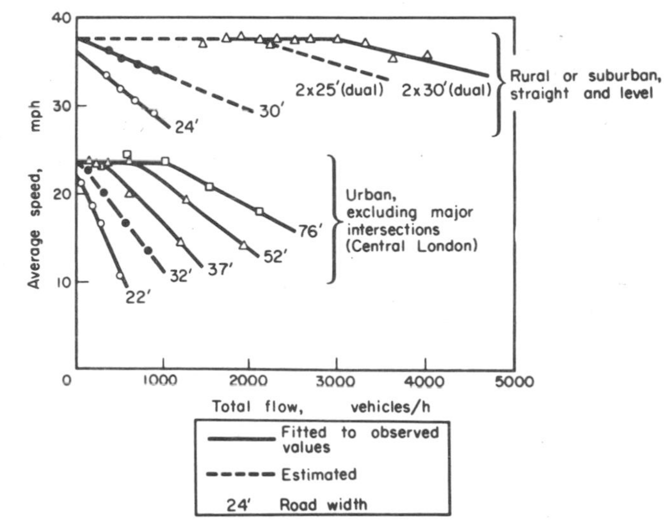

Fig. 3.14.(i) Capacities as observed in London.

(i) WARDROP and DUFF. *Factors Affecting Road Capacity*. London 1956.

This lack of clarity and precision is typical of Traffic Engineering as a science. In its primitive present stage constant definition is of the greatest importance. Conflicting figures, particularly with regard to urban traffic problems, are to be found. Contrary conclusions then spring from differing evidence. This should be borne in mind when studying the subject or using its contributions. Comparison of figures given by different authorities from different countries cannot be very effective. Methods of collection, criteria and even definitions vary. There is an urgent need for a comprehensive range of codes of practice covering the entire subject of traffic engineering. Only in this way can a sensible approach and analysis of any situation result. Dr. Smeed wrote in 1960:

"*A number of methods are used for deciding on the position of new or improved roads in existing towns, and for the layout of the road system in new towns. One method often used for existing towns is to improve or replace those roads that are obviously over loaded. This method, however, suffers from the serious disadvantage that it perpetuates an unsatisfactory method of routeing, and if there is no existing road between two points there never can be one because the necessity for it will not be realised.*"

"*Another method is to find the origins and destinations of traffic, and the journey times on the existing road system, and then to estimate the effect on the times of travel of a new or improved road in a given position, taking into account any expected changes in travel habits. This method has much in its favour but, as it is in practice not possible to find the effects of all possible new roads, it will not necessarily result in the best possible roads being built. It is probable that the best possible roads in an old or new town cannot be planned until a comprehensive theory of urban travel is available.*"(i)(ii)

III. 3.3. *Momentum and Danger*

The mass, multiplied by the speed of the vehicle, even in its weakest or slowest forms, gives it a momentum significantly greater than the pedestrian. This danger applied to horse-drawn carriages. When London had a fifth of its present population, 100 or about an eighth of the present traffic deaths were recorded. Day(iii) says this was the case in 1860, when there were far fewer vehicles, although traffic jams were common.

(The bicycle, non-powered, approximates the pedestrian.)

(i) Smeed R. J. *The effect of some kinds of routeing systems on the amount of traffic in the central areas of towns.*

(ii) "...we are still some way from an applicable theory of traffic movement and road capacity." From a review of a textbook by A. Proudlove, *T. E. & C.* August 1963.

(iii) Day A. *Roads.* London 1963.

Danger arises out of this momentum:

a) if traffic is brought to a sudden halt by stationary objects (lamp post, tree, etc.);
b) by collision of two vehicles, aggravating the momentum danger;
c) by collision between pedestrian, or bicycle, and vehicle, where the momentum of the vehicle results in death or injury of the former, unprotected or vulnerable.

Arranging and mixing all kinds of traffic in an inefficient manner results in high accident figures: alarmingly high compared with other causes of death; surprisingly low considering the chaos in towns. The old and young are killed and maimed far more often than the adult between 19 and 49. Society's attitude is contrary to Christian ethics which teach special care for the weak and vulnerable (e.g. in war). Doxiadis calls our traffic mixing "immoral".(i) (See also p. 68.)

Vehicles may well be regarded as "misguided missiles", aimed at pedestrians and each other.(ii)

Fig. 3.15. The Immoral Mixture, as suffered.

Fig. 3.16. The Immoral Mixture, as seen.

(i) Doxiadis C. *Ekistics.* Town & Country Planning Summer School 1959. Also *R.I.B.A. Journal*, September and October 1960.

(ii) Ritter P. *Man & Motor Don't Mix.* Nottingham 1960. (distributed to all M.P.'s).

Fig. 3.17. Cars in combat.

Some relevant figures for traffic accidents are given below:

"100,000 deaths on the roads of the world every year,...
1,000,000 serious injuries,...
3–4,000,000 light injuries,"(i)
and uncounted numbers of accidents leading only to material damage, plus all the unreported ones,
and an uncountable volume of anxiety and worry for the welfare of oneself and others.
All these are rising annually.

It should be borne in mind that "light" injuries are of a kind that would go unattended by a doctor if suffered in the home (in Britain).

Taking the U.S.A., the most fully motorized country to date, we can see in the following figures what threatens other countries if they motorize without proper provisions: in 1957,
deaths from all communicable diseases—24,000,
deaths from road accidents –39,000,
7% of all available hospital beds are taken up by men suffering from impact and physical damage by vehicles.
(Material costs are enormous: see economic section)

Pedestrian casualties only: (in Britain approx. $\frac{1}{4}$ of total killed). Of the total number of pedestrians killed and seriously injured, 22,539, about 8600 were under eighteen (2058 under four!)
10,000 over 50
and only 3400 adults between 19 and 49.(ii)

(i) *WHO*, Geneva 1962.

(ii) Figures taken from Ministry of Transport Road Accidents. H.M.S.O. 1961.

Fig. 3.18. Each symbol represents 1000 killed or seriously injured in Britain in 1961.

At a recent Road Safety Congress the Danish representative described their research as follows.

"*In order to get to the bottom of this whole problem of the very sad accidents involving small children, we have recently carried out an investigation, which was carried out by Sergeant Viggo Kirk, M.A., from the Traffic Department of the Copenhagen Police, comprised oj* 522 *accidents, which means all accidents registered by the police during the years* 1958, 1959 *and* 1960, *in which children from* 0 *to* 6 *years were involved.*

The investigation proved among other things:

1. *that the accident involved* 65% *boys and* 35% *girls;*
2. *that especially the oldest children were alone when they got involved in accidents;*
3. *that* 17%–18% *of the children were passengers and consequently personally not to blame for the accident. But our means of transportation are equipped for adult passengers only, e.g. the foot-holds of motor-cyles and the outlook from motorcars. This means that children do not sit securely in the first case, and in the other that they often stand at the front seat near the windscreen;*
4. *that* 53% *of the children came from the most densely inhabited districts;*
5. *that* 69% *of the children came from homes of two rooms or less;*
6. *that* 48% *had received traffic education, i.e. that they had been taught how to walk on the pavement, how to cross the street and how the traffic lights (or the traffic-controlling policeman) work;*
7. *that* 9% *of the accidents happened on Sundays,* 20% *on Saturdays and* 14% *on each of the other workdays;*
8. *that most accidents took place in April and August. In April perhaps because the children again begin to play out of doors after winter; in August perhaps because the children have just come home from holiday where they have been used to going about rather freely;*
9. *that* 42% *of the accidents happened very near home;*
10. *that* 93% *of the adults who were involved in the accidents were men; only* 7% *were women;*

11. *that 55 % of the children were used to going out on their own;*
12. *that 88 % of the children were used to behaving as they did when the accident took place;*
13. *that 41 % of the children had their playground near home.*
14. *that 65 % of the accidents were caused by children running into the carriageway, and in 53 % of these cases a stationary vehicle was blocking the view."*(i)

The attitude of the planner to these figures must be more positive than that displayed by road experts at Salzburg in 1962:

None of the many papers on this subject mentioned the right of a child to be safe. In fact life for a child, according to the German delegate, is not much fun:

"*A child of pre-school age should always have his hand held by an adult whose responsibility it is to see that the child does not leave go. Objects such as balls, that easily roll about, must be kept in a net or a bag.*"(ii)

Planning can eradicate the need for such edicts.

The same congress dealt with the nature of injuries caused by the impact of unnecessary sharp and hard protrusions on the car with the soft human body. This is a late realization but the positive conclusions to be drawn from it are quite clear: the law *must* intervene. Styling cannot be allowed to break bones.

"*We have found it impossible in every serious car–pedestrian collision to relate the point of the first impact of the car with a particular anatomical area of the patient.*"

"*In about two-thirds of these collisions, the front near side of the vehicle, either bumper, mudguard or head-lamp, made first contact with the pedestrian, in less than one-third it was the centre front or centre off side of the vehicle. In the remainder, the sides or rear of the vehicle made contact after the car had skidded, generally on a wet surface. The severity of the initial injury depends upon the force and the sharpness of the object first hitting the pedestrian. Thus, compound fractures of the legs were commonly caused by the sharp edges of metal front bumpers or their over-riders. Chest and abdominal injuries were often caused by the sharp projecting cowling over head lamps. There is no reason, other than styling, why any feature on the outside of a motor vehicle should project, and so increase the severity of the first impact injuries to pedestrians.*"

"*When the first impact is low on the pedestrian's body, and even when the speed of the vehicle conforms to city regulations, the pedestrian is often thrown high into the air to land either in the road or on the bonnet of the car. If the latter, the pedestrian then hits the windscreen or its frame, finally to be thrown forcibly onto the road. There to come to rest, or hit roadside furniture, or maybe to be hit by another road vehicle or to be run over. In short, a series of impacts account for the typical multiplicity of pedestrian injuries, and their severity depends upon the violence of the forces.*"

"*It is of interest to note that somewhat similar pedestrian injuries occurred during the early development of railways in the nineteenth century, and ceased only when railway tracks were segregated from all other traffic. Similar segregation is the solution for pedestrian/motor vehicle collisions, particularly in built-up areas where the road carries heavy volumes of motor transport and pedestrian traffic. Until this segregation is achieved, the severity of initial impact injuries would be lessened by spreading the blow over a wider body surface by the elimination of all sharp projections from the outside of all vehicles, and by lessening the fatalities from run over accidents by the provision of side and rear bumpers to all lorries with high ground clearance.*"(i)

Fig. 3.19. Fashion.

Accidents need to be reduced as an urgent measure, the most effective propaganda seems to be a proper combination of moderate realism, and positive suggestions of what people can do to protect themselves against accidents. Good examples of this are the short B.B.C. TV. items. Scare techniques or horror description, have proved worse than nothing, after the first application, satisfying only the consciences of people that something is being done. On examination they do not prove effective.(ii)

(i) GROES-PETERSEN E. *Road Safety Training of Pre-school Age Children in Copenhagen.* 6th International Road Safety Congress Report. London 1962.

(ii) MATZUTT, DR. MARIANNE Cologne University, Germany. *Road Safety Training of Pre-school Age Children.* 6th International Road Safety Congress Report. London 1962.

(i) GISSANE W. University of Birmingham. A Report for the Road Injuries Research Group. Birmingham Accident Hospital. England. 6th International Road Safety Congress Report. London 1962.

(ii) J. L. MALFETTI, Scare Techniques and Traffic Safety; *Traffic Quarterly*, April 1961. Eno Foundation.

III. 3.3.1. *Accidents in traffic segregated areas.* An analysis and comparison of two traffic segregated areas in Gothenburg has shed some light on details of planning necessary for safety.(i)

Kortedala, completed 1957, 24,000 inhabitants, was compared with Guldheden, completed 1952, 9000 inhabitants. All accidents between 1956 and 1960 were investigated.

The chief differences are as follows, though neither of these schemes seem particularly well planned:

(a) secondary feeder roads in Kortedala are in the main through roads with four-way junctions, housing strung along them, door opening onto them; in Guldheden, culs-de-sac with three-way junctions;

(b) There is a greater degree of traffic segregation in Guldheden with an underpass. At Kortedala paths cross secondary feeder roads. The estates have in common an inordinate amount of roadside parking which is a constant danger.

In Kortedala 17·5% of accidents (the largest single percentage) was due to vehicle–pedestrian conflict. In Guldheden, on the other hand, only 3%.

Most of the accidents to pedestrians in Kortedala were due to people rushing into the roadway between crossings. In 30% of these, vision was reduced by parked cars. Boys of 2–3 and 2–6 were particularly accident-prone. In Guldheden few accidents involved children. 2·7 times as many accidents occur to children between 2–15 in Kortedala, taking into account the relative child population of each, and 4 times as many accidents occur to children below the age of six. The highest accident rates occur in through-roads. However, even in culs-de-sac accidents do occur.

Fig. 3.20. Many accidents happen, especially with small children at Kortedala. Doors face onto roads and direct contact is lacking from flats to play areas within the block. These are entered, as can be seen, along the ramp on the left. Small children have a limited circle of action and so play mainly on the road and the parking areas immediately available.

(i) Gunnarsson S. O. *Traffic Accidents in Two New Residential Areas in Gothenburg*. Gothenburg 1962. Summary in English.

The following conclusions are drawn:

"1. *Consistent and, as far as expedient, complete differentiation of traffic leads to a lower frequency of accidents and of injuries to persons.*"

"2. *The design of the street system is of decisive importance for safety in traffic. Thoroughfares of low standard should be avoided. Streets of dwellings should be short, culs-de-sac with ample space to turn a car, or as horseshoe streets.*"

"3. *A system of streets with three-way crossings probably provides a higher degree of safety than one with four-way crossings.* (*Detailed investigations are in process.*)"

"4. *Parking space should be provided on special sites, away from the streets. Kerb parking should be prohibited.*"

"5. *Steps must be taken to ensure safety for children in traffic.* Entrance to dwellings should be towards the playgrounds. *The path between dwelling, playground and school must be made safe.*"(i) (See also p.34.)

Fig. 3.21. Underpass for pedestrian and cyclist at Guldheden illustrate the greater degree of traffic segregation.

(i) Ice-cream vans have proved a special danger in Britain.

Fig. 3.22. Kortedala. All accidents. Crossed circles involved injury to people:

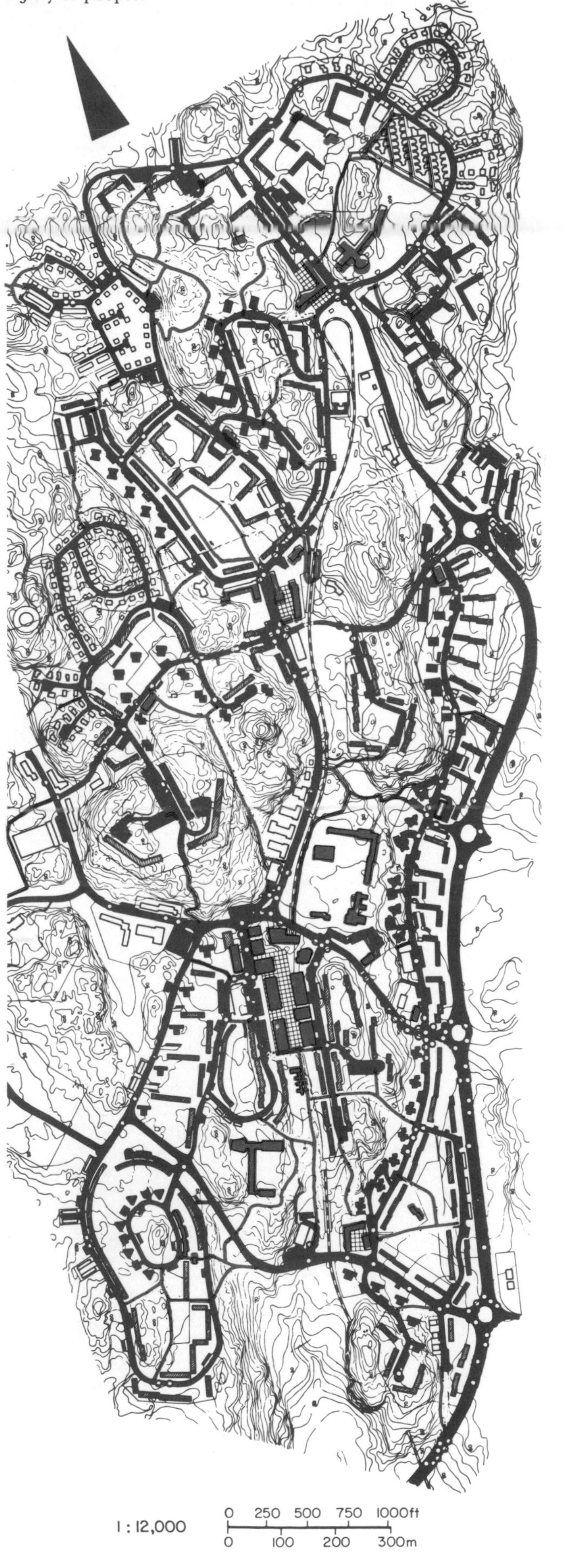

Fig. 3.23. Guldheden. All accidents. Crossed circles involved injury to people.

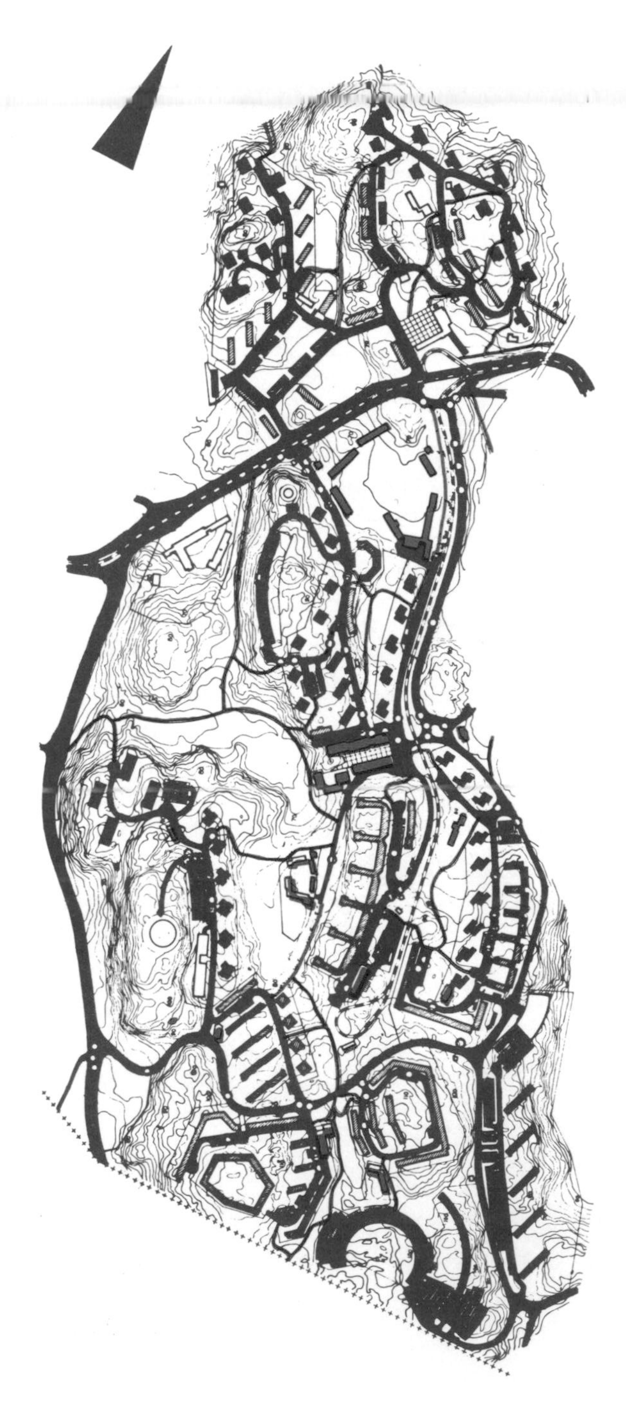

III. 3.4. *Curvature of Path*

The tendency to spiral movement is universal. This includes not only the movement of each organism as a whole but also of every liquid and solid within it.(i)

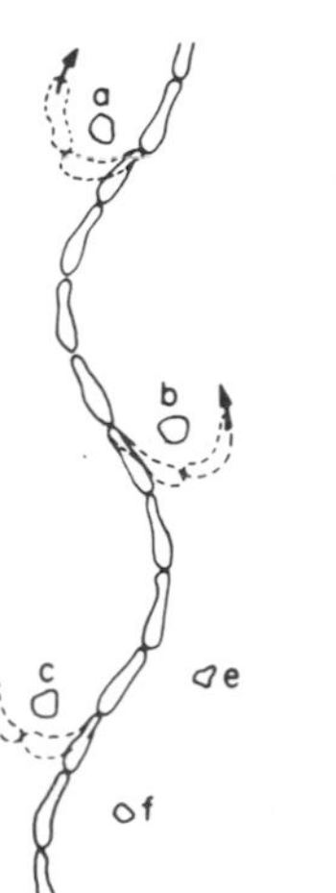

Fig. 3.24. Ameboid movement.

The degree of curvature is modified by the opposing tendency to reach a specific goal and the speed and impetus, i.e. momentum of movement.

The very gentle meander of a footpath worn into turf of a field by those who wish to cross it diagonally illustrates the tendency of the aim at a goal attempting to overcome the tendency towards spiral movement. A blindfolded person, walking in complete spirals, swimming in spirals (or if he cannot swim, sinking in spirals!) demonstrates the innate tendency unaffected by conscious goals or intent.

Walking along straight paths, and indeed driving along straight roads, can be satisfying only in that it sets up tension in contrast to the innate tendency to movement in spirals, expressed in meandering paths. Because of this, the effective length of straight paths and the frequency of their right application is limited. Similarly with roads. It is generally valid to plan to satisfy the meandering. The degree of curvature should be right for the speed, power and urgency of goal, of the travelling person or vehicle. Hence the decisive difference between the need of a driver and a pedestrian.

Pushkarev(ii) has clearly and fully illustrated the advantages, nature and technique of spiralling motorways, combining horizontal and vertical alignment, in contrast to such as are composed of segments of circles and straight lines.(iii)

(i) SCHAEFFER A. A. *Ameboid Movement:* Princetown University Press 1920. PETTIGREW J. B. *Design in Nature.* Vols. I, II, III. London 1908.

(ii) TUNNARD and PUSHKAREV, *Man Made America: Chaos or Control.* Yale University Press, 1963.

(iii) See Section V. 3.5.1. (p. 75).

A grid-iron, or other geometrically determined lay-out of straight lines, makes sense only as an intellectual triumph over nature. The small, old cities, designed in this way for reasons other than traffic, represented symbols of supremacy of man in small areas surrounded by wild, wolf-infested forests. The grid-iron lay-out of modern cities does not make sense and is not rational. It makes all movement, walking or driving, unpleasant. The straight line indicates an urgent goal, but there is often no goal visible, just weak endings to boring streets. It may be deduced, from general principles, that this is one factor influencing the dislike of walking in the U.S.A. This would mean that walking is less pleasurable than in Europe and that driving, in dead straight lines, is the lesser of the two unpleasant movements within a town. (At University campus and suburban shopping centres Americans walk.)(i)

The patterns of rhythm in movement are created not only by curvature itself but also by incidents which simulate curvature along the line of movement: alternate widening and narrowing of the volume along which one travels: vertically by bridges, roofs, etc., or horizontally by walls or hedges, coming right up to the path itself and then opening out again. This means rhythm can be experienced along a relatively straight path. It makes walking more pleasant, and so less arduous and seemingly shorter, and similarly with driving.

The curvature arrived at through the consideration of the two basic tendencies listed above must not be confused with the planning of roads in curves to slow down traffic in residential areas for the extremely dubious purpose of making them safer.(ii) Undue curvature makes walking unpleasant, particularly when the wish to get from *a* to *b* is uppermost. And the constant irritation and concentration enforced by artificially curved roads makes them less safe than straight ones. They tire the daily roundsman for whom they are mostly intended until he is more dangerous because more tired. Lack of sight lines makes far slower speeds dangerous. Short straight service roads are best. Enforcing slow speeds where necessary by vertical bumps at crucial points is far more sensible and effective.

Routes designed for man are not right for motor and vice versa. In addition the presence of fast motion alongside the pedestrian makes him feel his own movement as uncomfortably slow.

(i) Similar sentiments can be found in LONGSTRETH THOMPSONS' book *Site Planning in Practice.* London 1923.

(ii) TURNER R. *Garages in Residential Areas.* T.P.R. July 1959.

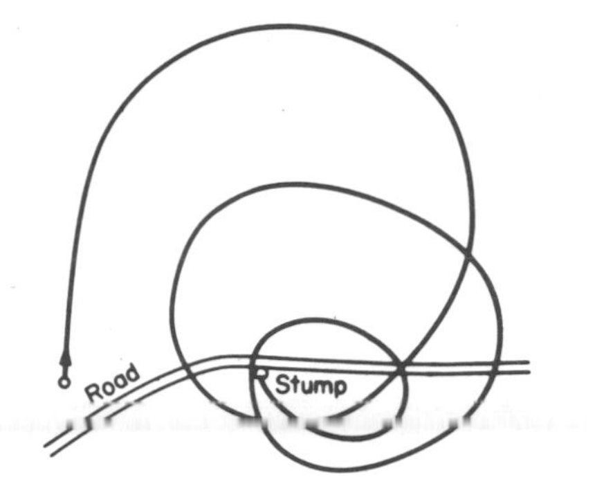

Fig. 3.25. Human being, blindfolded, i.e. goal excluded.

Fig. 3.26. The Medieval road and bridge.

Fig. 3.27. Motorway curve of high speed movement of vehicles. The Antony Wayne area on the Palisades Interstate Parkway.

III. 3.5. *Mutual Frustration*

Apart from danger and nervous tension, frustration can be isolated as a mutual effect man and vehicle have upon each other.

For the driver the frustration of standing and waiting, instead of moving smoothly, is bound up with pedestrian crossings, traffic lights and jay walkers.

For the pedestrian the frustration of the rhythmic stride of his walk, having to stop, if only to look right, look left and look right again, decisively spoils his pleasure. To have his rhythm broken at frequent intervals is galling. This is particularly apparent in the grid-iron cities of the U.S.A., where city walking, along small blocks, even along main streets, may be safer but is continually disrupted by crossings, and where disobedience of robots (traffic lights), "Don't Walk" signs, results in an immediate fine.

Fig. 3.28. Part of a cartoon by URS.(i)

Jane Jacobs(ii) in her shallow analysis of planning problems, orientated towards large cities of the U.S.A., argues in favour of many small blocks. She insists on "streets" and limits her illustrations to grid-iron Manhattan. She denounces the super-block, as if it had to act in a frustrating way to citizens on the move with its more extensive and traffic free path areas. She describes only bad use of this principle. "These streets are meaningless because there is seldom any active reason for a good cross section of people to use them." This merely means that a path system must make meaningful and plentiful connections with the surrounding areas, satisfying "desire lines"(iii) and giving opportunities for creative additions, citizen participation. (See also p. 236.)

It has become clear that a pedestrian street can contain more of the activities desired by Jacobs than a street where cars and pedestrians confront each other. It can already be seen in the pedestrian streets of Cumbernauld that the bustle of people is dominant even at this early stage.

(i) From Kurier November 1962.

(ii) Jacobs J. *Death and Life of Great American Cities*. London 1962.

(iii) See Section V.

III. 4. Ecological Considerations

III. 4.1. *Fumes and Poison*

Man, as an organism, can be ecologically in harmony with his planet. A number of his activities, however, have been disruptive of his environment. Relevant to this book is the effect of petrol as fuel resulting in fumes, smogs and the poisoning of the atmosphere with odourless, invisible, carbon monoxide.

The most infamous example of smog occurs in the hundred or so miles of coast line by Los Angeles where micro-climatic factors, such as the mountain range close to the coast, aggravate matters dramatically. Released through "3,000,000 cars burning 5,000,000 gallons (22,729,800 litres) of gasolin daily"[(i)] it appears in its more noticeable form on sixty days of the year. Even in its less noticeable form, concurrently with sunshine, it is announced, in the weather forecasts, as "weak eye irritation", contrasting with "strong eye irritation" on the sixty days.

Anti-fume devices are about to be legally enforced in new cars, as manufactured to be used in California, to reduce 600 tons (610 tonnes) of hydrocarbons deposited daily. This, however, does not remove the carbon monoxide poison nor the oxides of nitrogen of which 400 tons (406 tonnes) are released.[(ii) (iii)]

Gasoline or petrol smog is radically different from the more usual smoke smog. Sunshine activates it, smoke or fog does not. It has been studied with great concern in California and the studies proceed. To date, damage to 70 different species of plant have been confirmed and recorded. Damage to salad crops alone, from such smog, in the Los Angeles area was estimated at $5,000,000 (£1,666,000) in 1956.

Many animals and birds have been recorded to suffer from irritations and illnesses similar to those which afflict human beings in petrol smog. Mice have lost weight, compared with control samples.

Human beings already suffering from diseases which result in a decreased oxygen-carrying capacity of the blood react most sensitively to the direct poisoning effects of carbon monoxide, even in diluted quantities. People generally suffer from the effects of smog. The extent to which this results in deaths and hospital admissions seems as yet undetermined. But eye, throat and nose irritation are normal in areas of high concentration, i.e. West California. Children are reported as emotionally disturbed during periods of such "gasoline smog". Visibility is reduced but not to the degree of smoke smog. In Europe, the effects are less spectacular, but the micro-climate at one cross road in Vienna makes it necessary to replace traffic policemen frequently, since a number have been overcome by carbon-monoxide poisoning while on duty.[(i)]

Figs. 3.29. and 3.30. Los Angeles Smog (from *WHO*).

(i) *The Poison We Breathe*. World Health. Geneva. May–June 1961.

(ii) HEINEMAN H. *Effects of Air Pollution on Human Health.*
CATCOTT E. J. *Effects of Air Pollution on Animals.*
THOMAS M. D. *Effects of Air Pollution on Plants.*
All in: Air Pollution, *WHO* 1961.

(iii) MIDDLETON J. T. & DIANA CLARKSON, Motor Vehicle Pollution Control. Reprint *Traffic Quarterly*, April 1961. Eno Foundation.

(i) REICHOW H. B. *Die Autogerechte Stadt*. Ravensburg 1959.

Cancer-creating substances in exhaust gases represent the most vicious aspect of the ecological disharmony resulting from petrol and diesel engines.

The fuel used at present in almost all our vehicles, is, as Lewis Mumford pointed out quite clearly in *The Culture of Cities* in 1938, a grave pollution to the air we breathe.[i] This can only be regarded as a lowering of our standard of living in the biological sense. It is most undesirable, and the fuel urgently needs replacing. The powerful financial interests involved in such a change must be recognized and dealt with intelligently. They include both capital and labour.

It is certainly only the State which is in a position to initiate action. As the liberty-loving U.S.A. has shown that State action can be applied in California, there should be no difficulty for other Governments to act in the interest of the country as a whole in this very far-reaching matter, with the clear goal of clean air in mind.

The fumes and poison which are at present inevitably associated with vehicles are not a necessary by-product. One hopes that this will not prove to be the main reason for having given them their own environment, apart from pedestrians. Electric or other means of propulsion are bound to come sooner or later. At present it makes separation especially urgent, of course, and affects many details of design considerations (i.e. underground roads and garages).

Fig. 3.31. Diesel engines are the worst offenders. Ironically compulsory testing of vehicles in Britain does not apply to commercial vehicles.

(i) Hence "Infernal" as well as Internal Combustion Engine.

III. 4.2. *Noise*

Noises made by human beings, such as those of shouting children near the "night-shift" bedroom, can be a nuisance. Planning for the proper distribution of human sounds is possible.

Noise made by vehicles are disturbing and offensive to man. The Wilson report on Noise[i] collects together some telling evidence. The Building Research Station, testing at 540 points all over London (1961–62), found that at 84% of these points road traffic noise predominated.

> "*In the* 1961–62 *London Survey a sample of approximately* 1,400 *people were questioned about noise and its importance relative to various other factors. Table III.* 3. *summarises the answers given by these people to the question 'If you could change just one of the things you don't like about living round here, which would you choose'? and thus shows the relative importance, for Central London, of the various factors mentioned.*"

TABLE III. 3. *Relation of noise to other factors*

The one thing that people most wanted to change	% people who wanted to change it
Noise	11
Slums/dirt/smoke	10
Type of people	11
Public facilities/transport/council	14
Amount of traffic	11
Other facilities/shopping/entertainment	7
Other answers	1
No answer, or vague reply	5
Would change nothing	30

The kind of noises people found disturbing are given in Table III.4.

TABLE III. 4. *Noises which disturb people at home and outdoors*

Description of noise	Number of people disturbed, per 100 questioned	
	when at home	when outdoors
Road traffic	36	20
Aircraft	9	4
Trains	5	1
Industry/Construction works	7	3
Domestic/Light appliances	4	—
Neighbours' impact noise (knocking, walking etc.)	6	—
Children	9	3
Adult voices	10	2
Wireless/T.V.	7	1
Bells/Alarms	3	1
Pets	3	—
Other noise	—	—

(i) *Noise, Final Report.* Committee on the problem of noise. H.M.S.O. July 1963.

The report gives a comparison of 1948 and 1961 responses to noise disturbance.

TABLE III. 5. *Comparison of people's reaction to noise at home in the* 1948 *and* 1961 *Survey*

Individuals' reaction to noise	External noise (% of people)		Internal noise (% of people)	
	1948	1961	1948	1961
Those who are disturbed by noise	23	50	19	14
Those who notice but are not disturbed	19	41	21	14
Total of people who notice noise	42	91	40	28
Those who do not notice noise	58	9	60	72
	100	100	100	100

"*Further analysis shows that a high proportion of these, approximately* 30% *of the total sample, claim to be seriously disturbed by the one noise found most bothersome, generally traffic. Some measurements have been made of the noise climate in the immediate locality of the homes, of about half the people questioned, and it appears that whether people live in noisy or in quiet places does not affect the proportion of them who are seriously disturbed.*"

The report concludes that: "*road traffic is, at the present time, the predominant source of annoyance, and no other single noise is of comparable importance.*"

Sound increases are felt as a logarithmic increase of intensity, i.e. a variation of 10 phons up or down is felt as an increase or decrease of 100% twice or half as loud:

Normal conversation 50 phons.

Office background noise through traffic in side street 40–60 phons.

Office above, but in main street, 50–75 phons (window closed).

Motorbike 65–105 phons.

Heavy lorry in a wide road from 22 ft (6·7 m), distance, 85–100 phons.

It has been found that tall buildings do not escape sound as was once thought.

Statutory requirements are to be enforced in London with respect to noise levels, as permissible in various kinds of buildings, a recent remarkable advance.

As with air pollution through fumes and poison, so pollution through noise will be remedied to a great extent by the improvement of engines. But a rush of agitated air will always accompany fast, large vehicles, whatever the engine.

It is not possible for a town street to be planned for cars of today so that the human voice can still be an efficient tool in that same street.[1] The volume of noise of cars is such that only by shouting is communication possible in their presence. This represents a constant physical strain on the vocal chords and lungs and discourages conversation.

Because of their nature and association a large proportion of the noises, representing power, speed, aggression and hardness, are a nervous strain.[i] Open windows become a mixed blessing. The air is polluted with sound waves and psychological tension, incipient accidents, as well as fumes and poison.

Motorways create special problems.[ii]

Positive planning proposals have been made by Reichow on how to counter noise nuisance,[iii] and he shows among other things how general principles properly applied can obviate remedial measures.[iv]

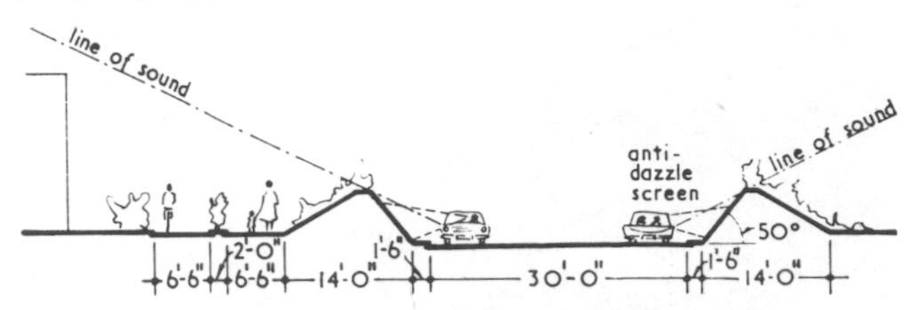

Fig. 3.31. a. Banks to counter noise nuisance from roads.

Fig. 3.32. When truth and surrealism come close to each other.

(i) Tooting of horns, screeching of brakes and tyres, shrill groan and whine of acceleration and deceleration, etc.

(ii) HUMPHREYS H. R. Noise and the Motorway. *A.J.* 11.7.62.

(iii) REICHOW H. B. Noise Abatement. *A.J.* 13.2.63.
REICHOW H. B. *Die Autogerechte Stadt.* Ravensburg 1959.

(iv) See also Section V and Vogler, Medizin und Städtebau, Munich 1957.

III. 4.3. *Vibrations*

The mechanical movement of vehicles creates rhythmic vibrations unlike those of pedestrians. Due to weight and insistence these have considerable effect on building fabric and represent a noise nuisance.(i) Protection may be built into the vehicle (e.g. the rubber tyres on the Paris Metro), the road surface, or by insulating or isolating the building.

An interesting parallel exists in Venice. The swell from the motorboats is washing away the stonework along the waterline which has withstood the gentle ripples of the gondolas for centuries.

Fig. 3.33. In Venice the path system combines with the canals. Photo by Chris Millard.

Fig. 3.34. The front door on the left would suffer from motor launches as countless front doors in Britain suffer from trunk roads skimming past their houses. Photo by Chris Millard.

Fig. 3.35. The movement of slow boats is inherently beautiful. Painting by Canaletto.

(i) B.R.S., Effect of Traffic Vibrations on Buildings.

III. 5. Sociological Considerations

III. 5.1. *Influence of Private Cars on Social Patterns*

III. 5.1.1. *Constructive effects.* Possession of private cars makes possible much more frequent contact between people who wish to be together but live far apart. It makes possible the attendance at gatherings of many people who would not be there without the use of their own car. These aspects are particularly important in low density areas, isolated farmstead development and suburbs. In medium and high density residential areas the car becomes less efficient, for there is less room for it to manoeuvre and less need.

It can also be said that in car parks, or any area where groups of cars park in proximity, during evenings and week ends, those working on them will be brought into social contact through common interest in their vehicles (observed at Willenhall Wood, Coventry).

III. 5.1.2. *Destructive effects.* Co-operation, in a functional and not a sentimental sense, is a rational and logical expression of basic human needs. Examples of such co-operation on a spontaneous basis are: mutual help in looking after small children and babies for the benefit of mothers, the offspring's social development and the relationship of mother and child; help during the crisis of a birth at home, increasing as the acceptance of natural childbirth principles grows; help during illness; help with simple short-term tasks too much for one; taking in of parcels during absence; keeping an eye on the house during holidays, etc.

This kind of co-operation can of course occur most conveniently and spontaneously between people geographically and functionally close.

The disruptive effect not only of the moving car but the very existence of a motor road has been isolated in an extensive research project carried out by the author.(i)

III. 5.2. *Research Project*

III. 5.2.1. *Design of research project.* The factor of access, i.e. the effect of the absence of motor traffic, was isolated by coupling and comparing pairs of areas of housing which were similar in as many respects as possible, except the manner of approach to the front of the house. This was by path in the one, by road in the other. Great care was taken in the selection of samples. Theoretically the same statistical results might have been collected from many single houses, paired as in the above way. But to examine compact areas gave better opportunities for observation and for presenting the findings in vivid sociograms.

To safeguard against the influence of other factors a large number of areas were surveyed, seventeen paired replicates, 504 houses in all.

The main part of the survey was composed of three groups:

1. Victorian. 2. Post 1914–18 War. 3. Post 1939–45 War.

The size of the pairs varied from 2 × 32 to 2 × 8. The older groups were conveniently available in Nottingham, where footpath access in neither type is exceptional. The contemporary pairs are from Radburn housing at Northampton, Wrexham and Sheffield.

Investigation techniques

a) *Interview*

The purpose of the interview was twofold:

1. To ask a prepared schedule of questions in each case.
2. To converse on relevant subjects to form a general impression.

b) *Observations*

The relevant areas were visited a great deal apart from the interview, at all times of the day and the evening, and all times of the week. The observations helped to decide to what extent objections and praise listed by the inhabitants were rational, to what extent subtle contacts, greetings, chatting in the street, waving from windows and talking across hedges corroborated or contradicted the statistics collected, and the information gleaned in conversation.

(For example, it was instructive that a child, whose parents had said that the road in front was *not* dangerous, had been observed as she tried to run into that same road in front of a car. The mother had administered, there and then, such a beating as might well have been the previous history of the little boy in another road, who, according to his mother, "preferred" to sit on the front door step to running around. (Like a pillar of salt, one might add.) Such behaviour, which people tend to put from their minds and forget, even in lengthy interviews, could be deduced, generally speaking, from the observations made inconspicuously.)

III. 5.2.2. *Presentation of results.* Information gathered was assembled in a way designed to give a clear and immediate comparison between individual pairs of path and street, and statistics and summaries describe the picture as a whole, in the following sequence:

(i) Ritter P. Thesis—submitted University of Nottingham 1957. See also *Archit. J.* 17.11.60. ATTBO 2 (1961), 3 (1961), Stockholm.

a) Each pair was described verbally, photographically and by a plan to the scale of 1/1250, with details of access, amenities and surrounding development.
b) On juxtaposed, diagrammatic plans to a scale of 1/500, there are entered on to each house respectively,
   1) the number of inhabitants.
   2) number of generations.
   3) number and ages of children.
   4) length of stay.
c) The same diagrammatic plans have sociograms superimposed to give the direction and extent of contact within each group, for each of the three main types of contact measured. The symbols of the sociograms are such that an immediate impression of the amount of "blackness" on the plan is indicative of the amount of social contact.
d) Subjective impressions, as recorded from the inhabitants, are reproduced verbatim.
e) Subjective impressions of the investigator, formed on the basis of all sources, are given separately.

Below some typical survey results for one of the Victorian pairs (i.e. a path and a street each lined both sides with a terrace of houses) and for one inter-war pair.

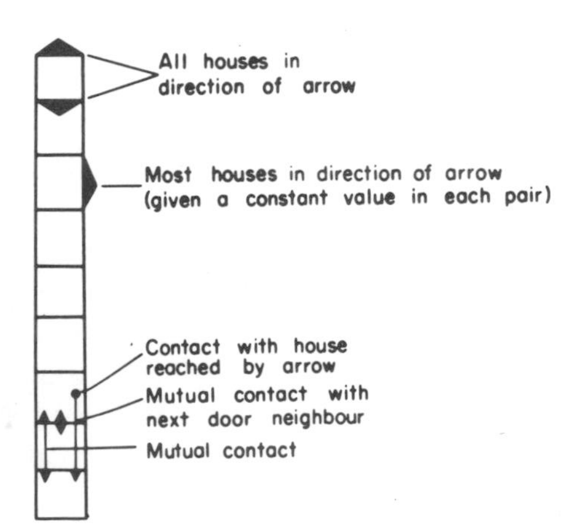

Fig. 3.36. Symbols on Sociograms.

A. Mid-Victorian Example

Fig. 3.37. Path sample.

Fig. 3.38. Road sample.

Fig. 3.39. Path plan and Road plan.

Opinions and Impressions of Inhabitants (logical and illogical)

Path: "Main road at end danger to children."
"No road an advantage." (Two with children.)
"Locality changed, different class of people now, children a nuisance, does not interfere with anybody if they don't interfere with her." (Lady of 70, resident 36 years.)

Street: "Like living in street."
"I was born in 'terrace' (path) but prefer street because the terrace is a throughway to the main road."
"Non-schedule buses use the road to avoid traffic lights; they speed."
"I never leave the children alone, they are always in the garden."
"I don't speak to anybody." (Fifteen years' resident.)

Observations of Investigator

Path: This is a connection between two roads, one of which is a main road. There is a very considerable flow of walkers through this area; it is more exposed visually and less enclosed. There is little feeling of privacy. From the main road this area is a green oasis in the stone and brick all around. Trees, too large, are pruned and deformed with periodic severity. There is no lighting. All houses use back and front doors.

There is a concentration of relative newcomers on one side and this seems to have had a disrupting effect. Nevertheless there is a friendliness and a sense of belonging apart from the statistical recordings. Children play in relative safety. Every one of the inhabitants is satisfied with the path access and the two with children stressed this with a rare emphasis. Three, of six families with children, can ask for help with these. Five families spend some leisure time with their friends in the area concerned and nine could ask to borrow a tool.

Street: All use back and front doors.

Although this is a well-defined area there was no feeling of identification with the street or the group of houses at all. One person would like to have changed to a path. Three are most vehement in their dislike of paths. They cannot explain this rationally and wrinkle their noses when talking about it as if touching something nasty. Nor could it be discovered why dogs were "bad" in the path area. The only theory to make rational sense of this might be that as the public area for dogs is smaller, the nuisance of dogs' faeces per square yard is greater. Nor is there a gutter, perhaps an important detail.

Both families with children can ask for help with them. Three people spend leisure time together and nine can ask to borrow a tool.

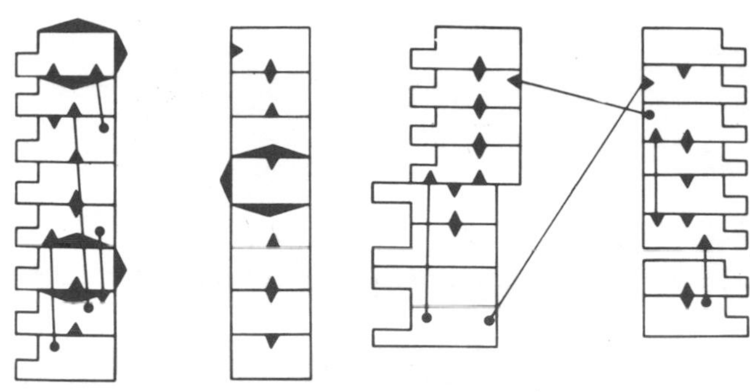

Fig. 3.40. Taking in a parcel.	II Path	IIA Road
Total of contacts	65	22
Houses in sample	16	16
Contacts per house	4·06	1·37
Houses in contact	14	13
Contacts per house in contact	4·64	1·69

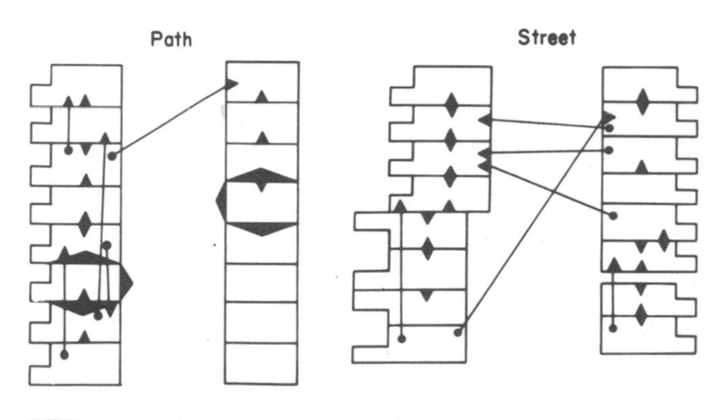

Fig. 3.41. Essentials in emergency.	II Path	IIA Road
Total of contacts	43	27
Houses in sample	16	16
Contacts per house	2·69	1·69
Houses in contact	9	13
Contacts per house in contact	4·77	2·07

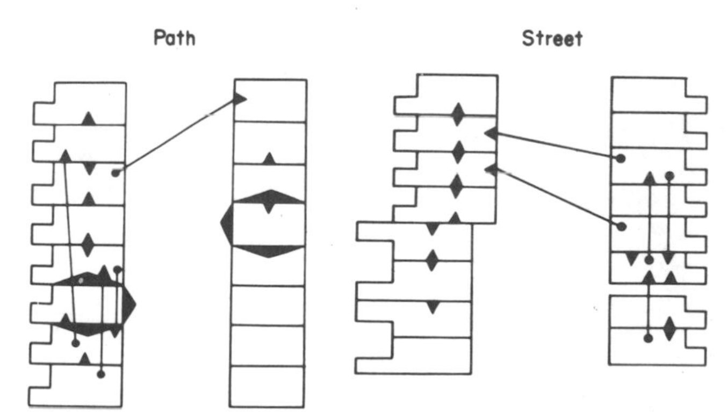

Fig. 3.42. Help in illness.	II Path	IIA Road
Total of contacts	45	20
Houses in sample	16	16
Contacts per house	2·81	1·25
Houses in contact	10	11
Contacts per house in contact	4·60	1·81

b) *A* 1929 *Example*

Fig. 3.43. Path sample.

Fig. 3.44. Road sample.

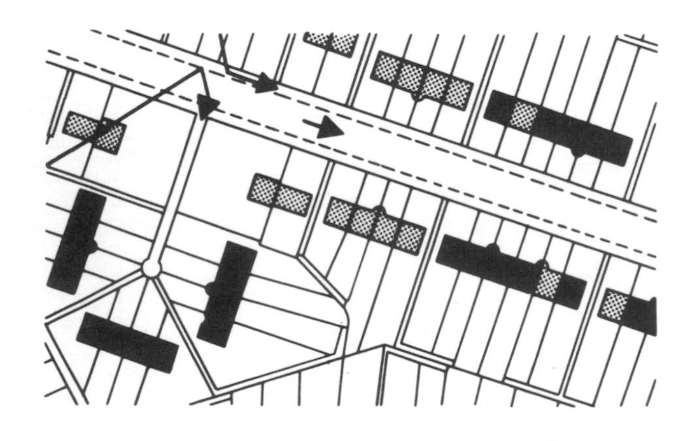

Fig. 3.45. Path plan and Road plan.

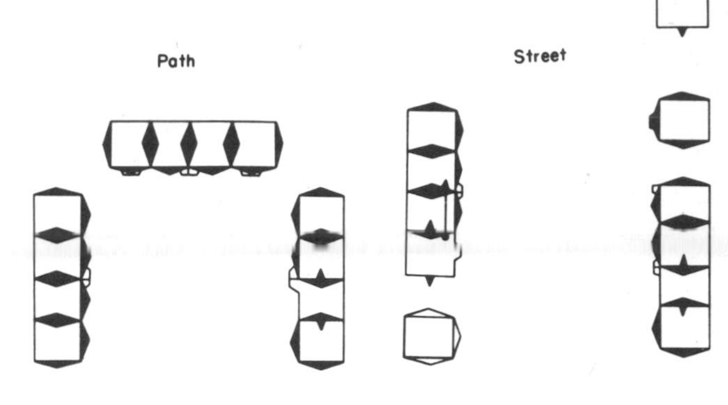

Fig. 3.46. Essentials in emergency.	X Path	XA Road
Total of contacts	122	94
Houses in sample	12	12
Contacts per house	10·17	7·83
Houses in contact	12	12
Contacts per house in contact	10·17	7·83

Opinions and Impressions of Inhabitants (*logical and illogical*)

Path: "Children congregate in here."
"It's more private."
"It's ideal for little children."
"Children like playing at the top."
"There's more interest in the road."
"I like it, it's quiet and there are no children."
"It's nice, away from traffic."
"Nice and private and no cars."

Street: "The road is safe." (Three.)
"It's lively and there is lots of traffic, it's not safe." (Two.)
"The close (path areas) are closed in."
"Living on the front ('roadfront') is better there (in path), the children play round lamps" (there are none!) "and fences."
"It's easier for callers on the road."
"I am frightened of the dark in the path."
"You are so shut off in the cul-de-sac." (From slums.)
"There are no buses here, I don't like the paths, I don't know why."
"All the traffic up and down makes you wonder with the children, safer in path."
"Children play in the road; greens round here not for playing on!"(i)
"I like to see a bit of life."

(i) Most greens in Nottingham are railed off or have notice boards forbidding play on them!

Observations of Investigator

Path: The statistically recorded co-operation in these twelve houses is the highest found anywhere. In relation to taking in parcels every possible contact is felt to be available, i.e. each one of the twelve people feels she could ask the eleven others. The feeling of belonging is very strongly developed; any disadvantages are regretted because the path is liked. There is strong feeling of privacy. Except for children, others do not venture up the cul-de-sac. It is this lack of life, this imposed privacy, that is felt and frequently expressed as a disadvantage. Surprisingly, none of the four families with children could ask anyone for help with them, only one person spends leisure time with another, or admits to doing this, but eleven feel they could talk over a personal problem. It may well be that in this area the children are rather old and so the mothers cannot see any reason to ask for such help. This is a factor that might explain what otherwise seems like an odd pattern.

Street: This is a straight minor road with traffic growing rapidly as cars become more frequent on the estate. The narrowness of the road and tree planting make it quite pleasant. With an absence of garages there is a parking of cars along the road which makes it into a particularly unsafe one. This is most pronounced in the evenings. Yet it enhances the children's play! (if not the cars.)

No one wishes to move into a path area and the objections to the path area are, as generally found, vague and irrational.

One of five houses with children can ask for help with them (three children under six in the sample).

III. 5.2.3. *Summary of whole survey.* The general statistical conclusions of the whole survey of 17 pairs are as follows:

a) The total number of contacts (help with essentials in minor emergency, through taking in parcels and in illness) is significantly greater in the path areas. p is less than 0·01 in each case.[(i)]

b) The number of contacts per house is likewise significantly greater.

c) The total greater extent of contact is not only due to greater extent of contact across the path but also along each terrace, separately.

d) The greatest difference is found in contact through help in illness, the most intimate of the three main types of contact measured.

e) The extent of "all-round contact" with everyone in the group, as compared with pure "next-door-neighbour-only" contact, is also instructive: "next-door-neighbour-only" contact is found in roughly a third of the households along paths, but in half the households along roads. All-round contact on the other hand occurred almost exactly twice as often in paths as in roads.

f) Roughly 40% of path households spent some leisure time with others in the sample compared with 30% in the roads.

g) Sixty-two per cent of path people thought the area friendly, only 42% of street households expressed this opinion.

h) Fifty-two per cent of path households with children can ask for help with them, but only 37% of road dwellers are in that position.

i) Twice as many households with children are found in those path areas in existence over 25 years, where free choice has been operative for long enough to make this extremely significant. Knowingly or unknowingly people have preferred such areas for children, it seems.

j) The percentage of households without any contact, in asking to borrow essentials and to take in parcels, is similar in path and street. The differentiation is found in the total number of contacts found in each group, the number of all-round contact households and the extent of contact in help during illness, which is a more intimate relationship.

k) Taking, then, help in illness, we find that 11% of households in paths could not ask anyone for help. In the roads this disturbingly high number soars to 17%, one in every six households.

Thus it is clear that the statistics indicate to planner and architect that households along paths gain, sociologically speaking, to an extent that is quite decisive.

The information is directly relevant design data required for housing lay-out, and to go contrary to the evidence, unless specially justified, is obviously perniciously wrong.

Where walking becomes more frequent, whether it is by old people, mothers with prams or children to school, friendships start more frequently and loneliness is likely to be a rarer event.

(i) Probabilities obtained by ranking paired replicates; see Wilcoxon F. *Some Rapid Approximate Statistical Procedures.* New York 1959.

III. 5.2.4. *Survey at Stevenage.* Finally, a similar sample survey carried out for the author at Stevenage. The sociologist who carried out the survey wrote:

"*The social contacts are useful ones, they include help at childbirth, baby minding, help in sickness, sharing a washing machine, running a clothing club, paying each others' rent etc.*"

Fig. 3.47. Path sample.

Fig. 3.48. Road sample.

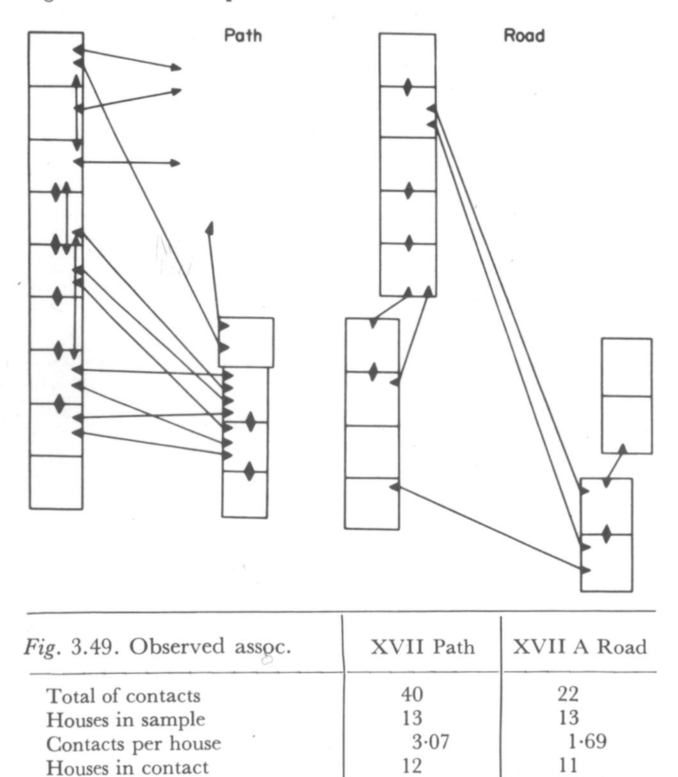

Fig. 3.49. Observed assoc.	XVII Path	XVII A Road
Total of contacts	40	22
Houses in sample	13	13
Contacts per house	3·07	1·69
Houses in contact	12	11
Contacts per house in contact	3·33	2·00

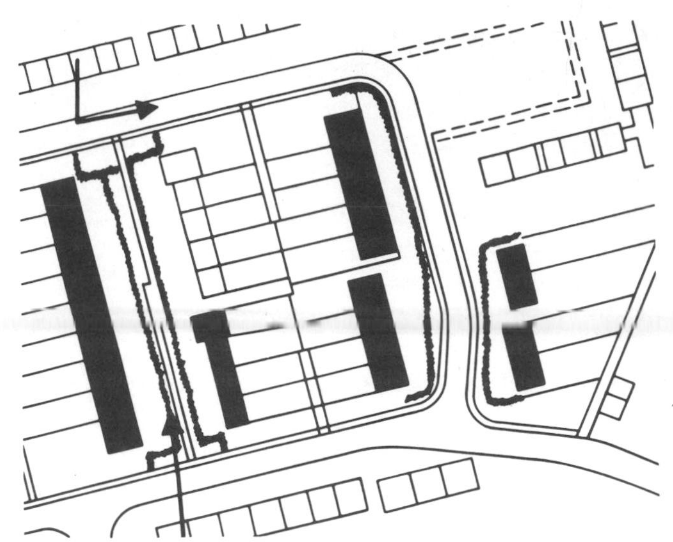

Fig. 3.50. Plan of path and road.

III. 5.3. *Social Aspects of Public Transport*(i)

The nature and quality and location of public transport determines its value as a potentially fruitful time for meeting, conversation, discussions and other socially positive phenomena. In designing public transport vehicles the wish to converse, as indeed the wish to be left to read, or think quietly, should be borne in mind.

For the young, the meeting of others through the use of public transport is a positive asset not gained through daily commuting by private car (how many spouses met at bus stops?).

Useful and pleasurable associations result from repeated coming together. This means the travelling time is positively used for more than the purpose of travel, particularly for regular travellers to work, school and shops.

An outstanding example of this is The Debating Society formed on the commutors train from Brighton to London recently, asking for extra members in *The Times* through an advertisement.

If the positive aspects of communal travel are borne in mind then the design of stations and the means of conveyance themselves will develop according to more than the purely economic criteria now used.(ii)

III. 5.4. *Moral and Legal Considerations*

The proposition for a separate path system conjures up in the minds of many people only the path systems they know. These are often in parks, dark, unlit, and sparsely used, lonely and the scenes of sexual crimes. The latter are exploited to the full

(i) LIEPMAN K. *The Journey to Work.* London 1955.

(ii) At the back of some models of the Boeing 707 there are some seats which face each other across a table top. The popularity and attractive nature of these places compared with the rows of tourist or first class seats illustrate the point.

by the press to give the aura of danger to these paths even where this is not rational. There are areas, such as Central Park, New York, where it is rational to be afraid however.[(i)]

It is therefore important to clarify the point. First of all one must distinguish between that which is criminal and violent and that which is merely sexual, and perhaps offensive to some people, but not illegal and indeed, as Lewis Mumford stresses,[(ii)] is an important human function for which the path system may rightly provide.

Secondly, the motor car is in greater use for an extension of sex-life, legitimate or illegitimate, than any path system of the future is ever likely to be.

Thirdly, police arguments which say that paths cannot be patrolled by vehicle, that criminals cannot be properly pursued if they run on to path systems, and that paths plus roads necessitate a doubling up of police duties, must be analysed.

It emerges then that paths planned as an integral part of housing are much more the concern of the inhabitants than the normal road in front of houses so that policing becomes unnecessary. Emergency 'phone boxes are all that is required. More efficient pedestrian-free roads need policing only by cars. To plan cities from the point of view of ease of pursuit of criminals by the police is absurd and must not determine a policy contrary to many primary requirements of people. (It reminds one of the planning of Paris by Hausman for ease of military access.)

Screams can be heard in a quiet path environment, and, because of their feeling of belonging, inhabitants take immediate notice of it, unlike the "no-one's concern road".

A path system should be and can be well lit with windows giving more light and a "safe" feeling, quite unlike the park paths imagined by those who cannot go beyond the limits of their own experience.

Finally, the emphasis laid on sexual crimes and attacks tells us more about the speaker than about the facts. It has a sobering effect to inform people that 90% of children murdered are murdered by their parents!

In Montreal a footpath system of the Cité Jardin[(iii)] has no public lighting whatsoever. In twenty years there has been no attack of any kind. The children of the area do not regard the dark paths as areas of danger. This is due to the feeling of belonging. The areas which include paths regarded as extensions of the gardens, are sometimes separated by fences. These occur, for example, where families

(i) JACOBS, J. *Life and Death of Great American Cities*. 1961.

(ii) *Town Planning Review*, Vol. XX. 1941.

(iii) See example.

have had to cut off their swimming pool from the public path—so that it does not represent a danger to young children.[(i)]

In terms of legal requirements, it is plain that if the needs of pedestrians and vehicles are considered specifically, then the bye-laws and regulations for one will be inappropriate to the other.

The law could help positively: to quote from a propaganda booklet published by the author and distributed, among others, to all M.P.s in 1961:

"It is now high time to recognise as criminal the building of roads for the monstrous immoral mixture of traffic that causes accidents, and pollutes both the driver's realm and the pedestrian's. This would put legitimate pressure on professionals to inform themselves of up to date (even if they are thirty years old!) techniques and to apply them with gusto."

"It would further force the private developer to cease feeding the lethargic, ignorant public with lethal, 'lollipop' lay-outs. After all, lollipops dangerous to health are forbidden by law, so is insanitary drainage, and unsafe construction and the belching of smoke;"

"Why not the more spectacularly harmful mixing of mutually irrating mutilating traffic?"

"Make traffic 'pollution' illegal—and let us have the lovely alternatives."

In summer 1963 C. D. Buchanan was reported in the Sunday Telegraph as hinting that some legal enforcement of traffic segregation might be necessary.

Fig. 3.51. Basildon, Essex. Who would attack anyone here? And who would not come out to give protection?

III. 5.5. *Crowds of People and Crowds of Cars*

Vehicles have no social life of their own, and even where they are gathered together, as in open-air

(i) "*Monks at the Franciscan priory at Palmerston North fear that a proposed path around the Centennial Lagoon would destroy their privacy. The priory joined a deputation to the city council when it was asserted that a path on the eastern side of the lagoon would present an opportunity for immorality.*" *New Zealand Herald*, quoted in *Punch*, London, 11 January 1961.

cinemas in California, this is a factor which, except for the relationship between occupants of each car, is not conducive to sociability.

This is important in the consideration of shopping streets which are made unpleasant for social intercourse by the presence of groups of cars. When they are gathered in huge car parks surrounding the shopping centre there is additional danger of isolating the centre from the people and their dwellings and paths round about, socially, by the great desert of metal and asphalt.

III. 5.6. *Psychosocial Factors*

Jackboots make us less sensitive to the environment and also, symbolically and really, to other people. Encased in "steel armour" almost every driver changes from his pedestrian self into a far more aggressive personality. This is well known. In this manner it is interesting to note that not only does the car have antisocial effects, but once tied to the car, people themselves become less sociable, co-operative, rational, considerate and kind. The car has only one way of communicating by sound: "Get out of my way!"

Fig. 3.51a. Half of a cartoon by ROSNER R., from J.R.I.B.A., February 1961

It is often argued that children brought up in a path system will not learn to negotiate roads safely when they have to. This is not the case. The fact is that a child can easily learn to take care on specific occasions. But in an environment of traffic roads only we are asking for constant vigilance. This no healthy young child can manage: it does not allow it to dream, or run, or let go.

So the educational argument is truly this, that, given a path system, there is a reasonable basis for teaching children that roads and cars are dangerous.

III. 6. Planning Solutions

III. 6.1. *Functional Traffic Arrangements*

Man must have a worthwhile environment. Vehicles are a wonderful aid and should be well provided for. The problem the planner cannot solve is the absolute evil of poisonous petrol fumes. Beyond this the basic and thorough examination of physical, ecological and sociological factors lead to the conclusion that the conflict between man and vehicle can be resolved if each is given a separate, fitting environment. This traffic separation, or segregation, is one fundamental factor in future planning. It can be achieved in a number of ways—used singly or combined—both in old and new areas.

There are three basic ways in which pedestrian and vehicles can be given a separate environment.

a) Horizontal segregation.
b) Vertical segregation:
   i) Pedestrian above vehicles, pedestrians on ground or on deck.
   ii) Vehicles above pedestrians, cars on ground or on deck.
c) Segregation by time.
   Vehicles (or pedestrians) allowed at certain times only.

These possibilities will be discussed in the contexts of hundreds of examples. (For analysis see p. 157.)

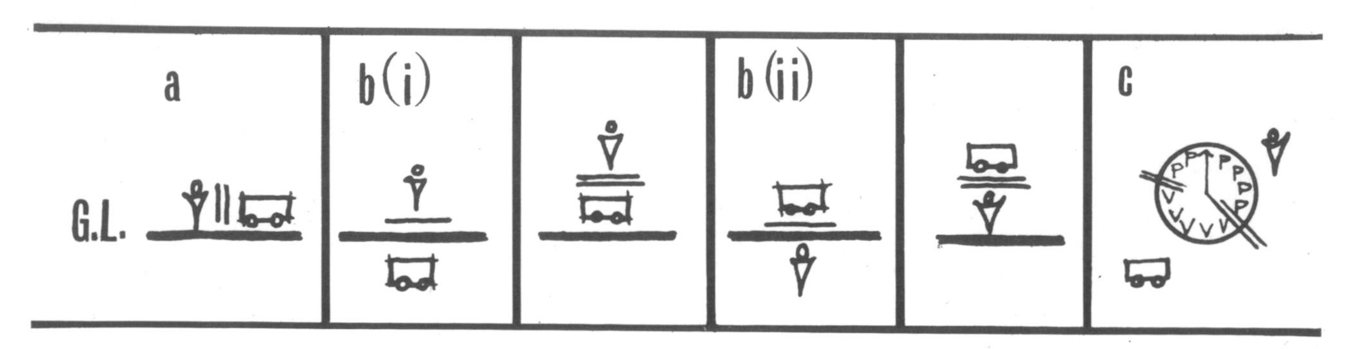

Fig. 3.52. Diagram of traffic segregation possibilities.

III.* 6.2. *Character of Roads and Paths*

The path system should reflect the climatic, micro-climatic, cultural and local patterns of living and building. This theme is referred to throughout the book. Road surface will tend to show far greater similarity the world over. But local building, plants and contours should be in specific harmony.

Fig. 3.53. The grandiose scale of the main road network to satisfy the growing private car ownership. The great width distinguishes it decisively from railway lines. The tighter bends reflect the greater flexibility of road traffic.
(Boston, Mass, U.S.A.)

Fig. 3.54. Grandiose shapes can have a fine sculptural value, truly to car scale. However, in Los Angeles, where this photo is taken, one seems to spend much time driving under bridges, or so it seems.

Fig. 3.55. Pedestrian traffic has come to be taken for granted as merely a matter of "sidewalks" to the road. But once a path system is perceived as a separate entity from the roads the imagination should adapt itself. Possibilities increase as new techniques of surfacing become available. Many small areas of different materials can occur in the path system, fulfilling many functions. (Greenbelt, Maryland, U.S.A.)

Fig. 3.56. The ingenious stair on the left has ramped tracks for donkey carts. This form is found in other towns of Italy and some have steps carefully geared to donkeys in the middle strip and to human beings on the outer strip. Photo, Millard.

Fig. 3.57. Democracy in action. And where things get desperate the children take matters into their own hands: a "Kids not Cars" demonstration in a street in the East End of London. Other placards read "Play-street not Park-street", "Keep out cars or else", "Wake up Council", "Play the game it's our play-street", "We want a park", "Cars mean death". A later picture showed the children pushing a car bodily out of "their" street.

IV. Needs of Man as an Organism

IV. 1. Biological Function of Walking. **38**

IV. 2. Qualities Required in the Pedestrian's Environment. **39**
IV. 2.1. Man Sized Spaces. 41
IV. 2.2. Detail Matters. 43
IV. 2.3. Endless Variety. 44

IV. 3. Study of an Urban Pedestrian Piazza. **46**
IV. 3.1. Minute by Minute Survey. 46
IV. 3.2. Visual Analysis. 47
IV. 3.3. Explanation why Piazza works so well. 50

IV. 4. Convenience. **51**

IV. 5. Mechanical Aids. **52**
IV. 5.1. Pedal Cycles. 52
IV. 5.2. Pedestrian Conveyors. 53
IV. 5.3. Escalators and Lifts. 54
IV. 5.4. Trolleys. 54
IV. 5.5. "Pedestrains". 54
IV. 5.6. Carveyors. 55

IV. 6. Information. **55**

IV. 7. Comparison of Path Systems. **55**

IV. Needs of Man as an Organism

IV. 1. Biological Function of Walking

Today, towns discourage walking. Standards of safety are low. There is little which is picturesque. The air chokes and poisons. Car noises deafen and drown voices. Scale, surface, speeds are unsympathetic. However, the use of limbs and lungs is basic to healthy development of the organism. It is therefore a necessary quality of town-planning that it attracts people to walk in cities, residential areas and centres.

Doctors, again and again, recommend exercise and deep breathing of fresh air as a remedy for many ailments and as a prophylactic measure. They have to prescribe as a medicine what should be part of the pleasure of existence itself, because to breathe deeply in our streets today is to savour poisons and fumes. Shallow breathing is the natural defence against such displeasure, as is also the change from walking to driving and riding as passenger.

Examination of 7000 children showed that 58% from the U.S.A. failed in a fitness test whereas only 9% of European children did so. Lack of exercise (and walking was taken as one of the most important forms) was isolated to be the reason for the low U.S.A. standard.(i) Children who have had exercise learn better at school, sleep better at night and feel fresher the next day. The play-activity essential to all young animals involves running and jumping to a great extent. These are not separate activities for "play-grounds". This is activity indigenous to the whole life of the child—it is not practicable along roads, as it is along a path, playing all the way to and from school: catch games—ball games—singing and shouting—telling yourself stories as you go along. We need "linear children's play grounds". The paths to school are the most important. Not only can healthy children cover incredible distances with great pleasure if the route is inviting, and learn a lot on the way, but safety allows children to linger on their way home. There need be no worry about traffic accidents, which is now sadly rational, as soon as anyone is late.

Chatting to friends is a joy while walking, unlike conversation in bus, train or underground against roars, puffs and whistles. Contact with trees, flowers, the sky, birds, squirrels and sheep can be a daily routine. There need be less urgency to spend week-ends driving into the "country", desperate to get into a little bit of green (surrounded by fellow seekers). Taking the dog for a walk can be a pleasure.

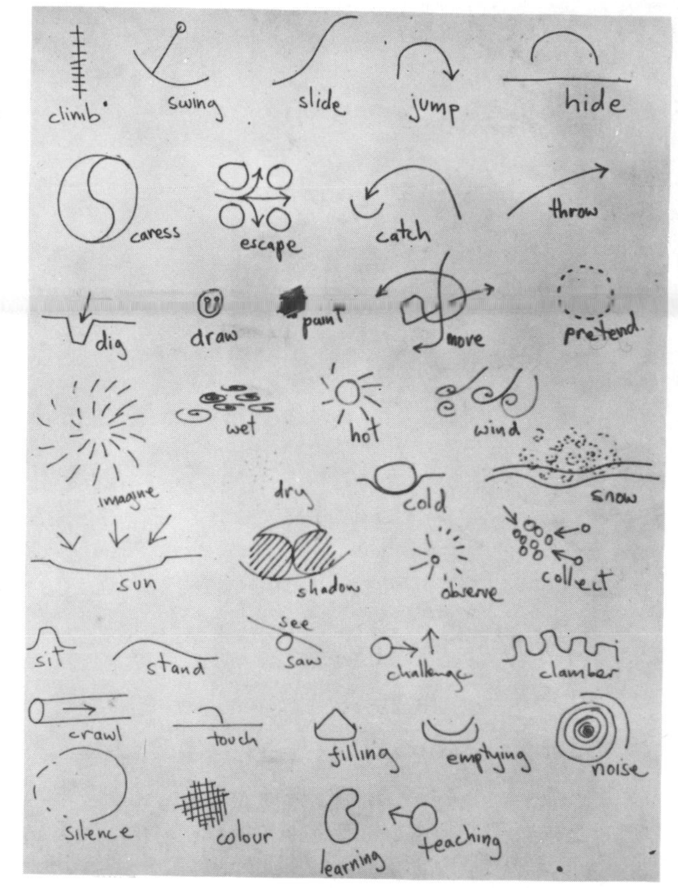

Fig. 4.1. Play-movements drawn by Tony Gwilliam.

Old people can have an inviting new dimension to their life. People will choose to walk farther and more often.

Those who claim that the American tendency of more cars and less walking is irreversible do not see the whole picture. U.S. citizens can be seen to walk considerable distances, hilly and exposed, on the University campus, without noticeable grumbles.(i) The suburban shopping centres mean a great deal of walking for the thousands who flock to them: from the parked car, as much as 600 unpleasant feet (183 unpleasant metres) to the entrance of the centre, and inside, probably under cover, as much as 2000 convenient feet (610 convenient metres). Then back the 2000 feet (610 metres) and the 600 feet (183 metres). Where walking is pleasant, where the environment is planned for walking, even U.S. citizens will flock to walk, and like it.

It is useful to describe in some detail what attracts people to be pedestrians. Once one plans for a purely human environment, the designer can attune far more sensitively to the individual need and character of each path, lane, square, court, piazza, alley, parade, bridge, underpass, avenue, precinct, green, row, back, twitten, gannel, snicket and vennel.(ii)

(i) New York University Project, reported by Vogler and Kühn in *Medizin und Städtebau*, Vol. 1. p. 484. Munich 1957.

(i) See also Reston New Town, U.S.A.

(ii) From Farehoe and Browne, *Architectural Review*.

The tactile quality of floors varies a great deal. The springy lawn, the compacted earth, even tarmacadam are gentler to the muscles than stone slabs or concrete. However, rubber and composition shoe soles and heels can transform hard impact also. Shoe fashions, like fashions in cars, act against efficient pleasurable movement, encouraging change for the sake of profits. Variety in floor texture, keeping comfort for pram wheels in mind, should be contrived where possible.

To show what the new environment might look like helps a great deal towards the attainment of it. Men have forgotten that it can be a joy to walk to work, smelling freshly cut grass, noting the seasons, falcons circling, streams gurgling, children running: why not return to sanity?

There is in Gothenburg an exquisite experience for those who leave their cars. From the splendid new Student's Hotel, Volrat Tham (for both sexes), a footpath leads with delightful informality. It invites, and eight photographs on the next page will show the route.(i)

IV. 2. Qualities Required in the Pedestrian's Environment

Volumes, areas and lengths must be geared to the space–time scale of walking. "Man-size" spaces, to use Sir Hugh Casson's vivid phrase, have full effect. Differentiation between areas of travel and areas of arrival in microcosm and macrocosm become a key to the design theme. This can be experienced in Cumbernauld already.(ii) The innate tendency to meander, and the need for the shortest routes for the pedestrian, "desire lines", are taken into account.

Contrasts, giving rhythm to walks, can be gained in a number of ways:

Change of direction: sharp corners, with surprise and promise beyond, are possible when car sight-lines do not dominate design (see the Gothenburg example).

Change of size or height: a straight path leading through a variety of spaces is not felt as a straight line. Experience of a route is an interaction between the pedestrian and the environment. Enclosing elements of the path coming near and going out simulate a meandering path. Ceiling variation range from the sky, to trees, roofs of covered ways to give shelter from weather, underpasses and archways.

(i) All photos by Nadja Wiking and kindly sent by Professor Sune Lindström from Gothenburg after I had experienced the path and wanted to write about it. P.R.

(ii) See example p.117.

Fig. 4.2. Canopy enhances long view with tighter volume. Caversham, drawing by Roy Gibson.

Fig. 4.3. Underpass contrast by volume, sound and light. Stockholm.

Change of texture of floor, walls or roof: differentiation in paving can add pattern, walls proper may allow views through or hedges provide contrast. Green and urban character may follow one another. External walls must be regarded as having a multiplicity of functions on the outside. Just as the inside wall serves not only to keep out the weather but also to hang pictures, lamps, provides for recesses, flues and windows, forms bays and niches.

Fig. 4.4. Hoops are fine on footpaths.

Fig. 4.5. From the really dense bushes and trees...

Fig. 4.6. ...you emerge onto a small but delightful park.

Fig. 4.7. Past the lake and past a school playground, informality remains with greenery close around.

Fig. 4.8. Up some steps and tall walls rise full of promise.

Fig. 4.9. Small cracks between them allow a little set of steps to lead through and down.

Fig. 4.10. At once one has a breathtaking change of scale and formal urbanity replaces the paths and tracks.

Fig. 4.11. Moving out from among the impressive buildings the piazza and vista in front of the large pedestrian square are really effective.

Fig. 4.12. The powerful sculpture and its fountain command even this great space. Beyond it, turning around one has the full view of the Museums, the Opera House and the Concert Hall.

To make the path environment really interesting requires mixed land-use, so that not only the colourful, useful detail of 'phone boxes, post boxes and posters are encountered, but also shops, slot machines of endless kinds, factories, markets, stations, hot dog stalls, cinemas, theatres, churches, offices, etc. The view of the vehicles, at a different level or clearly, horizontally segregated, for controlled sections, also adds interest. Such limited association with vehicles is quite a different matter from the present constant, direct, compulsory association. Similarly, rapid transit in the form of electric trains, their viaducts and movement, enlivens the environment, gives it meaning and colour.

Lighting should come from standards of a delicate size or brackets. Warm, plentiful light should not disfigure nor make a hypocrite of the moon-light lover who has to tell his loved one she looks heavenly, even when sodium light makes her look like death warmed up.

It is an axiom in the age of relativity that a pleasant walk is a shorter walk than a tedious one.[i,ii]

Because of their width, roads are very obvious routes to take. When a path system is introduced to people who are used to walking along roads, they may find it difficult at first to follow paths. We visualize the route we use whether at the outset of a journey or while directing someone else. Paths present a different problem from roads in this respect. Thus, it is important to bear this in mind and design and identify routes with marks of recognition that allow easy visualization and easy orientation both for oneself and the directing of others. The architecture and the floor treatment are the most obvious means for this, but where difficulties arise paths could be signposted fully and if it is done with taste and possibly a touch of humour (not practical jokes) it will add to the scene. The above applies to large residential areas and new towns in particular. And if it is not borne in mind people may use roads even though they are the longer way round (see also Section VIII).

(i) BEAZLEY E. *Design and Detail of the Space Between Buildings.* London 1962.

(ii) CULLEN G. *Townscape.* London 1961.

IV. 2.1. *Man Sized Spaces*

The many examples in this section, showing a great variety of pedestrian-only areas, urban or green, should kill the fallacy that such areas must be just this or just that, that only the vehicle can bring sufficient interest and life to the traffic routes we use for the greatest part of our lives. Venice manages to be interesting and vital without cars, with its own specific traffic segregation of water-ways and land ways. All the examples in this book show further variations on the theme of paths and spaces for man.

Fig. 4.13. Patio housing, Germany. Professor Rainer.

Fig. 4.14. Split level housing, Cumbernauld Development Corporation. A space with promise beyond.

Fig. 4.15. Path between Patio Housing, Copenhagen, Eske Kristensen. Only access to dwellings.

Fig. 4.16. Patio housing, Copenhagen. E. Kristensen.

Fig. 4.17. Housing in Baghdad. Doxiadis Associates.

Fig. 4.18. Housing, Carbrain, Cumbernauld New Town Development Corporation.

From the above emerges the new pattern of alley and "little place", "man-sized space" point of arrival.

Fig. 4.19. Farsta Shopping Centre. Safe and exciting on the way home from school.

Fig. 4.20. Brightly lit path at Hemel Hempstead, obviously married to houses and safe.

Fig. 4.21. Greenbelt, Maryland, hedged gardens and the minimum of maintenance area.

IV. 2.2. *Detail Matters*

Fig. 4.22. Rails are fun.

Fig. 4.23. Blackeberg shopping precinct, fountain with many uses.

Fig. 4.24. Walls as play equipment; and fences: you run and rattle sticks. Cumbernauld.

Fig. 4.25. Bollards in Coventry's centre.

Fig. 4.26. Detail, minute detail matters.

Fig. 4.27. Willenhall Wood I, Coventry—small wall, bench, flowers, rail, dog and lots of children.

Fig. 4.28. Hard wearing surface used in a lively way. Cumbernauld.

IV. 2.3. *Endless Variety*

Fig. 4.29. A mediaeval town street for man and slow small carts. The stone slabs in the centre denote an early appreciation of the needs of man on foot. (After Reichow.)

Fig. 4.31. Covered pedestrian bridge between separately owned bank and shop, Minneapolis, U.S.A.

Fig. 4.32. Continuous deck access used by milkman, Parkhill, Sheffield. Photo by Roger Mayne.

Fig. 4.30. Fontenay aux Roses, flats by Lagneau, Weill, Dimitrijevi, and Perrotet. Photo by Biaugeaud.

Fig. 4.33. Goods lift entrance recess—the only recess on deck at Parkhill, so used for playing "house" and by lovers.

Fig. 4.34. Mediaeval Chester, The Rows, Vertical Traffic Segregation. Although today both "Row" level and street level are shops, the picture seems to indicate service and supply at street level, and shopping in the protected, clean upper decks.

Fig. 4.35. The city wall walk in Chester crosses the main roads and gates on arches in a number of places. The famous clock is on one of the gates.

Fig. 4.36. The moat in romantic setting and the Cathedral close give a very fine quality to the Chester Wall Walk.

Fig. 4.37. Covered pedestrian bridge, from mediaeval Lucerne.

Fig. 4.38. Covered paths link blocks of dwellings with shopping and bus stops. Coventry, Hillfields.

Fig. 4.39. Cluster of Colleges, Oxford. Cloisters provide covered routes.

IV. 3. Use Study of an Urban Pedestrian Piazza

Christopher Millard has made a minute-by-minute study of the Piazza del Palio in Siena with its shell-shaped, central, pedestrian "campo" approached by eleven, narrow, mainly pedestrian streets. (Survey May 1963, photos January 1963.)

IV. 3.1. *Minute by Minute Survey recording People Entering and Leaving.*

A small sample from 1050 items recorded, from 5 a.m. to 10 p.m., to show the variety of happenings and the kind of observations which pick out significant detail from what is poetically termed "the pageant of life". Fundamental is the point that only in a pedestrian environment could these things happen and seem significant.

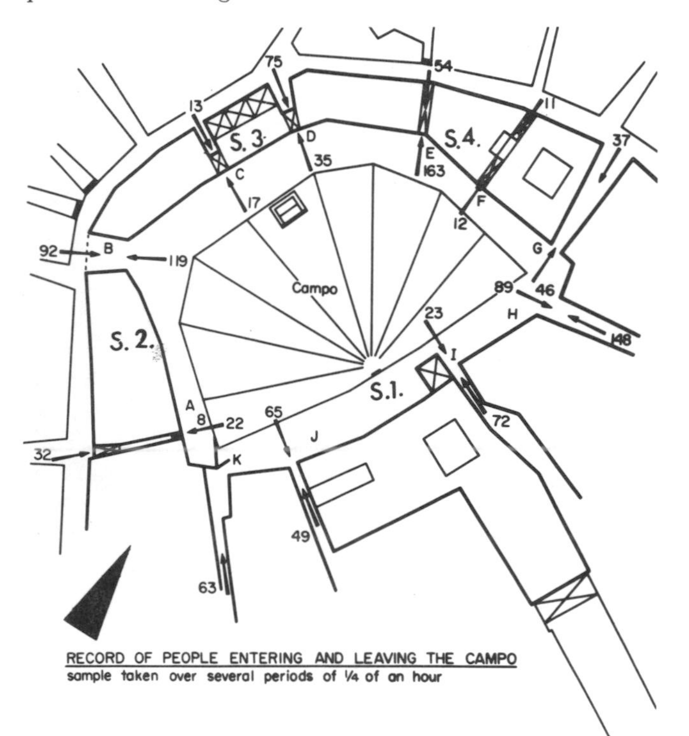

Fig. 4.40. Plan of Piazza del Palio with record of numbers of people arriving at and leaving campos.

To quote from Millard's report: (items selected)

"5.00 *Sun striking buildings on sides 2 and 3.*

5.10 *Practically no activity, a few people crossing to fruit and vegetable market.*
Road sweepers now active.
Swallows flying in great numbers twisting and weaving at a lower level, twittering and screaming all the time.

6.00 *Banana stall wheeled onto side 4 from I.*
Man carrying bundle of newspapers disappears up E from J.
Man with dog enters Campo from I with dog on lead, releases animal which promptly performs in centre of shell much to owners satisfaction.

6.30 *Woman walking with bicycle from which hang three milk churns; passes through piazza from K along 1 to H, very gently all the way.*
Shopowners sweep outside their individual premises.
Man leans on bollard then decides that the one he has chosen for meditation does not provide him with a good view moves four to the right to rectify situation.
Workman pushes wheel barrow along 1, up to K and along top of campo to A, barrow hauled backwards up steps.
Blue nun (colour is remarkable in early sun) crosses from B to I.
Old man leans on railings of Fonte Gaia and watches the water.
Several elegantly dressed girls, one carrying floral decoration, emerge from K and walk along 1 to H.
Two workmen carry on long distance conversation as they approach one another from opposite sides of the piazza.

7.00 *Dog chases pigeons.*
Man studies wall map of city.
Two dogs romp together (tremendous gyrations).
Chairs being arranged outside bars on top of Campo.
40 *people in or crossing Campo.*
Policeman arrives on Campo from Palazzo Publico watches water in Fonta Gaia.
Man leans on bollard shading sun from eyes side 2.

7.30 30 *people in the piazza, most in transit on way to work.*
Milk being delivered to buildings looking onto the Piazza.
Large party of youths and girls, 30+ *stand about Fonta Gaia having emerged from D, group photographs being taken.*
Man fills watering can from tap of Fonta Gaia.
White dog, apparently no owner, wanders about Campo.
At least 2 *stationary discussion groups are in progress on the piazza now, several pairs, mainly men.*

7.45 60 *people on Campo.*
Old man sitting in chair on top of Campo surveys the activity thereon.
Boy carries large box of vegetables across Piazza.

8.00 *School children still much in evidence crossing in all directions.*
Tourist in alarming pink dress has appeared from B, stands outside bar on side 2.
At least a dozen chairs outside the bars are occupied at the moment.
2 *sinister looking youths sitting on wall of Fonta Gaia.*
About half the shops round the Campo are open.
Piles of rubbish swept up earlier are now being collected.

8.30 40–50 *people on Campo.*
2 *men push mobile tourist stall into place on top of Campo.*
Several men leaning on bollards, some reading, some watching what little activity there is at the moment, two old men talk, 2 *men now cleaning out their pipes under water tap of Fonta Gaia.*

9.00 *Several youths apparently with nothing to do watch girls carrying water from Fonta Gaia taps.*
Woman throws bread to pigeons, on one side of piazza, then all take to the air and fly to another spot where someone is apparently producing more interesting food.
Quite a trickle of people crossing from E to I.

9.30 *Shaggy black dog chases pigeons.*
130 *people on Campo.*

9.45 *Priest rests briefcase on top of bollard and muses in the sunshine.*
Woman being photographed feeding pigeons.
Small girl stands and admires the crowd at the bottom of the Campo then swings on railings of Fonta Gaia.
4 *Indians,* 2 *of them women; the women in beautiful saries, cross from tourist information to B.*

10.00 230 *people now in piazza.*
2 *women push pram, sun shades raised from G to B.*
Small girl leaps in amongst pigeons and scatters them.
Small boy relieves himself under mother's guidance in centre of shell.

Tremendous sense of movement now, people seem to be walking everywhere, the piazza is alive with movement and purpose.
Small girl chases pigeons, called to order by man.
Man carries potted plants from E to J.

10.30 *Youth carries aluminium pipe from J to B.*
6 *bollards being leant on or talked round at the moment.*
Small boy plays on railings of Fonta Gaia.
Small girl relieves herself on steps in front of Fonta Gaia.
4 *people sitting on walls of Fonta Gaia.*
250 *to* 300 *people on Campo.*

11.00 *Tourist Bus turned off shell by policeman.*
Hearse passes down side 1.
Young man has difficulty controlling small boy who insists on looking down drain of Campo.
6 *people at souvenir stalls at the moment.*
Postman crosses from E to J.
300+ *people on Campo.*
Not so many on the shell; most people are avoiding the sun-baked centre of the piazza and are walking round the top of the Campo. The buildings along side 2 *are casting a shadow now.*

11.30 *Small boy trains binoculars upon windows of buildings flanking piazza (fruitless, most have shutters up at the moment).*
Girl dragged by small but eager dog into centre of shell then up to top and out of A.
Second girl with a more restrained animal crosses from B to I.
First girl and dog return, dog still has boundless energy apparently.
Very old lady with stick eases her way down B.
Girls in bright summer frocks, usually tourists cross to and fro all the time; local office girls and shop assistants usually dressed more formally, skirts and suits.

12.00 *Man carries pile of new shirts from C towards I, changes his mind and veers towards J.*
Caged birds sing all the time.
Woman takes over running of pigeon food stall from man.

12.30 7 *men talk in two groups round bollards at head of B.*
Very hot, 3/8 *or more of Campo bathed in sunshine.*
Practically all tables in restaurant on side 2 *are occupied,* 50 *seats of other bars and restaurant side* 3 *are in use.*
200+ *people on Campo now of these only* 22 *are actually on the shell and most of these are in transit across piazza, only* 6 *are stationary,* 4 *by Fonta Gaia, other* 2 *taking photographs in centre of shell.* (IV. 3.1. cont. page 50.)

Fig. 4.41. One of the small walls of the fountain, (Fonta Gaia) elderly woman selling sweets central.

Fig. 4.43. Approach B on plan.

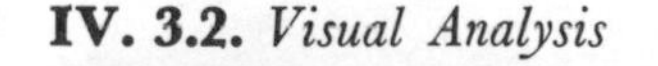

IV. 3.2. *Visual Analysis*

Fig. 4.42. Approach C on plan.

Fig. 4.44. Approach D on plan.

Fig. 4.45. Approach A on plan.

Fig. 4.47. Approach E on plan.

Fig. 4.46. Approach J on plan.

Fig. 4.48. Approach F on plan.

Fig. 4.49. Approach I on plan.

Fig. 4.51. Approach G on plan.

Fig. 4.50. Approach H on plan.

Fig. 4.52. Approach K on plan.

1.00 *Pigeon food venders have retreated before the sun.*
Small child in bright red pedal car drives round top of Campo watched by father.
Man leaves Chemists shop on side 2 *to watch child in car.*
Clothes stalls still open in F, very few people use this access point for entering or leaving the piazza.
150 *people in piazza at the moment, only* 17 *on shell.*
3 *discussion groups in progress now; one of* 3 *people, the other two composed of pairs of men, all are gathered round bollards at top of Campo.*

1.30 *Two men in blue overalls cross from E to J, friend carrying metal blocks catches them up.*
One pram now being wheeled round top of Campo from E towards H, turns about and returns along top of Campo.
Policeman returns across Campo to I.
200+ *people in the piazza, greater proportion are on the shell.*
Woman empties bag of food for pigeons; all flock to investigate, half return on foot to food venders stall.
175–200 *people in the piazza mostly round top of Campo, many sitting at bar tables.*

2.30 11 *bollards in use.*
Woman leads very large dog across the piazza from side 1.
Man admires baby in pram; small boy watched by parents feeds pigeons.
Several pairs of office girls or shop assistants walking to and fro along the top of the Campo.
6 *youths display with large yellow plastic ball in centre of shell.*

3.00 *Very old woman carrying an assortment of bags and old paint pots, puts down her burden and pauses for a rest in the centre of the shell en route from I to B.*

3.00 *Hand cart loaded with something indescribable is pushed onto Campo from H.*

3.30 *Small boy on bicycle stops to admire footballers in centre of shell.*
Group of 4 *eligible young ladies walk arm in arm from B to E, paralysis of footballers.*
Small boy falls, is picked up, tears follow.
Football game broken up by policeman.
Small boy succeeds in tipping bin from dust cart, tremendous clatter, pigeons rise in mass in alarm.

4.00 200 *people on campo.*
4 *boys gather round a fifth on bicycle in centre of Campo, small boy on bicycle speeds away across the shell.*
Man filming Palazzo Publico.
3 *curio stalls, operational.*
Caged birds call to one another.

4.30 *Small boy bursts balloon, very surprised, two other small boys stand in the centre of the shell wondering what to do.*
3 *girls sucking lollipops masquerade as troops using the lines painted on the Campo pavement for yesterday's ceremonies.*
Small boy demonstrates clock-work car to another admiring but sceptical child.
Banana seller still doing a brisk trade in his stall side 4.
Funeral procession of 15 *black bearers, priests and banners, crosses from G to B, few people notice.*
280 *people in the piazza.*
Man attempting to teach small girl to ride a very small bicycle ends up by performing on it himself.
One baby summarily dumped on pavement of shell sits and contemplates.

5.00 *Group of three small girls organise a game, one is hauled away to have a second jumper put on, game developes into tick from line to line of the paved courses leading to the fulcrum of the shell.*
Policeman wanders about on the shell.
350 *people in the piazza now, of these* 250 *are on the shell, no determined movement in any direction though a few are coming and going.*

6.00 *No sunlight strikes pavement of piazza at all now, reflected from facades round piazza.*
8 *prams on Campo at the moment.*
Small boy sits on paving and examines something minutely.
Girl with ridiculous heels hobbles from E to B via a large gathering of mothers in the centre of the shell.
4 *soldiers walk in line abreast across Campo and up B.*
Road sweeper is working on the shell.
500+ *people in the piazza now, probably nearer* 600, *about* 1/3 *seem to be in transit.*
Large party of small boys now chasing group of small girls.
Fisherman with all his gear crosses from D to H.

7.00 *Waiters stand outside Restaurants and boss hoping to attract custom.*
20 *bollards or more are now in use as leaning posts or focal points for discussion groups.*

7.30 *Pigeon food vender now packing up her wares.*

8.45 *Seats in bars occupied.*
Several people sit outside restaurants having full evening meal.
Light all round the piazza, bar lights and subdued street lights.
A few small groups are scattered over the unlighted shell.

9.30 *About* 100 *people are in the piazza at present most are gathered on top of the Campo.*
Only about 14 *people are on the shell.*
2 *dogs play in the dimly lighted centre of the shell otherwise there is no movement in the centre of the piazza. All movement is on the lighted periphery.*
Caged birds still sing. More audibly now for there is little sound other than a few voices and footsteps.
Lights on the buildings surrounding the piazza are subdued but ample.
Youth and girl, arms round one another admire Fonta Gaia.

10.00 *Only one man leaning on a bollard at the head of B.*

IV. 3.3. *Attempt to Explain why Piazza works so well.*

Again we quote from Millard's report:

"(*a*) *Temperament of the People:*

The Italians are naturally gregarious; they enjoy being in places where there are many other people.

The way of life and working hours all contribute to this.

The way of living whereby people are for the most part flat dwellers means that open spaces are used to a greater or lesser degree in much the same way as an English park.

(*b*) *Position of Piazza in relation to physical properties of ground:*

The Piazza is on plan at the junction of three ridges.

Vertically the piazza is not at the highest point of the slope of the ridges but is some 15–30 *feet lower.*

The piazza is lower on the sheltered side of the ridges also it is protected by the highest point of the town, the hill upon which the Cathedral stands.

(*c*) *Position in relation to main routes and thoroughfares:*

The piazza is enclosed round its curved sides by the main route through the city, but, this is sufficiently narrow for one way traffic only for the most part, and the traffic is forced to move very slowly because there are several sets of traffic lights to slow it down. Pedestrians and traffic mingle fairly safely in these streets. All the ways out of the piazza, which lead up onto the main street, are provided with zebra crossings on the main street.

(*d*) *Position in relation to functions within the city:*

The piazza occurs between several main features in the city, the market, the Cathedral, the town hall, the regional Administration, and together with the fact that the street running round the curved sides of the back of the piazza is the main shopping area, gives it a very good position.

(*e*) *Orientation:*

High buildings round the curved sides protect it from Northern and Eastern blasts from the snow fields of the mountains. The inner faces of these buildings reflect the sun all day, in the early morning the light catches buildings at one end of the curve, the other end is in shadow; in the evening vice-versa; at mid-day sufficient shadow on one side or the other for people to walk in a certain amount of shade. The flat side of the palazzo is exceedingly good at keeping mist out of the Campo when it rises up the valley beyond the market. (On several occasions I have seen the Market piazza full of mist (or not seen of course) whilst there has been clear air in the Campo).

(*f*) *Aesthetic:*

The buildings all round are tall and enclosing. The dished central portion is very restful to the eye, textures are rich and colours warm. (A proper aesthetic analysis of the Campo is being done in connection with the scheme and will follow when complete)." (At the British School in Rome.)

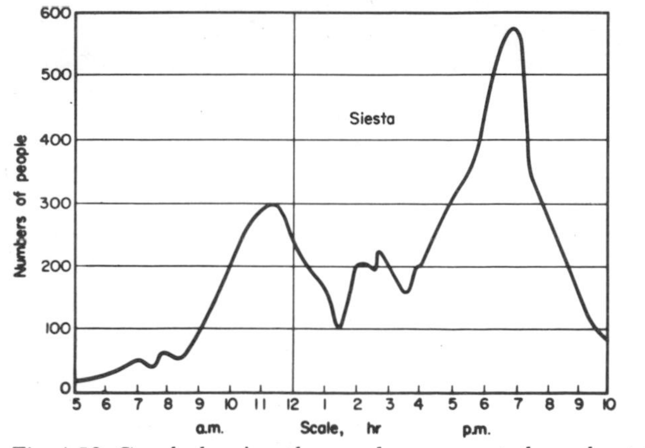

Fig. 4.53. Graph showing the numbers present throughout the period of observation of Piazza del Palio.

IV. 4. Convenience

The convenience of paths can be measured by the directness and ease with which they provide access to all the places of gathering or communal value, from swimming pool to post box. Most important of all, in most instances, are the public transport stations: these will have an increasing importance as public transport develops in quality.(i)

The study of directness of paths is the study of desire lines:(ii) the straight line between home and goals should be approximated as often as possible for as many people as possible and touch as many goals as possible. Underpasses and bridges must not demand unreasonable climbs for mothers with prams and old people. Rails should accompany steps.

A combination of direct paths to stations and good public transport or mechanical aids can extend "convenient" distances considerably for the pedestrian. Dennis Chapman has shown this in his study of Convenience.(i) 89% of 455 people who were questioned at Middlesbrough, England, found Albert Park "convenient", living within *half a mile* (800 metres) of it. But so did 96% of those who lived within *12 minutes travelling time* from it, combining walking and a bus ride.

To the factors listed above, affecting convenience, it is essential to add protection from rain, snow, sun, wind, ice, depending on climate, through roofs, walls, heated floors, complete enclosure, canopies.

Fig. 4.54. Trees as roof.

To summarize:

Convenience of a path system depends on four factors:

- the walking distances, directness and identification of routes;
- the connections with public transport and the efficiency of this;
- the pleasantness of the route, making it an end in itself;
- the protection from weather, making it seem shorter.

(i) Convenience, a desirable quality in town-planning, studies by Dr. D. Chapman in *Human Relations*, LII, 1, February 1950.

(ii) MOORE R. L. Psychological Factors of Importance in Traffic Engineering. *R.R.L.* 1956.

IV. 5. Mechanical Aids

IV. 5.1. *Pedal Cycles*

The bicycle, non-powered, is the oldest of the minor mechanical aids available to man. It is remarkably efficient, takes up little room, gives the rider healthy, pleasurable exercise, is simply stored, does not foul the atmosphere, is easy and cheap to run. Its momentum, normally ridden, is not great, and so danger from it is very limited. This also means that it can be safely used from a very early age, particularly for the journey to school. Weather protection has not yet been developed—beyond oilskin clothes.

The proportion of people using cycles in Holland is very great, but even in central Copenhagen, with the present discouraging traffic and no special provision for the cyclist, approximately 20% of peak hour travellers are on pedal cycles. Taking an average of six towns near London, 35% of the work journeys were on cycles. In Crawley a survey in 1957 showed that although this new town was not providing for this type of transport 25% of the people went to work that way.(i)

Cambridge Massachusetts and Cambridge England and Universities the world over demonstrate the popularity of bicycles for students.

The use of the bicycle extends the "convenient" distance, so that non-motorized access is attractive to many more people. At Copenhagen airport the free use of rubber-wheeled non-powered scooters by the staff, taking them and leaving them at random, has a relevance to town centre planning.

In assessing the information, that the trend is away from the use of cycles, we must note that the roads are continually filling up with cars, which discourage cycling even more directly than they discourage walking. The stench and danger make the last woman cyclist in New York a headline item and do indeed decrease numbers in most places.

So we have less cycling because there is less opportunity to cycle due to inadequate planning. The deduction should be that planning adequately can easily reverse the trend, as cycling is a healthy and pleasurable activity for a great range of age-groups. Stands at shopping centres, right in the precinct, are possible: the cycle can be wheeled, taking shopping and baby.

The bicycle does not fit quite nicely into either the vehicular or pedestrian category. Planners have been tempted to forget it and rationalize it out of existence. It is easy to do this if "trend planning" is used: as the use of bicycles is decreasing one may accept this as an inevitable trend and plan less and less for them. Such an argument has led Jellicoe to ban it from his theoretical scheme Motopia, saying without evidence, that "it is an anachronism in the modern world" (like walking!).

More seriously, the plan for Hook, not to be built either, makes little provision for cycles. Flat sites like Motopia's albeit a theoretical exercise make the banning of cycles poignantly wrong. Cumbernauld, with its steep hills, ignores the needs of the cyclist on more rational grounds.

It is quite obvious from the above that it needs routes more akin to the pedestrian than the powered vehicle. It may share routes with either, given special cases, but where there are petrol fumes it is not reasonable to plan for the deep breathing that goes with effort. Sharing paths with pedestrians leads to somewhat slower speeds, and design and marking of paths which are divided for use. But all main paths should have, running alongside, a cycle track physically separated. A hedge is ideal. The construction of a light cycle track which might easily carry the equivalent of a lane of motor traffic of those who would otherwise abandon the cycles is an economically attractive proposition. It is important to distinguish quite clearly between the powered cycle or scooter and the non-powered. It is a crucial difference.

Fig. 4.55. Shared footpath and cycle track at Kiel, with sign indicating the arrangement.

(i) *The Planning of a New Town*, L.C.C. London, 1962.

Fig. 4.56. Bicycles for students, Cornell University, Ithaca, U.S.A.

Fig. 4.57. In Holland bicycles really matter.

IV. 5.2. *Pedestrian Conveyors*

In 1893 a moving pavement was an attraction at the Chicago World Fair. In 1962 a moving pavement, one of perhaps 50 in the world, took passengers up to the Monorail station at Seattle World Fair. The slow growth of this aid to the pedestrian is due to its overlapping the function of the motor-car, as long as it was possible to park at the kerb and enter wherever one wished to go.

The number of moving pavements is likely to increase rapidly in the near future, although their application should be limited to special cases. They have an urgent application to underpasses and overpasses, where gradients allow this and flow is great. They have a place in pedestrian centres where public and private vehicles are banned from large areas. Their capacity for traffic flow is very large, as people travel standing, closely packed together or add to the speed by walking at the same time. It seems feasible and desirable to invent and incorporate flap seats so that at least some of the passengers can rest while on the conveyor.

The first moving pavement in Europe was installed by Otis Elevator Company at the Bank Station in London for 1960. It is 300 ft (91 m) long and consists of two 4 ft (1·2 m) wide belts. They travel at 2 mph (3·2 km/hr), 180 ft/min (55 m/min) and it is possible to walk on them making a total speed of something like 5 mph (8 km/hr). They negotiate an incline of 10 degrees, have an hourly capacity of over 20,000 people. The cost for the installation and mechanism was £500 ($1500) per foot approximately.

For window shopping, speeds of 50 ft/min (15 m/min) are recommended. Conveyor belts can negotiate inclines of up to 15 degrees, one in four, and curves down to 22 ft (6·7 m) in diameter.

The crossing of conveyors raises difficulties and sets limitations.

Other makers of passenger conveyors are: Stephens, Adamson and Hewitt, Robins Incorporated, all in U.S.A.

Fig. 4.58. Travolators taking passengers up the Town Station of the Seattle Monorail installed for the World Fair 1962, and now the city's property.

(i) McElroy, J. P. Pedestrian Conveyors, *T.P.R.* XXXII, 2. July 1962. Goss, A. Moving Pavements, *A. J.*, 5.1.61.

IV. 5.3. *Escalators and Lifts*

In hilly towns, the use of mechanical help to overcome heights will be essential to make public transport termini in valleys or on hills attractive and to join residential areas with shopping in a convenient way. The cost of the mechanical help is tiny compared with the cost of road construction.

The mechanical aids have special relevance to the multiple use of land. Parking above or below ground garages can be directly connected with shopping areas. High or low level public transport is already connected in this manner in every underground railway system.

At sea-side resorts, on cliffs, lifts to the beaches are common. In Stockholm, the lift taking people from sea-level to the height of a densely inhabited rock with many professional offices is used by almost everyone, adults and children, when laden. A small fare is paid each time.

The matter-of-course installation of vertical mechanical aids in the townscape is a direct outcome of the primary consideration of the pedestrian and the realization that private cars can be admitted to towns only in limited numbers and then only to the "gates" of the centre. Public transport becomes more attractive and convenient when linked with travelling pavements, lifts and escalators.

Fig. 4.59. Lift for pedestrians walking from the centre of Stockholm and the old town to the cliff-like Slussen. A low fare is paid.

Fig. 4.60. "Pedes-taxi" at Seattle World Fair.

IV. 5.4. *Trolleys*

Trolleys to take shopping and baby, as already used in many supermarkets, may become as common as milk bottles. Their universal use from shop to home could be financed by the stores in the same way as in fact milk-bottles are; and once one store does this all may have to follow. For the planner this means avoiding steep gradients, and "ups" and "downs" (steps) and crossing roads. Where a path must cross a road the kerb should become a little ramp. We make provision for cars in this way where drives occur, why not the same comfort for prams and trolleys? (At such places cars must be slowed down by ridges on the road.) Their use at the residential end of the journey, from car or public transport to home, and also by those who deliver post and all other things along paths, is becoming more frequent and will become universal, like the coal bag or petrol can.

The postman uses a specially designed 3-wheel trolley at Mies van de Rohe's Lafayette, Detroit.

IV. 5.5. *"Pedestrains"*

The word, coined by Victor Gruen, is meant for a train of carriages, so low, so slow that they can be easily mounted and dismounted. Such transport has been provided at most of the world fairs. It is used at McGill University, Montreal, for people to view at the open day, at the Tivoli Pleasure Gardens at Copenhagen, and at Disneyland, California, to bring people from the vast car park to the entrance. The idea is that if they move slowly they will be right in the pedestrian environment of precinct and shopping promenades.

They are obviously useful in many cases. Perhaps one of the nearest approaches to them are the cable cars in San Francisco which have been working in a similar way, although in motor roads and on rails for over sixty years.

Such vehicles can certainly give convenience, colour and movement to busy shopping areas without making them dangerous, but they must be run with the greatest care and discretion.

Fig. 4.61. Pedestrain as used at McGill University open day as envisaged by Victor Gruen for Fort Worth and Fresno. (See also p. 157.)

IV. 5.6. *"Carveyors"*

Victor Gruen's proposals for East Island, New York, includes another original idea for conveying passengers. Here we have the nearest thing to a moving pavement with seats, to take its place with his previous idea for "pedestrains": the "carveyor". It would run along a covered concourse with stations 900 ft apart. The idea is for individual platforms with seats which move quickly between stations and slow down at boarding islands. This system is intended to make unnecessary all private cars restricting vehicular traffic on the island scheme to emergency services.(i)

Fig. 4.62. Section through "Carveyor" running in Victor Gruen's proposed concourse on East Island, New York.

IV. 6. Information

Public relations need to be highly developed and applied in many effective ways when new ideas on planning are implemented. It is absurd to expect the public to "buy" new architectural or planning products, without something to inform them of the virtues and values of the new as compared with the old. This is taken for granted with all other products. The success of the new depends upon its proper use.

The need for good public relations is therefore axiomatic. Information on a local basis, personally, with each specific occupation of a new scheme, dwelling, shop or office, must be supplemented by education on a broad level. Planners and architects have a very great deal to learn in this field which is hardly ever touched upon.

The reversal of trends, as has been mentioned (IV. 4.1), certainly needs powerful communication, even propaganda, i.e. the use of existing trends to reverse others. For example, an enlightened government, caring for the health of its citizens and the survival of our city centres and the appreciation of new kinds of residential lay-out with traffic segregation, might use the fashion for slimming to encourage walking: "Walk to Health" ("and Slim"). The truth, vividly presented, acts as powerful propaganda. "Want to Slim? Why not Walk?" ("with Him") could replace the "Is Your Journey really Necessary?", of the war years in Britain.

IV. 7. Comparison of Path Systems in New Towns and Villages.

(Plans to similar scale).

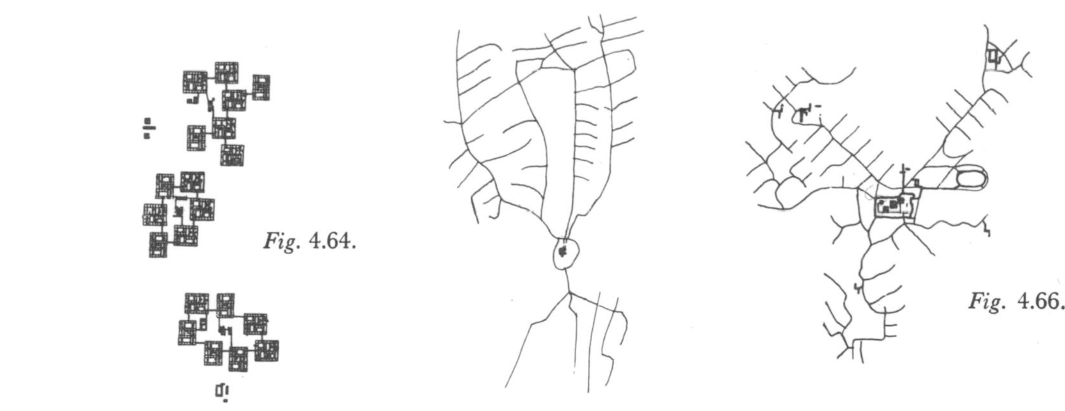

Fig. 4.63. Cambridge Village, 3500 inhabitants (see p. 147).

Fig. 4.64. Erith, 25,000 (see p. 124).

Fig. 4.65. Sennestadt, 24,000 (see p. 136).

Fig. 4.66. Sabende, 20,000 (see p. 135).

(i) *Interbuild.* London, June 1961.

Fig. 4.67. Hook, 100,000 (see p. 121).

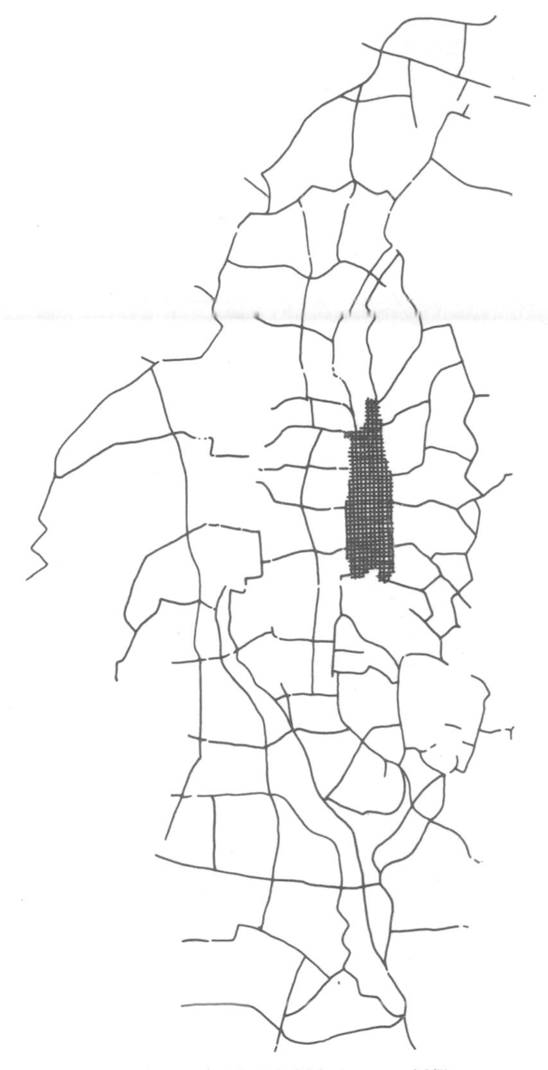

Fig. 4.68. Cumbernauld, 72,000 (see p. 117).

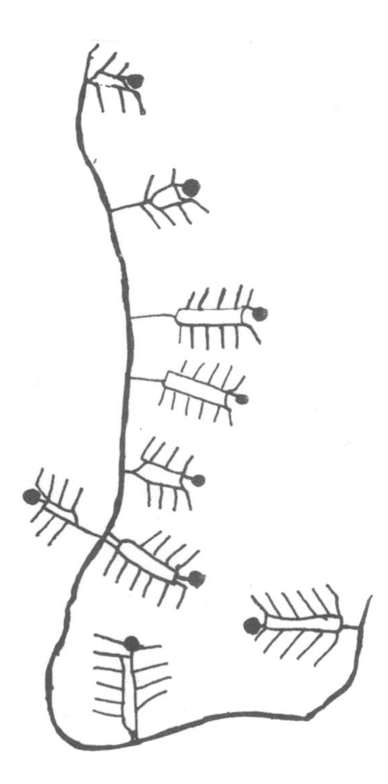

Fig. 4.69. Cologne, 100,000 (see p. 133).

Fig. 4.70. Toulouse le Mirail, 100,000 (see p. 126).

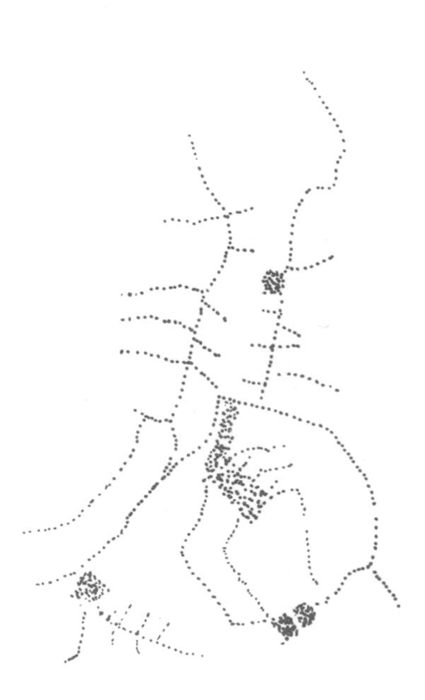

Fig. 4.71. Sputnik, 100,000 (see p. 146).

V. Needs and Nature of Vehicles

V. 1. Man as Driver. **58**
V. 1.1. Perception. 58
V. 1.2. Road Furniture. 59
V. 1.3. Training. 59
V. 1.4. Emotional State. 59

V. 2. Motorization Forecasts, General. **60**
V. 2.1. Transportation Studies and Surveys. 62
V. 2.2. Vehicles and Town Space. 65
V. 2.3. General Economic Consideration. 67

V. 3. Routes of Vehicles. **68**
V. 3.1. Administration of Road Planning. 68
V. 3.2. Road Space, Shape, Noise and Danger. 68
V. 3.3. Capacity of Roads. 70
V. 3.4. Speed. 73
V. 3.5. Types of Roads. 75
V. 3.5.1. General. 75
V. 3.5.2. Motorways and Freeways. 75
V. 3.5.3. Main Roads. 79
V. 3.5.4. Service Roads. 79
V. 3.5.5. Summary. 79
V. 3.6. Junctions, General. 79
V. 3.6.1. Some examples of Junction Flow. 81
V. 3.6.2. Junctions of Three Routes. 81
V. 3.6.3. Junctions of Four Routes. 82

V. 4. Parking. **84**
V. 4.1. General and Basic Requirements. 84
V. 4.2. Parking Lots, Ground Level Only. 84
V. 4.3. Multilevel Parking Garages. 85
V. 4.4. Mechanical Parking. 88

V. 5. Functions of Private Vehicles. **88**
V. 5.1. Journey to Work. 89
V. 5.2. Schools and Universities. 89
V. 5.3. Shopping. 89
V. 5.4. Visiting. 90
V. 5.5. Sport and Cultural Events. 90
V. 5.6. Week-ends and Half Day Holidays. 90
V. 5.7. Long Holidays. 90
V. 5.8. Interpretation of Statistics. 91

V. 6. The Form of Private Vehicles. **91**
V. 6.1. Factors Affecting Form. 91

V. 7. Commercial Vehicles, Vans and Lorries. **93**
V. 7.1. Numbers and Growth. 93
V. 7.2. Size, Speed and Flow. 93
V. 7.3. Effect on Flow. 93

V. 8. Public Transport. **94**
V. 8.1. General Attitude. 94
V. 8.2. Principles of Public Transport Systems. 95
V. 8.3. Types of Rapid Transit. 95
V. 8.4. Comparison of Typical Features. 95
V. 8.3.1. Buses. 95
V. 8.3.2. Duo-rail systems. 96
V. 8.3.3. Monorail. 96
V. 8.3.4. Underground Railways. 98
V. 8.3.5. Self-guided Bus Trains. 99
V. 8.4. Comparison of Typical Features. 100
V. 8.5. Examples. 101
V. 8.5.1. Philadelphia. 101
V. 8.5.2. St. Louis. 102
V. 8.5.3. Boston. 103
V. 8.5.4. San Francisco. 103

V. 9. Glossary of Terms. **105**

V. Needs and Nature of Vehicles

V. 1. Man as Driver

V. 1.1. *Perception*

The joy of driving in a vehicle lies in its comfort and speed although "comparatively few passenger car drivers desire to travel at speeds which equal or approach the potential speeds of their vehicles even under the most favourable highway and traffic conditions".(i) The pedestrian relates directly to the townscape or landscape. The interior of the vehicle moves with the driver, as a space about him, and his passengers. Partly for that reason, each person when driving tends to be less considerate, more aggressive and foolhardy. He is not really involved with his fellow men outside his own box. Appreciation of and sensitivity to surroundings changes with speed. Detail observations should not be required of the driver. This eliminates undue changing of the focus of the eyes. It is reported from the U.S.A. that average urban drivers spend 1·2 sec looking at average girls along their routes.(ii) At urban speed, this means that for 50 ft the driver is not in control. Diversion of attention is likely to be a far more important factor influencing motor accidents than ever admitted, as such admission would be bound up with deep guilt feelings. The more attractive the girl, the greater the menace. This also applies to advertisements, particularly the kind that is not scaled up for the motor speed. If they compel attention they have a vicious influence. In Michigan statistics showed a correlation of accidents and commercial diversions to the driver. Clear roads showed 3·7 accidents per million vehicle-miles, those with more than 4 road side developments per 1000 ft averaged 13·5.(iii) This is a serious reason why adverts and pavements should be taken off the sides of urban roads, and pedestrian traffic, with its pretty girls, segregated. It is further an argument against bridges which allow the motorist a free view of those walking across it, in built-up areas.

Hamilton and Thurstone(iv) formulated five principles on the effects of speed on vision of the driver.

As speed increases:

(a) concentration increases,
(b) the point of concentration recedes,
(c) peripheral vision diminishes,
(d) foreground detail fades increasingly,
(e) space perception becomes impaired.

(i) *Highway Capacity Manual.* U.S. Department of Commerce. Washington 1950.
(ii) Stopwatch on Roving Eye. *Daily Telegraph*, 22 Nov. 1962.
(iii) McMonagle J. C. *Traffic Accidents and Roadside Features.* Washington 1952.
(iv) Hamilton and Thurstone, *Human Limitations in Automobile Driving.* U.S.A., 1937.

Given a design speed, Fig. 5.1 shows the distance at which the driver will naturally focus on an object, and so the kind of detail which is suitable.

Glare, monotony and regular, rhythmic stimuli should be avoided. The last leads to sleep, which is dangerous when driving! They also decisively affect reaction time. On the "normal" duration of this

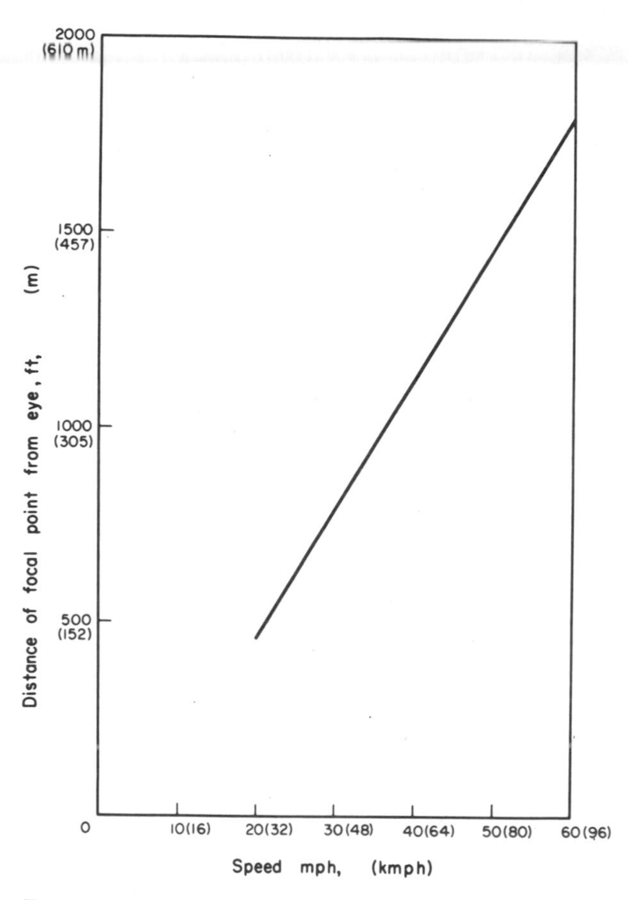

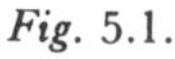
Fig. 5.1.

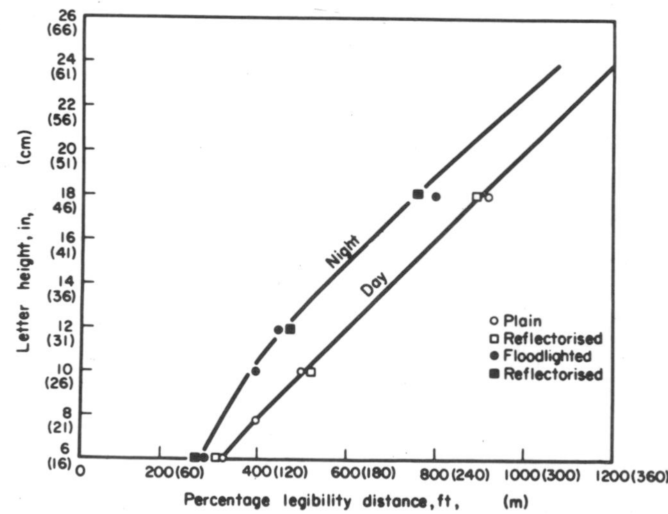

Fig. 5.2. Letter heights at varying distances legible to motorists (clear Gill Sans type). With speed, the angle of vision decreases from as much as 120° to 40° at 60 mph (96·6 km/hr).(i)

(i) Hamilton J. R. *Safe Driving.* New York 1937.

Fig. 5.3. Manahawkin Bay Bridge in New Jersey. Light Fittings in the bridge rail included for better fog lighting. Aesthetic advantages are obvious. However, some sense of organic rhythm should be incorporated into lighting design or surroundings to give scale and bearings. This does not mean the monotonous rhythm of poles.

psychologists and traffic engineers are not agreed, yet it forms a vital factor in formulae affecting almost every aspect of design for practical road capacity.

V. 1.2. *Road Furniture*

Good communication with the driver must take into account not only the speed at which he moves, which determines the size of signs and lettering, but also the fact that intuitive, felt orientation is preferable to intellectual indications. "No need for signs" is the best principle for design, but obvious symbols are better than writing. The extent of writing for motorists, up to 48 words on lamp posts with four different signs, in Washington, for example, can extend beyond all reasonable limits. The use of coloured road surfaces, shapes and symbols, and overall simplification is required in every country. Reichow, in his design for the Sennestadt in Germany, differentiated between classes of roads by the colour of road lighting. In Britain, recommendations of a definite kind for road furniture have recently been drawn up by a committee set up to study the field.(i)

The techniques of road lighting so far developed are elementary.(ii) Lighting from continuous railings rather than posts has been employed in a few places, i.e. Germany and U.S.A. Whether this or other ideas prove to be an optimum solution it is too early to say. What can be ascertained is that lighting from lamp-posts, whatever the colour or nature of the light, is a crude idea, which will be superseded. The separation of pedestrian traffic from vehicular traffic makes possible developments not practicable in the existing streets with mixed use.(i)

(i) Worboy Committee Report, H.M.S.O., London 1963.

(ii) Berry and Robinson, Engineering for Traffic Conference Papers, London 1963,

V. 1.3. *Training*

The training and passing out of drivers is at a ludicrously low standard in many countries. You need not even be able to reverse into a parking space in Britain. Driving should be taken more seriously and the training should be far more thorough and systematic.

V. 1.4. *Emotional State*

The emotional state of a driver is a very important factor. Not only do we wish to have people good tempered because it is pleasant, but they also prove to be safer. Thus the road itself should contribute by making driving as pleasurable as possible.(ii) (See also p. 14.)

(i) See also Section V. 3.2.

(ii) Tunnard and Pushkarev. *Man-Made America: Chaos or Control.* Yale, 1963 (and p. 75)

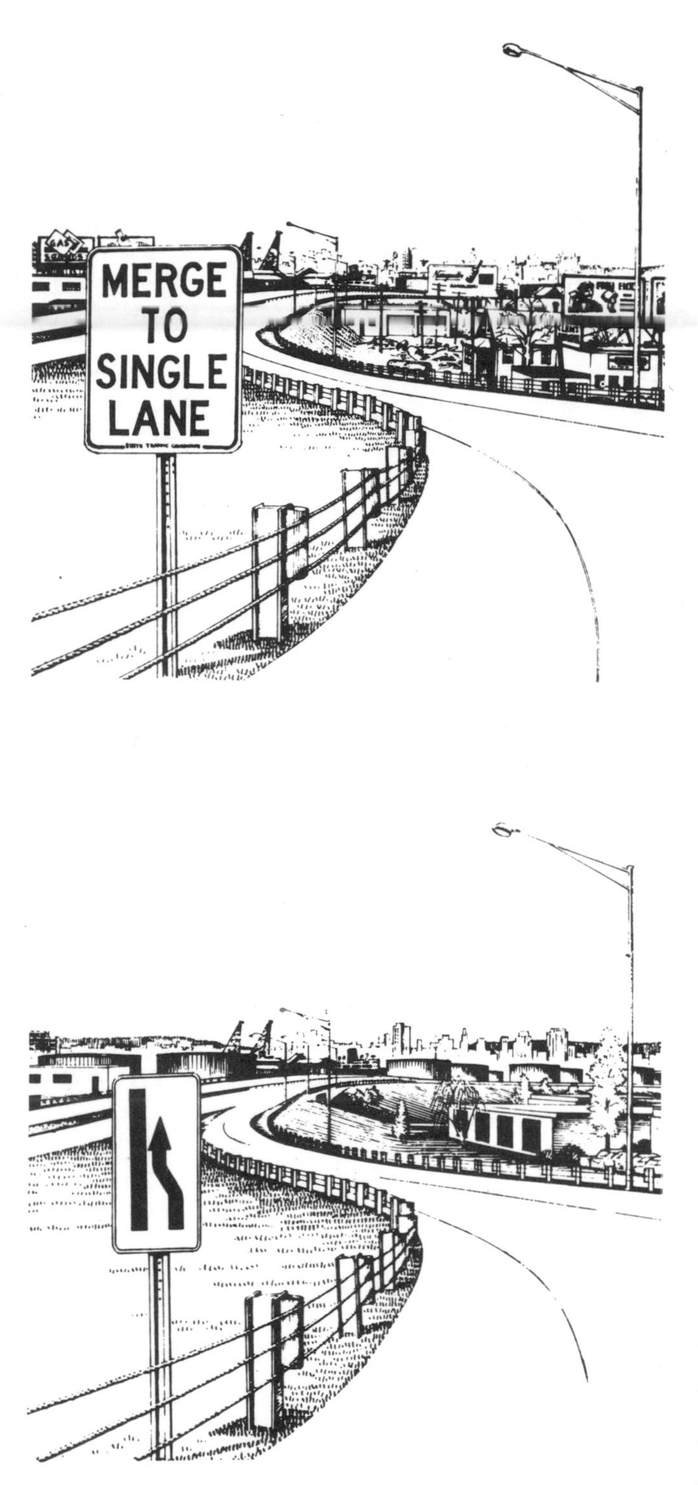

Figs. 5.4. and 5.5. Example from *Man-Made America: Chaos or Control*(i) showing the advantage of symbols

(i) TUNNARD and PUSHKAREV. *Man-Made America: Chaos or Control.* Yale University Press, 1963.

V. 2. Motorization Forecasts, General

The basic dilemma of traffic in towns was shown clearly by an often reproduced diagram from Le Corbusier's *Un ville contemporain.* It contrasts road space existing mainly on the outskirts with the maximum need in the centre.

His proposals mirror his conclusions—the first proposal for urban motorways (see Fig. 10.30).

The rate of growth of the vehicle population and the saturation level of any given area are of prime interest to the planner.

Before figures are given, it must be said that to cater for statistics which generalize and project trends unreservedly contradicts two basic concepts of planning. First: planning should determine trends, at least in part, and not merely follow them. Second: there are important differences within a statistical generalization. Local differences may be decisive for the planner (0·25 cars per person in the City of Chicago; 0·41 in adjacent urban areas).

Another point which affects a planner's calculations for future provision of roads, or parking, is the growth or otherwise of the habit of walking and the use of public transport. This may mean that although car ownership grows, the use of the added numbers becomes selective, i.e. cars may tend to be used for out-of-town journeys mainly, if public transport in a town is excellent and walking pleasant and convenient.(i)

Even today cars cannot be used as they were a few years ago: in most central areas you cannot drive to any actual shop and park outside. With growing ownership, more and more restrictions become axiomatically enforced, not just by law.

Forecasts of car ownership have in the past been ludicrously and repeatedly low. In the forties the Ministry of Transport acted on the assumption that traffic would not increase more than 100% beyond pre-war volumes. In 1958 this level was passed. It is only recently that American ownership and economics have been used as a realistic basis.

In calculating with exceptional care (quite fresh for British new towns) the parking needs of Cumbernauld, Geoffrey Copcutt lists the Road Research Laboratory's forecasts(ii) according to curve-fitting of post-war trends, according to increases in income and according to production capacity. He then has to guess a national average level and adds:

(i) "one aspect of affluence has already been forsworn by Mr. Krushchev—private car ownership, ... the official line is that private cars are a nuisance, and in the abundant society of the future Soviet citizens will be able to hire cars at short notice." *Observer*, 10 March 1963.

(ii) CHANDLER K. N. *Traffic Trends.* R.R.L. Slough 1956.

"... *future trends are almost impossible to predict, being largely dependent on government policy* ..."

He is dealing with a small New Town where full motorization demands can be met. He assumes 10,000,000 cars in 1980 as a minimum in the country and 20,000,000 cars as a likely maximum, taking into account American trends. This would mean 0·7 or 1·4 cars per family respectively. (At present small U.S. towns go up to the maximum already.)

Taking total vehicle numbers, the Road Research Laboratory has predicted the long-term growth of the vehicle population again, in its 1960 report.(i) The Ministry(ii) in its 1963 report, taking into account the R.R.L.'s figure and those of the National Institute of Economic and Social Research as well as data on income and production, come to similar results.

TABLE V.1.

These approximate forecasts were made:

	By the R.R.L.	By the Ministry Group
1970	17 million vehicles (12 million cars)	12–13 million cars
1980	25 ,, ,, (18 ,, ,,)	16–18 ,, ,,
1990	30 ,, ,, (22 ,, ,,)	
2000	34 ,, ,, (24 ,, ,,)	
2010	36 ,, ,, (26 ,, ,,)	

In 1962, the Ministry of Transport gave a 160% increase by 1980 as an official recommendation to local authorities for their road calculations. In Britain saturation level is taken as 0·4 private cars per person and 0·55 of all vehicles. At the end of 1962 there were 0·53 vehicles (0·46 private cars) per person in California. Local differences are tremendous, particularly those between large city centres and their very low density suburban areas as found in the U.S.A.

While it may be right for New Towns to deduce needs in this way, and try to satisfy them fully, even in the central areas, it is unrealistic and fallacious to work in this way in assessing structure and changes of existing towns. At Liverpool, Shankland is taking 0·3 as the ultimate number of cars per person for his central area plan (see Section VII).

Indeed, even in Cumbernauld the figures for car usage are modified by the assumption that to walk to the centre will be far more common there than in the much less dense new towns of the earlier type in Britain. It is possible to plan for such modifying factors, and it is desirable and increasingly important to take them into account.

To summarize, it can be said that a 200% or 400% increase in cars determines a change in their use. Increase in numbers itself modifies the use pattern. It is the duty of the planner to foresee this and plan for it. Then, the town can become convenient and cope well with the life within, though it may not cope with a theoretically projected car population seen as a *laissez-faire* invasion of an inevitable force.

Of all factors in planning, I believe this to be one of the most important: that the car is assigned a new, efficient place in the town as a whole, and that the function it has hitherto fulfilled, possible only for a small minority, is replaced by carefully planned alternatives for everyone's use.

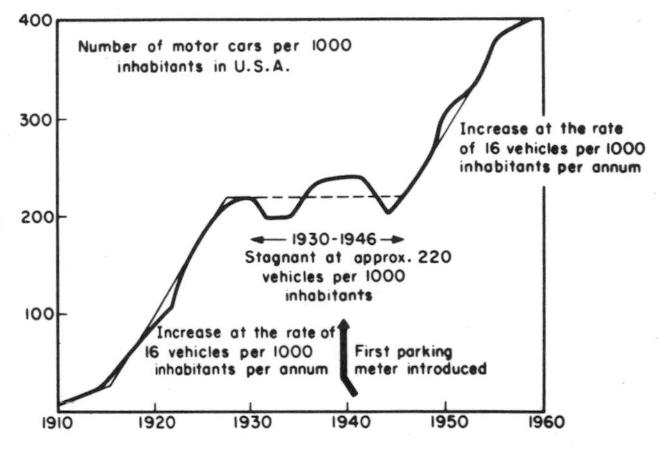

Fig. 5.6. Number of motor cars per 1000 inhabitants in the U.S.A. showing growth from 1910 to 1960.(i)

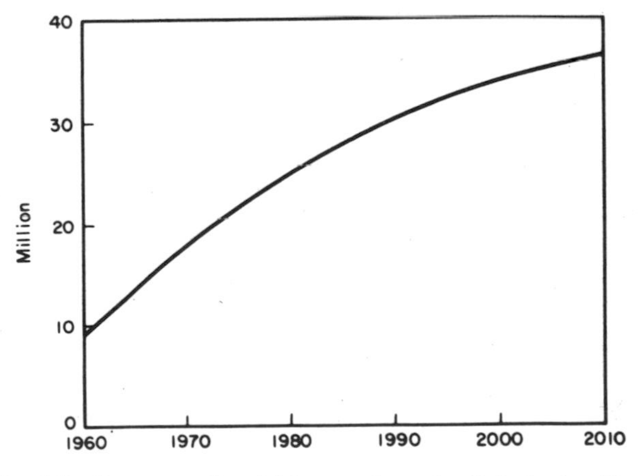

Fig. 5.7. Forecast of vehicle numbers in Britain by Road Research Laboratory, up to 2010. (Report of the R.R.L. H.M.S.O. 1960.)

(i) *Report of the Road Research Board*. H.M.S.O. 1960.
TANNER J. C. *Forecasts of Future Numbers of Vehicles in Great Britain*. R.R.L. September 1962.

(ii) Min. of Transport, *The Transport Needs of Great Britain in the Next Twenty Years*. 1942.

(i) SMITH W. and Associates. *Future Highways and Urban Growth*. New Haven 1961.

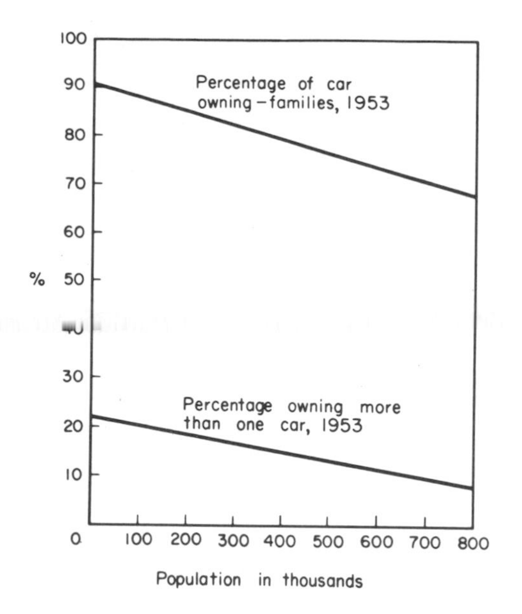

Fig. 5.8. Car owning families in the U.S.A. plotted from 32 cities. Percentage of car owning families and of families owning more than one car in cities with a population from 20,000–8,000,000, showing tendency for more cars in smaller towns and the tendency for more multi car families in the smaller towns.[i]

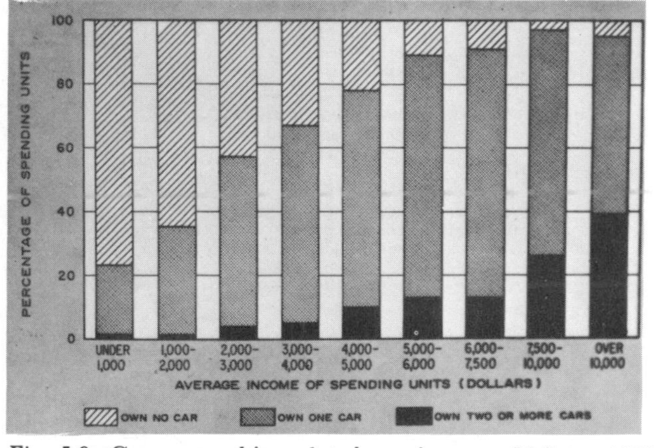

Fig. 5.9. Car ownership related to income, U.S.A., 1958. From *Future Highway and Urban Growth*. Note that beyond $10,000 income the number of families not owning a car drops.

Fig. 5.9a. "Into the seventies?" road shortage affecting motorization forecasts.,[ii]

(i) Bendtsen P. H. *Town and Traffic*. Copenhagen 1961.

(ii) JAK, *Highway Times*, London, August 1963.

V. 2.1. *Transportation Studies and Surveys*

Transportation studies pioneered in the U.S., in Chicago, Detroit, San Diego and other cities, have come to replace the limited traffic "surveys". The latter lead to piecemeal remedies, usually aggravating a neighbouring problem (see Hyde Park Corner Improvement, London). They could not rationally assess all the journeys of the inhabitants according to the character, capacity, finance and facilities of any given town.

The creation of a new all-embracing body like the London Traffic Management Unit, is necessary before reason can combine the many parties involved in any large town.

Simply explained, transportation studies start with all required journeys, péople and goods, found through house-to-house origin and destination surveys and are backed up by many other refined enquiries. (See also pp. 198, 199.)

The proposals for the town development are then projected and an outline transportation system evolved which is related to growth factors. The criteria for this are economic and developmental. The possibilities for public transport are balanced as a factor at this point. For the passenger, convenience is closely bound up with the duration of any journey. Finally, value judgements govern the art of town design though these are based on factual data.

The point of a transportation study is to assess the total number of journeys and give this data. The value judgements relate to the capacity of a town limited by its character and economy. They decide what the forms of transport shall be, what kind of parking (commuter or shopper) shall be provided. It is not simply a matter of assessing how many private cars will want to enter a town and park there, and then to design for it. Much of the work of traffic engineers is still based on the ludicrous assumption that if only you count the cars it will be known how many there are, and then you can provide for them, or, at best, multiply the result by one and a half, according to the Ministry figures, and provide for such a future number, without regard to the many other factors: notably, for example, that an improved road attracts traffic (immediately and continually as the car population rises,) and that multi-storey car parking remains empty, in spite of "desperate need", if placed and priced wrongly.

The city of Leicester's Study Programme for Transportation Patterns (City Planning Officer, W. K. Smigielski) designed by their traffic engineer, Middleton-Harwood, is a good example of modifying the U.S. studies to British towns. Here, as elsewhere (Coventry, London, Newcastle, Liverpool), computers are used to find possible combinations. (See Appendix.)

For Official Use	Card Type (1)	Zone No. (2 3 4)	Sample No. (5 6 7 8 9)	Person No. (10 11)	Person Identification	Occupation and Industry .... (12 13 14 15 16 17 18)

All information below is for **TUESDAY, MAY, 1963**

Please record ALL JOURNEYS you made on this day

ONE of these forms should be completed for EACH member of the household 3 YEARS AND OVER

Please read explanatory notes on the bottom of this form before filling in your answers.—Please do not mark numbered boxes

	WHERE DID THIS JOURNEY BEGIN? Write below ADDRESS of each place Use BLOCK CAPITALS Be as precise as possible; give number in street. The starting place of the second and later journey is always same place as end of previous journey.		WHERE DID THIS JOURNEY END? See Note 6. Name the PLACE VISITED—NOT where you left your vehicle		WHEN? Write below the PRECISE TIMES when you left on each journey and when you arrived. Put a.m. or p.m.		HOW DID YOU TRAVEL? Mark "X" below See Note 7											PURPOSE OF JOURNEY Please state below what was the purpose. FROM one of the following TO one of the following See Note 8 0. Work 1. Firm's business 2. Personal business 3. Shopping 4. Have meal 5. Social or recreational 6. Pick up or set down passenger 7. Catch train or bus 8. School 9. Home		IF YOU DROVE A VEHICLE On all or part of your journey								
	DISTRICT	NUMBER and STREET or NAME of well-known building	DISTRICT	NUMBER and STREET or NAME of well-known building	LEFT	ARRIVED	0. Drove car or light van	1. Passenger in car	2. Taxi passenger	3. Drove heavy commercial vehicle	4. Passenger in commercial vehicle	5. Bus passenger	6. Drove motor or power cycle	7. Passenger, motor or power cycle	8. Pedal cycle	9. Train	10. Walk	From	To	WHERE DID YOU PARK IT? At the end of the journey. Give name of street or name of car park	WHAT KIND OF PARKING? Mark "X" below: Private Car Park	Public Car Park	On Street	Not Parked	WHAT TIME DID YOU PARK VEHICLE? Put a.m. or p.m.	FOR HOW LONG DID YOU PARK IT? Precise Time: Hours	Minutes	HOW MANY PASSENGERS? Including driver
First Journey																												
	19 20 21		22 23 24		25 26						27	28						29	30	31 32 33		34			35 36	37	38	39 40
Second Journey																												
	19 20 21		22 23 24		25 26						27	28						29	30	31 32 33		34			35 36	37	38	39 40
Third Journey																												
	19 20 21		22 23 24		25 26						27	28						29	30	31 32 33		34			35 36	37	38	39 40
Fourth Journey																												
	19 20 21		22 23 24		25 26						27	28						29	30	31 32 33		34			35 36	37	38	39 40
Fifth Journey																												
	19 20 21		22 23 24		25 26						27	28						29	30	31 32 33		34			35 36	37	38	39 40
Sixth Journey																												
	19 20 21		22 23 24		25 26						27	28						29	30	31 32 33		34			35 36	37	38	39 40
Seventh Journey																												
	19 20 21		22 23 24		25 26						27	28						29	30	31 32 33		34			35 36	37	38	39 40
Eighth Journey																												
	19 20 21		22 23 24		25 26						27	28						29	30	31 32 33		34			35 36	37	38	39 40
Ninth Journey																												
	19 20 21		22 23 24		25 26						27	28						29	30	31 32 33		34			35 36	37	38	39 40
Tenth Journey																												
	19 20 21		22 23 24		25 26						27	28						29	30	31 32 33		34			35 36	37	38	39 40
Eleventh Journey																												
	19 20 21		22 23 24		25 26						27	28						29	30	31 32 33		34			35 36	37	38	39 40

Notes for Guidance

Note 1 **A Journey means a movement from one place to one other place.** Most people will record at least two journeys, a journey out from home and a return journey back to home.

Example 1 If you went to work morning and afternoon, returning home for lunch and after work, record this as four journeys.

Example 2 A trip to Granby Street shops, then to the De Montfort Gardens, then back home. Record this as three journeys.

Note 2 **In the case of two members of a household making the same journey together, enter each persons' journey on a separate form.** For example, for a child taken to school by an adult on the way to work, enter the child's journey on the child's form and the adult's journey on the adult's form.

Note 3 **If you are a driver of a Taxi, Public Service Vehicle or Commercial Vehicle, or if you are a "Driver's Mate",** DO NOT record the journeys you travel during work this information is being obtained in other ways. DO record your first journey from home to work and your last journey from work to home, and all other personal journeys.

Note 4 **Night workers** should record the journeys they make on this date only, from midnight onwards.

Note 5 Work includes attendance at University, Commercial College, etc.

Note 6 **Where did this journey end?** If you visited several shops in one place, record this as one journey and give the name of the shopping area only. If you went outside Leicester and district give the name of the TOWN (etc.) only. In other cases be as precise as possible, e.g. by giving the Street name and number, or well-known place or building, or name of Shop and Street.

Note 7 **How did you travel?** If you went by train, record each journey to and from the station in the Leicester area, and each journey on the train. Walking journeys should only be recorded only if they are TO or FROM work or TO or FROM bus stop, railway station or parked vehicle.

Note 8 **Purpose of journey.** Examples of typical journey will be FROM home TO work, FROM work TO shopping, FROM shopping TO eat meal. The purpose being entered under the appropriate FROM TO column. The TO Purpose of the previous journey will always be the same as the FROM purpose for the next journey.

Fig. 5.10. The Home questionnaire designed and used by the City of Leicester Planning Officer for an area beyond the city boundaries, in collaboration with the County Planning Officer and the County Surveyor. (See also Appendix.)

It is absurd to project provision for car use on opinions collected from drivers today. Yet this is what the Ministry Group were looking for to make their statement on the needs of the next twenty years. Indeed Birmingham University School of Traffic Engineering with the Birmingham School of Planning are collecting just such information on a great scale. Such data may not be entirely useless but there is a great danger that the statistics culled from the statements of the mass of people will be used to project needs. This would be nonsensical. The increase in car-ownership requires positive planning for optimum use—as derived from first principles by careful judgements, not an approximation to the average opinion of the ignorant. Imaginative, bold provision for vehicles will stem from creative original thought on the basis of fundamental facts. In planning for motor cars sensibly, past or present wishes cannot be the criterion. The wishes change as the number of cars increase and as the environment changes. For example, given an uncongested road, 100% of car owners may say they want to ride to work. Given a congested road they may say that they like efficient rail transport.

Once it has been accepted that sensitive, expert and informed value judgement will fix the number of cars an existing town can absorb, and that there will always be pressure to exceed this by factions of society, the point of many of the demand surveys becomes very much less urgent and sensible. Cheap parking will, inevitably, be taken up if the roads can cope with the traffic at rush hour, and road improvement will usually bring more cars, not less congestion. But planning can relocate industry and can create counter-attractions. This changes origin and destination figures. Planning does not merely mean following trends and satisfying data based on peak hour needs. These can be decisively changed also if arrangements and slight re-organization spreads the peak load over a longer time. This simple step will surely be taken quite soon, with the encouragement of local authorities and Government. It is the easiest, cheapest way of easing congestion, or at least allows more cars the congestion now enjoyed by the few.

One other general principle can be stated. Where rail communication is good, the need for roads is less than along those desire lines which are not so served. The reverse holds true. Rail communications in urban areas should provide routes not satisfied by roads. This principle is valid locally and nationally and pin-points the mistaken positioning of motorways. British Railways' figures show that their most profitable routes are parallel to busy roads. This indicates efficient use which ought to be extended.

The relative volume of road and rail passenger and goods traffic over the whole of Britain have been tabled by Day(i) and the Ministry of Transport.(ii) Though both acknowledge the same sources the figures are different in part, which is typical of the vagueness and uncertainty that permeates past statistics and dominates future forecasts. The Ministry report is "trend" planning, almost entirely extrapolation and projection rather than creative planning.

TABLE V.2. *Inland passenger travel by land (hundred million passenger miles)*

Public Transport	1952		1958		1960	
	Day	Min	Day	Min	Day	Min
Road	51·1	59·1	44·1	43·4	41·0	43·9
Rail	24·0	24·0	25·0	25·5	24·3	24·8
Private cars and taxis	37·4	35·1	67·6	61·5	77·7	74·8
Total	112·5	109·2	136·7	130·4	143·0	143·5

Source: Estimates based on Public Road Passenger Statistics, British Transport Commission Reports and Road Research Laboratory data.

TABLE V.3. *Inland goods transport (thousand million ton miles)*

	1952		1958		1960	
	Day	Min	Day	Min	Day	Min
Road	18·8	18·3	23·1	23·1	28·1	26·6
Rail	22·4	22·4	23·1	18·4	18·6	18·7
Total	14·2	40·7	46·3	41·5	46·7	45·2

Source: Estimates based on Ministry of Transport British Transport Commission and Road Research Laboratory data.

The Ministry Report says:

"*All this tends to confirm our view that the present inter-urban railway movement of some* 16,000 *million passenger miles per annum is unlikely to change appreciably. This is on the assumption of no significant change in the cost of rail transport relative to other prices. Rail fares have now caught up with the general rise in retail price levels; and the response of rail carryings to price changes in recent years suggests that there is very little room for further increase without provoking resistance of an order which so reduces passenger journeys that receipts are not likely to rise.*"(ii)

Dr. Beeching's British Railways report,(iii) a month later, makes clear that there will be a

(i) DAY A. *Roads*. London 1962.

(ii) Ministry of Transport. *The Transport Needs of Britain in the Next Twenty Years*. H.M.S.O. 1962.

(iii) British Railways, Dr. Beeching Report. 1963.

considerable rise in fares. This policy in a Welfare State must be compared with United States' policy described in Section V. 8.1.

It is important to realize that large scale railway closing down proposed by Dr. Beeching's Report, 1963, is not because passenger traffic has fallen significantly, but so the railways can make money. This is a false criterion for a nationalized industry leading to short-sighted "business methods" for a small service.

With planning powers in Britain it would be feasible and sensible to plan industry and settlements so that under-used lines of communication became better used, rather than accentuate a number of very serious trends which, on the one hand, lead to depopulation of large tracts of the country and on the other overload the transport system of the areas already most seriously congested.

Only comprehensive planning can save Britain from disastrous trends about to be accelerated by "trend planning" of the worst kind, with money profit as the false criterion.

V. 2.2. *Vehicles and Town Space*

(Space requirements of "Motorization" as compared with public transport.)

Dr. Smeed[(i)] has worked out some interesting correlations of types of transport, working area and resulting size of town, for populations from 10,000 to 5,000,000. Although his calculations are purely theoretical and restricted to commuters, leaving out of consideration parking demands of shoppers, the figures clearly demonstrate the overwhelming effect of universal private car usage, as compared with public transport. This effect increases with the size of population. He notes that multiple use of land for parking, roads and buildings is essential if large cities are to be fully motorized. This requires great expense and exceptional levels of co-operation.

Dr. Smeed takes the ground space per worker as 100 ft². Car parking space in multi-level garages as 20 ft² (for a 16-storey car park) and on one level 200 ft². Occupancy rate of 1·5 persons per car. The contents of these tables teach many valuable lessons on careful perusal.

Dr. Smeed's parallel calculations for theoretical residential areas, with 10 workers to the acre, show less dramatic results. Extra road space required for car journeys to work, only slightly affects the town radius. However, close to the centre, a similarly immense demand on area for roads is found as in the calculation for the centre, up to 56% in a town of one million. Parking or garaging is not taken into account in the residential calculations.

TABLE V.4. *Traffic and Town Space for* 10,000–50,000 *inhabitants.*

Commuter's transport	Nature of parking	Parking per head	Town of 10,000				Town of 50,000			
			percentage				percentage			
			town r	roads	parking	work	town r	roads	parking	work
Urban railway	—	0	0·11 miles (0·18 km)	0·2	0	99·8	0·24 miles (0·39 km)	0·3	0	99·7
Bus 44 ft road	—	0	0·11 miles (0·18 km)	1	0	99·0	0·24 miles (0·39 km)	2	0	98·0
Bus 24 ft road	—	0	0·11 miles (0·18 km)	1	0	99·0	0·24 miles (0·39 km)	3	0	97·0
Car Urban motorway	multi-level	13·3 ft² (1·2 m²)	0·11 miles (0·18 km)	2	12	86·0	0·26 miles (0·42 km)	4	11	85·0
Car Urban motorway	ground level	133 ft² (12·4 m²)	0·16 miles (0·26 km)	1	56	43·0	0·37 miles (0·6 km)	3	55	42·0
Car 44 ft urban road	multi-level	13·3 ft² (1·2 m²)	0·12 miles (0·19 km)	5	11	84·0	0·27 miles (0·44 km)	11	10	79·0
Car 44 ft urban road	ground level	133 ft² (12·4 m²)	0·17 miles (0·27 km)	4	54	42·0	0·38 miles (0·61 km)	8	53	39·0
Car 24 ft urban road	multi-level	13·3 ft² (1·2 m²)	0·12 miles (0·19 km)	9	11	80·0	0·28 miles (0·45 km)	19	10	71·0
Car 24 ft urban road	ground level	133 ft² (12·4 m²)	0·17 miles (0·27 km)	6	54	40·0	0·39 miles (0·63 km)	14	49	37·0

(i) SMEED R. J. *The Traffic Problem in Towns.* Manchester Statistical Soc. 1961.

TABLE V.5. *Traffic and Town Space for* 100,000–500,000 *inhabitants.*

Commuter's transport	Nature of parking	Parking per head	Town of 100,000				Town of 500,000			
			percentage				percentage			
			town r	roads	parking	work	town r	roads	parking	work
Urban railway	—	0	0·34 miles (0·55 km)	0·6	0	99·4	0·76 miles (1·22 km)	1	0	99
Bus 44 ft road	—	0	0·34 miles (0·55 km)	2	0	98	0·78 miles (1·26 km)	5	0	95
Bus 24 ft road	—	0	0·35 miles (0·56 km)	4	0	96	0·79 miles (1·27 km)	10	0	90
Car Urban motorway	multi-level	13·3 ft² (1·2 m²)	0·37 miles (0·6 km)	6	11	83	0·86 miles (1·38 km)	13	10	77
Car Urban motorway	ground level	133 ft² (12·4 m²)	0·53 miles (0·85 km)	4	55	41	1·21 miles (1·95 km)	9	52	39
Car 44 ft urban road	multi-level	13·3 ft² (1·2 m²)	0·37 miles (0·6 km)	16	11	73	0·97 miles (1·56 km)	31	8	61
Car 44 ft urban road	ground level	133 ft² (12·4 m²)	0·55 miles (0.89 km)	11	51	38	1·32 miles (2·12 km)	23	44	33
Car 24 ft urban road	multi-level	13·3 ft² (1·2 m²)	0·42 miles (0·68 km)	26	8	66	1·11 miles (1·79 km)	48	6	46
Car 24 ft urban road	ground level	133 ft² (12·2 m²)	0·57 miles (0·92 km)	19	47	34	1·45 miles (2·32 km)	37	26	27

TABLE V.6. *Traffic and Town Space for* 1,000,000–5,000,000 *inhabitants.*

Commuter's transport	Nature of parking	Parking per head	Town of 1,000,000				Town of 5,000,000			
			percentage				percentage			
			town r	roads	parking	work	town r	roads	parking	work
Urban railway	—	0	1·08 miles (1·74 km)	2	0	98	2·44 miles (3·93 km)	4	0	96
Bus 44 ft road	—	0	1·11 miles (1·79 km)	7	0	93	2·60 miles (4·18 km)	16	0	84
Bus 24 ft road	—	0	1·15 miles (1·85 km)	13	0	87	2·80 miles (4·51 km)	27	0	73
Car Urban motorway	multi-level	13·3 ft² (1·2 m²)	1·25 miles (2·01 km)	17	10	73	3·16 miles (5·09 km)	34	8	58
Car Urban motorway	ground level	133 ft² (12·4 m²)	1·75 miles (2·82 km)	13	50	37	4·24 miles (6·82 km)	25	42	23
Car 44 ft urban road	multi-level	13·3 ft² (1·2 m²)	1·48 miles (2·38 km)	41	7	52	4·48 miles (7·21 km)	67	4	29
Car 44 ft urban road	ground level	133 ft² (12·4 m²)	1·96 miles (3·15 km)	31	39	30	5·47 miles (8·80 km)	55	25	20
Car 24 ft urban road	multi-level	13·3 ft² (1·2 m²)	1·79 miles (2·88 km)	60	5	35	6·35 miles (10·22 km)	84	2	14
Car 24 ft urban road	ground level	133 ft² (12·4 m²)	2·24 miles (3·61 km)	47	30	23	7·18 miles (10·56 km)	74	15	11

(In 53 central cities in the U.S.,[(i)] 28% of total developed area was taken up by roads, and similar percentages in satellite towns and urban areas.)

(i) BARTHOLOMEW, HARLAND. *Land Uses in American Cities*. Harvard City Planning Series, Vol. XV. Harvard University Press, 1955.

In a subsequent study[i] Dr. Smeed calculated the numbers of workers who could approach a city by car at peak hour given 20% of the town occupied by roads and parking of six-storey height. He compared direct, radial and ring routeing.

The density of workers is according to London figures, and peak period is extended to last two hours morning and evening. Car occupancy is taken as 1·5. The speed is taken as 10 mph and road capacity calculated accordingly (a very high capacity in London). If they are to travel at 15 mph, the speed taken by the Road Research Laboratory normally as the reasonable minimum speed to aim at, the numbers have to be divided by two. Road surfaces are taken to serve in their normal direction only, even at peak hours.

Dr. Smeed is careful to point out that the road space must of course be available exactly where it is wanted for the calculation to be applicable.

The working populations which Dr. Smeed calculates are immense and his conclusions seem to be contrary to the experience of large American cities which have found that they have to rely on public transport to a very substantial, if varying, extent. However, the colossal width of the ring road in his case and many other questionable assumptions, the lack of economic considerations and the theoretical nature of the exercise, leave it quite incomplete as a piece of evidence. What the tables suggest could not actually be achieved, even if roads are perfectly designed and distributed. Again, as with other work, the shopping parking is not taken into account nor a variety of other volume consuming factors, such as commercial traffic which takes 24% of peak period traffic.

Thus, though the figures quoted in the table seem misleading, if not considered in their entire complex context, the chief contribution of Dr. Smeed with this study lies in the clear evidence of the different efficiencies of road systems in towns. The tangential system, recently incorporated in a number of plans, would be an interesting fourth comparison. His general conclusions are:

"(*i*) *For traffic routes, narrow roads are wasteful of ground space.*
(*ii*) *High-capacity radial roads continued to the centre of a town are likely to encourage people to travel by routes which pass through the centre of the town. A system of such roads in a large town is therefore likely to cause congestion in the centre, unless a large area in the centre of the town is devoted to roads.*

(i) SMEED R. J. *The effects of some kinds of routeing systems on the amount of traffic in the central area of towns.* R.R.L. 1962.

(*iii*) *Provided they have frequent access points, high-capacity ring roads attract traffic with no business to carry out in the inner part of the town. They, therefore, tend to result in reduced congestion in the centre. Their disadvantages are that they result in increased distances of travel and in increased total area required for roads.*
(*iv*) *Provided the central area of a town has an appropriate road system, including a ring road of adequate capacity, which is used by all drivers in such a way as to make their journeys within the central area as short as possible, and provided the journey to work is spread over an adequate period, and provided too many drivers do not travel alone, and if multi-storey car parks are used for parking, the calculations suggest that it might be possible for everybody working in the central area of a fairly large town with a density of workers similar to that of Central London to travel to work by car.*"
(See also p. 152.)

TABLE V.7. "*Maximum numbers of persons who subject to appropriate conditions, can work in a suitably designed town, when they all travel by car (carriageway and parking spaces limited by 20% of ground area.)*"[i]

Routeing	Case 1 (uniform density of destinations in central area) Number of Persons	Case 2 (density of destinations inversely proportional to distance from centre) Number of Persons
Direct	95,000	103,000
Radial	44,000	54,000
Ring if the circumferential road is not included in the 20%	1,119,000	487,000
Ring if the circumferential road is included in the 20%	34,000	28,000

(i) An increase of x% in the area of roads and parking increases the number of workers that can travel by car by $2x$%. (U.S.A.)

V. 2.3. *General Economic Consideration*

It is cheaper to drive a car round city streets than park it. This anomaly brings to the fore the odd assumption that every one has a right to bring his huge metal box on wheels into a city, poison the citizens, and then drive out again without charge. But if he stops he is charged or fined, whether outside his house or not! To the man from Mars it becomes odder still when he sees the citizen pay for toll roads and bridges outside cities, where there is no shortage of space. Something is topsy-turvy, quite obviously.

A statement by the Minister of Transport of Britain may become historic. It said:

"*Alternatives may have to be considered to prevent private cars unduly inflating peak hour traffic flows.*"(i)

This, the last sentence of his 1961–62 report, may be the first of a new chapter on car use in cities. Here is an indication that, apart from and beyond restricting parking, or making people pay for it in meters, roads may be regarded as something too scarce, in the centre of the city, to be left for one and all to help themselves to 200 ft^2 at a time, or much more while moving (see also p. 152).

Although isolated ministries or spokesmen may make such statements there is little known of which trends are the right ones to encourage and what principles should govern redevelopment and new development. The research for this has only just begun(ii) and is to be undertaken, seemingly, without the Universities and without specific participation of the ministry responsible for town planning, which once upon a time had its own research staff under Holford:(iii)

"*In its Annual report the Road Research Board of the DSIR says that the panel on urban roads, which it set up two years ago, has called attention to the urgent need for research to establish the facts, techniques and principles upon which to base the redevelopment of new towns.*"

"*We have accepted the panel's recommendation to investigate those aspects of town planning that affect the generation and movement of traffic; the aim will be to identify and measure the social and economic forces that are operating and to establish relations between traffic and such factors as land use, population densities and cost of travel. To assist this research we have recommended that a working party be formed, to include members from the Road Research Laboratory and the Building Research Station of the DSIR and representatives of other interested departments.*"

(i) *Roads in England and Wales*, 1961–62. H.M.S.O. London 1962.

(ii) *Annual Report* (1962) *of the Road Research Board of DSIR.* H.M.S.O. 1963.

(iii) Seymer N. Regional Planning and Transport. *A.J.* London 21.8.1963.

V. 3. Routes of Vehicles

V. 3.1. *Administration of Road Planning*

The technical and administrative alternatives are many, depending on the political system, the size of the town and other factors. The 31 steps which have to be taken in Britain before a road can be built indicate complexities involved. (See Table p. 69).

V. 3.2. *Road Space, Shape, Noise and Danger*

In rural areas the delight and elegance of driving lies in the smooth, gentle, organic rhythm of vertical and horizontal meander, right for the speed of the road, and giving contrast in terms of large elements, such as woods, lakes, fields, walls, bridges, viaducts, tunnels. In urban areas the small elements facing roads today need tying together into large visual units along all classes of roads, except in the very low-speed, car-port or service lanes with direct access to doors. The restriction on a multiplicity of openings on to roads helps functionally with the segregation of pedestrians from vehicles. Future planning should have walls, fences, embankments or cuttings, even of a modest foot or two, as the standard edge to roads, in units of several hundred foot lengths, giving the right scale, as well as safety at speed. Where embankments are higher, they serve the additional valuable service of sound barrier between road and residential or business areas, as Reichow has demonstrated.(i) Adequate banks can be built from 20 in. of road excavation plus site subsoil from foundation work. Alternatively, the road may be excavated and raised, excavation giving the soil for the banks where raised. There are further considerations regarding the edge of traffic routes.(ii)

A considerable proportion of accidents to cars arise out of crashing into lamp-posts, telegraph poles, trees, bus stops and other vertical items right beside the road. Although this is plain it nevertheless seems extraordinary that few steps are taken for the removal of these death traps. This becomes particularly relevant when roads are planned for vehicular traffic pure and simple. In studying statistics, it is important to realize that, though this type of accident may be a low percentage of the total, the frequency with which it leads to death is high. From a simple, logical standpoint it is absurd to make the first thing a car hits when off its track, as little as one foot at times, an object as disastrous as can be. (If it is not a pedestrian on the present system.) The frequency and ubiquity of this condition has obscured its illogicality.

(i) Reichow H. B. Noise Abatement. *A. J.* 13 Feb. 1963. (See also p. 25.)

(ii) See also Section on Capacity, V. 3.3.

TABLE V.8. *The 31 steps. Even when a road scheme has been put forward these steps have to be taken in Britain. They often cause delays of three to seven years. Some have to be repeated if one step fails. "An inquiry into the 31 Steps is not only necessary, it is long overdue," Mr. J. Rawlinson, C.B.E., Chief Engineer, London County Council.*

	1	2	3	4	5	6	7
	Ministry request the local authority to prepare 1 (2) order details.	Local authorities circulate appropriate authorities and other bodies to obtain observations.	Local authorities submit to Ministry draft Order details.	Ministry headquarters prepare draft Order.	Draft order placed on deposit (3 *months*).	Objections received by Ministry.	Inquiry to discuss objections (*if not resolved*).
15	14	13	12	11	10	9	8
14 (1) Order made (*if not amended by findings of inquiry —in which case matter reverts to item* 10).	Inquiry to discuss objections (*if not resolved*).	Objections to (2) received by Ministry.	15 (1) Order placed upon deposit.	Ministry headquarters prepare draft 14 (1) Order.	Local authorities send, in first place, 14 (1) Order details to Ministry.	Ministry invite local authorities to prepare land plans 14 (1) Order details and scheme details.	Order made (*if not amended by findings of inquiry—in which case matter reverts to item* 1).
16	17	18	19	20	21	22	23
Local authority submits land plans to the Divisional Road Engineer of the M.O.T.	Divisional Roads Engineer submits land plans to Ministry headquarters.	Ministry headquarters submits land plans to Chief Valuer.	Chief Valuer forwards to District Valuer	Local authority forwards scheme details to Ministry.	Ministry approve or amend scheme details.	Local authority prepares draft contract.	Ministry approve or amend draft contract.
31	30	29	28	27	26	25	24
Contractor commences work on site.	Clerk of local authority prepares contract.	Local authorities obtain Committee approval to tender.	Ministry approve tender.	Tenders to Ministry for approval.	Local authority advertise contract and receive tenders.	Ministry give clearance to item 24.	Local authority awaits clearance of land acquisition and authority to commence work.

Based upon a Paper by Col. S. Maynard Lovell, O.B.E., T.D., County Engineer and Survey for the West Riding of Yorkshire, presented in 1957 at the conference "Highway Needs of Great Britain" organized by the Institution of Civil Engineers.

Fig. 5.11. Glancing off device might have saved a life.

The solution lies in the reduction of vertical elements to the absolute minimum *and* arrangement for cars to glance off them if they do hit any, or a safety region, a resilient fence or the "multi-flora rose" which, given an 8 ft depth, can take vehicles at 60 mph and halt them comparatively gently. The even more obvious need for barriers between opposing flows of traffic is still not universally answered. Such vertical elements as have to be supplied should be, like "keep left" signs in Britain, easy to knock over. Where such signs protect pedestrian islands the principle must be reversed. Here the protection is specifically necessary, because of the mixing of man and motor, and the vehicle should glance off any barrier. In such areas speeds should be slow.

The reports from the 6th International Safety Congress confirm the above points:

"*The number of single vehicle accidents amounts to about 30% of the total number of road traffic accidents in Sweden. About half the number of single vehicle accidents involve collisions with roadside features of various kinds.*

"*Since single vehicle accidents constitute such a predominant type of accident, it is necessary to make thorough analyses so as to elucidate the causes of these accidents. The effects of roadside features should also be closely studied in this connection.*"(i)

The French speaker at the Congress gave figures showing the danger of closely spaced tree lining to avenues, and reported that, as a result, many avenues were to be thinned out.

Figures were given showing the importance of obstructions due to the presence of street furniture in single-vehicle accidents and Moore, from Britain, added that:

(i) Bruzelius N. G. Single Vehicle Accidents in Relation to Roadside Features. *6th International Road Safety Congress Report.* London 1962.

"*The danger of lining fast routes with a palisade of strong and massive lighting columns is emphasized, and some experiments on the relative lethality of lighting columns when struck by vehicles are described. It is pointed out that lighting columns must be capable of shearing at the base if this danger is to be minimized.*"(i)

The Italian figures (which, incidentally, unlike the other countries included "running over pedestrian" as "single-vehicle accidents") showed that "collision with stationary objects" accounts for 2·5% of all accidents only, but for 10% of the death role. This type of accident, in other words, is four times as lethal as the average.(ii)

V. 3.3. *Capacity of Roads*

Whether routes remain in the form of roads in the present sense, or whether hovercraft tracks are built or cleared in other places, the same sort of relationship between ground and vehicle, via gravity, will persist. The capacity of the routes will be related to speed, efficiency and reaction time.

"*It would be incorrect to state that the possible capacity of a two-lane highway is 2000 vehicles per hour. If, however, the statement be amplified to say that the possible capacity of a level, tangent, two-lane highway with a 24-foot surface, free from lateral obstructions within 6 feet of its edges, and with no major intersections at grade, is 2000 passenger cars per hour, then it is substantially complete and correct.*"(iii)

In a table W. R. Bellis(iv) lists the full number of factors limiting the capacity of roads:

"*The largest lane-capacity known to have been experienced is 2437 cars in one hour. Let us assume that a larger volume could be obtained such as 3700 cars an hour a lane shown in 'Elements of Capacity Diagram'. Elements limiting capacity to a volume less than 3700 are listed at the left of the diagram. Each of these impeding elements reduces the capacity by a degree peculiar and variable depending on the element. Most of these elements have not been isolated so that their effect can be measured. A few have been measured, and, in the case of freeways, the combined influence of a group on impending factors can be measured since these factors have been eliminated.*"

(i) Moore R. L. Single Vehicle Accidents in Relation to Street Furniture. *6th International Road Safety Congress Report.* London 1962.

(ii) Sturni G. P. Road Accidents Involving a Single Vehicle in Relation to the Nature of the Roads. *6th International Road Safety Congress Report.* London 1962.

(iii) *Highway Capacity Manual.* U.S. Department of Commerce Washington 1956.

(iv) Bellis W. R. *Traffic Quarterly.* ENO Foundation, Saugatuck, January 1959.

"The effect of some elements can vary from zero to 100%. Some such elements are pedestrians and weather. When any one element reduces the capacity by 100% or more, the effect of the other elements is meaningless. In the diagram, capacity is shown as varying from zero to 3700. A capacity is shown at 600 to represent city streets but some streets have less and some have more, depending on the effect of the impeding elements."

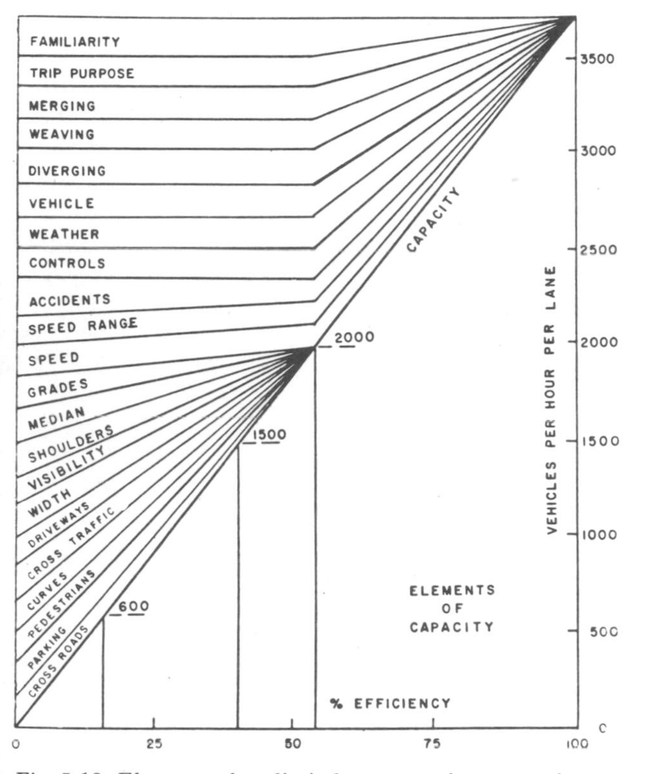

Fig. 5.12. Elements that limit lane capacity on a city street.

The Road Research Laboratory has assessed the factors affecting road capacity and the following table shows the conclusions for rural and urban roads. For the latter a minimum acceptable speed of 15 mph is taken as the basis for their calculations.(i)

Unlike U.S. figures the flow here increases as speed decreases down to 10 mph for urban traffic.(ii) (See Fig. III.9.)

TABLE V.9. *Capacity of streets at various running speeds.*(i)

Running speed mph	Total flow (vehicles per hour) for carriageway-width of: 20 ft	30 ft	40 ft	50 ft	60 ft	70 ft
20	—*	350	700	1000	1350	1700
15	250	700	1200	1700	2150	2650
10	450	1100	1700	2350	2950	3600

* Extrapolation is unreliable here.

(i) WARDROP and DUFF. *Factors Affecting Road Capacity.* Road Research Laboratory. London 1956.

(ii) See also 3.11; 3.12.

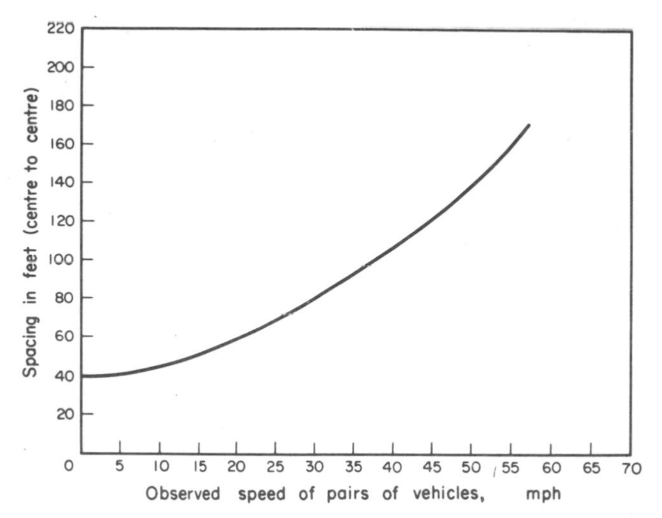

Fig. 5.13. Spacing between vehicles on Rural U.S. roads. Minimum spacings allowed by the average driver when trailing another vehicle at various speeds (two-lane highway).

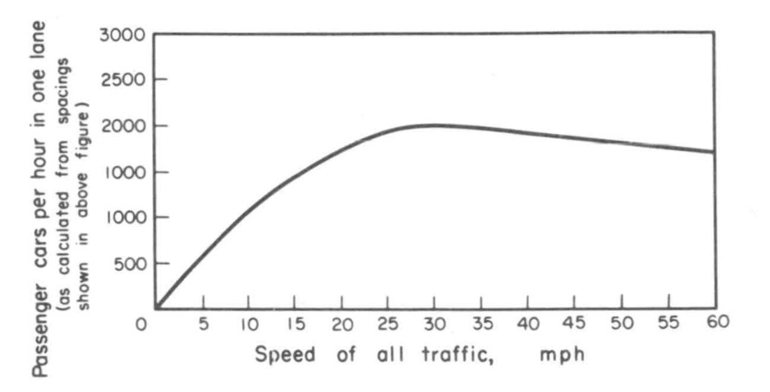

Fig. 5.14. Capacities of lanes at varying speeds. Maximum capacity of a traffic lane, based on average spacings between pairs of vehicles travelling at the same speed (two-lane highway).

Roads are most efficient and take most traffic per hour when marked lanes are used evenly at something like 30 mph. This comes as a surprise to many. It is a crucial point to bear in mind when the planners of old cities, with a minimum area to give over to the car, work out the number of lanes. It means that to be able to keep up a speed of thirty is important and to limit speeds likewise. This conclusion is based on the observed distances between vehicles on U.S. rural two-lane highways.

At that maximum 30 mph or higher for multi-lane roads (35 mph–40 mph) a lane will take 1500 vehicles per hour whether one of two lanes or one of four lanes. At a speed of 45 mph to 50 mph this is reduced to a thousand or 900 for the two-lane road. Add one lane and the extra capacity is only 350 vehicles per hour (discrediting the lethal 3-lane highway on economic grounds).

This is of course a theoretical capacity assuming uninterrupted flow. However, junctions of any kind have speed limits. At low speeds of 35 mph ramps at junctions need not slow down traffic.

The very high speed lanes of roads, like the third lane of the M1 London–Leicester, at 120 mph, will take only 120 vehicles for its 46 mile length in one hour, so great are the distances observed between fast-moving cars.

The capacities of most existing urban roads are crucially affected by junctions as well as parked vehicles and many of the factors of V.9. This brings about the great variation of flow as recorded in III. 1.3. Again, sophisticated metropolitan drivers use road space to the utmost and the spacing between vehicles at low speeds is much less than the U.S. figures for rural roads, giving a considerably higher capacity for British urban roads, particularly in London (see Wardrop and Duff, R.R.L., London 1956).

TABLE V.10. *Capacity of uninterrupted streets. Comparison of Road Research Laboratory results with those of Normann* (U.S.A.)

Number of 12 ft lanes (two-way roads)	Carriageway widths, ft	Capacity per 12 ft lane	
		R.R.L. London traffic (15 mph) vehicles per hour	Normann 20 per cent heavy lorries (35–45 mph) vehicles per hour
2	24	220	620
3	36	340	560
4	48	400	1250
6	72	550	1250

TABLE V.11. *Increase in traffic speed and capacity due to a reduction in parking level of* 10 *vehicles per mile.*

	Carriageway width, ft	Increase in speed at constant flow, mph	Expected increase in capacity at 15 mph, per cent
Central London streets	30–45	½	4–5
Suburban streets	20–40	¼	2–4
Three London sites (detailed study)	32–38	¾	7
Hounslow High Street	30	¼	2½

One of the capacity reducing elements, studied in U.S., is analysed below. Most of the others have been studied in one way or another more or less conclusively. (See also pp. 15, 159.)

TABLE V.12. *Capacity expressed as a percentage of the capacity of two* 12*-foot lanes with no restrictive lateral clearances. Combined effect of lane width and edge clearances on highway capacities.* (i)

Clearance from pavement edge to obstruction	Obstruction on one side				Obstruction on both sides			
	12-foot lanes	11-foot lanes	10-foot lanes	9-foot lanes	12-foot lanes	11-foot lanes	10-foot lanes	9-foot lanes
	Possible Capacity of Two-Lane Highway							
Feet								
6	100	88	81	76	100	88	81	76
4	97	85	79	74	94	83	76	71
2	93	81	75	70	85	75	69	65
0	88	77	71	67	76	67	62	58
	Practical Capacity of Two-Lane Highway							
6	100	86	77	70	100	86	77	70
4	96	83	74	68	92	79	71	65
2	91	78	70	64	81	70	63	57
0	85	73	66	60	70	60	54	49
	Possible and Practical Capacities of Two lanes for One Direction of Travel on Divided Highways							
6	100	97	91	81	100	97	91	81
4	99	96	90	80	98	95	89	79
2	97	94	88	79	94	91	86	76
0	90	87	82	73	81	79	74	66

(i) *Highway Capacity Manual*, U.S. Department of Commerce, by the Committee on Highway Capacity, Department of Traffic and Operations Highway Research Board. Washington 1950.

V. 3.4. *Speed*

Vehicles rarely move at the speeds for which they are designed. Average car speeds on motorways in Europe vary only from 52 mph to 54 mph. In many areas of the United States, notwithstanding the existence of motorways and powerful cars, absolute state speed limits of 65 mph or thereabouts are enforced. In fact it is surprising to Europeans to what a considerable extent much slower speeds are prescribed and obeyed on many of the new U.S.A. roads. High minimum speed regulations can guarantee great flow capacity. Smooth, uninterrupted journeys are a better aim than interrupted high speeds.

On rural roads South Carolina has maximum and minimum speed limits. Two signs on some posts limit drivers to between 45 mph min. and 55 mph max.

Speeding over short distances between the slower vehicles is tempting, but very rarely effective in making journeys significantly quicker. This is combined with great danger. Speed of cars in unrestricted free areas is something quite different from speed on the other kinds of road where everyone is tied to an average. Propaganda could make this point clear to many impatient people, now creating danger for themselves and others to very little purpose.

It is not recognized and so ignored that at higher speeds less time is saved for a standard increase. The longer the distance, the greater the saving in every case.

It should be made clear how little there is to be gained by the extra speed attempted by many drivers at high speeds, beyond what feels safe to them. They are likely to relax and become better drivers.

Speed is largely determined by the road and the character of the driver, within certain practicable limits.

Speeding is often cited as largely contributory to accidents.(i) Evidence from the time of the imposition of 30 mph limits in urban areas in Britain bears out this argument: accidents dropped dramatically.(ii)

(i) COBURN T. M. Speeds on European Motorways. *Internation Road Safety and Traffic Review*. VII, 4. Autumn 1959. NEWBY R. F. Effect on Accidents in London Area. *Traffic Engineering and Control*. July 1960.

(ii) SMEED R. J. *The Influence of Speed and Speed Regulations on Traffic Flow and Accidents*. Road Research Laboratory. 5th International Study Week in Traffic Engineering. Nice 1960.

TABLE V.13. *Time saved with speed increase of 10 mph at 10 mph*

Distance (miles)	10 mph	20 mph	Time saved 100%
10	1 hr	½ hr	½ hr
50	5 hr	3½ hr	2½ hr
100	10 hr	5 hr	5 hr

TABLE V.14. *Time saved with speed increase of 10 mph at 100 mph*

Distance (miles)	100 mph	110 mph	Time saved 10%
10	6 min	5·5 min	½ min
50	30 min	27 min	3 min
100	1 hr	55 min	5 min

However, statistics lead to vague generalizations. There is evidence that enforced slow driving, for those who have quick reactions, leads to a lack of concentration in their driving which is a greater danger than the fast speed with concentration. Research bears out that this is a valid theory and should modify the principle of "slow is safe".(i)

> "*Figures based on reports from nineteen states show the approximate speed of vehicles involved in fatal accidents during* 1955, *with the following percentages of vehicles in four speed ranges:* 0–20 *mph*, 17 *percent;* 21–40 *mph*, 30 *percent;* 41–60 *mph*, 35 *percent;* 61 *mph and over*, 18 *percent. Our results suggest that individuinals the kind of population considered* (346 *Californian students*) *who report consistently higher driving speeds than average have, in reality, traffic records free of accidents as often as other drivers.*"

Summarizing his paper on the influence of speed and speed regulations in a number of European countries, Dr. Smeed includes the following points:

"(*i*) *A high proportion of drivers do not limit their speeds as they should do in order to accord with legal requirements.*

(*ii*) *In nearly all cases for which we have satisfactory data, a speed limit had a marked effect in reducing the higher speeds.*

(*iii*) *In other cases, the distribution of the speeds was identical before and after the imposition of the limit.*

(*iv*) *Speed limits seem to have a marked effect in reducing fatal accidents in urban areas. They have much less effect on slight or damage-only accidents.*

(i) STEWART R. G. Are We Over-Emphasising Speed as an Accident Cause? *Reprint Traffic Quarterly*. Saugatuck October 1957.

(*v*) *There is some evidence that motor cyclist and pedal cyclist fatalities are especially affected by speed limits.*

(*vi*) *The imposition of speed limits on a number of main roads, including motorways, seems to have had a favourable effect on road accidents.*"

But he adds:

"*No attempt has been made in this report to discuss some aspects of the subject. Thus there can be no doubt that drivers should adjust their speed according to circumstances. A speed limit cannot—in practice—be put low enough to make it a suitable speed in bad conditions, such as fog, and on the other hand it loses much of the value if the limit is put at a value suitable for especially favourable conditions such as early in the morning on a good summer day with no pedestrian or other traffic on the road. If the great majority of drivers could be relied upon to adjust their speed properly according to circumstances, this argument against the use of speed limits would be very strong. Some information on the possibility of using other methods for controlling speeds would be very useful.*"

Fig. 5.15. Junction showing variety of curves in the design.

The speed for vehicles on roads (or in the case of hovercraft, just above them) is limited not by the power of the vehicle but by the space taken up by speed. Between the front and back of a fast moving car there is, for safety and braking reasons, such a large gap that the capacity of a lane becomes very small. Overtaking at such speeds takes up relatively more space still if both cars are very fast.

Curvature relating to speed and possible superelevations of road have been plotted by Reichow.[(i)]

The figures to the outside of the spiral refer to the permissible radii of highway curves in metres and are constant between the white rectangles superimposed upon the black spiral. The small trapeziums indicate on the top side the speed in kilometres per hour, and the figures below the trapezium relate to the percentage slope of superelevation. Thus, to determine the degree of superelevation for a particular curve at a predetermined designed speed, study of the spiral will produce the answer. Superelevation, radii or permissible speed can be readily determined.

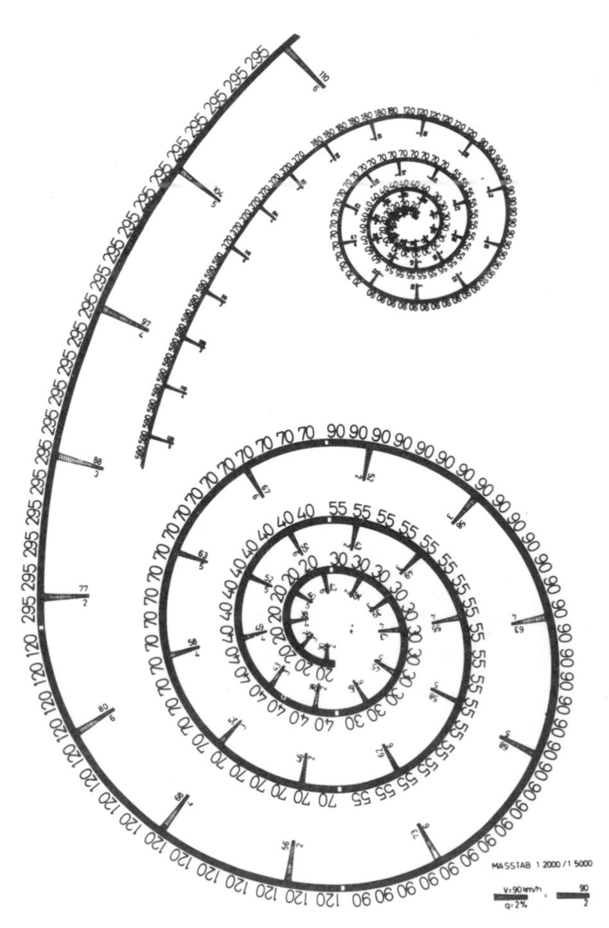

Fig. 5.15a. Reichow's spiral diagram for the calculation of superelevation and curvature permissible at various speeds.

(i) REICHOW H. *Autogerechte Stadt.* Ravensburg 1959.

V. 3.5. *Types of Roads*

V. 3.5.1. *General.* All roads should have this much in common: they are specifically, and for the most part exclusively, used by powered vehicles. Heavy and light traffic should be segregated where high speeds or inclines occur.

Roads are designed for:

a) certain maximum speed.

b) certain maximum flow capacity.

The proper use of any facilities still depends on the driver.

The faster, safer, high capacity road has fewer access points. This means it provides best for long and stereotyped journeys, and it is crudely restricting. It makes the motorist less like the free agent pictured by some, who say that with his car he can go anywhere. In reality he can't go to the city because there is no space. And when a network of motorways has been provided it becomes increasingly difficult to reach a point you can see, a hill you might want to reach. It takes a long time and some very long returns from wrong turnings once on an urban motorway, to reach a goal which may be so near but yet not on a motorway. Only a minute number of goals can be signposted or even listed on a map to help avoid this.

So the efficiency of a road cannot simply be measured by the flow per lane or the journey time of those who use it. The access it does not provide and the access it prevents must also be taken into account.

V. 3.5.2. *Motorways and freeways.*

A.: *Rural*

Few access needs make rural motorways a logical solution for heavy volumes of traffic. They take the through traffic from the local road system. The capacity of a motorway is very high so that together with the local system it may satisfy vehicular needs for saturation level if combined with rational use. Design speeds should be high, up to say 100 mph for a third lane. There is no conclusive evidence that shows 90 mph to be more dangerous than 65 mph on roads designed for such speeds.

However, U.S. rural freeways are designed for speeds between 50 mph and 70 mph.

The shape of motorways in the townscape and landscape is a vital aspect of environment which is sadly neglected. Those making the decisions are people not trained in aesthetics, overwhelmingly concerned with economics and influenced by the ease of pegging out by traditional methods. Pushkarev, Sylvia Crowe and Lorenz have written clearly on the principles for good motorway design from this point of view.[(i)]

Pushkarev shows that there are three possibilities both vertically and horizontally:

(a) straight lines ("tangents") joined by arches or circles,

(b) straight lines joined by spirals,

(c) continuous curvilinear alignment, spirals or spirals and arches.

In design the first two are arrived at by the designer drawing the longest straight lines he can on any site and then connecting them by curves. The approach in the third category is quite different. Optimum arches are established and these connected by sections of spirals. Extra costs may be involved in providing the most beautiful route. This is not an outrageous idea. Pushkarev notes that of 910 articles in the Proceedings of the Highway Research Board of the U.S.A., in twenty years, only eight dealt with aesthetic aspects.

Pushkarev, under sections dealing with the internal harmony and external harmony of the motorway, shows how the consideration of the advantage of spiral alignment vertically and horizontally leads to helical forms, and that a proper sculptural design of motorways can only be achieved by three-dimensional planning. This is the exception at the present day. Evidence is submitted that pleasant motorways prove safer than those less carefully designed, albeit theoretically as technically efficient. Other evidence shows that straight roads ("tangential") create more accidents than curving roads.

> "*Unless it is aimed at some landmark, it is aesthetically uninteresting, since it is totally predictable: it is monotonous and fatiguing, since the view is completely static; it encourages excessive speeds, since the driver tries to 'get it over with'.*"

From this one can conclude that the spiral planing for roads, even in completely flat areas, is a necessary creative aspect of road design. To quote Pushkarev:

> "*The designer has to treat the twin ribbons of pavement (kerb) as a sculptural form in its own right. This form, taken as a plastic abstraction, derives its beauty from four elements: first from the harmonious rhythm of its curves—their form, their scale, and their co-ordination in three dimensions; second from the proportions of the shapes it encloses as seen from the driver's seat—those that the varying median strip forms in*

(i) Tunnard and Pushkarev. *Man-Made America: Control or Chaos.* Yale University Press, 1963.

Crowe, S. *The Landscape of Roads.* London. The Architectural Press, 1960.

Lorenz, Hans. *Aesthetische und optische Linienführung* der Strasse, XI Internationaler Strassenkongress, Rio de Janeiro, Sektion 1, 1959.

perspective and those between the horizon and the pavement; third, from the way it clings to hills, jumps over valleys, winds along bodies of water, or pierces steep rock barriers; fourth, from the vistas it offers the traveler —broad panoramas from hilltops, dramatic industrial complexes, memorable landmarks, broad expanses of water. (These, in general, cannot be moved to suit the highway, but the highway can be orientated to bring them into view.)"

Fig. 5.17. The straight lines joined by arches.

Fig. 5.16. The tangent or straight line.

Fig. 5.18. The arches joined by spiral segments, continuous alignment as recommended by Pushkarev. Most pleasant to look at from the air, from road by the driver, and to the onlooker from ground level. It is sympathetic to contours.

It is calculated that the net savings in constructing the U.S. system (12 billion per year, and 9000 lives) will total more than the cost of the motorways. By 1972 it will have saved 75,000 lives. A three year delay from 1972 to 1975, would, it is calculated, result in a loss of $10 billion and about 10,000 lives. On a mile per mile basis urban motorways will provide five times the benefits of rural ones.

The new system to serve 20% of all traffic is adding 5% to the total of land used for roads in the U.S. to date, i.e. 1,150,000 (350,000 are on existing rights of way) to 22,000,000 acres, (460,000 hectares to 8,900,000 hectares 141,000 hectares on existing rights of way) an average of 6 miles (9.7 km) of road per acre.

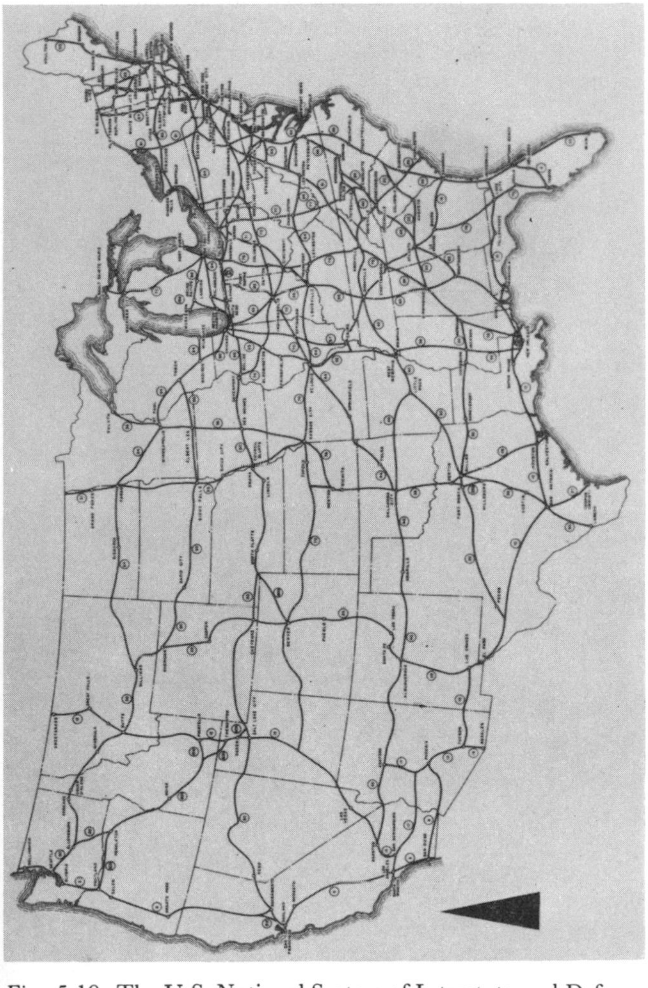

Fig. 5.19. The U.S. National System of Interstate and Defence Highways is based on the Federal Aid Highway Act of 1956. It is at present aimed at 41,000 miles (64,000 km). Interstate System planned to be completed by 1972, connects 42 capitals and 90% of towns with more than 50,000 inhabitants.

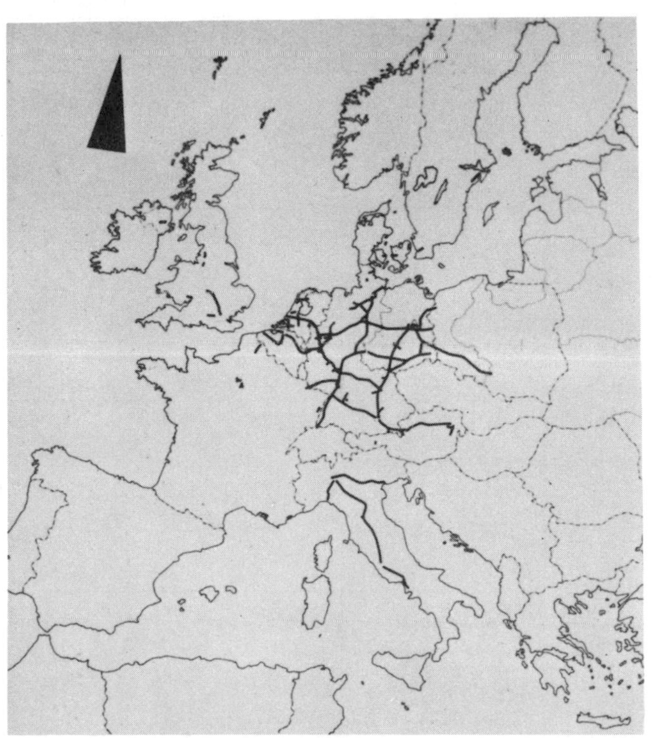

Fig. 5.20. 3400 miles (5400 km) of motorway completed and under construction in Europe, 1962. Motorway mileage planned are; Germany: 3000 (4828 km). Italy: 3000 (4828 km). France: 1235 (1987 km). Britain 1000 (1609 km); all for the early 1970's.

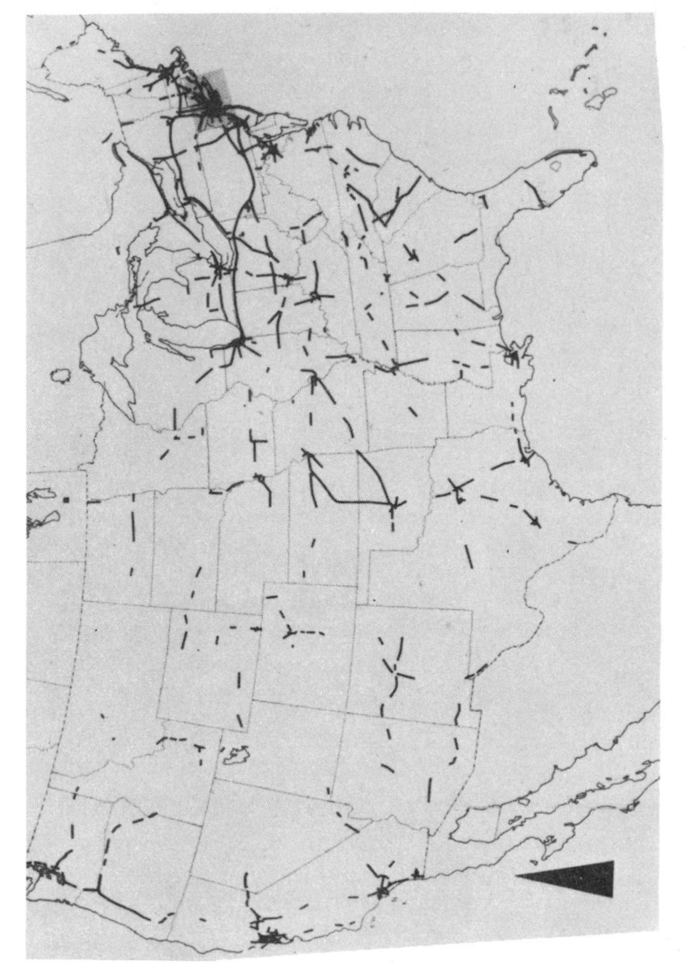

Fig. 5.21. 6700 miles (10,700 km) of freeways completed and under construction in 1962 in the U.S.

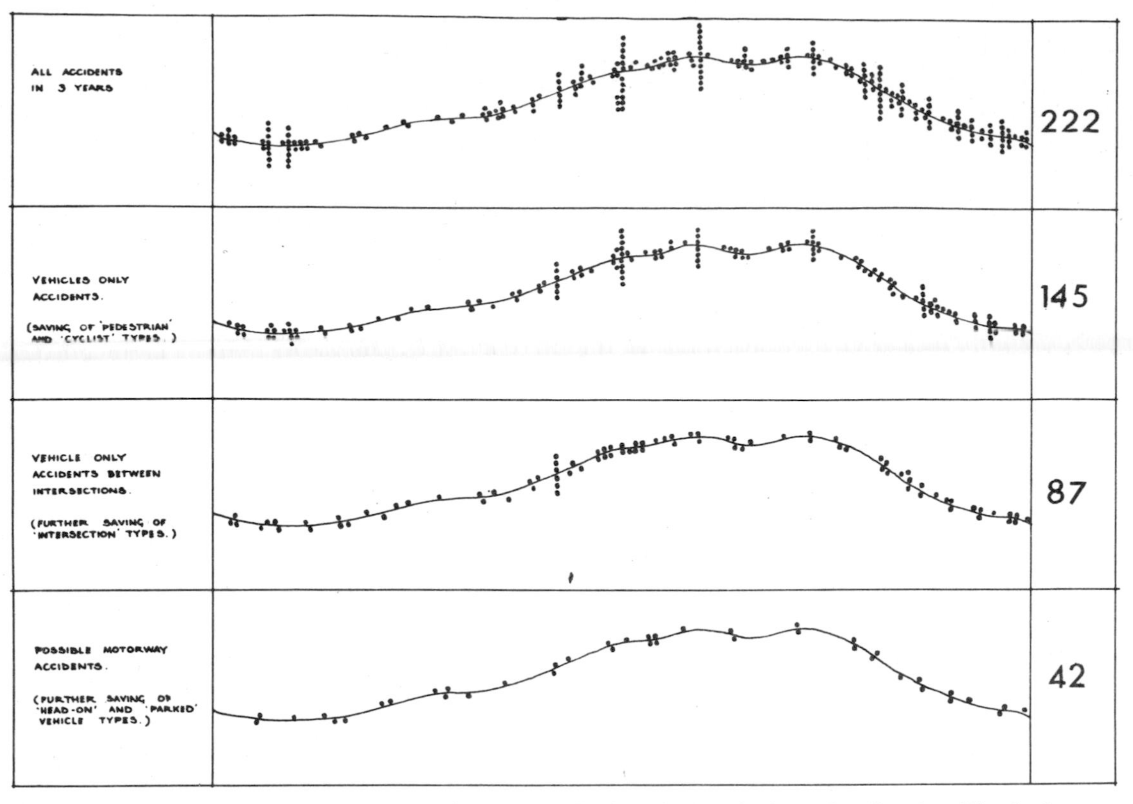

Fig. 5.22. Potential saving of accidents by conversion to a motorway of the East Lancashire Road linking Manchester and Liverpool. It must be noted that by present British standards this is already a very good road with lengths of dual carriageway.

It is calculated that the London–Birmingham motorway will have more than repaid its cost of £21,000,000 in ten years, calculating savings in accidents, working time, fuel, and wear and tear.

B. *Urban:*

The urban motorway suffers, basically, from a contradiction. Motorways can have few access points, as each requires a long slipway, loops under and so on, and in urban areas a motorway with few access points has a limited use. It is a hollow triumph if an immense freely flowing number of cars on a new motorway merely represent those who now have a specially created, longer, if somewhat safer, journey.

On balance there are likely to be many urban routes which are best as motorways. If only the petrol engine was not so dirty and poisonous one could welcome wholeheartedly the through traffic of motorways skirting quite close to the city. In many this is logical. Since a clean non-poisonous fuel is surely coming, it is probably right (if optimistic) not to exclude volumes of traffic on this account. This means "tangential" motorways.

Motorways on pure commuter routes make less sense, particularly if densities are not low and good park-and-ride public transport can be provided.

Fig. 5.23. Terracing a freeway in Brooklyn, N.Y. scaling down and integrating the powerful horizontal element into the urban fabric and the contours. The waterfront is directly on the right.

The success of urban motorways depends on adequate reservoir capacity of related parking and the design of efficient junctions with the existing road system. Urban motorways will be designed for between 35 mph and 50 mph, as this allows maximum flow, useful speed and fairly tight curves.

The "ring", right round a town, has been found a false concept in many cases. Some sections of it are heavily used, others only lightly, even where provided. Plotted desire lines show the functional disadvantages of the ring-road clearly. A motorway system may at first glance look like a ring. But the provision at junctions show the "tangential" character: connections are provided in the major directions only.(i) (See also Urban Roads, VII. 3.)

V. 3.5.3. *Main roads.* Urban or rural main roads should have no development along them, nor pedestrians, unless there is a clear physical barrier.

Access should be controlled and marked. Many more urban roads can be made into such main roads, allowing urban motorway speeds for clearly marked stretches, possibly one way with progressive traffic lights. The major need is for proper provision along these roads. They cannot merely be "designated", but must be created. Thirty mph limits have no useful function on such roads. A much finer distinction of speeds suitable for varying stretches, going up well beyond 30 mph is reasonable. "Distributor" roads can often fall into this category.

V. 3.5.4. *Service roads.* Urban and residential service roads are those which have direct contact with buildings and people. It is this which limits the speeds severely. As on every factory gate, there should be a clear sign for traffic to slow down to between 15 mph and 3 mph. These speeds must be rigorously enforced and engrained by local and national government propaganda. Such roads should be kept to the absolute minimum as they are inefficient for car and man.

The efficient use of road space lies in the clear differentiation of functions. To have fast routes faster and unobstructed and slow routes slower. The more sophisticated traffic becomes, the closer the spacing at slow speeds. (See also p. 240.)

On every road the traffic stream as a whole should move at the optimum to avoid overtaking. This is a dangerous and, as far as time-saving is concerned, overrated sport in urban areas, on all but the clearest, steepest stretches where heavy vehicles slow down a great deal.

(i) See Liverpool, p. 180.

V. 3.5.5. *Summary.* If public transport and motorways are provided judiciously, street parking and other avoidable flow stoppages eradicated, and proper time and space use encouraged through differential charges, thus lengthening peak periods, there is no reason why our towns should not regain their good intimate quality and provide well for the driver. The key lies in efficient multi-storey parking and its direct connection with the chief roads at many points. A road system very largely independent of the pedestrian will result. (See also Section VIII and pp. 55, 56.)

V. 3.6. *Junctions, General*

The science and art of junctions is already vast and still developing; but one important first principle is the steep increase in danger points with the number of traffic streams. Particularly noteworthy is the difference between three two-way streams and four two-way streams.

Fig. 5.24. Four-way and three-way junctions, 16 and 3 danger points respectively. The road-works required to allow smooth similar flow in all directions is much less with two three-way junctions than with one four-way one, as is the number of danger points if the intersections are not level or grade separated. (See example of roads at Toulouse le Mirail and Sennestadt.)

The above point makes nonsense of any new proposal which creates cross roads. Chandigarh, on this count, is not rational; grid-iron plans generally complicate traffic unecessarily.

Reichow, who saw this point clearly in the forties, developed the "organic flow" concept of roads. This anticipated many of the sane and sensible road designs now generally advocated, and there is in this concept an idea which, though not always applicable, makes good sense: Reichow assumes that there is and should be planned for, a *preferred direction.* He recommends that all minor roads should be channelled into this at an oblique angle. Taken to extremes, however, his "branching" arrangements bring too much traffic together unnecessarily. The oblique angles ensure that if cars collide the damage and impact are minimized. These junctions have recently been advocated by Dr. Smeed, R.R.L., for Cumbernauld.(i)

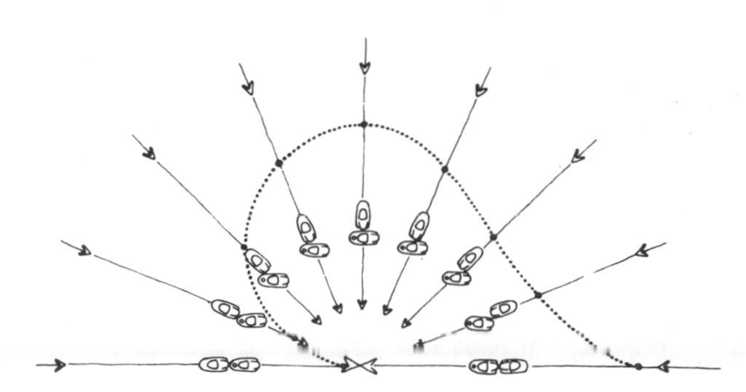

Fig. 5.25. Reichow demonstrated lessening of danger in colliding by making fusion of traffic streams at an acute angle.

Fig. 5.26. Traffic accidents for six months plotted on a map of Stockholm. Note the concentration at cross-roads. (See also pp. 199, 307.)

(i) Cumbernauld Development Corporation. *T.E. & C.* March 1963.

V. 3.6.1. *Some examples of junction flow, driving on the right*(i)

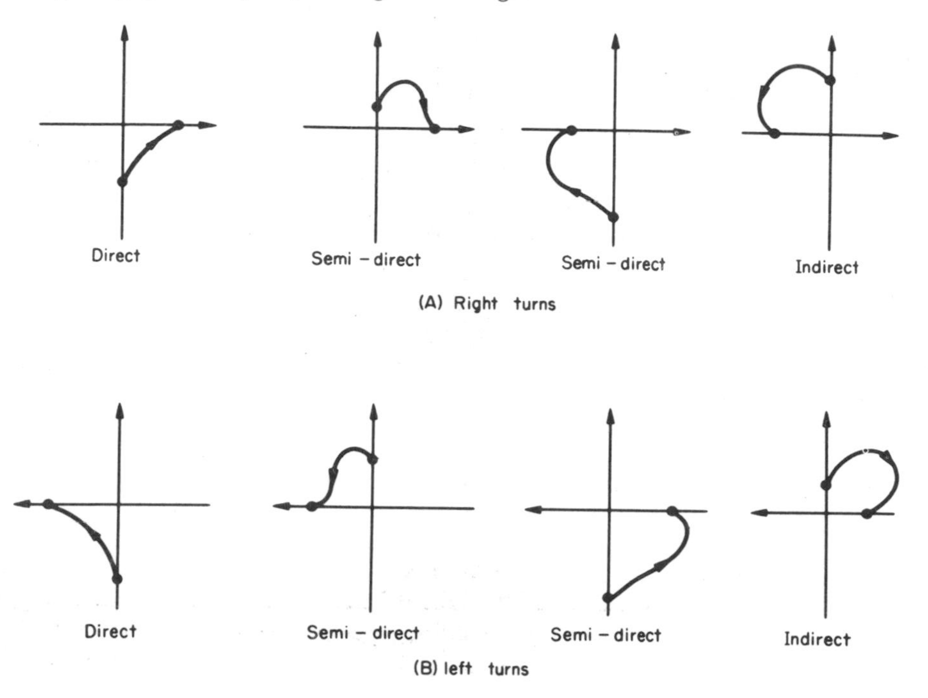

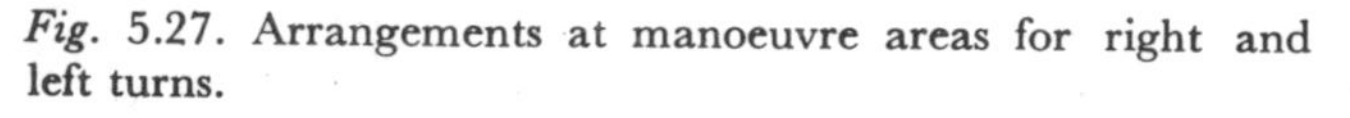
Fig. 5.27. Arrangements at manoeuvre areas for right and left turns.

V. 3.6.2. *Junctions of three routes*(i)

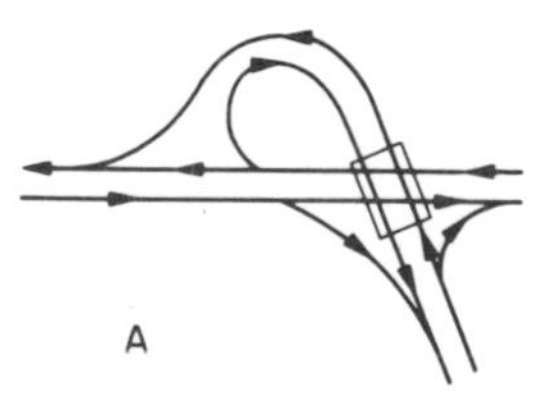

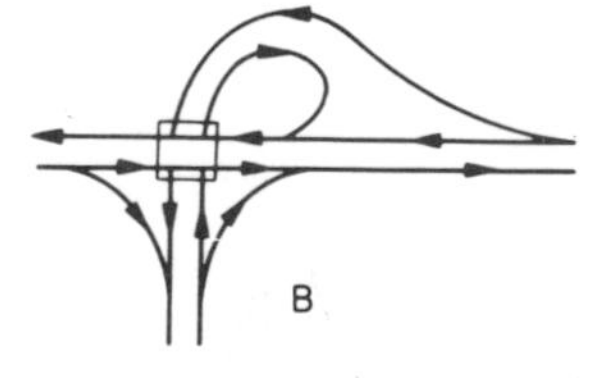

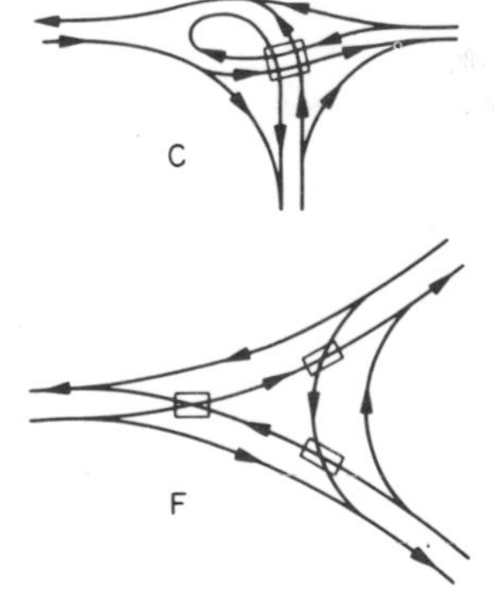

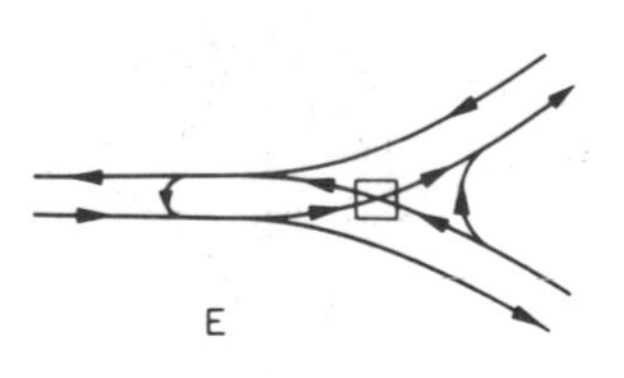

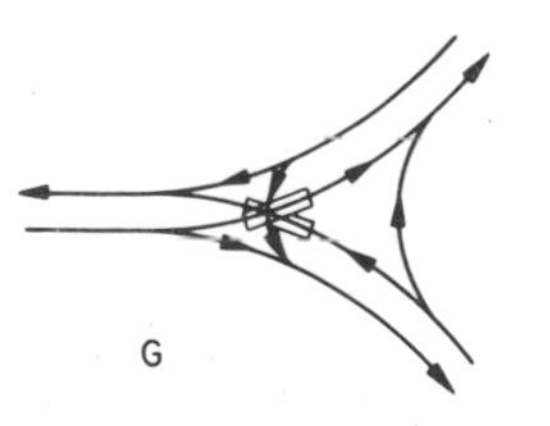

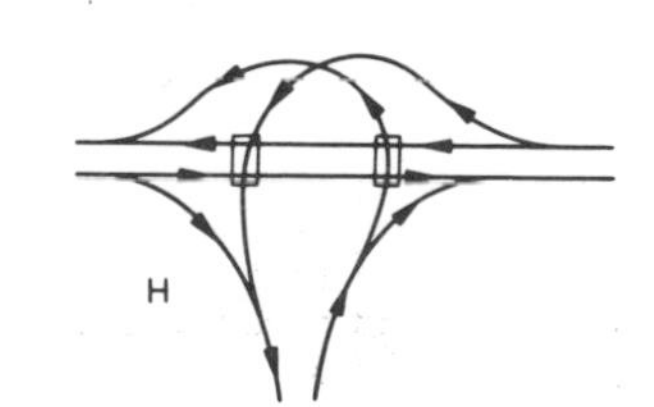

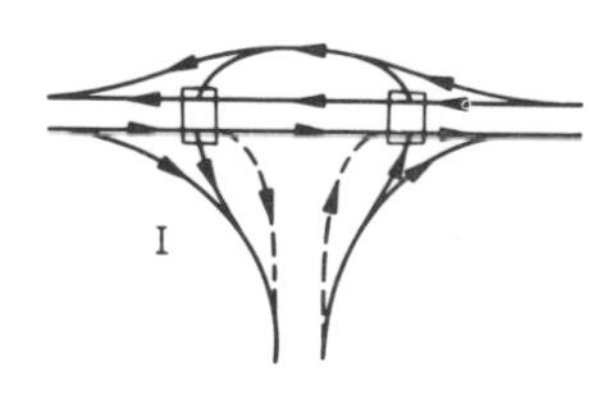

Fig. 5.28. Three major routes require one bridge only to keep smooth non-crossing flows in all directions.
A. Trumpet—one bridge, semi-direct and indirect left turns.
B. T—one bridge, semi-direct and indirect left turns.
C. T—one bridge, semi-direct and indirect left turns.
D. T—one bridge indirect left turns.
E. Y—one bridge two direct left turns.
F. Y—3 bridges, direct left turns.
G. Y—tri-level bridge, direct left turns.
H. T—3 bridges, semi-direct left turns.
I. T—2 bridges, semi-direct left turns.

(i) MATSON, SMITH, HURD. *Traffic Engineering*. McGraw-Hill (1955).

V. 3.6.3. *Junctions of four routes*[i]

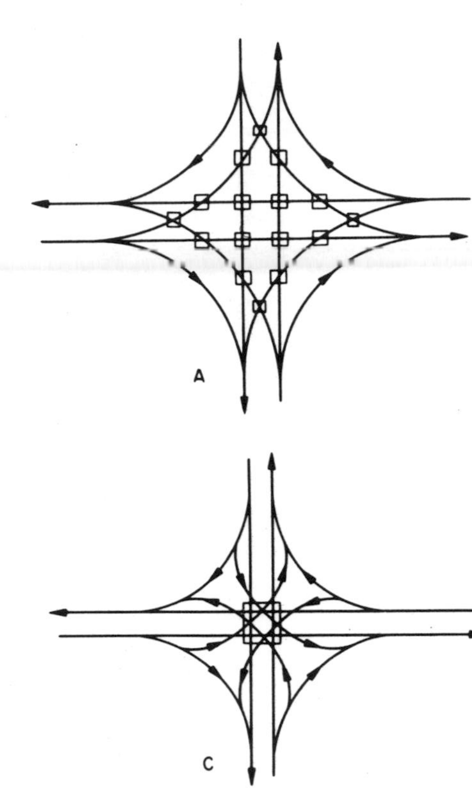

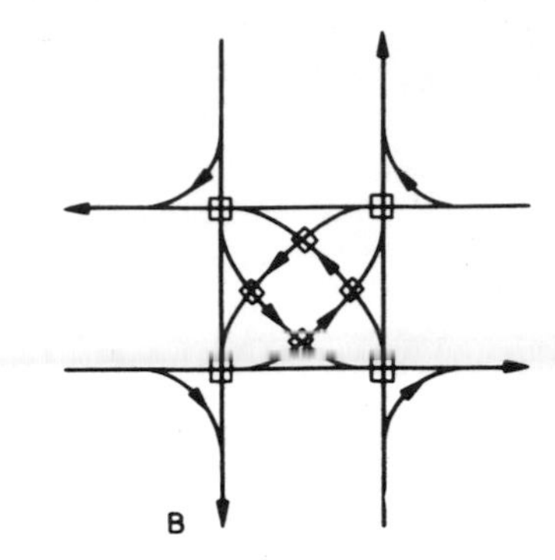

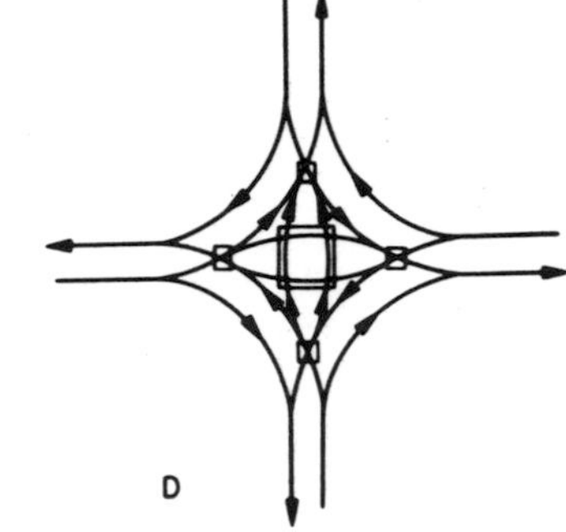

Fig. 5.29. Diagrammatic four route interchanges. Basic types.
Four leg interchange—direct left turns.
A. 16 bridges.
B. 8 bridges.
C. 4-level bridge.
D. 5 or 8 bridges braided.

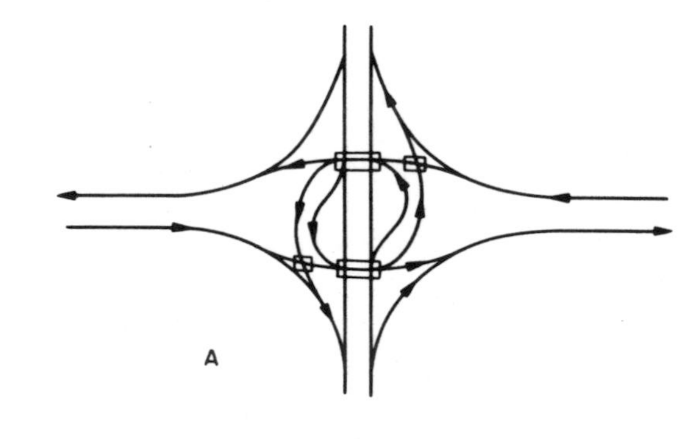

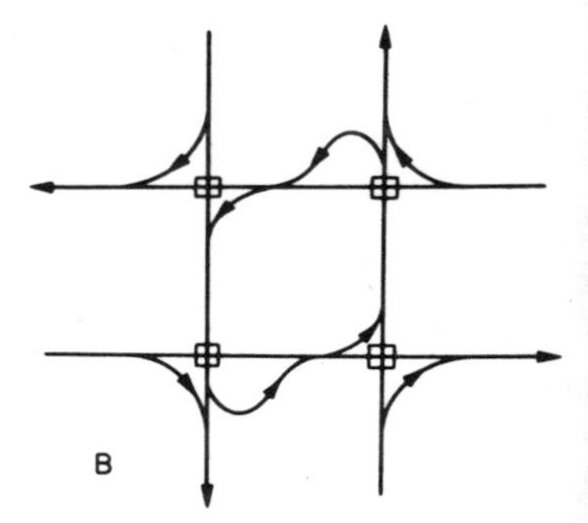

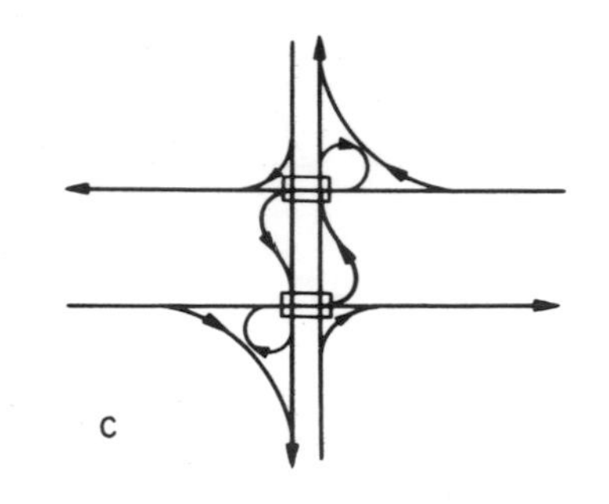

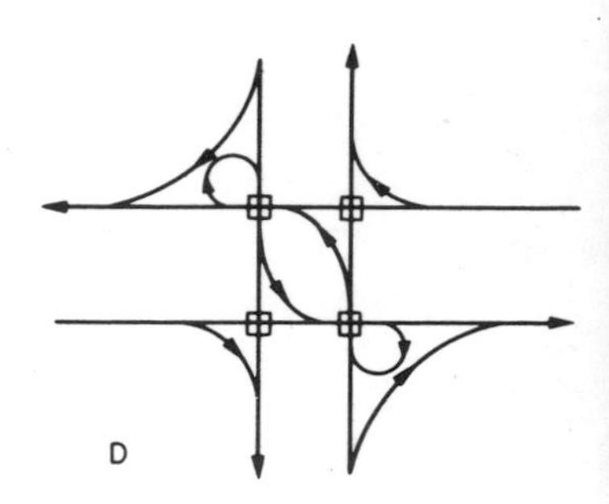

Fig. 5.30.
A. Four leg interchange, semi-direct left turns.
B. Four leg interchange, direct and semi-direct left turns.
C. Four leg interchange, semi-direct and indirect left turns
D. Four leg interchange, indirect and direct left turns.

(i) MATSON, SMITH, HURD. *Traffic Engineering*. McGraw-Hill (1955).

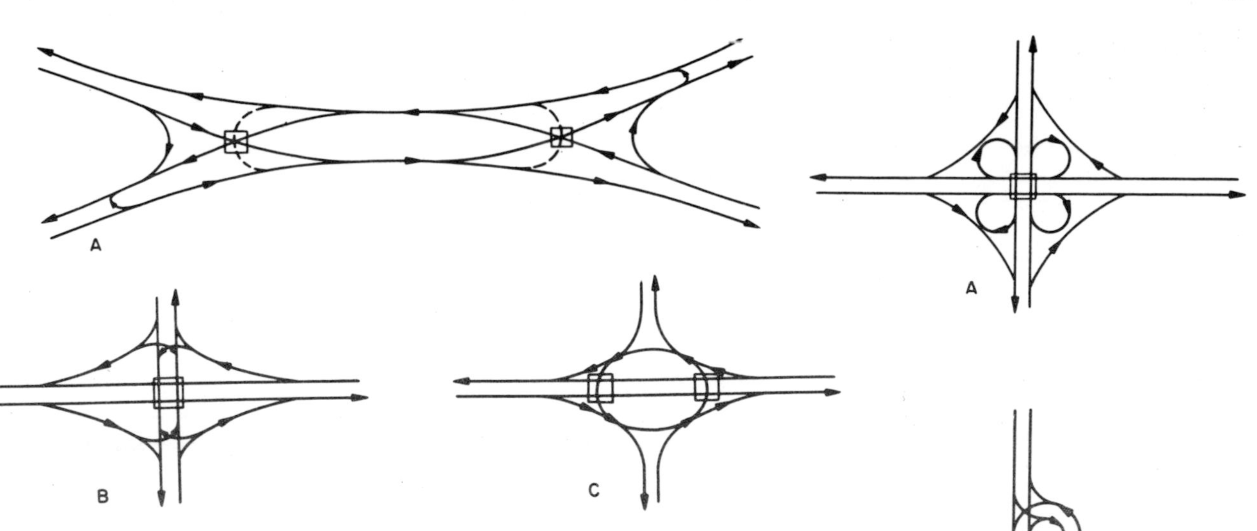

Fig. 5.31.
A. Four leg interchange double-Y—2 bridges, two direct left turns.
B. Four leg interchange, diamond one bridge, semi-direct left turns.
C. Four leg interchange, bridged rotary, 2 bridges, semi-direct left turns.

Fig. 5.32.
Four leg interchange—indirect left turns.
A. Complete cloverleaf, one bridge.
B. Partial cloverleaf, one bridge.

Fig. 5.34.

Fig. 5.33. Four-level interchange, Los Angeles.

Fig. 5.35. Clover leaf vintage 1961, Word-Road U.S. 80 freeway[1] $\frac{1}{3}$ mile diameter earthworks. (0·53 km). (See also pp. 22, 74, 76, 152, 154, 162, 165, 169, 180.)

(1) *California Highways & Public Work.* May–June 1961.

V. 4. Parking

V. 4.1. *General and Basic Requirements*

Although it is cheaper to drive a car round town than park it, the vast majority of vehicles are parked most of the time, at their home or their destination.

There are two approaches to parking. One can develop the idea that city space is precious and that a car parked there all day is bad policy. This would lead to an increase in the use of taxis and in hire cars which might, with some sort of slot meter, stand at defined places in any city and which could likewise be left at similar, clearly marked places on arrival. Secondly there is the idea of parking itself which can be developed. "Park-and-ride", combining public transport and private car, by leaving the car at the station, is a very important variant described later.[i]

It has been shown that though a city seems to be desperately in need of parking or garaging, people are not willing to pay for it, at the home end or the goal, beyond certain cost limits. Further, unless parking is strategic it is not fully used (see Section VII). The deeply engrained habit of regarding the road as a stopping and parking place cuts across much rational thought on the subject and prevents action.

There can be no figure in numbers of feet defining how far car parks can be situated from goals in residential or central areas. Given the mechanical aids to the pedestrian, park-and-ride, a roof over his head, and an interesting route, it is not the number of feet, but the total convenience of the whole arrangement that counts.

Simple single use of land for parking should only be allowed as a temporary measure. Such areas are nearly dead land. It is wasteful for it to have a regularized long life, even in many residential areas. Multiple use of areas should at least include two levels of cars or better residential, office, educational, commercial or industrial use on other levels. (See also pp. 13, 90, 109, 130, 152, 162, 177, 221, 235.)

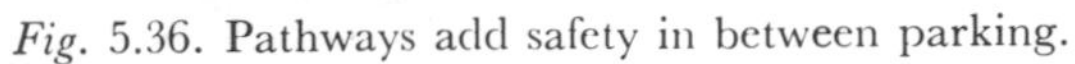

Fig. 5.36. Pathways add safety in between parking.

(i) See section on public transport.

Fig. 5.37. Diagram of parking possibilities.

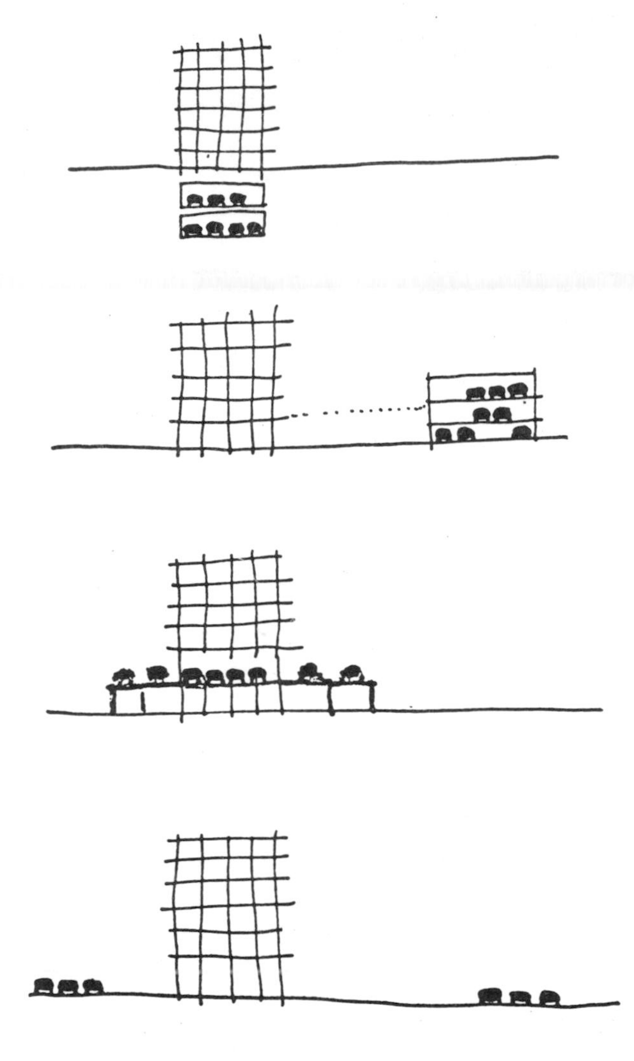

V. 4.2. *Parking Lots, Ground Level Only*[i]

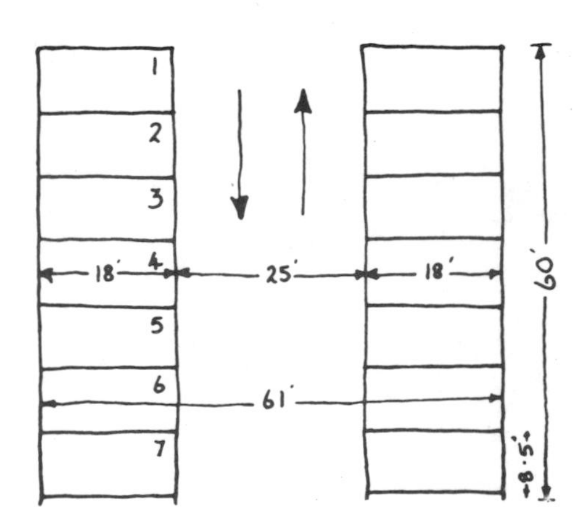

Fig. 5.38. Ninety degree parking, two-way traffic.

(i) Adapted from BURRAGE and MOGREN, *Parking*, Saugatuck 1957.

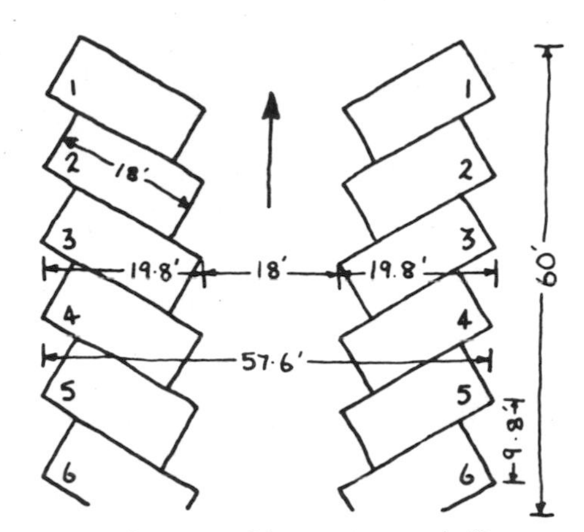

Fig. 5.39. Sixty degree parking, one-way traffic.

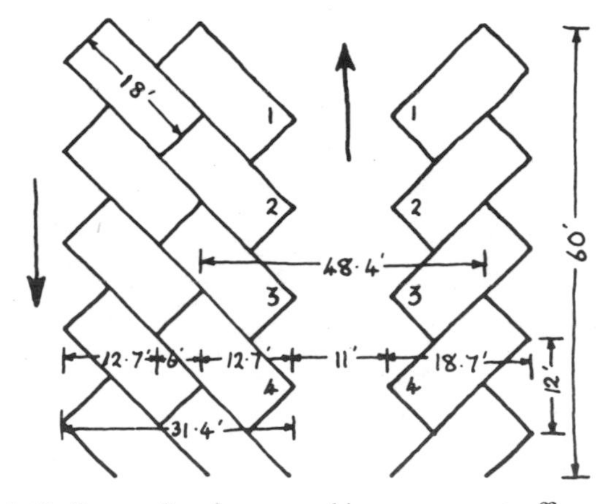

Fig. 5.40. Fourty-five degree parking, one-way traffic.

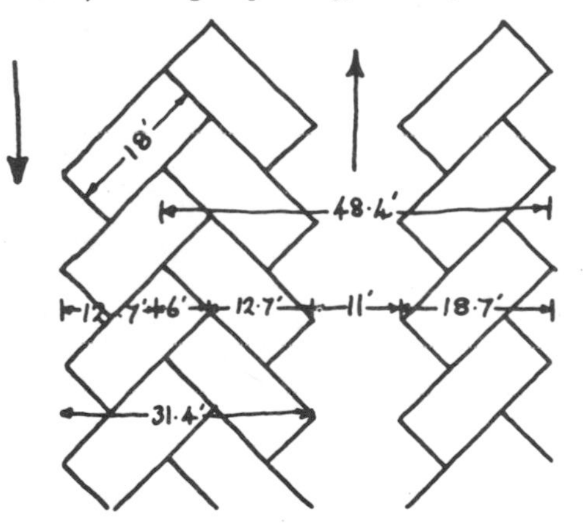

Fig. 5.41. Fourty-five degree parking, one-way traffic.

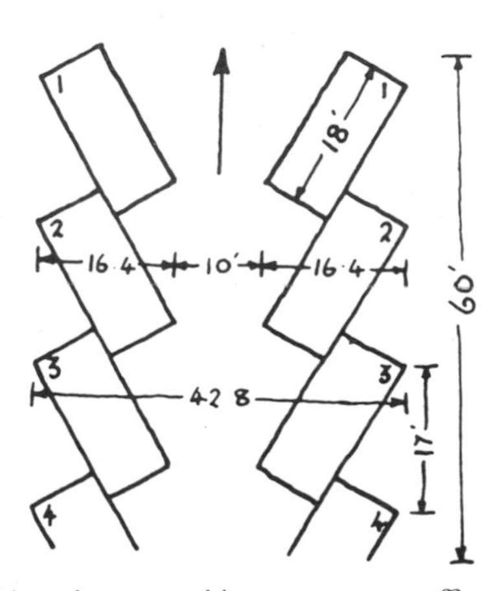

Fig. 5.42. Thirty degree parking, one-way traffic.

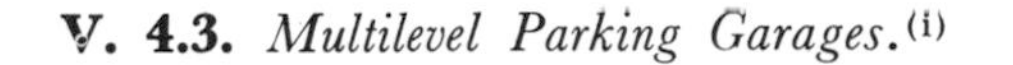

V. 4.3. *Multilevel Parking Garages.*(i)

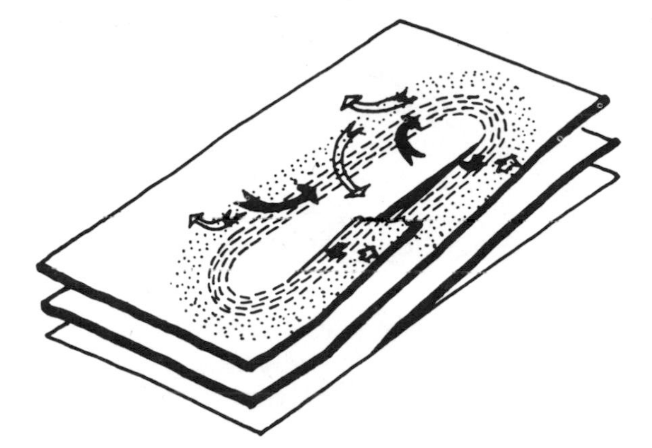

Fig. 5.43. Two-way Straight Ramp System.

Fig. 5.44. Parallel Straight Ramp System.
Measured in an attendant-parking deck on a ramp 65 feet long with a banked 180° turn at the top of the ramp and a partially banked 90° turn at the bottom, the up movements ranged from 5 to 9 sec, the down movements from 4 to 7 sec. The greater time required for the up travel may be accounted for by the approach to the blind turn at the top of the ramp; for down travel, this turn has been negotiated before the car reaches the ramp. Such physical differences in ramps will greatly affect operating time on the ramp.

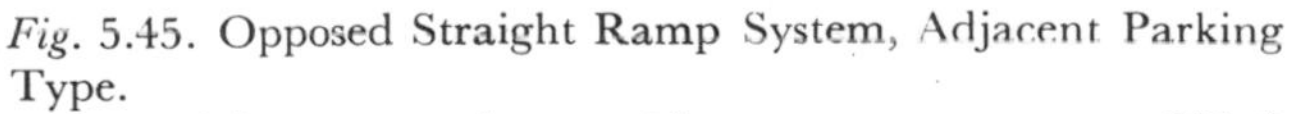

Fig. 5.45. Opposed Straight Ramp System, Adjacent Parking Type.
Measured in an attendant-parking garage on a ramp 100 ft long with a partially banked 90° turn at the bottom and 180° turn at the top. The range of travel times per floor was 13 to 19 sec, and for the ramp 4·5 to 6 sec. The down ramp in this garage was of a different type.

(i) Most types after RICKER R. E., *Traffic Design of Parking Garages.* Eno Foundation, Saugatuck 1957.

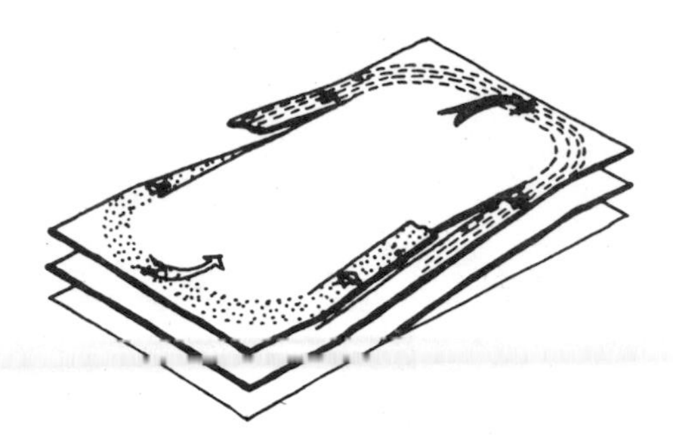

Fig. 5.46. Opposed Straight Ramp System, Clearway Type.

Fig. 5.47. Two-Way Staggered Floor Ramp System.

Fig. 5.48. Parallel Circular Ramp System.
Measured for 30 cases of up travel, with a range of 11 to 13 sec, and for 35 cases of down travel, with a range of 7 to 10·5 sec.

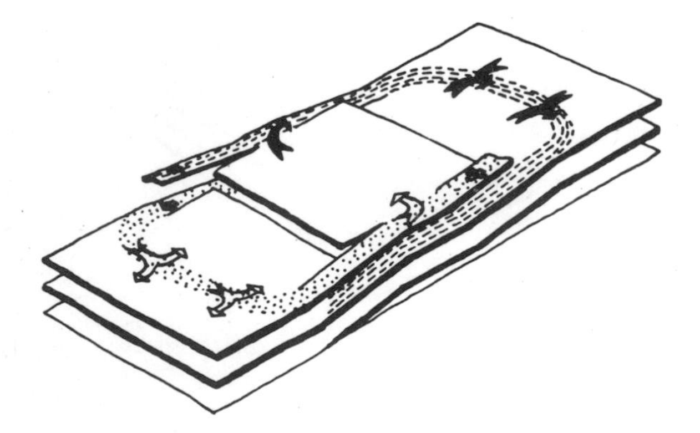

Fig. 5.49. Opposed Circular Ramp System.
Measured on a ramp with an inside radius of 18 ft and travel paths of 150 ft and 95 ft for up and down movements, respectively. The range of travel times for up movements was 18 to 31 sec for 35 cases, and for down movements 11 to 20 sec for 44 cases (per floor).

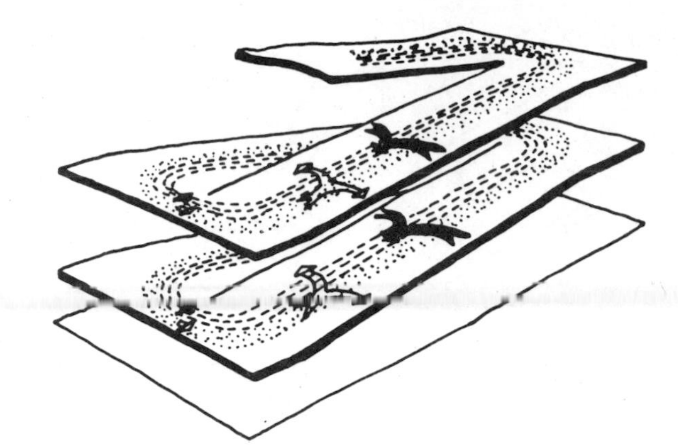

Fig. 5.50. Sloping Floor System.

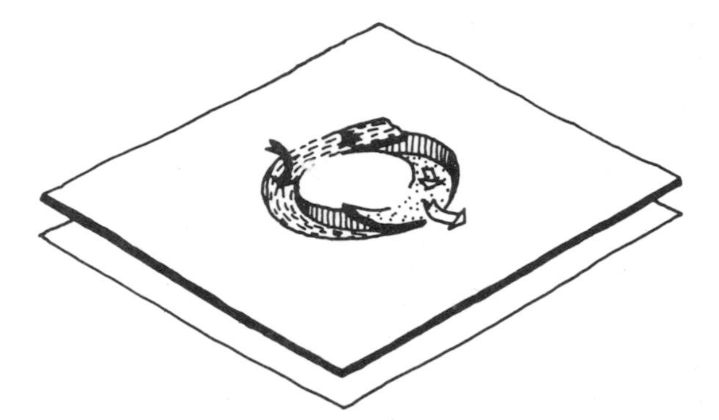

Fig. 5.51. Semi-Circular Ramp System.

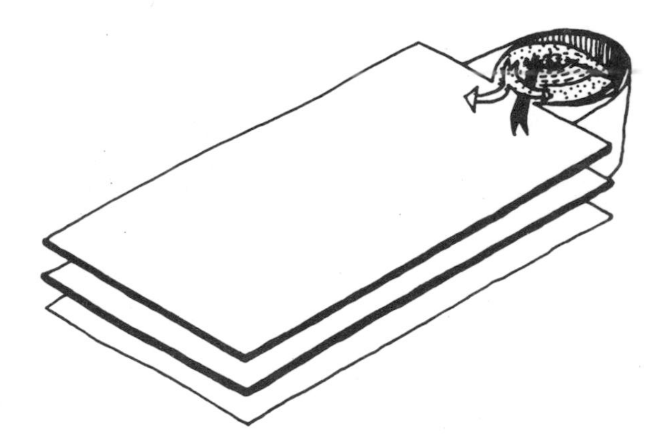

Fig. 5.52. Three-Level Staggered Floor Ramp System.

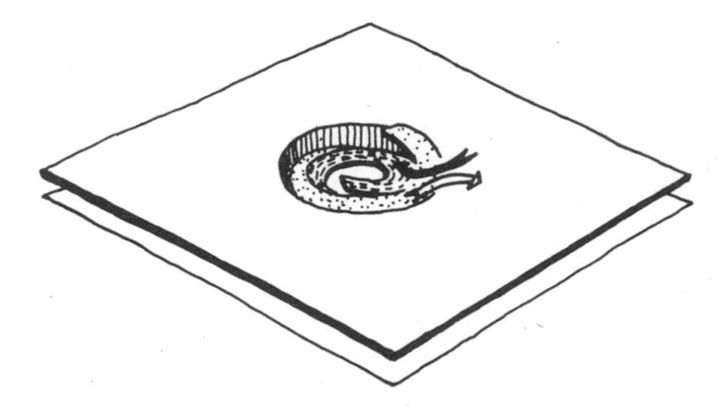

Fig. 5.53. One-Way Tandem Staggered Floor Ramp System.
The theory of probability cannot be applied to this type of delay because cars are not spaced on the ramps at random. Drivers tend to travel in groups, because of delays on the ramps, the grouping effect of passenger elevators, and conditions on the main floor.

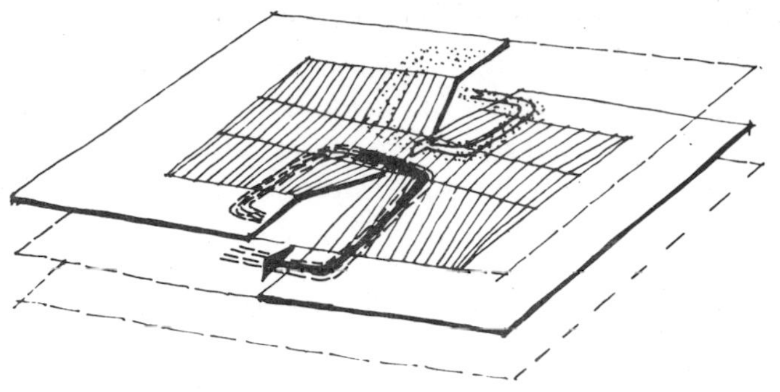

Fig. 5.54. The ramp is a warped floor which is split in the middle. Cars can pass down and up by passing through the split. Mr. Khoury estimates that this makes a 5% saving on space on the most favourable of the remaining versions of ramp. Other advantages are that, as the peripheral floors are horizontal, they could be used for office space; also that the floor–ceiling height is the same at every point in the floor. As the ramped parts of the floor can be constructed in straight sloping runs (like a hyperbolic paraboloid roof) construction would not be much more expensive than that of a conventional structure.

Fig. 5.55. *"The 'Silo' Garage. The general form of the Silo is characterized by an access-ramp in the form of a rising helicoid with parking space on both sides, turning about a central axis, and concentrically surrounding a helicoidal exit-ramp which in turn surrounds the central opening. There is at least one point of tangency per floor (or turn of the helicoid) between the access and exit ramps, thereby making it possible for departing automobiles to turn on to the exit ramp at every complete turn of the access helicoid. The exit ramp can be either continually contiguous to the access ramp, or it can turn in the opposite sense."*

"Entering autos maintain a strict one-way circulation pattern as they drive up the access ramp, and departing cars a strict one-way circulation down the exit ramp. In this way, it is possible to achieve a complete separation between entering and departing automobiles and thereby to create a smooth traffic operation, especially important at rush-hour periods. A major defect of sloping-floor garages is thereby eliminated, that is, the congestion caused by two-way traffic which occurs particularly at the lower levels."(i)

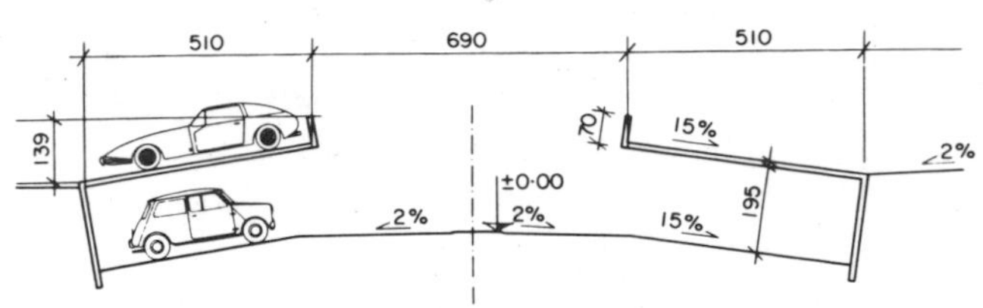

Fig. 5.56. *"The 'Park-a-Back' System. The 'Park-a-Back' system is ideally suited for large open parking lots, such as to be found at shopping centres, large industrial plants, commuter railroad stations, airports, and fringe lots in medium-sized cities. The system offers the parking-lot operator a new way of increasing the capacity of a lot at a minimum of expense."*(i)

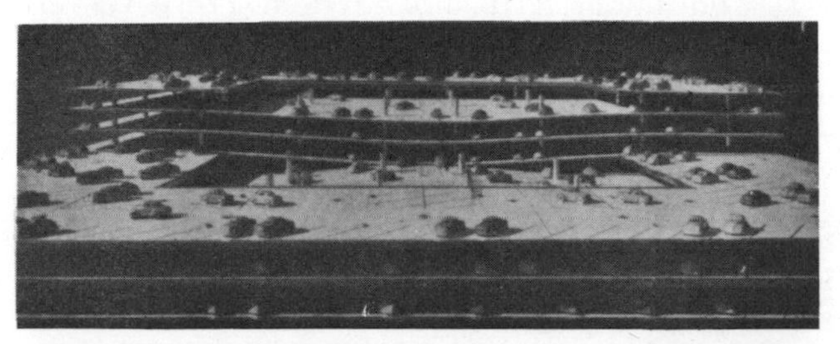

Fig. 5.57. Section through warped floor system of Fig. 5.54.

Fig. 5.58. Parking and ramp approaches at bus terminal, New York.

Fig. 5.59. The scale of Victor Gruen's garages, planned for Fort Worth, make them effective entrances to the city.

(i) GLECKMAN W. B. Two New Parking Systems. Reprint *Traffic Quarterly*, Saugatuck July 1961.

V. 4.4. *Mechanical Parking*

Basically the following methods have been used:

a) Hydraulically raised platforms.
b) Mobile platforms on fork-lift trucks.
c) Simple lifts feeding floors.
d) Lifts combined with dollies, mechanical or manual.
e) Lifts which move horizontally after or before raising the car, combined with dollies.
f) Lifts which move horizontally while simultaneously raising the car.
g) Continuous vertical belt system (Ferris Wheel).
h) Continuous vertical belt combined with dollies.

All the above are more costly than non-mechanical multi-storey parks. They tend to absorb cars more slowly, which accentuates the need for reservoir space at the entrance. (The slow dense flow of the ramped garage represents a very considerable reservoir space.) The Type (f),[i] is the most advantageous from this point of view.

Because of cost and their economy in space and volume requirement mechanical garages are most suitable on tight city sites. (Floor to ceiling height 5 ft 3 in. in the example illustrated.)

Fig. 5.60. Showing one of the two lift towers each with two lifts and the trolley for horizontal movement. Single row of parking on each side of the lift well into which the car is "posted". The lifts move 75 ft (25 m) per minute, vertically; 300 ft (100 m) per minute horizontally. Building completed 1963. Designed by Howard Lobb & Partners for the Parcar Ltd., in the City of London, part of a comprehensive development scheme.

(i) *A. & B.N. London*, 13 February 1963.

V. 5. Functions of Private Vehicles

The use of cars has been measured in relation to all trips taken, in relation to trips to central areas, and those specific to certain functions.[i]

Results in twelve study cities are as follows:

In an old city (Chicago) 75% of all trips taken are taken by car.

In a small new city (Reno) the figure is 98.5%.

For work trips alone the percentages are relatively 67%–97%.

For social trips 82%–99%; for business and shopping trips 87%–99%.

In the old large towns there is a considerable percentage of poorer people who walk, in high density areas.

The very high figures, advanced as inevitable, good trends by some (i.e. American Automobile Manufacturers Association), cannot apply to old cities in Europe and are in fact countered by planning in the recent U.S. new town of Reston.[i]

In United States suburbs the proportion of journeys has been given as 30% of the trips are to work, 20% shopping, 10% school, 22% recreation, 18% others. (See also p. 152.)

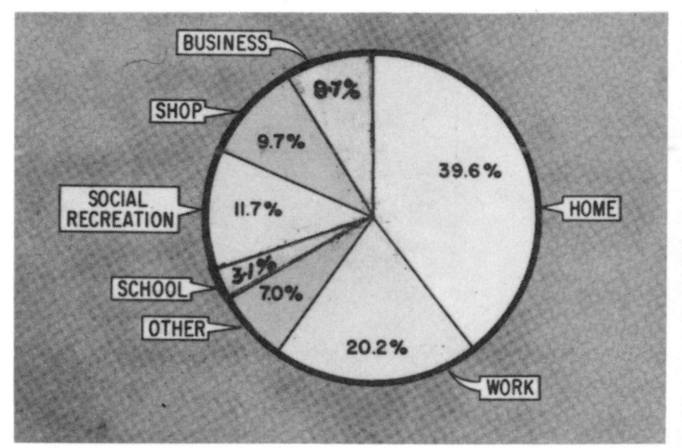

Fig. 5.61. Trip percentages, all trips.[ii]

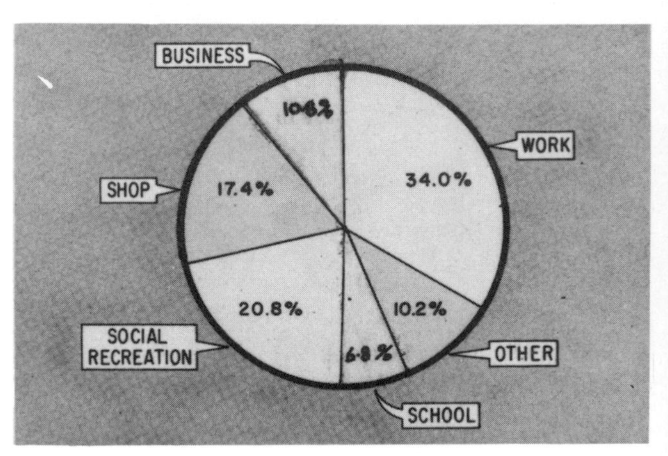

Fig. 5.62. Trip percentages, home-based trips.[ii]

(i) See Section VI. 9.17.

(ii) Smith W. and Assoc. *Future Highways and Urban Growth.* New Haven 1961.

V. 5.1. *Journey to Work*

The convenience of setting out from one's own home and arriving right at the place of work becomes possible less often with growth of car numbers. To begin with many garages are some hundred feet from the home, particularly in cheaper housing, and when the place of work is in a central area, as most places of work still are, there is an acute lack of parking facilities. This leads to undue expense, or undue distances to walk after abandoning the car, or both. It is usual to find a far greater percentage of people going to work by private car in the small towns of a fully motorized country like the U.S.A., and much smaller, though widely varying proportions, in large cities. Road capacity, and so the speed of the journey to work, are also against those wishing to travel to work at rush-hours. The percentage of total numbers entering a city is smallest for private cars during rush hours, and greater when shoppers make their journeys.

The convenience of public transport and the density of any particular residential area also determine the degree to which the journey to work is taken by car. In Detroit 60% of people in the C.B.D. at 3 p.m. used public transport, and 40% of all the people coming to the centre. In Chicago the figures are respectively 88% and 68%, Los Angeles 52% and 34% and Copenhagen 61% and 53%.(i)

The journey to work, the same each day, the vehicle idle all day, lends itself better than any other to be changed to public transport.

"*Studies of commuters' attitudes based on questionnaire surveys in Washington, Los Angeles, San Francisco, and South New Jersey, have indicated that about one third of all auto drivers would ride improved rapid transit, but that another third would not be potential riders because of their dependence on, and conditioning to, auto transportation. The proportion of auto drivers diverted to existing rapid transit facilities as shown by experiences in several cities, has been less.*"(i)

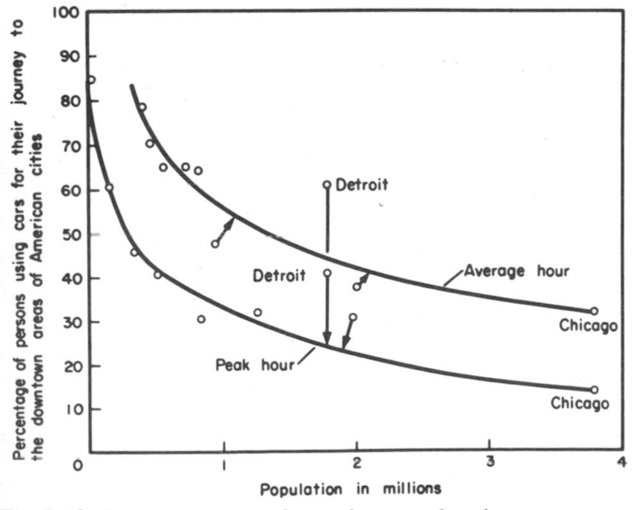

Fig. 5.63. Journey to central area by car; showing greater use in smaller towns and smaller use during peak periods. (U.S. 1950.)(ii)

(i) See also the section on Public Transport.

(ii) *U.S.A. Public Administration Service Procedure Manual 2A*, page 41.

V. 5.2. *Schools and Universities*

Many children are taken to school by their parents in cars. A considerable proportion of these are on their way to work. In dangerous areas, and that is most areas, with an environment not designed for walking and even less for cycling, the temptation to take children to school by car is great. Many more children find their own way home by public transport or walking, because the end of their day does not coincide with that of father; more likely with mother shopping, especially in two-car families.

It is obvious that the use of cars by pupils will have to be limited in many instances, judging by U.S. experience. Not only do "high schools" have serious parking problems but most universities do not, cannot, allow students to use their cars in the campus at all. A great deal of walking is to be found even in the hilly universities.

V. 5.3. *Shopping*

Going shopping is not as standardized a journey as the one to work. It entails the bringing home of goods in many instances. It may also mean the transport of small children. All this is not only simpler in a car, where it can be used to good effect, but is encouraged by the unpleasurable journeys now offered by our streets and towns to mothers with children on foot and in prams, and people with small trolleys "upping" and "downing" at each crossing. Public transport does not yet cater for trolleys and push-chairs.

Hence the emergence of the suburban shopping centres of the U.S.A. The point of these, it must be clearly understood, is not to cut out walking. But the women who take their children and prams here know they will find a parking place and that they will be able to walk and shop with their youngsters in comfort and with pleasure, once they have passed through the car park.

In Europe, suburban shopping centres are likely to be less frequent, as strenuous efforts are made to resist the corrosion of the cores of cities. Out-of-town shopping facilities sap the commercial life of city centres.(ii)

(i) Smith W. and Assoc. *Future Highways and Urban Growth.* New Haven 1961.

(ii) Bendtsen P. H. *Town and Traffic.* Copenhagen 1961.

Increased delivery services from shops are one solution to the problem of carrying, solved by the motor vehicle performing a communal function.

Since the shopper parks for short periods, the space he occupies is, from a commerical point of view, more fully used than the parking space occupied by the man at work all day. Parking for shoppers should always be at a premium near city centres. It must be reserved in some way or workers' cars will take up the spaces before shoppers arrive, albeit illegally.

V. 5.4. *Visiting*

Both professional (doctor) and social visits mean many different and various journeys for which the advantages of the private car are unparalled. In this sense of greatly increased mobility the car is contributing to social life of people, particularly in low density areas. These journeys are made at off-peak hours.

V. 5.5. *Sport and Cultural Events*

When 100,000 people are gathered together and one in two has come by car the parking area for the cars alone, not counting the coaches, would be twelve and half million ft^2 (46,000 m^2) or two thirds of a square mile! With vast road provisions necessary such meetings in central areas are impossible even for lesser numbers. At an off-rush-hour-period opening of the Lincoln Center in New York even the most select Rolls-Royce owning clientele was forced into the very dirty and, compared with those of Europe, sub-standard, Underground Railways (Sub-ways), leaving their limousines and chauffeurs stuck in hours of chaos.

Although commutors' day-time parking is useful for evening occasions it becomes clear that only in special circumstances can large-scale gatherings reasonably depend on a great proportion of private cars, unless it is right out in the country where cheap land can be obtained and a local bus service or "pedestrain" is installed to bring people from the edge of the parking area, i.e. three quarters of a mile away, to the entrance. This is the solution at Disneyland near Los Angeles. Such colossal areas of parking cannot, in most cases, be landscaped satisfactorily.

V. 5.6. *Week-ends and Half Day Holidays*

One of the most pathetic spectacles, shown in Mumford's film *The City* in 1939, is the week-end rush out of the city. So many dash out that the countryside and the escape they seek is debarred them by sheer numbers. So they end up in their thousands, just along the roads, breathing the carbon monoxide they wished to escape and hearing the very noise which during the week distracts them.

The short term escapes from home are largely due to the lack of fresh air, swimming pools and lack of worthwhile activity near home; lack of space to run in and play with one's children, lack of walks and the distance from the countryside. Individual small gardens have severe limitations. All these shortcomings are likely to become less through good planning so that one may foresee a decline in the short term tracks away from home. The use of public transport brings out even more people, but at least they do not clutter up the country with their vehicles.

Fig. 5.64. Disneyland parking, Los Angeles.

The solution in the long run can only be the creation of more areas where it is worthwhile to stay put, particularly in dense countries. The rigid dividing line between holiday and work will disappear, as prosperity increases which will cut the pathetic spectacle at week-ends. Yet this trip is, and will remain, really worthwhile for many. The air in our cities will be well worth escaping for some years. But we will have to share the view with our fellows.

Several decks of car parking on piers might solve the need for those who like to go to the sea and merely sit in the car, a need obvious in most resorts. This might help to keep other resorts, or parts of resorts, free from cars. Devon, England, is planning for such a situation and is to prevent actively the motorization of every spot of coastline.

V. 5.7. *Long Holidays*

Whether touring, or travelling a long distance to a destination, the private car, fully used, is incomparably more economical than rail or combined rail and air travel today. As these trips tend to send people to certain spots at certain times the car parking problem at the terminus can be extreme.

Countless little villages all over the world, particularly on the coastlines, i.e. Cornwall in Britain, have been completely spoiled. The atmosphere is changed by the predominance not only of the moving car which literally blocks the urban streets most of the time, but also by vast areas of parked vehicles in the wrong places.

The long holiday is the most likely use of the private car to be developed in the foreseeable future. The road capacities will stand this if traffic is carefully organized, as staggering is comparatively easy. The provision for the vehicles at destinations is crucial, so that holiday makers do not despoil what they have come to enjoy.

V. 5.8. *Interpretation of Statistics*

Statistics have to be very carefully read not to mislead.

The point can be made particularly well by quoting some figures for the use of cars in towns:

Taking the City of Chicago, taking figures all relating to the same period, private car use could be expressed in any of the ten statements given below:(i)

a) 12% of persons present in the CBD at 3 p.m. come by car.
b) 32% of all persons entering the CBD in a week-day come by car.
c) 41% similarly, from an alternate source.
d) 75% of all trips in Chicago Urban Area ar made by car.
e) 67% of work trips are made by car.
f) 86% of social trips are made by car.
g) 81% of all shopping and business trips are made by car.
h) 3·5% of all auto trips in Chicago are to and from the CBD.
i) 2·6% of all trips in Chicago are car trips into the CBD.
j) 2.3% of all work trips in Chicago are work trips into CBD.

Read in isolation such figures, selected to make a point, can be false evidence, so if used as argument such statistical figures must be carefully examined before their relevance to any case in point is granted. For, even beyond the above variables, CBD's are variously defined, urban areas likewise, and the same unreliability is attached to much other basic datc.

The simple and valid conclusion, as that the greater the city the smaller the percentage of CBD trips, can be gathered from a few minutes contemplation with maps.

(i) (a) and (b) From Bendtsen, P. H. *Town and Traffic*. Copenhagen 1961. (e)-(j) From Smithward Assoc. *Future Highways and Urban Growth*. New Haven, Feb. 1961.

V. 6. The Form of Private Vehicles

Private transport in the U.S.A. today is almost 100% by car. The high proportions of mopeds, motor-bikes, motor scooters, motor-bikes with side cars, bubble and miniature cars, make the road picture in Europe, let alone Africa and Asia, different in many respects. Even the family car means a different dimension in Europe and in the U.S.A. And given a system of the right expressways the scale towards much larger commercial vehicles becomes feasible.

V. 6.1. *Factors Affecting Form*

A. *Fashion:* In a capitalist society, geared in many countries to the sale and manufacture of cars, there is great pressure to sell new models as often as possible. Thus, change becomes a need in itself for commercial attitudes. The less accountable whims of people also matter. Whereas the motorbike was the most popular symbol of manhood for countless Swedish youths only some six years ago, today it is the large, second-hand American car. This is not purely due to a rise in the standard of living (See also p. 92.)

B. *Prestige:* Through advertisements vehicles have become status symbols, in the form of the largest, latest car, the fastest car, the loudest roaring motorbike. Grandiose shapes and chromium encrustations stem from the wish to impress. The insecurity of people is compensated by the power and pomposity of their cars. Unable to love living things, they lavish enormous care, pride and attention on their machines.

C. *Lack of road and parking space:* Comparatively recently there has been a very strong tendency in Europe towards miniature cars and the adoption of scooters and mopeds on a large scale. As such cars can easily carry as many people as the average family vehicle normally does (namely between 1 and 2 persons) and taking up much less road and car park, the logic behind this invention is clear. In the U.S.A. this has not swayed the public. A timid move towards the "compact" car, which was the size of the European family car, was soon reversed by such absurd, contradictory advertising as the "largest of the compact cars", as a virtue to be extolled. Two-wheeled variations are a mixed blessing. It is possible to weave in and out and travel more quickly in heavy traffic but these same manoeuvres prove very dangerous.

D. *Convenience:* Large families, and those who negotiate long distances with much luggage, may well choose a larger moving space on strictly rational grounds. Many other individual needs are satisfied by particular forms of vehicles.

E. *Running cost:* The great majority of small cars use less fuel than the larger ones and are indeed cheaper to buy and keep their value better. These are important factors in determining choice in Europe.

F. *Wind resistance:* In detail design, streamlining is determined by the speed of the vehicle, and is a direct aid to economic and quiet running.

G. *Wheels:* With the advent of vehicles that rise on air cushions, cutting out friction (hover-craft), the wheel may not remain a determining factor.

H. *Manufacture:* Pressed steel bodies dominate car design, but the future of glass fibre and other plastics may have an influence on form.

I. *Safety:* In European countries pointed extremities are not allowed on vehicles (see III. 13), charging like multiheaded rhinoceroses about the streets. Models in 1961 and 1962 showed that this very obvious danger was allowed in the U.S.A., and the number of impalements must have been high (see also section on Danger—III. 2.3.).

J. *Material damage:*(i) The change from framed bodies to pressed bodies made a radical difference to the extent of repairable damage on cars. The effect to owner was indirect in most instances because of compulsory insurance.

There is a weak and inefficient attempt in terms of overriders, bumpers and rubber extremities to lessen the superficial damage done in the vast majority of minor collisions. This factor, which with a little uniformity, standardization and common sense might help a great deal to raise the morale of drivers and the appearance of cars, has hardly been exploited in a way which might be beneficial to drivers.

K. *Engine:* The nature of the internal combustion engine or any electric or other device, has to be accommodated to its fuel, and so affects the form.

L. *Anthropometric factors:* Position of persons in vehicles vary from sitting almost upright to the nearly lying position, dictating greater or lesser height.

That bad design of vehicles contributes substantially to accidents is suggested and explained by G. Grime who divides limitations found in vehicles into four classes: seeing and perception, communication and information from vehicles, control of the vehicle and protection of occupants when accidents occur.(i)

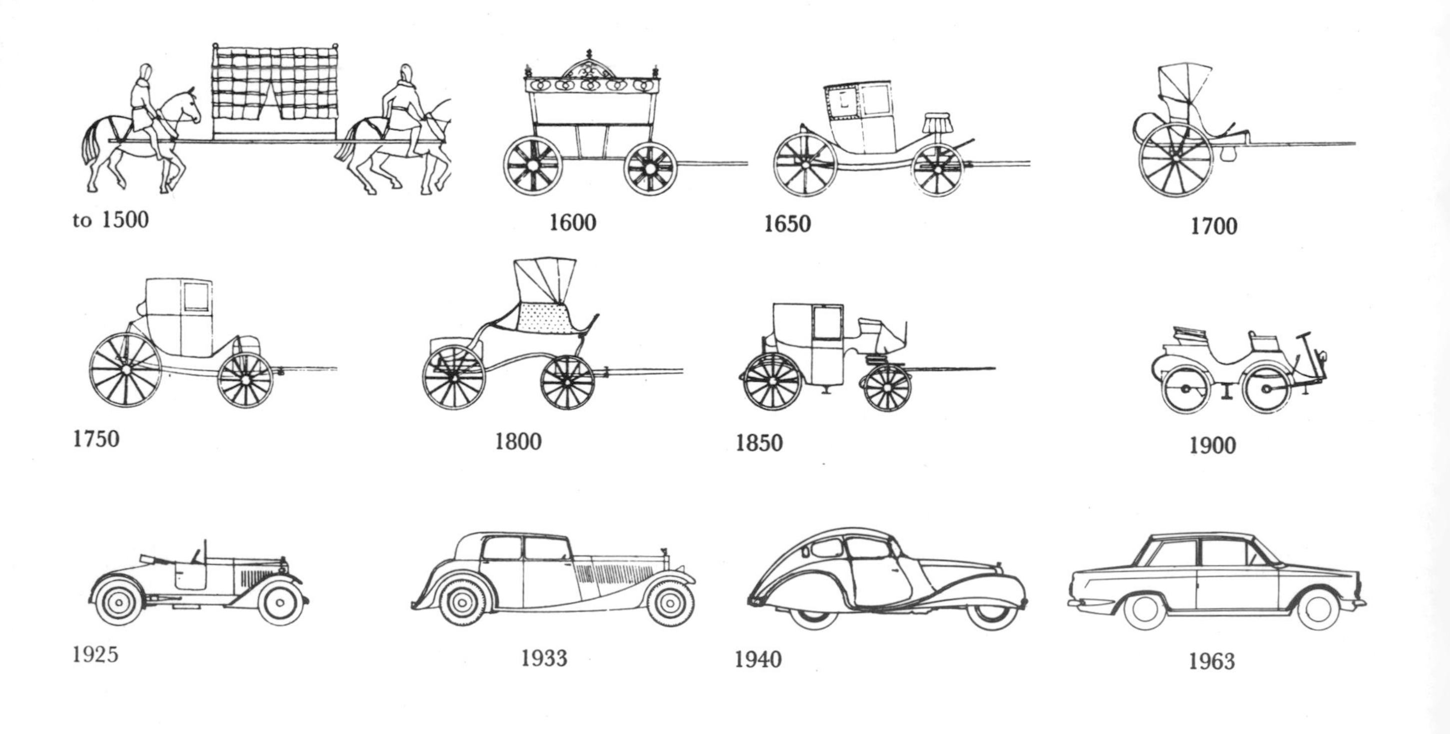

Fig. 5.65. Development of the private carriage.

(i) The Minister of Housing said in the Commons that "Councils could use their powers to make special request collections which could be used to help solve the problem of derelict cars being dumped around cities". We must avoid the U.S. state in this matter.

(i) GRIME G. *Car Design and Operation in Relation to Road Safety.* R.R.L. Reprint from *The Practitioner*, April 1962.

V. 7. Commercial Vehicles, Vans and Lorries

V. 7.1. *Numbers and Growth*

The proportion of commercial vehicles is higher in Britain than in other countries.

TABLE V.15.(i)

	Commercial vehicles	Total vehicles
Germany (1961)	736,000	9,718,000
Britain (1962)	1,460,000	10,050,000
U.S.A. (1960)	12,300,000	73,010,000

In Germany 30% of motorway traffic is by lorries, a much higher percentage than that for all roads. Special long trailers are used but to relieve weekend travel heavy trucks are forbidden on the Autobahnen. Some recreational routes built as motorways in the U.S.A. are out of bounds to trucks altogether. *Growth*: The growth of commercial vehicles is usually taken as different from the growth of private car numbers. Commercial vehicles form a larger percentage when numbers are low and subsequently the increase lessens so that the percentage drops. Miles per vehicle are much higher for commercial vehicles than for private cars so that the percentage of commercial vehicles on the road at any one time is higher than the percentage of commercial owned.

The ultimate percentage of commercial vehicles was taken as 10% in the prediction of Glasgow Traffic,(i) but as much as 20% is also suggested in other quarters. The use of railways and their development may affect inter-urban traffic development.

V. 7.2. *Size, Speed and Flow*

The Highway Capacity Manual of the U.S. divides trucks into "light" which take 13·4 ton loads and "medium" which take 17·8 tons. The R.R.L., on the other hand, divides commercial vehicles equivalent to private cars which weigh themselves, without load, up to 1½ tons, medium up to 3½ tons, unladen, and heavy over 3½ tons. (1 ton equals 1·016 tonnes.)

V. 7.3. *Effect on Flow*

In assessing the reduction factors of commerical vehicles on road capacity the Highway Capacity Manual figures show that 10% of "dual-tyred" trucks reduce capacity as calculated for 100% private car use, by 9% on the flat and 23% on rolling terrain, and 20% of trucks reduce capacity by 17%

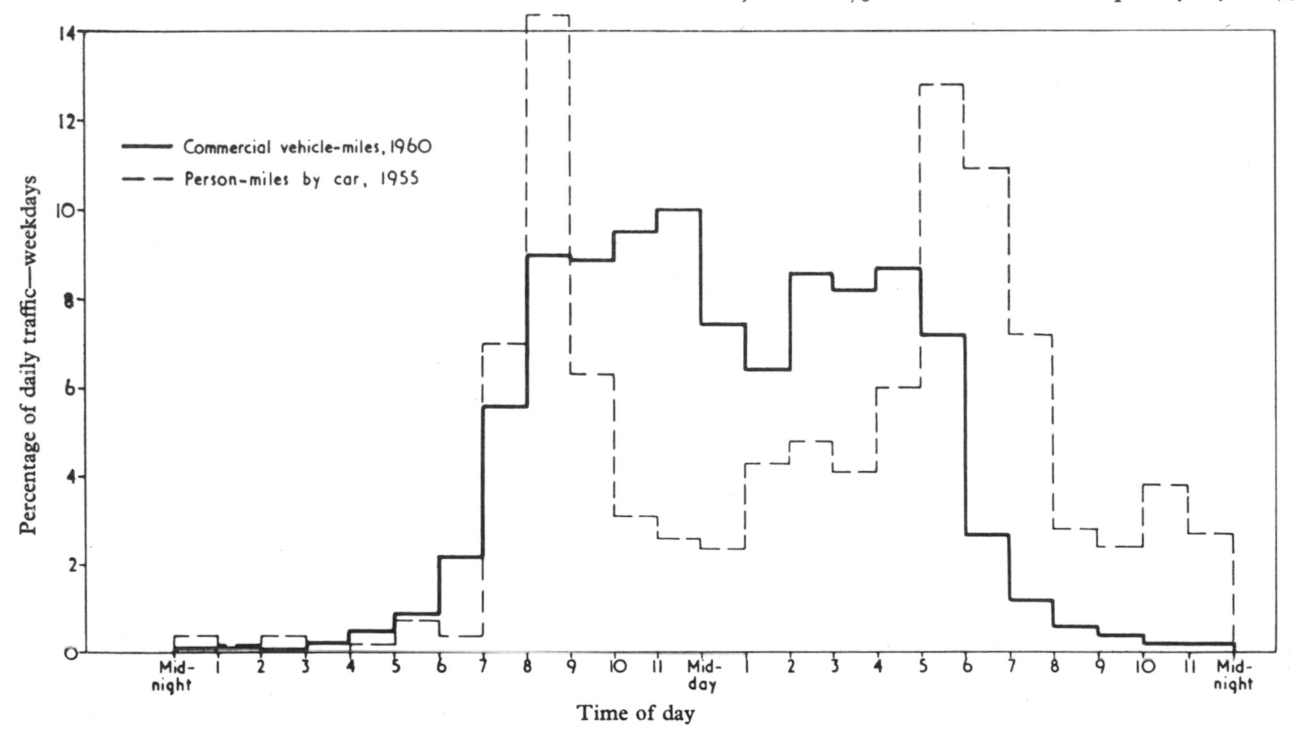

Fig. 5.66. Hourly distribution of commercial vehicle-miles compared with distribution of person-miles per car.(ii)

(i) Figures supplied in part by SMEED. R. J. R.R.L., letter, 19 March 1963.

(ii) DAWSON R. F. F. Survey of Commercial Traffic in London. *T. E. & C.* August 1963.

(i) HODGEN R. Prediction of Glasgow's Future Traffic Pattern. *T.E. & C.* March 1963.

and 37% respectively. Three-lane roads should have two lanes for traffic uphill to make overtaking safe and easy. Many opportunities for this exist.

The British R.R.L. suggests that in calculations of "moderate curvature and gradient", medium and heavy goods vehicles, such as buses and coaches, should be taken as equivalent to three private cars. This reduces a two-lane carriageway maximum of 4500 (per 16 hour day) to 3250, 2900, 2600 vehicle capacity if 15%, 22½%, and 30% of medium and heavy goods vehicles are calculated for, and dual three-lane from 25,000 to 19,000, 17,000 and 15,000 respectively, again for 15%, 22½%, and 30% of total traffic.

As can be seen, this aspect of traffic engineering is far from developed or refined. There is agreement on the following general principle:

The reduction factor increases with
a) steepness of gradient,
b) slowness of the initial speed at the foot of a hill,
c) increase of average speed of all traffic.

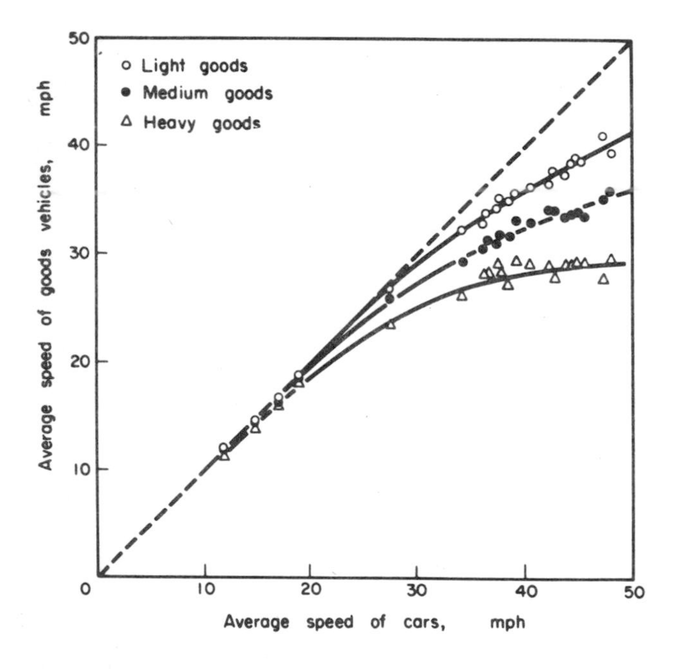

Fig. 5.67. Relation between goods vehicle and car speeds on level roads. R.R.L.

All this does not account for the very serious disturbance of traffic which is created by lorries. Local variants can aggravate the effect of any of these factors decisively.

Commercial traffic has not been subject to much research. The entire Bibliography of the R.R.L. has only one reference particularly concerned with commerical vehicles. *Traffic Engineering*, the comprehensive U.S. textbook, makes only two brief mentions of it.

V. 8. Public Transport

V. 8.1. *General Attitude*

Some kind of public transport must inevitably remain the major way to our cities, particularly the large ones and particularly during peak hours. Even in the most fully motorized part of the world, Los Angeles, with enormous road systems and little concern in the past for Public Transport, more than half the people arrive at work by this means. Yet in the C.B.D. 20% to 30% of the families own more than one car. In places like New York and London it is more like 90% who commute by rail or bus.

It is a mistake to note that public transport has declined in favour of the private car, regard this as an inevitable trend and plan for it by reducing public transport further, at the same time increasing unduly facilities for the private car. It is a mistake because this trend, as the American cities have shown, can only go so far. Then, inevitably, the private car traffic grinds to a halt and subsequently stabilizes at a rate and convenience level that makes public transport attractive by comparison for many. But if public transport has previously been cut according to the trend, then it will offer less, the tendency to use it will be less, and the whole city traffic suffers. Life in all its aspects is reduced in the centre. The easy way out through suburban commercial development is not practicable in Europe, and is not by any means the answer to most problems, even in the States.

From the desperate attempts many U.S. cities are now making to revitalize their public transport, and bring it up to date, we can learn this lesson: it is good planning in every way to keep railways in good trim and buses running well. The policy, which from a short-sighted profit-seeking viewpoint increases prices for suburban trains and shuts down lines in 1963, is tragic and disastrously uneconomic in the long run. Roads, assessed in the same way, don't pay either!

Britain should emulate the attitude of the Federal Government of the U.S.A.:

"*Mass transportation, it is believed by those of us who are responsible for the administration of this program, must be viewed as an essential public service and not solely as an enterprise from which a maximum financial profit is to be derived. Efforts to cover total costs, including those of attracting and supporting large capital outlays, have often proved self-defeating when fares—generally the sole, or principal source of revenue—have been increased. Those who can afford to pay the higher fares have been impelled to turn to private transportation; while those heavily reliant upon public transporation—the so-called 'captive transit rider' in the low income group, and the non-drivers, mostly the youngsters and the oldsters—are unduly penalized by being effectively restricted in their urban travels.*"

"*Balance in local transportation is thus destroyed. The federal efforts seek to restore that balance by inspiring and assisting communities in working out their own solutions. It will neither nationalize the local transit industry nor displace local initiative and control. It expresses legitimate United States interest in urban transportation.*"(i)

Financial support for experiment and long range programmes available in the U.S.A. are described in the next section. With nationalized transport the problem can be solved more easily if European countries choose to do so.

It is clear that in countries where the railways are nationalized the rationalization is much simpler. In fact, state–local authority co-operation is, at least in Britain, at a disastrously low level, largely non-existent, not even thought of. Decisive chances are thus missed. Government instructions at the highest level are urgently needed to integrate activities of British Railways and local transport authorities with national and regional planning.

V. 8.2. *Principles of Public Transport Systems*

In cities, Public Transport must be cheap, frequent, quick, easily accessible, clean, with attractive termini and polite staff. But paying should be mechanically regulated or the service ought to be free, like so many other services absorbed in the rates. This in itself would logically differentiate "public" transport from the private car in town and release a vast pool of labour from a pretty sterile job.

U.S. public transport(ii) shows a significant correlation between increase in use, density, and number of people per car. This can be applied in a creative way: to plan at considerable density around public transport routes will reduce the need for cars and their use for standard trips.

Beyond such general points it is important that the stations are strategically placed. In new systems, such as the one in Stockholm (see example), the stations are placed by the neighbourhood centres, with mutual benefit. There is not enough parking in some of these stations. Park-and-Ride has been developed in many U.S. cities. (Cleveland offers about 5000 such parking places.) However, by American standards the housing is of high density around Stockholm and most people live within walking distance of these new centres and stations.

When existing lines are to be promoted to major importance it is vital to supply local shopping provisions and car parking by the stations. But it must not be forgotten that to develop proper pedestrian routes, the centring on such stations is a vital part of their success. In fact, the placing of routes may be influenced more by the position of the path system than the road system where new housing estates are to be connected with central areas.

Central termini, however, must be truly central and combine with many other services, amenities, parking and the main footpath network.

In central areas, the rapid transit will be overhead, or underground, or on its own separate track, along routes reserved for it. This would be on one side of a motorway perhaps, with underpass connection from one side and no crossing of the road whatsoever. Within the centre itself very slow public transport can be envisaged as entering the pedestrian malls, respecting the prominence and prime rights of the people on foot. This requires careful handling. Little experience exists.(i) (See p. 157.)

V. 8.3. *Types of Rapid Transit*

V. 8.3.1. *Buses:* The cheapest form of public transport to be turned into rapid transit is the bus. In the U.S.A. it has been introduced on to freeways where its speed is comparatively high. The advantage of the ready-made track, the mobility in being able to turn into minor roads, and the comparatively cheap vehicle are counterbalanced by the shortness of life, the high labour costs in traditional staffing, particularly in Britain, with driver and conductor. While the bus is oil or petrol driven it has the added disadvantage of being the worst polluter of the atmosphere.

In some places special lanes for rapid bus routes are being provided side by side with the existing free-ways (Stockholm, Chicago). In the U.S.A. special lanes to allow buses only to go contrary to other traffic in one way streets retain their convenience at a maximum compared with private cars (Baltimore).(ii)

Buses stopping at frequent stops, as in most British towns, and as trams and trolley buses do on the continent, as buses still do in most U.S. towns, become inefficient amongst the peak traffic. Their stopping and starting is a major contribution to congestion along roads. They really are part of the anachronistic mixture of traffic which has to be sorted out in each and every town. In centres such vehicles are obviously more useful, for the room they take up. Roads might be reserved for them.

(i) WEAVER R. C. The Federal Interest in Urban Mass Transportation. *Traffic Quarterly*, January 1963.

(ii) Smith W. and Associates. *Future Highway and Urban Growth.* New Haven 1961.

(i) LIPP M. N. Lincoln Road Mall. *Traffic Quarterly*, July 1961.

(ii) LLOYD F. J. Traffic Priorities Keep America's Buses Moving, *T.E. & C.* London, January 1963. Also to be found in Sienna, Italy.

The adaptability of buses is demonstrated by the recent introduction of a fleet of five to take 23 cycles below and passengers on the upper deck through the Dartford–Purfleet Tunnel (Britain).(i)

Again, at slow speeds of up to 15 mph, trolley-buses might share the pedestrians' realm in many places; as they do not pollute the atmosphere with noise or poison. The scrapping of many trolley-buses seems short sighted and will be regretted in years to come. They are less efficient than buses, given today's criteria. But criteria are changing in a predictable way, making trolley-buses better than diesel buses for central, pedestrian-dominated areas. It is not the trolley bus which is the anachronism—it is the attitude to the private car.

V. 8.3.2. *Duo-rail systems:* As with buses, so with rail transport: we can distinguish between rapid transport, through routes, and the local and stopping varieties. Trams are the oldest variety of widespread mass transport and in their up-to-date versions on the continent they have much to commend them. They do not pollute the air and their route is clearly seen by pedestrians, so lessening danger. This makes them suitable as a possible mass-transit means in an otherwise pedestrian mall. For this they would be slowed down of course. They have the same high capital investment cost and long life normal for rail transport (Chestnut St., Philadelphia).

Trams have been ousted from most British cities, and will be from the remaining ones, it seems, because they conflict with motor traffic in the inadequate streets of towns. This may, as with trolley-buses, be a shallow reason for rejecting them, tearing up the rails, and wasting the stock. If the air pollution by diesel engines is taken seriously as a carcinomous agent, this point may become clearer still.

Suburban trains and local trains serving cities are abandoned as private cars become popular. But as the roads get blocked by the rising numbers the importance of trains emerges once again. In fact, modern town planning practice shows that railway lines into the central stations of cities are an extremely important asset and investment. These latently highly efficient means of communication should be run well and their existence is a good reason for building up the areas around the out-stations densely and providing them with ample car parking, particularly in low density areas such as are found in the U.S.A.(ii) In Chicago, the Congress Street Line, since it was modernized and opened in 1958, has attracted 32% more passengers than the old line, mostly from private cars. (See Fig. 5.68.)

Thus, the reverse of the very process which is happening in many countries, including Britain, is the right one. Railway lines should not be closed down merely because they become unprofitable in a very narrow sense. They are a precious national and local investment which should be surrounded by planning to make them work. They save the construction of roads, which are vastly more expensive than the deficit of the railways, and much more antisocial, economically wasteful with ground, less efficient in fog and snow, creating more accidents and unable to enter big towns or old towns in great numbers.

Fig. 5.68. Chicago's Congress Expressway, the first in the U.S.A. with rapid transit rail in its median strip. During rush-hours this railway carries more passengers than do the cars in the eight lines of the road.

V. 8.3.3. *Monorail:* Since the installation of the Monorail in Seattle for the 1962 World Fair, this means of rapid transport has been widely discussed. For three years previously the 2/3 full size Monorail had been running in Los Angeles, in Disneyland Fairground, working very well. They also have single lines in Tokio and Turin. The main advantage of the monorail seems to be its use of the existing roads for its supports, thus not disrupting the traffic of the usually overcrowded streets below either during building or once they are running. The ride is not in any way quicker or more comfortable than in ordinary trains (the author has used the two U.S.A. prototypes). The Alweg type Monorail (described above, developed in Germany) has the advantage over the new French suspended type in that its supports are less high, and it can run on ground level where possible. The French type developed by Sauffege at Château-neuf-sur-Loire, runs more quietly however.

(i) *Daily Telegraph*. London 9.5.63. (Picture and description.)

(ii) See Philadelphia, Urban Renewal section, in City Centres example.

Fig. 5.69. Wupperthal suspended Monorail.

Fig. 5.70. Not without appeal. Alweg Monorail, Seattle.

There will be places where, considering the various factors, monorails might be the right answer. But in the city they clutter the streets only slightly less than the overhead railways just removed from the street of New York. This resulted in an increase of amenity and property values. In rural areas the Alweg monorail type can run on the ground. The type which is suspended cannot make such economies possible. However, "the daddy of them all", the Wupperthal monorail, opened in about 1900 is suspended over the town river and is an interesting special case. (It has been denounced as grossly inhuman in the townscape by Reichow.)[(i)]

Fig. 5.71. Alweg Monorail, Seattle, obstructing the view in the street.

Fig. 5.72. Alweg Monorail over lake at Turin to avoid purchase of building land.

(i) Reichow H. *Die Stadtbaukunst* Braunschweig 1948. Dark J. W. Why not the Monorail? *J.T.P.I.* London, February 1963.

V. 8.3.4. *Underground railways:* Suburban lines which, running on the surface, go underground when approaching the centres of big cities relieve the transport problems of places like London, Paris, and Stockholm enormously. The uncivilized depressing effect of long stays underground (however attractive the shopping) must be taken into account. Improvement of environment underground may range from chandeliers (Moscow), to especially sexy posters (London), and rubber-tyred coaches (Paris). The micro-climate could be improved in most instances. This is promised for the New Victoria Line in London.(i)

Leslie Tass(i) stresses that the routing of new underground railways, particularly in the centre of cities must follow desire lines and not existing road routes. In Toronto the latter mistake was apparently made in the most recently completed underground railway. The new London Line, begun in 1962, on the other hand, will serve a very much needed desire line not at all well catered for by roads. Some points of interest on this, the first Underground line to be built in London since 1907, are:

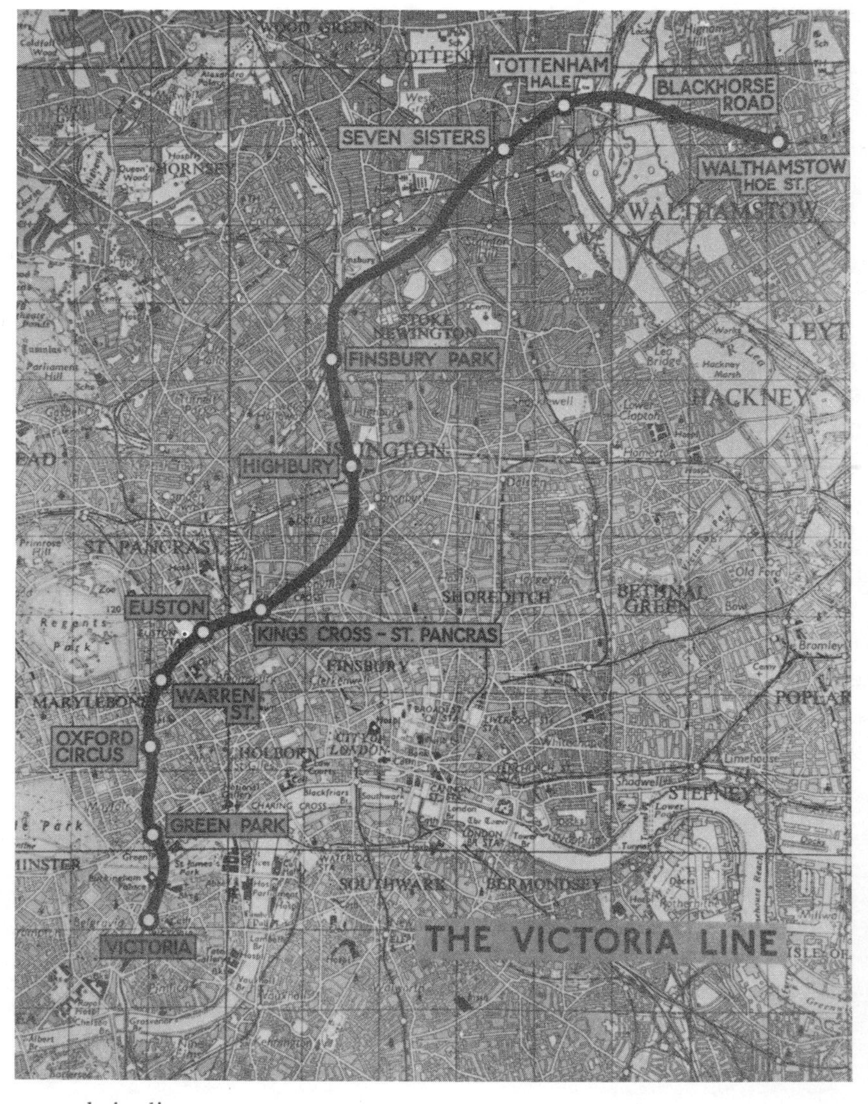

Fig. 5.73. New London underground route serves desire line not served by roads (by permission of London Transport Executive).

(i) *The Victoria Line.* H.M.S.O. 1960.

(i) TASS L. New Rapid Transit Planning. *Reprint Traffic Quarterly.* October 1961.

"Quicker, shorter, and more direct daily journeys for those crossing central London or travelling from the northern and southern suburbs. Victoria station will be linked for the first time direct with Green Park for the Mayfair area, Oxford Circus for the shops, and given a much quicker route to Euston, St. Pancras and King's Cross for the main line stations. Quick interchange with eleven existing Underground and British Railways stations, at some important stations by merely crossing the platform.

Reduction in street congestion by motorists leaving their cars at home and making a quicker journey by tube to Town.

Spacious stations designed for easy interchange and passenger movement between platform and street, based on research into passenger flow. Escalators will lead direct to platforms wherever possible.

Closed-circuit television at central area stations to assist in controlling and speeding up the movement of passengers at peak-hours.

Two-minute rush-hour services with a capacity of 32,000 *passengers an hour in either direction—the capacity of fourteen motorway traffic lanes.*

New ventilation techniques to cut out draughts on platforms, and to keep the tube cool.

Smoother and quieter rides for passengers due to long-welded rails, special 'fins' in the tunnels to reduce noise, and the minimum of curves.

The trains may be fitted with automatic driving control, with which London Transport is now experimenting.

Automatic signalling controlled by 'programme machines' interpreting a 'timetable' of punched paper rolls into signalled train movements."(i)

V. 8.3.5. *Self-guided bus trains:* The possibility of self-guided bus trains was first suggested publicly in New Orleans on 15 October, 1958, when the General Manager of Chicago Transit Authority, Walter J. McCarter, spoke at the annual meeting of the American Transit Association.

"Studies have been made of the practicality of combining rail rapid transit with other expressways of the comprehensive system programmed for Chicago. It was determined that rail rapid transit would be justified in two of the expressways, the Northwest and the South, but that it would not be warranted in the Southwest Expressway, which is to serve an area of relatively low population density."

"This finding led to CTA's consideration of self-guided bus trains for the Southwest Expressway as a means of obtaining in substantial measure the advantages in efficiency, economy and increased passenger-carrying capacity that are derived from combining rail rapid transit with an expressway."

(i) *The Victoria Line.* H.M.S.O. 1960.

"Actually there would be some important additional benefits. One is a low operating noise level. The bus trains would travel along their reserved lanes quietly and smoothly on pneumatic tyres. Another is flexibility, a desirable quality that is lacking in a fixed rail facility."

"In regular operation, the bus trains would possibly be capable of attaining speeds of 60 *miles per hour, or more. One operator would be assigned to each train. He would control the motive power and braking facilities of the buses in the train from his position at the wheel of the front bus."*

Unique Operating Procedures:

"Operating procedures would be unique, and especially suited to the requirements of areas of relatively low population density. In the outlying sections, the buses would be operated manually as single units over a number of scheduled routes on surface streets. Operators would drive their buses to an assembly point alongside the highway. There buses would be coupled into trains, and each train would proceed on its run to downtown with only one operator aboard."

"At a point near the central business district, the bus trains would leave the highway and proceed to a terminal alongside the expressway. Here the bus trains would be separated into single units again, and each bus would then be operated manually over diverse routes, providing fast, convenient delivery to passengers."

"On the out-bound trip, the manually operated single units would come from their diverse central business district routings to the staging area alongside the expressway where the buses would again be assembled into trains for the trip to the outlying staging point."

"Mechanisms for control of the motive power and the brakes of all of the buses in a train from the operator's position in the front bus have been developed and over-the-road vehicles with such equipment are now being sold commercially."

"The guidance system that would function when the buses are coupled together in a train uses proven techniques that have performed successfully for years in the operation of self-guided industrial trucks."

"With the co-operation of Barrett Electronics of Northbrook, Ill., Chicago Transit Authority recently completed a series of preliminary tests of the feasibility of adapting the industrial truck guidance system to the operation of buses in trains."

"An industrial truck was equipped with the electronic sensing device and the necessary guidance cable was intalled in the floor and pavement area of a section of the shops."

"On this test course, the industrial truck operated without any manual controls whatever, making pre-programmed stops for a predetermined period of time at each stop."

"In the opinion of Chicago Transit Authority, the outcome of these preliminary tests justifies proceeding with the final series of tests—the equipping of transit buses with the electronic devices and installing the guidance cable in a section of highway where test operations at highspeed, simulating express bus-train operation, could be conducted."

"Financing of the final series of tests from funds appropriated to the Federal Housing and Home Finance Agency for aid to urban mass transit is amply justified because the benefits to be derived from the operation of self-guided bus trains would be widely shared."

Fig. 5.73b. Lane reserved for buses, Sweden.

V. 8.4. *Comparison of Typical Features*

TABLE V.16. *Urban Transportation Facilities: A Comparison of Typical Features*(i)

Urban facility type (a)	Number of lanes or tracks (b)	Typical frequency of access or stops (c) (in feet)	(in metres)	Peak hour peak direction: Person-trip capacity (d)	Typical speed (e) (in mph)	(in km/hr)	Typical full capital cost for one urban route-mile (f)	Cost/capacity ratio (f)/(d) (g)
Automobiles								
Freeway	8	6000	1829	9000	32	51·5	$15,000,000(ii)	$1670
At-grade expressway	8	2000	609·6	6000	25	40	3,000,000	500
Arterial	6	600	183	3000	20	32	*	—
Present transit								
Express subway	2	9000	2743	50,000	33	53	22,000,000†	440
Local subway	2	3200	975	50,000	20	32	22,000,000	440
Express elevated	2	9000	2743	50,000	33	53	7,000,000	140
Local elevated	2	3200	975	50,000	20	32	7,000,000	140
Freeway express bus	2	4000	1219	8000	22	35	*	—
Arterial limited bus	2	2000	609·6	7000	14	22·5	*	—
Local surface bus	2	600	183	7000	10	16	*	—
Limited tramline								
—In available private rights-of-way	2	2000	609·6	34,000	25	40	2,000,000	60
—In arterials	2	1200	365·8	34,000	17	27·4	1,000,000	30

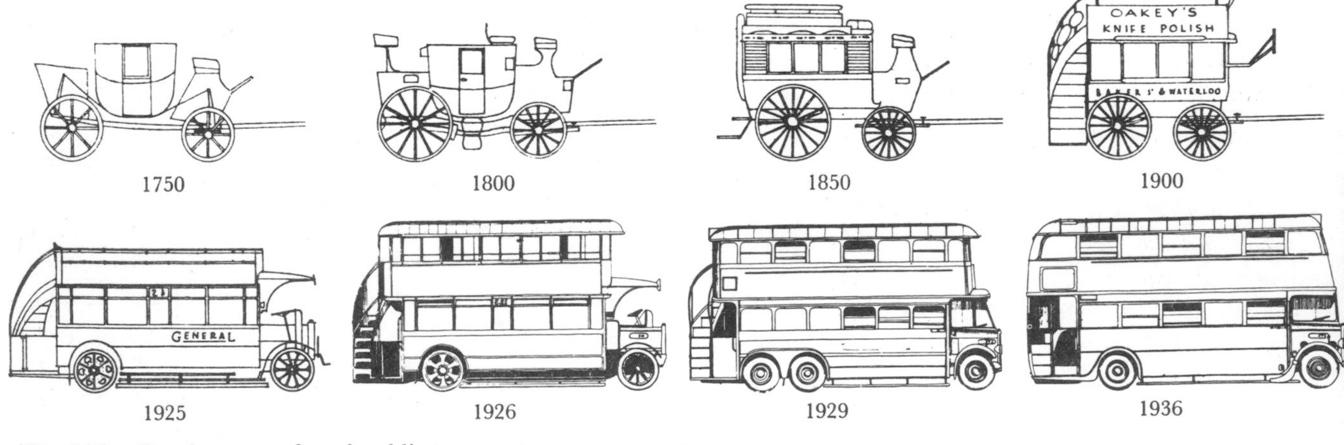

Fig. 5.73a. Development of road public transport.

(i) QUINBY H. D. New Concept In Transit Facilities, *Traffic Quarterly*, April 1962.
* Travelled way assumed already available.
† The New Victoria Line, London, costs $16,000,000 per mile including rolling stock (author's note).

Metric dimensions added by the author.
To change from dollars to pounds sterling (approximately) divide by three.

(ii) Liverpool Urban Motorway is estimated at £6,000,000 per mile, inclusive of land cost, ¼ of the total (6 lanes.)

V. 8.5. *Examples*

V. 8.5.1. *Philadelphia:* The Passenger Service Improvement Corporation was set up in Philadelphia in 1960. It is the body which carries on the work done previously by the Urban Traffic and Transportation Board, born out of the research and work begun in 1953. Of all the revitalizing programmes in the U.S. it is the most effective and is therefore described in some detail.

The above bodies worked in close contact with the City Planning Commissions and the Federal Highways Departments, as well as the local and state public and private companies involved in Transport. To quote from a progress report by the PSIC:

"*Twelve commuter railroad lines and two subways form the backbone of local transportation.... The railroads in recent years had been in trouble. Ridership had been dwindling alarmingly. Rising costs pushed fares up and forced the removal of trains.*"

"*At the same time automobiles flooded the streets, strangling traffic, and sapping the public purse with costly road building programs, traffic control installations and increased police services.*"

"*To save the dying commuter railroad network, the City of Philadelphia acted in concert with the Pennsylvania and Reading Railroads on October* 26, 1958. *'Operation Northwest' was born.*"

"*Aim of the program was to make rail travel attractive so that motorists would leave their cars at home. Both the Pennsylvania and Reading dropped fares on their routes to Germantown-Chestnut Hill to a flat* 30 *cents—about half the former cost for many users. They also scheduled additional trains. The City Government helped underwrite some of the costs.*"

"*The 'Operation' efforts have produced encouraging increases in ridership as millions of people have taken advantage of the cheap, comfortable and convenient transportation offered on these trains.*"

"*Latest figures show that* 6.2 *million are riding the 'Operations' yearly. This is a* 44% *gain in ridership over 'pre-Operation' years.*"

"*The upsurge in ridership is particularly remarkable when compared with the continuing decline of commuter networks elsewhere in the country.*"

"*The 'Operations', have attracted nationwide attention and are serving as models for similar programs.*"

"*PSIC stemmed from the efforts of Mayor Richardson Dilworth and City Council of the City of Philadelphia. It was set up in the summer of* 1960 *as a non-profit corporation to take over the management of the 'Operations' already in existence and to develop similar programs in other areas.*"

"*Making policy is a* 15-*member non-salaried board of civic, business and government leaders, railroad and railway union officials. Serving the board is a technical staff responsible for conducting studies and recommending action.*"

"*For* 1961, *City Council appropriated* $1.5 *million with which PSIC purchases commuter rail services from the railroads. Council and the voters also have approved expenditures of some* $8.7 *million for important capital improvements including new, air-conditioned cars, and the U.S. Federal Housing and Home Finance Agency may furnish an additional* $11 *million. Fares collected from riders wholly or partly will repay capital investments.*"

"*A committee of PSIC directors and the technical staff studies operating deficits stemming from the reduced fares, additional trains, and modernization programs. Negotiations with the railroads follow. Ultimately a figure is recommended to City Council. The subsequent payment to the railroads covers their losses only in part.*"

"*Without aid, the commuter rail system—so vital to the entire region—simply would dry up, as has happened elsewhere. Without rail and transit service, our highways would be paralyzed by autos at peak periods; business activity would be stifled; huge tax expenditures would be consumed by the losing battle to move motor vehicles; business and industry would flee the region.*"

"*All business and industry in a metropolitan area must be accessible to both their employees and their customers. Rail offers the only means for moving masses of people through a built-up metropolitan region rapidly and economically, especially at morning and evening peak travel hours. Meanwhile necessary motor traffic which must be on the highways at peak periods enjoys reduced congestion.*"

"*Expressways for motor vehicles—even if enough could be provided—are fantastically costly. Contructing highways and furnishing traffic control devices and police services gulp tax dollars. In* 1960, *Philadelphia paid* $50 *out of general taxes for each auto registered in the city, but only* $2.50 *for each resident for subways (railway) or commuter trains. Expressways, in addition, destroy tax-producing real estate.*"

"*Citizens overwhelmingly want rapid and inexpensive tranportation. Each day some* 216,000 *persons arrive in downtown Philadelphia during peak periods by public transportation including commuter trains—four times as many as come by auto.*"

"*Since the chartering of PSIC, passengers have increased so rapidly that railroads are pressed to make available enough equipment to provide seats for every passenger. As a result, the City is now arranging to purchase, for lease by the railroads on a self-liquidating basis,* 26 *new air-conditioned commuter cars. Eight older cars already have been rehabilitated for the comfort of riders.*"

"*PSIC is working with neighboring counties in the concerted attack on traffic congestion and better com-*

muter service. Officials in sub-urban counties adjoining Philadelphia realize that without quality transportation their own development and growth can be impeded."

Bucks, Chester and Montgomery Counties already have joined Philadelphia in the Southeastern Regional Compact, a co-operative program designed to achieve outstanding transportation for the entire area,(i) and the demonstration program started September 1962, to run for three years. The Federal Government is paying two-thirds of the costs, as this is a "demonstration". To obtain two-thirds federal funds for this venture it had to be a demonstration. The application for the federal funds starts with the following ten points, quoted because they summarize the basic principles:

"*We expect to demonstrate that:*(ii)

1. *Improved rail transportation service, properly priced, can draw new riders, especially from highways, and that this can reduce the cost of transportation facilities for a major metropolitan region.*
2. *Innovations such as free parking, zone fares, improved schedules, and integration of feeder buses are attractive to potential riders.*
3. *Reverse riding return tickets can be increased on suburban railroad lines.*
4. *Riders who must come by automobile can be attracted to parking facilities located away from the traditional town centers if they are spacious and free.*
5. *Residents in low-density, suburban areas can be persuaded to use mass transportation facilities.*
6. *Such mass transportation service can be adjusted to meet the needs of citizens at a particular time and varied as their needs change.*
7. *Riders can be attracted between suburban areas if the proper service is provided.*
8. *Such a program will reduce the deficits operating railroads now face and preserve the commuter railroads to serve transportation needs of suburban counties as well as central cities.*
9. *Several governments can not only plan for transportation facilities, but operate a program co-operatively.*
10. *Homebuilders, residents, and planners can make decisions favourable to the future of such facilities if their quality is improved and their life assured (i.e., in buying or building houses or apartments, decisions on car ownership or two cars per family, zoning, etc.).*"

(i) *Meeting the Transportation Crises*. Passenger Service Improvement Corporation, Philadelphia 1959.

(ii) Proposal for a commuter transportation demonstration grant made to the Administrator, Housing and Home Finance Agency, Washington. 4 May, 1962.

The increase in ridership as experienced in Philadelphia was forecast by the Market Research Service Incorporated, employed by the Urban traffic and Transportation Board in 1958. Their enquiry found out other points of interest. Those who average $4500 (£1500) income per annum were *more* likely to take to train riding than those who averaged $2900 (£970).(i)

In projecting Philadelphia's traffic to 1980 it is interesting to note that the number of private car drivers and passengers is increased from 20% of the total central area population to 27% only. Public transport will however increase by 20%. This must be compared with a reduction in the use of public transport by about 25% between 1947 and 1955.

In listing the gains made by the rail experiments it is relevant to note that over 60% came from previous private car users and others from increased trips to town, a point that encourages down-town business activity and life in general. Few came from the buses.

How important public relations are was born out by a survey. It was found that people did not use the trains because station car parks were full so that they could not leave their cars to board them. But they did not know that special bus services had been started to collect people from near their homes for just such cases, to take them to stations.

The City of Philadelphia has started a programme of obtaining thousands of extra parking places in the cheap areas surrounding the stations in out-lying districts. (See also Philadelphia p. 176.)

V. 8.5.2. *St. Louis:*(ii) In 1953, the St. Louis Public Service Company, in an effort to win people from their cars to Public Transit, started to operate bus routes to town from car parks about quarter of an hour's ride away from the centre, before the driver meets the congestion of the central area. The experiment was a success and the service was extended. It had an indirect influence also. People copied the idea of leaving their cars and changing to public transport, even where this was not specially provided for.

Most of the people using the new service had formerly driven all the way to town and were male workers. Good publicity was an important aspect of the success of the experiment.

(i) Market Research Service Inc. Market Potential for Pennsylvania Railroad. Philadelphia 1958.

(ii) Rexford C. W. Park-ride in St. Louis. Reprint *Traffic Quarterly*. Eno Foundation, Saugatuck, April 1955.

V. 8.5.3. *Boston:*[(i)] Investigation in Boston has shown that the Highland Branch Rapid Rail Transit catered for 35% of previous car drivers. It released 5%, i.e. 1300 of the central parking spaces. The line was particularly heavily used at rush-hours, of course, and took the trade away from the private bus company so that some routes had to close down. The typical rider was in the higher income bracket, $10,000 (£3300) or more, owned one car or, in the case of 32%, two, and was a professional man. This is the kind of person normally expected to shun mass-transit in the U.S.A. The line was built at a tenth of the cost per mile of the Boston Central Artery Freeway, which had cost $10,000,000 a mile (£3·3 million).

The latter could carry 8000 passengers an hour, but the rail line could easily carry 24,000 passengers with improved facilities.

V. 8.5.4. *San Francisco. Bay Area Rapid Transit:*[(ii)] On 6th November 1962 Bay Area voters approved $792 million (£264 million) of general obligation bond issue for the construction of the Three-County Plan Rapid Transit.

Eleven years after the Californian Legislature created a special Commission to study the Bay Area transportation problem, the three major ones, of the nine counties originally involved, are at the point where actual construction is planned to start in January 1964. The first trains are to run in 1966. In 1968 the system is to be 80% complete and in 1971 the entire 75 miles (120 km) of the three counties plan, will be ready.

Lightweight electric trains will run at 70 mph (112 km/hr), stop 20 seconds at stations, to give average speeds of 50 mph (80 km/hr). Trains and fare collection will be regulated by computor. After examination of all other systems, the area project chose duorail which will travel above, below and along the ground and through a tube on the bottom of San Francisco Bay. 40,000 parking places will be provided at stations along the route. By 1975. The New Rapid Transit is forecast to increase its percentage of peak hour traffic to 42%, from 28% in 1959.

In explaining the need for the Rapid Transit Lines the liberally documented report makes the following statement, encouraging voters just before the November 1962 poll:

(i) Schneider L. M. Impact of Rapid Transit Extensions on Suburban Bus Companies. Reprint, *Traffic Quarterly*, January 1961, Eno Foundation.

(ii) *Rapid Transit for the Bay Area, a summary of Engineering, Financial and Economic Reports submitted to the San Francisco Bay Area Rapid Transit District*, Stone & Youngberg 1961. Also *Fact Sheet* 111. Progress Summary, 1962.

"*The only alternative will be the involuntary addition of countless new freeways, bridge crossing, and parking facilities throughout the Bay Area.*"

"*The motorist and taxpayer would bear not only the cost of constructing these new automobile facilities, but also would have to absorb the tremendous extra costs incurred by removing vast amounts of land from the tax rolls. Such land would be lost to potential use for homes, businesses and industries.*"

"*Extra taxes also would be required for local traffic controls, for smog reduction, for accident-prevention, and for additional utility construction.*"

"*Reliance upon an 'all-automobile' solution, according to the economic experts, will be twice as costly as the construction of a new regional rapid transit system.*"

"*By comparison, a rapid transit line can carry five times as many commutors, and yet occupy only one-fourth the amount of right-of-way necessary for a modern six-lane automobile freeway.*"

"*And even the State Division of Highways acknowledges that it cannot build auto facilities fast enough to cope with the future growth of automobile congestion.*"

"*Benefits of Rapid Transit:*[(i)]
Reduced automobile congestion.
Improved property values.
Reduced travel times.
Access to new jobs.
Smog reduction.
Lower commute costs.
Greater residential availability.
Greater cultural accessibility.
New leisure opportunities.
Expanded sales markets.
Lower insurance rates.
Auto accident reduction.
Expanded labor markets.
Preservation of scenic beauty.
New civil defense facilities.
Lower freight costs.
Better city planning.
Wider access to schools."

It is regrettable that car parking seems not to be an integral part of the station architecture and on more than one level. Pictures show cars surrounding stations and, whereas the aim to give good architectural quality to the stations is specifically stated, the cars are left to look after themselves. This is poignant, particularly where high level stations would cut out the ascent for all those parking on the first or second level up.

These four examples of a very different kind are symptomatic of the tendencies now emerging in the U.S.A., tendencies quite contrary to the "trends" as forecast.

(i) Taken from *Bay Area Rapid Transit District*. 1962.

Most American cities, particularly the very large ones, realize, as each makes its own study, that public transport, in terms of rapid, grade-separated vehicles of one kind or another, are essential to the survival of cities. This lesson, learnt late in the New World, should be heeded by Europeans in good time.

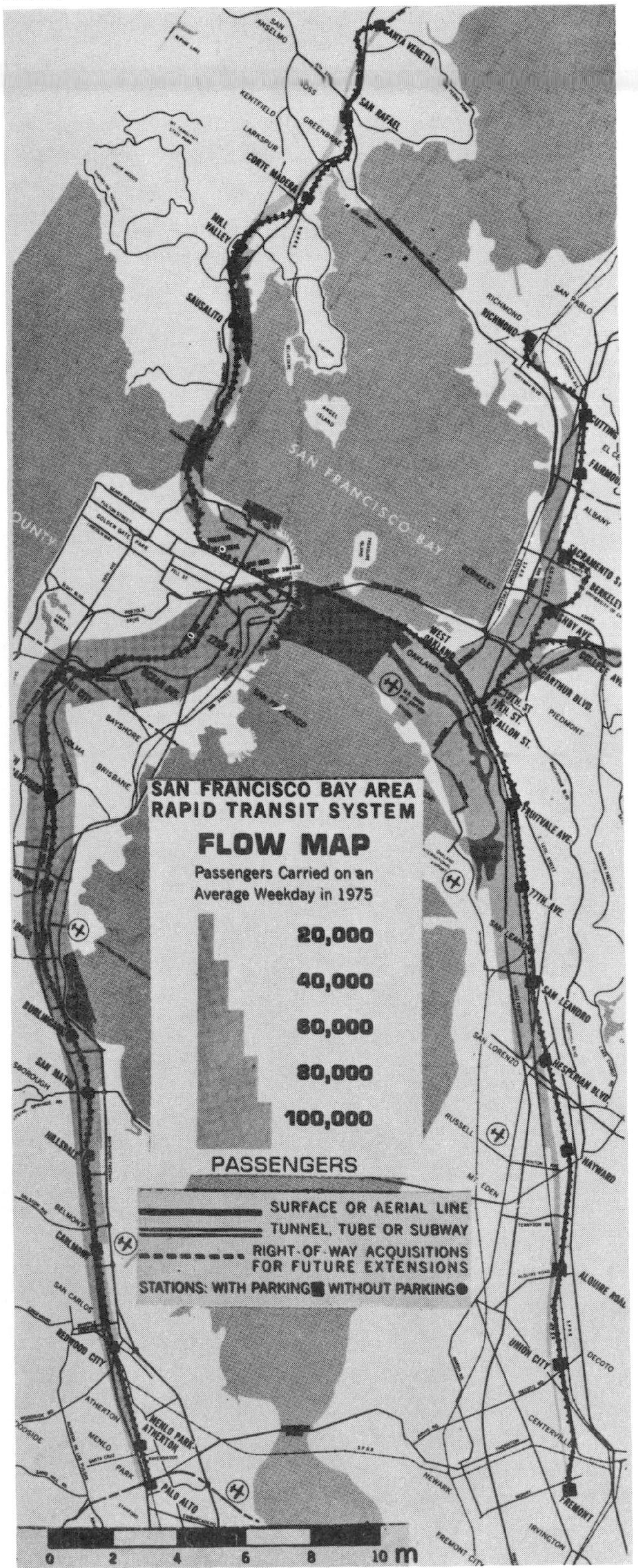

Fig. 5.74. San Francisco rapid transit proposals, showing park and ride provisions.

Fig. 5.75. Walnut Creek Station.

Fig. 5.76. Ocean Avenue Station San Francisco using the median strip of the Southern Freeway.

Fig. 5.77. Rapid Transit Trains and street cars sharing Market Street Subways in San Francisco.

V. 9. Glossary of Terms

Arterial Highway. A general term denoting a highway primarily for through traffic usually on a continuous route.

Basic Capacity. The maximum number of passenger cars that can pass a given point on a lane or roadway during one hour under the most nearly ideal roadway and traffic conditions which can possibly be attained.

Capacity. The ability of a roadway to accommodate traffic in given circumstances.

Critical Density. The density of traffic when the volume is at the possible capacity on a given roadway. At a density either greater or less than the critical density, the volume of traffic will be decreased. Critical density, occurs when all vehicles are moving at or about the optimum speed.

Density. The number of vehicles occupying a unit length of the moving lanes of a roadway at a given instant. Usually expressed in vehicles per mile.

Design Speed. A speed selected for purposes of design and correlation of those features of a highway, such as curvature, superelevation, and sight distance, upon which the safe operation of vehicles is dependent. It is the highest continuous speed at which individual vehicles can travel with safety upon a highway when weather conditions are favourable, traffic density is low, and the design feature of the highway are the governing conditions of safety.

Desire Line. A straight line between the point of origin and point of destination of a trip without regard to routes of travel (used in connection with an origin and destination study).

Dual Carriageway. A road in which there are two physically separated carriage ways reserved for up and down traffic respectively.

Expressway. A divided arterial highway for through traffic with full or partial control of access and generally with grade separation at intersections.

Flow Line. A line on a map joining the origin and destination points of a trip, following the best route available.

Freeway. An expressway with full control of access (100% grade separated).

Highway Grade Separation. A structure used to separate vertically two intersecting roadways, thus permitting traffic on the one road to cross traffic on the other road without interference.

Inner Loop. A ramp used by traffic destined for a right-turn movement from one of the through roadways to a second when such movement is accomplished by making a left-exit turn followed by a three-quarter round left-turn manoeuvre and a left-entrance turn.

Interchange, Traffic Interchange. A system of interconnecting roadways in conjunction with a grade separation or grade separations providing for the interchange of traffic between two or more roadways or highways on different levels.

Major Road. A road which has, or which is assigned, a priority of traffic movement over that of other roads.

Motorway. A highway for the exclusive use of mechanically propelled vehicles (a British Freeway).

Optimum Speed. The average speed at which traffic must move when the volume is at a maximum on a given roadway. An average speed either appreciably higher or lower than the optimum will result in a reduction in volume. Also known as critical speed.

Origin-Destination ("O and D") Study. A study of the origins and destinations of trips of vehicles and passengers. Usually included in the study are all trips within, or passing through, into or out of a selected area.

Parkway. An arterial highway for non-commerical traffic, with full or partial control of access, and usually located within a park or a ribbon of parklike development.

Passing Sight Distance. The minimum sight distance that must be available to enable the driver of one vehicle safely and comfortably without interfering with the speed of an oncoming vehicle travelling at the design speed, should it come into view after the overtaking manoeuvre is started.

Possible Capacity. The volume of traffic that cannot be exceeded in actuality without changing one or more of the conditions that prevail.

Practical Capacity. The maximum number of vehicles that can pass a given point on a roadway or in a designated lane during one hour without the traffic density being so great as to cause unreasonable delay, hazard or restriction to the drivers' freedom to manoeuvre under the prevailing roadway and traffic conditions.

Ramp, Access Ramp. An interconnecting roadway of a traffic interchange or any connection between highway facilities of different levels, on which vehicles may enter or leave a designated roadway.

Ring Road. A highway roughly circumferential about the centre of an urban area and permitting traffic to avoid the centre of such area.

Tidal Traffic. Traffic on a two-way road proceeding predominantly in one direction or the other according to time or other recurrent circumstances.

Time Contour Map. A map on which are drawn lines joining points which can be reached in the same length of time from a given point or a given zone.

Traffic Flow, Traffic Volume. The number of vehicles, persons or animals passing a specific point in a stated time in both directions unless otherwise stated.

Traffic Lane. A strip of roadway intended to accommodate a single line of moving vehicles.

Through Street, Through Highway. Every highway or portion thereof at the entrance to which vehicular traffic from intersecting highways is required to stop before entering or crossing the same and when stop signs are erected.

VI. New Towns with Traffic Segregation

VI. 1. Basic Requirements. **108**

VI. 2. Motor Traffic in New Towns, General. **108**

VI. 2.1. Parking in New Town Centres. 109
VI. 2.2. Cumbernauld, Research into C.A./Retailing and Parking. 110

VI. 3. Fully Motorized Town Centre for 250,000. **111**

VI. 4. Town Design, Based on Walking Distances. **111**

VI. 4.1. Density and Town Size. 111
VI. 4.1.1. Town size, population and density study. 111
VI. 4.1.2. Town density diagrams. 112
VI. 4.2. Comparison of Towns and their Walking Distances. 113
VI. 4.3. Town Shape. 114

VI. 5. Land Use. **114**

VI. 6. Open Space. **114**

VI. 7. The Neighbourhood Concept. **115**

VI. 8. The Principles of Growth. **115**

VI. 9. Examples. **117**

VI. 9.1. Cumbernauld, Scotland. 117
VI. 9.2. Hook, England. 121
VI. 9.3. Erith, England. 123
VI. 9.4. Toulouse le Mirail, France. 126
VI. 9.5. Vällingby, Sweden. 128
VI. 9.6. Farsta, Sweden. 130
VI. 9.7. Cologne, Germany. 133
VI. 9.8. Sabende, Guinea. 134
VI. 9.9. Sennestadt, Germany. 136
VI. 9.10. Neu-Winsen, Germany. 138
VI. 9.11. Alkmaar, Holland. 139
VI. 9.12. Chandigarh, India. 140
VI. 9.13. Kitimat, Canada. 143
VI. 9.14. Tapiola, Finland. 144
VI. 9.15. Sputnik, Russia. 146
VI. 9.16. Cambridge Village, Cambs. England. 147
VI. 9.17. Reston, Washington, U.S.A. 148

VI. New Towns with Traffic Segregation

VI. 1. Basic Requirements

When the planning of New Towns takes for granted traffic segregation, buildings, roads and footpaths must be combined and integrated into one environmental whole. The routes offered to people to get to work, school, shops and recreation must all be pleasant in themselves. Journeys, walking or driving, become ends as well as means. Staying at home, for old people and mothers with young children, should also become far more attractive.

This requires particularly sensitive attention to the path environment. It has been shown that detail is of great importance and path convenience is vital. This is an outlook diametrically opposed to the old and current one, where, particularly in housing, roads are laid out and built, housetypes are stuck alongside, and then landscaping is added as a cosmetic to try and make the thing appealing.

Not only visually, but functionally, the climatic protection to paths brings them closer to the buildings, makes them a real extension of the living space. The transition from residential to central areas, both with their mixed uses and special character, must be developed to the full, giving excitement, attraction, civic design.

It takes a great stretch of the imagination for those steeped in traditional and still usual attitudes of town-planning to picture the kind of planning, and the kind of result, obtained when the whole environment is designed three-dimensionally, or space in time—four-dimensionally.

VI. 2. Motor Traffic in New Towns, General

The motorway and rail transport must likewise become part of the environmental concept. They are a threat to the environment because of their scale, if not properly integrated. Functionally too, it is of the greatest importance that termini, car-parking and routes are strategically positioned. Rail transport must not be neglected merely because of the present trend towards private vehicles.

The provision for private cars must be worked out on the basis of projected ownership figures and, far more important in many ways, projected car usage, which will differ as the planner determines it. Thus, ownership may be largely beyond the planner's influence, but the use of those cars within the town need not follow trends, but is subject to planned capacities, subject to town design. The attraction of paths and public transport may go hand in hand with much smaller private car provision in the town centre than that calculated for Hook or Cumbernauld. This possibility is general.

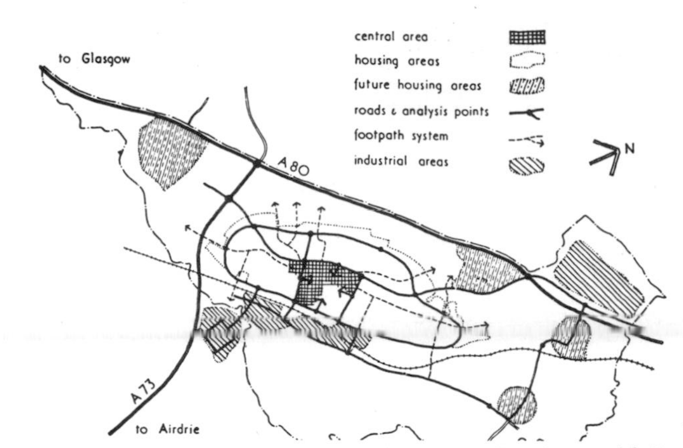

Fig. 6.1. The original proposed road pattern, Cumbernauld.(i)

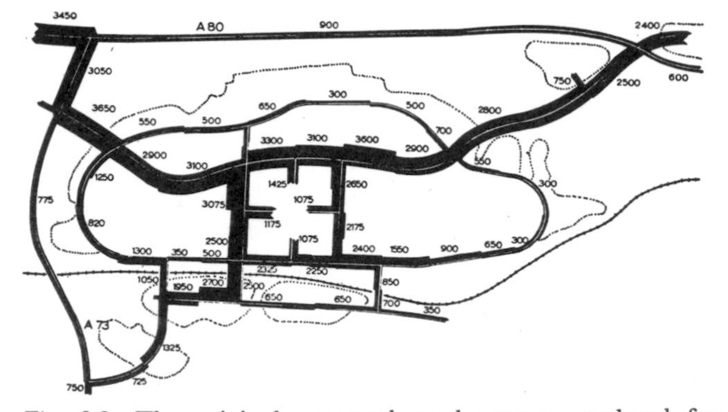

Fig. 6.2. The original proposed road pattern analysed for peak hour traffic volumes, numbers indicate total flow in mixed vehicles per hour, dividing strip indicates proportion in each direction, trunk flows not included. Cumbernauld.(i)

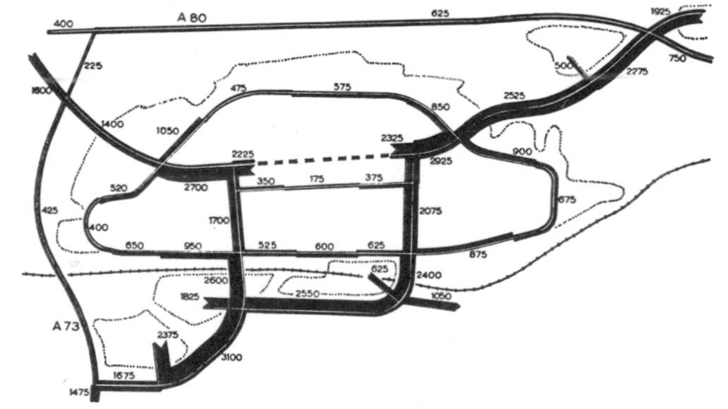

Fig. 6.3. Peak hour flow of revised road pattern. (See also p. 117 for road plan.) Cumbernauld.(i)

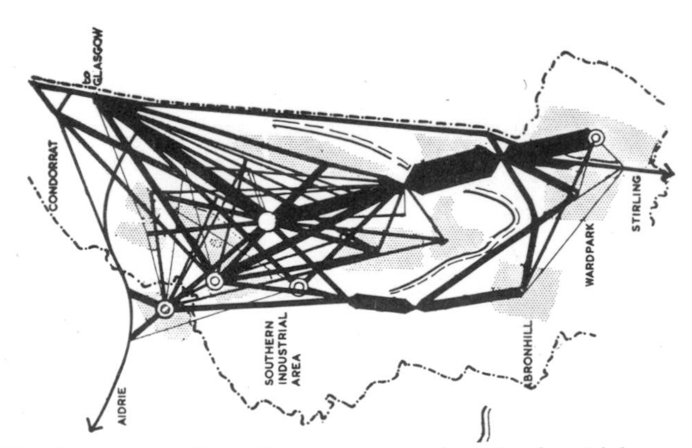

Fig. 6.4. Desire line diagram shown in mixed vehicles per hour. Cumbernauld(i).

(i) Logie G. Cumbernauld Road Lay-out, *A. & B.N.*, 29 March 1961. Also *A. D.*, London, May 1963.

VI. 2.1. *Parking in New Town Centres*

The number of parking spaces required in a new centre is very difficult to calculate and forecast.(i) At Vällingby the need was grossly underestimated and much of the pedestrian areas have had to be passed over to car parking. The Mark I British New Towns have shown no reliable or systematic method. In an answer to a questionnaire sent out by Cumbernauld Development Corporation, in their study of the problem, the following answers were obtained:(ii)

In Corby, 400 places (population 40–55,000), less than one per hundred population, was based on experience in neighbouring towns.

Hatfield, with 550 places (one per 50 persons), merely said "The town planner was well informed in these matters."

Peterlee, with 510 (one per 60), used the area available after the building lay-out had been catered for.

East Kilbride, with 1000 (one per 53), intended to provide according to the availability of space.

Harlow stated 1200 (one per 66), based on 20 cars per 100 ft (30·48 m) of shopping frontage but doubted whether this would be achieved.

Basildon is to provide 2400 (one per 33) based on 39 cars per 100 ft (30·48 m) shop frontage or 28 per combined 100 ft (30·48 m) office and shop frontage.

Hemel Hempstead planned 2000 (one per 30), one car per 400 ft^2 (37·16 m^2) of shop area, one car per 1000 ft^2 offices, one car per flat, one car per hotel bedroom, one car per 30 cinema seats.

Welwyn gave 1750 (one per 29) based on a census of parked cars.

Bracknell assumed one car per two families in ten years and provided spaces for 25% of the total cars in the town.

Crawley makes provision for 2700 cars (one per 18) using three separate bases:

a) 1 car space per 100 ft (30·48 m) shop frontage, plus one space for every 100 workers in the centre.
b) 1 car space per 800 ft^2 (74·32 m^2) shop and office area.
c) 1 space to every four of the population.

The Area Ratio Method, developed in the U.S.A., depends on the correct prior judgement of the applicable ratio between gross floor area of retail space and gross floor area of parking space. This may vary between 1:1 and 1:4 (the latter being the parking). For Cumbernauld this meant 8–32 acres (population 70,000).

(i) BURRAGE and MOGREN, *Parking*. Eno Foundation, Saugatuck 1957.

(ii) Cumbernauld Development Corporation, Report on Central Area, April 1960 (compiled by G. Copcutt.)

By the Unit Sales Method (dividing annual gross sales by product of: average unit sales × customers per car × minimum car turnover per space per day × no. of selling days p.a. × fraction of customers arriving by car) gives 1480 car spaces for Cumbernauld.

Cumbernauld's own calculations are based on two predictions of car ownership by 1980, for 10,000,000 cars for Britain and for 20,000,000, and allow for many factors including alternate day and evening use of parking space, greater use of footpaths, attraction of good provisions for parking, relative distances of inhabitants from centre, uses of cars in family, and Saturday and Friday peaks. Results from the long multiply-modified calculation, gives the need for 2050 parking spaces for 10 million and 3360 for 20 million (16 acres or 24 acres, 7·47 ha or 9·71 ha). These calculations are for shoppers alone and 750 car spaces for employees (or 1500, given the 20 million estimate) need to be added for 4720 employees.

Thus the minimum short term estimate for Cumbernauld is 2750 and the maximum 4500 (one per 25, or one per 15 inhabitants).

Cumbernauld and the scheme for Hook cater for all but about 20% of the population going shopping to the centre by car.

Hundreds of cars belonging to people not more than ¼ or ½ mile (400–800 m) distant from the centre are allowed for. This may not be reasonable where public transport is also available, directly above and below the centre and very near one's home, reached through sheltered lanes.

The great expense of 3000 or 8000 parking spaces for a town of 73,000 (Cumbernauld), or 100,000 as proposed for Hook, respectively, 56 acres (23 ha) for latter, may not justify repetition in similar circumstances.

The old version of new towns in Britain, and the low density new towns in Sweden, and existing towns, have taught the planners a lesson: that without adequate parking a centre does not thrive. But this may not apply to the new kind of New Town centre where the "bulk of the population" live within easy and convenient walking distance and local public transport is particularly convenient. It is surely quite plain that parking in towns which are not dense, and which have not been built for walking, is a far more urgent and widespread need than in towns where density and convenience have been borne in mind as attracting people to walk to the centre. There is the danger that planners wish to make doubly sure they are successful, which is an expensive luxury.

A multilevel centre tucks its several thousand

cars underneath out of sight and they do not form the barrier they represent at Farsta or Stevenage. Nevertheless there remain the huge junctions necessary to cope with the flow, allowed to continue that mere extra half mile right under the centre, one car to every twelve inhabitants. Which means, incidentally, that filled, these cars would hold 60,000 of the population of 100,000 in Hook. Taking the space between the cars, the whole 56 acres can be envisaged as accommodating the population of the whole town and more besides.

Perhaps the only justification for the provision of such parking space is an extraneous but perhaps realistic one; the pleasantness of the new pedestrian centres and the parking facilities may bring people from a far greater area than the normal commercial catchment areas calculated (largely within walking distance) and the centres may act as does U.S. out-of-town shopping.

High level or low level roads can take great car populations without allowing them to dominate and cut up the townscape with broad, barren, horizontal bands. Character should stem from the buildings of the town. We may become used to the idea of slicing cities apart by wide roads and freeways, but this need not be so, particularly in new towns. The solution is largely to be found in the unification of the now separated functions of road engineer and planner-architect. There is a truly difficult administrative problem where roads are built by state or other funds whereas the developer is private or the local authority. This must be resolved. It is indeed one of the specific functions of the planner to make obvious the absurdity of administrative split-mindedness.

VI. 2.2. *Cumbernauld Research into C.A./Retailing and Parking*(i)

To estimate the area required for retail and service trades. Known factors:

(1) The location of the town.

(2) Population of 70,000 (comprising 50,000 on hilltop site and 20,000 in satellite villages).

(3) Overall design of the town.

To solve this problem it is necessary to find some relationship between the area of shops to be provided and the size of the town's population.

1. A sample of 100 towns (populations 25,000–100,000) was chosen, each town sited within a 20 miles radius of one of the nine largest cities in Great Britain (excluding London). The object of this choice was to reflect the location of Cumbernauld close to a large city and the consequent effect on shopping within the new town due to this proximity.

2. Scatter diagrams were plotted and coefficients of correlation calculated for each of the Census trade categories relating sales to population size.

3. The average sales per head in each trade category for the 100 towns was calculated from the Census for the year 1950. This was adjusted to 1959 values and sales per head were multiplied by 70,000 to represent sales at that size.

4. Conversion factors to translate sales into area required were selected by empirical enquiry.

5. This estimate was adjusted to the circumstances of Cumbernauld.

" 'Sales conversion factors' are the equivalent of the American 'business capacity per unit area'. They have no absolute value that this space could bear. In practice this pursuit of the ultimate use by increases in efficiency is asymptetic in nature—there is no foreseeable limit to the advance distributive machinery might make. Even with the present restricted opening hours, however, the amount of sales made in a given space if plotted at different times of the day or week would produce peaks and troughs. The difference between the average slack day and the highest seasonal peak is the amount by which a current so-called 'efficient' conversion factor could be increased *without making any change at all* beyond a more even loading by the public."

(i) Copcut G., *A.D.* London May 1963.

TABLE VI.1. *Town population are varied according to densities.*

PARKING/ SATURDAY	→a.m. 8–9	9–10	10–11		→p.m. 11–12	12–1	1–2	2–3	3–4	4–5	5–6	6–7	7–8	8–9	9–10	→p.m. 10–11	
(1) Retail shop employees	750	750	750	—	750	750	750	750	750	750	750	—	—	—	—	—	
(2) Offices (25% weekday)	75	75	75	75	75		—	—	—	—	—	—	—	—	—	—	
(3) Residents (reserved)	300	300	300	300	300	300	300	300	300	300	300	300	300	300	300		
(4) Shoppers (maximum)	—	—	—	868	—	—	—	—	—	1740	—	—	—	—	—	—	
(5) Entertainments	—	—	—	—	—	—	—	—	—	—	—	500	—	—	—	—	
(6) Others	—	—	—	—	—	—	—	—	400	—	—	—	—	—	—	—	
TOTAL (max. possible)	1125	2433	2433	—	2433	2433	3230	3730	—	3730	3730	3730	1240	1240	1240	1240	1240

VI. 3. Fully Motorized Town Centre for 250,000

Full motorization, the town where every one drives his or her private car into town, is a concept elaborated in Professor Bendtsen's book, *Town and Traffic* (1961). Unlike the calculations of Dr. Smeed(i) (see Tables p. 65, 66) this similar exercise, starts not only from the use of cars for work but also for shopping.

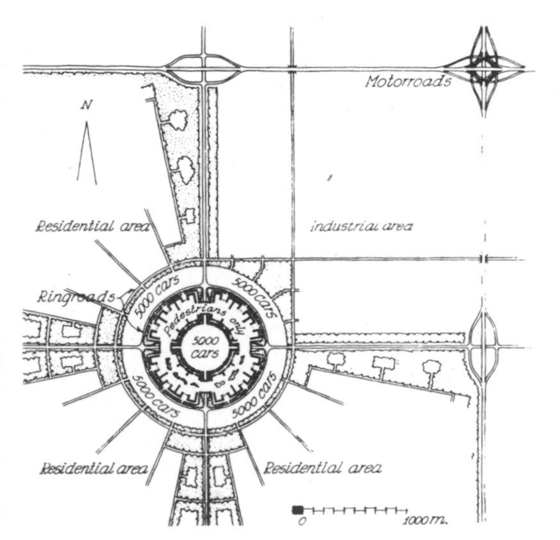

Fig. 6.5. Centre of an all motorized city with a population of 250,000.

Professor Bendtsen's assumptions are:

a) the site of the city is in an angle of two motorways
b) outlet capacity of these is 4 × 3000 = 12,000 vehicles/hr,
c) outlet capacity of arterials to residential areas etc., is 12 × 2000 = 24,000 v/hr,
d) half the traffic is taken as through traffic,
e) so maximum for central area is 12,000 v/hr. Thus, accordingly, by rule of thumb,
f) total parking accommodation is taken as 25,000,
g) "fully motorized" is taken to be one car for 2 persons,
h) 20% of cars in town should be able to park at once.
i) From this the size of the town population is calculated:

$$25{,}000 \times \frac{100}{20} \times \frac{2}{1} = 250{,}000.$$

i.e. the size of the fully motorized town is based on parking capacity at centre which is based on capacity of arterials leading from centre to residential areas.

Bendtsen suggests further that in addition to the 25,000 parking places another 3000 may have to be added if the shopping centre is to give the ease of parking which the American suburban centres do.

It is of interest that given architect Bonneson's plan (Bonnesen worked with Professor Bendtsen) to suit the needs of a fully motorized city of 250,000 on one level, the city centre sits isolated from any residential area by a 1000 ft (300 m) ring of parked cars. Its hollow core is filled with 5000 cars, a circle 1500 ft (450 m) in diameter.

Central area buildings are on a 1½ mile (2·5 km) pedestrian only ring. The area of this is 175 acres (70 ha) and the parking area would be 39 acres (17·7 ha).

Bendtsen himself suggests that such a city is not likely to be built. But he is of the opinion that this is because it is too "utopian"(!)

In fact this theoretical exercise shows clearly that old, two-dimensional concepts will not do to plan an adequate city in the future.

(i) SMEED J. R. *The Traffic Problem in Towns*. Manchester 1961.

VI. 4. Town Design Based on Walking Distances

VI. 4.1. *Density and Town Size*

VI. 4.1.1. *Town size, population, density study.* The relationship implications of these have been studied by M. Skrzypczak-Spak(i) in his thesis, published in Hanover in 1961, and the following figures and points are based on his work.

He assumed:

a) 30 minutes *maximum* convenient walking time.
b) 15 minutes to primary school and local shops.
c) walking speeds of:
 3 mph (5 km/hr) for workers,
 2½ mph (4 km/hr) for adults,
 1½ mph (2·5 km/hr) for the old, mothers with prams and children.
d) a concentration, centrally, of amenities, housing surrounding this, and all space consuming institutions on the periphery of this area (as in Cumbernauld or Hook).
e) a town centre provision of 21·5 ft² (2 m²)/person i.e. 5 acres (2 ha) to 10,000 inhabitants approximately.
f) centre car parking at the rate of one space to forty inhabitants (town designed for walking).(ii)
g) industry 54 ft² (5 m²)/person or 12·5 acres (2 ha) to 10,000 inhabitants.
h) public amenities 108 ft² (10 m²)/person or 25 acres (10 ha) to 10,000 inhabitants.
i) footpaths and greens 140 ft² (13 m²)/person or 32·5 acres (13 ha) to 10,000 inhabitants.
j) vehicular traffic 108 ft² (10 m²)/person or 25 acres (10 ha) to 10,000 inhabitants.
k) a circular town pattern.

From these assumptions (variable of course by local conditions) Skrzypczak-Spak produces a table demonstrating how varying densities affect the population number given a constant area of circular shape of 5000 acres (2000 ha), as that which a child can cover at a speed of 1½ mph (2·5 km/hr) and maximum time of 30 min.

From the table, the optimum size of 250,000 inhabitants shows a balance between residential and other uses in area.

If we now replace the assumption, made for convenience only, that the centre and city is circular, with an assumption that the town and its centre can be elongated, the number of people living within 1½ miles (2·5 km), 30 min walking distance,

(i) SKRZYPCZAK-SPAK M. *Der Fussgängerverkehr in den Städten und seine Erschliessungsmöglichkeiten.* (Town Structure, related to Pedestrian Traffic.) Thesis published, Technical High School, Hanover 1961.

(ii) This is of course very low compared with Hook, or Bendtsen's Town. Hook has 1:12 inhabitants, Bendtsen 1 :9 inhabitants.

TABLE VI.2. *Town sizes and population are varied according to densities.*

(A) Densities (metric)	50	100	150	200	250	300	350	400	P/ha
a. Total non-residential areas excluding agricultural	340	560	740	880	1000	1080	1150	1220	ha
b. Residential areas	1680	1440	1260	1120	1000	920	850	780	ha
a.+b. Total area	2020	2000	2000	2000	2000	2000	2000	2000	ha
Inhabitants in thousands	84	144	180	225	250	275	295	310	persons
Agricultural and open land at periphery added	2700	4400	5800	68,000	7500	8300	8800	9400	
(B) Densities (British Units)	20	40	60	80	100	120	140	160	P/acre
a. Total non-residential areas excluding agriculture	850	1400	1850	2200	2500	2700	2875	3050	acres
b. Residential areas.	4150	3600	3150	2800	2500	2300	2125	1950	acres
a.+b. Total area	5000	5000	5000	5000	5000	5000	5000	5000	acres
Inhabitants in thousands	84	144	190	225	250	275	295	310	persons
Agricultural and open land at periphery added	6750	11,000	14,500	17,000	18,750	20,750	21,750	23,500	acres

increases. Skrzypczak's 250,000 city grows to approximately 400,000 if increased in the manner illustrated in the Hook report. (See Section on town shape.)

In using the figures given it must be borne in mind that the assumptions set out are only one set. Many variations are possible, notably the planners of Cumbernauld and Hook set out with maximum walking distances of 15 min. Skrzypczak-Spak bases his figures of 30 min on the study made by Baniewicz[(i)] on Public Transport systems, which showed that the assumed reasonable travelling time for a commuter is 30 min, and that he spends only 10 min of this in the vehicle. The rest is the journey to and from the stop, and waiting.

Distance or time taken to be "convenient", quantity, can be considerably affected by the quality, the pleasant, sheltered, aided, or unpleasant, exposed nature of any given route.

VI. 4.1.2. *Town density diagrams.* Diagrams showing effect on city size by decreasing density can be very misleading.[(i)]

Increase in residential density seems to make a considerable difference to the area seen as rectangles. Seen as circles the increase is much less obvious. Walking distances on the circular town shape would show a small increase only. Thus to change the density of the residential area of a town of 250,000 from 40 ppa (100 ppha), to 100 ppa (250 ppha), would lengthen the maximum walking distance from periphery of residental area by only 500 yds (457 m), 19% of the radius of a circle as compared with a 42% increase in the area of the town. But, and this is the rub, the extra 500 yds would have to apply to an ever greater proportion of the total population, as at least 42% of the people would live in the added ring (given an even density).

Fig. 6.6. A main radial path at Cumbernauld leading under the ring road by flatted factories.

(i) From a regional study in Upper Silesia. *Industrial Areas*, by T. BANIEWICZ.

(i) Diagrams after *The Planning of a New Town*. L.C.C. 1962.

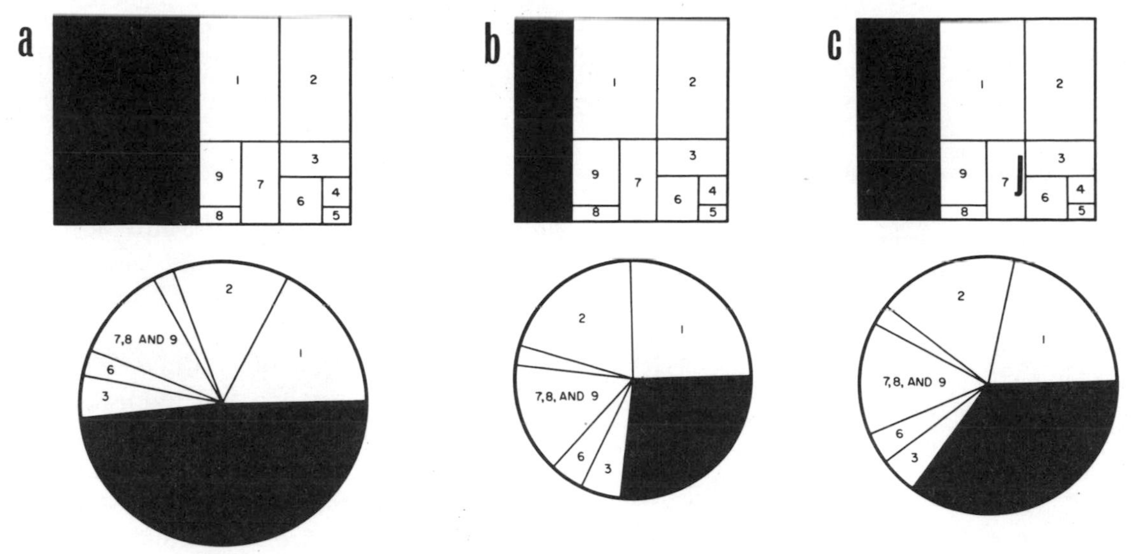

Fig. 6.7. Housing black. Other land uses numbered as below.
1. Industry 900 acres, (364 ha).
2. Open space 730 acres, (215 ha).
3. Main roads 200 acres, (81 ha).
4. Public Utility Services, 50 acres, (20 ha).
5. Hospital 25 acres, (10 ha).
6. Central area 160 acres, (65 ha).
7. Secondary Schools
8. Further Education 535 acres, (216 ha).
9. Primary 2600 acres (1062 ha).

a) Total area of town 3600 acres, (1457 ha). Residential area at 100 ppa net, 1000 acres (250 ppha) (405 ha)
b) Total area of town 4029 acres, (1619 ha) Residential area at 70 ppa net, 1429 acres (175 ppha) (580 ha)
c) Total area of town 5100 acres, (2064 ha) Residential area to 40 ppa net, 2500 acres (100 ppha) (1012 ha)

VI.4.2. *Comparison of towns and walking distances.*

TABLE VI.2. *Comparison of towns and walking distances.*

		Inhabitants	Area	Av. net density	Res. area	Walking distance
Harlow		80,000	6300 acres (2550 ha)	50 p/a (125 p/ha)	1600 acres (646 ha)	neighbourhood centre only within $\frac{3}{4}$ mile, ($1\frac{1}{4}$ km).
Cumbernauld		30,000	4150 acres (1699 ha)	85 p/a (212 p/ha)	600 acres (242 ha)	$\frac{1}{2}$ mile, (804 m), for most people to main centre.
Hook		100,000	7526 acres (3035 ha)	70 p/a (175 p/ha)	850 acres (344 ha)	$\frac{1}{2}$ mile, (804 m), for most people to main centre.
Farsta		35,000	—	—	—	$\frac{1}{2}$ mile, (804 m), for most people to main centre.
Vällingby		23,000	—	—	—	$\frac{1}{2}$ mile, (804 m), for most people to main centre.
Radburn		25,000	1280 acres (526 ha)	—	—	3 neighbourhood $\frac{1}{2}$ mile, (804 m) radians, $\frac{3}{4}$ mile, ($1\frac{1}{4}$ km) or 15 min to central points.
Cologne		100,000	16,500 acres (6600 ha)	—	—	20 min to work, 12 min maximum to station.
Skrzypczak-Spak's proposals	A	250,000	18,750 acres (7500 ha)	100 p/a (250 p/ha)	2471 acres (1000 ha)	$\frac{3}{4}$ mile, ($1\frac{1}{4}$ km), to centre.
	B	84,000	6750 acres (2700 ha)	20 p/a (50 p/ha)	4200 acres (1680 ha)	
Tolouse le Mirail (little industry)		100,000	1800 acres (720 ha)	very high	—	$\frac{5}{8}$ mile, (1 km), for most people to main centre.
Reston		75,000	6,800 acres (2,750 ha) 4,100 acres (1,660 ha)		60, 14, and 3 ppa (11ppa overall) 150, 35 and 7·5 ppha (27·5 ppha overall.)	Village centre and local recreation and some schools.

VI. 4.3. *Town Shape*(i)

Diagrams illustrating New Town concepts often take a circle as a model, for ease of explanation. The other school takes a long line. Many more people can be arranged within a given distance of a line than a point. This principle has an obvious relevance to towns designed for walking, where elongated centres will be within easier reach to more people. But, once within the centre, distances are of course theoretically greater and the centre less convenient, less compact. But public transport can be closer to more people along a given route also in an elongated area and to more with an elongated centre.

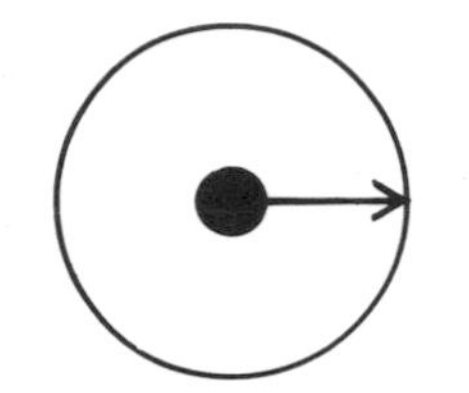

Fig. 6.8. Diagram of town shape and centre, circular.

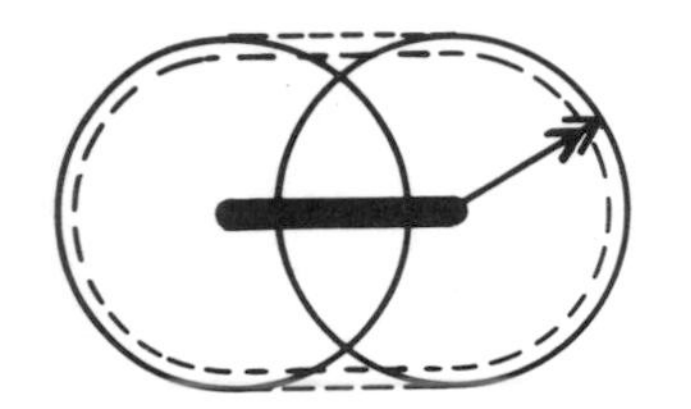

Fig. 6.9. The diagram used in the "Hook Report" mistakenly takes the greater area shown. As the central area is likely to be narrow if elongated, the smaller area is more diagramatically correct.

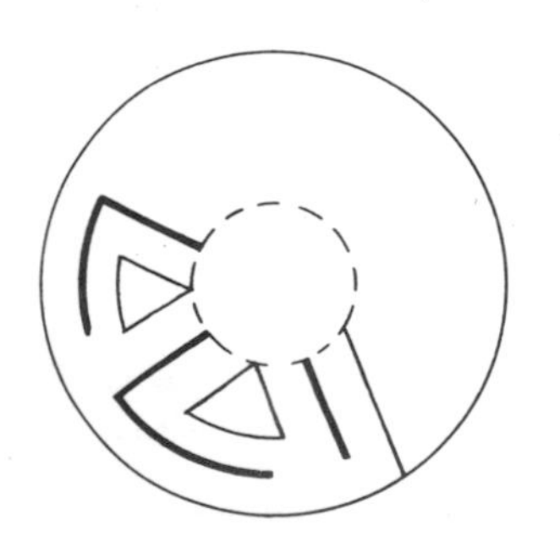

Fig. 6.10. Circular shape leaves areas not served by public transport.

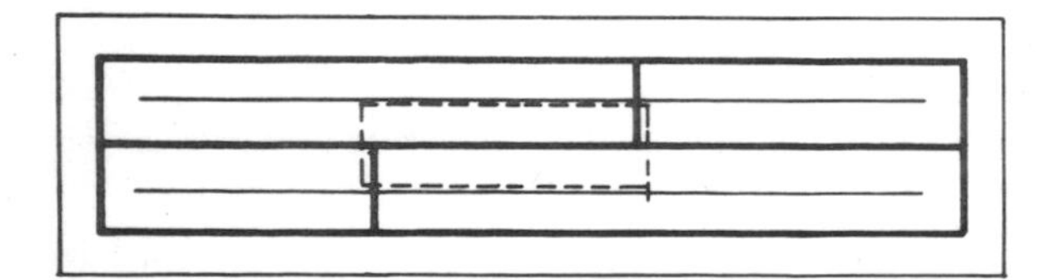

Fig. 6.11 Elongated or linear town shape is more evenly served.

(i) See also correspondence by KEEBLE and SCHOON in *A.J.* May 1963.

However, whether circular or elongated, it is a fact that any existing town centre represents a number of arms which link in the centre itself. Thus, a starfish shape, round or elongated, is almost invariably found.

In New Towns, where pedestrian routes should draw the pedestrian into the centre, it is desirable to create just this shape and not the centre isolated from the residential area, even if it includes dwellings.

With the growing importance of walking distances it has become fashionable to draw circles, at quarter mile (400 m) distances of radius apart and to stipulate that people within the areas have only that distance to walk to their goals. These diagrams assume that people can proceed as the crow flies, along their ideal desire lines. This can very rarely be the case and it underlines the need for a realistic and detailed, rather than a diagrammatic, appraisal of walking distances and desire lines.

VI. 5. Land Use

Mixed land use, reducing distances between work and, home is now recognized as a generally desirable idea. A minority has advanced it for a long time.(i) It comes very much into its own with a path system producing attractive walking and cycle connections. This idea, therefore, can be worked in with specifically decisive benefits in traffic-segregated new towns. Minor works by dwellings have been incorporated at Cumbernauld. (See Fig. 6.6, page 112)

VI. 6. Open Space

Unlike Swedish practice, in Britain huge areas for playing fields are statutorily reserved about each secondary school. At Cumbernauld it was first realized that, because of this, densities are affected decisively and so walking distances. Thus, taking the schools or their playing fields to the periphery of neighbourhoods, or towns, allows shorter distances within. It helps to bring the countryside nearer to the centre. Contrast with urban areas is bold.

Ideally the school playing fields should have a wider use than merely during school hours and school days. This artificial use barrier is due entirely to the departmentalized running of the services of our towns; the education and the parks department have separate staff, accounts, regulations, and policies.

(i) KLEIN A. Solving the Traffic Problem. Saugatuck, July 1949.

VI. 7. The Neighbourhood Concept

The sociological idea behind the neighbourhood can be successful. It depends on the planner. To date, planning has ignored vital functions so that any idea, good or bad, has not had a reasonable test. It must be borne in mind that there was a strong, supercilious resistance, until very recently, to the Radburn Idea, merely because some early examples of it had not pleased sophisticated, shallow minds who threw out the baby with the bath-water in condemning the basic idea.

Just as the lack of parking spaces in the first new towns in Britain has led to what may well be an overestimate of desirable car use in the present series of quite different new towns, so the failure of the neighbourhood principle, in past examples, is leading to a wholesale attack and abandonment of the idea, on principle. This is premature and unwise. The reason for the failure of the neighbourhoods so far, is likely to be the whole unimaginative process of creation, the lack of density within the neighbourhood, and the lack of footpath systems which throb with life and attract people, particularly old people and women with young children all day long. Hitherto, "the neighbourhood" has meant a tangle of roads with houses strung along them. Somewhere in the middle, only recognizable on the coloured plan, were the shops and the school, and possibly a hall.

A new "walking" town may be planned successfully with neighbourhoods, or without; it may be large or small. But if densities are not great and the town itself is large, it is necessary to have neighbourhood centres with specially convenient connections to public transport. As densities go up so does the practicable size of the town with one main multi-level centre rather than the subdivision into neighbourhoods.

Neighbourhoods, first pinpointed as a planning concept by Perry,[i] were seen in terms of an area by-passed by heavy traffic with a circulation of its own within.

Willmot[ii] considers that social ties occur in smaller units than any of these neighbourhood units and Doxiadis believes that "community blocks" will be the primary elements of a city. These are superblocks in fact, and again the size is variable.

Willmot has no doubt observed social ties, as has the author, which relate to small groups. But if we bear in mind that our larger units have not been planned in such a way as to encourage a functional distinction and so a feeling of neighbourhood, then it is easy to see why it cannot be observed.[i] The danger is to draw general conclusions from a special series of existing cases. Once it has been analysed why we would not expect neighbourhood identity in certain towns there is every reason why a general condemnation of the idea should not be deduced from the very partial evidence we are able to collect.

(i) Perry C. *The Neighbourhood Unit.* New York 1929.

(ii) Willmot P. Hook Plan. Did the Team Think? *A. J.* 7 February 1962.

Public relations have been notoriously bad, too often the attitude of "we" and "they" destroy identity with the town, although adversity, an enemy, does bring people together. If the growth of neighbourhood, beyond its use, is associated with the people in an imaginative way, then it will have identity and meaning.

A feeling of belonging, whether to a small group or a large group arises either out of the feeling that this group is making or suffering together, or from functional contact that is real and takes place daily. The former can occur in any size of group. The latter will involve more people if the density is high i.e. 50,000 in Cumbernauld, because walking distance is a crucial aspect of it. Unless there are either of the above two conditions it makes little sense to "plan" neighbourhoods on paper.

VI. 8. Principles of Growth

Failure of New Towns may be due in part to the idealistic concept of a final form to be attained. This has meant, in many instances, meaningless collections of houses which continue for years and years with much social misery.

From this and economic points of view it is desirable to have a principle of growth for every new town which, as in the organic field, carries with it a meaningful pattern from the beginning, throughout its growth and development, right for each particular case in point.

This line of thought can be seen in the work of Doxiadis (Ekistics), the plan for Hook and the plan for Le Mirail. We are concerned here with patterns and direction that give social and visual meaning to small groups of houses, patterns and direction which lead with increased meaning to larger organic wholes in which the previous groups of houses form a social function. This is more than a matter of the first few inhabitants having shops. It involves the growth of footpath systems so that their use is spontaneous from the beginning. And direction means that social attractions grow by the housing as this is built along the pedestrian environment. This may mean a larger initial outlay in terms of road construction.

(i) A notable exception is Park Hill, Sheffield (see section VIII).

Fig. 6.12. Stages at which Hook was to be built, central area increases with each stage. (See also section VI.9.2. Hook for completed town.)

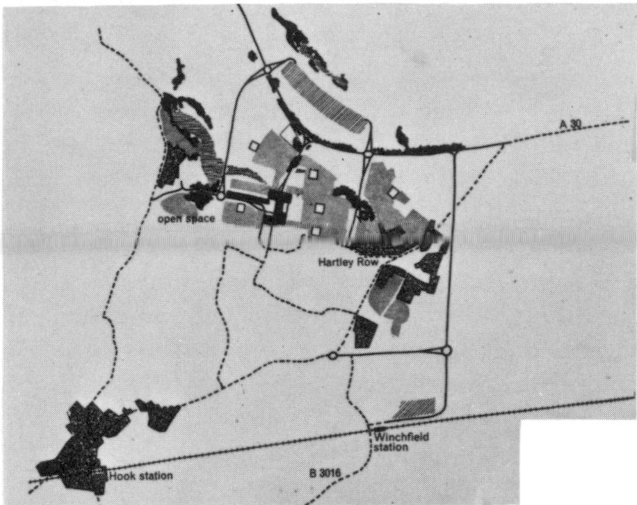

Fig. 6.12a. pop. 16,000

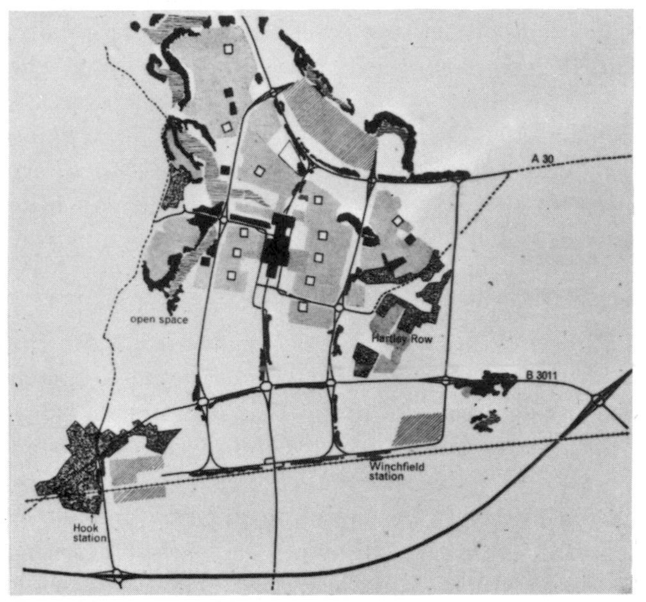

Fig. 6.12b. pop. 43,000

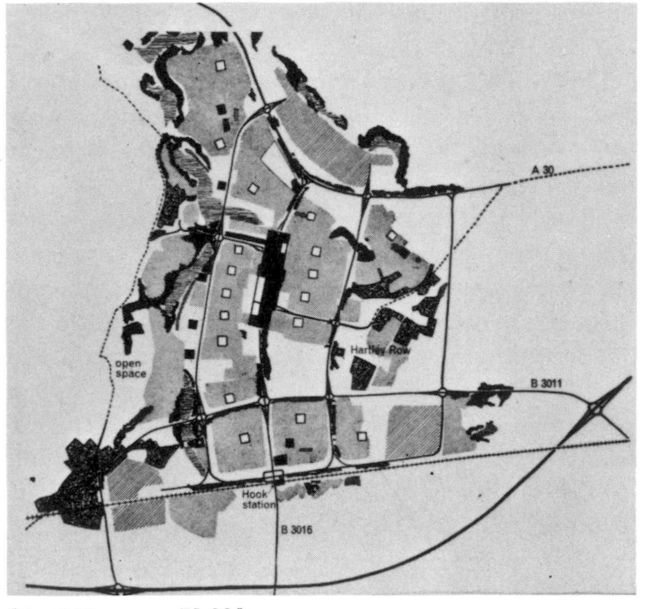

Fig. 6.12c. pop. 70,000

residential areas	329 acres
special residential areas	0·33 acres
central residential areas	16 acres
inner residential areas	181 acres
outer residential areas	131 acres
neighbourhood centres	— acres
central area	13·2 acres
retail	2·6 acres
public buildings, offices, etc.	2·8 acres
secondary schools (1)	4 acres
residential	— acres
remainder	3·8 acres

industrial areas	116 acres
Northern industrial area	83 acres
South Western industrial area	— acres
South Eastern industrial area	33 acres
Primary secondary school sites	4 acres
secondary school playing fields	20 acres
public playing fields	98 acres
hospital	9 acres
lakes	
existing development	395 acres

Year 5: Population 16,300

residential areas	919 acres
special residential areas	1·4 acres
central residential areas	50·5 acres
inner residential areas	414 acres
outer residential areas	453 acres
neighbourhood centres	8·5 acres
central area	43 acres
retail	9·5 acres
public buildings, offices, etc	6·4 acres
secondary schools (2) and technical college	14 acres
residential	0·7 acres
remainder	12·4 acres

industrial areas	289 acres
Northern industrial area	161 acres
South Western industrial area	68 acres
South Eastern industrial area	60 acres
Primary secondary school sites	16 acres
secondary school playing fields	60 acres
public playing fields	131 acres
hospital	15 acres
heliport	5 acres
lakes	
existing development	395 acres

Year 10: Population 43,700

residential areas	1520·5 acres
special residential areas	5 acres
central residential areas	92·5 acres
inner residential areas	537 acres
outer residential areas	886 acres
neighbourhood centres	15 acres
central area	69·2 acres
retail	14·8 acres
public buildings, offices, etc.	14·6 acres
secondary schools (3) and technical college	18 acres
residential	2 acres
remainder	19·8 acres

industrial areas	387 acres
Northern industrial area	195 acres
South Western industrial area	118 acres
South Eastern industrial area	74 acres
Primary secondary school sites	28 acres
secondary school playing fields	100 acres
public playing fields	422 acres
hospital	30 acres
heliport	5 acres
lakes	
existing development	395 acres

Year 15: Population 70,000

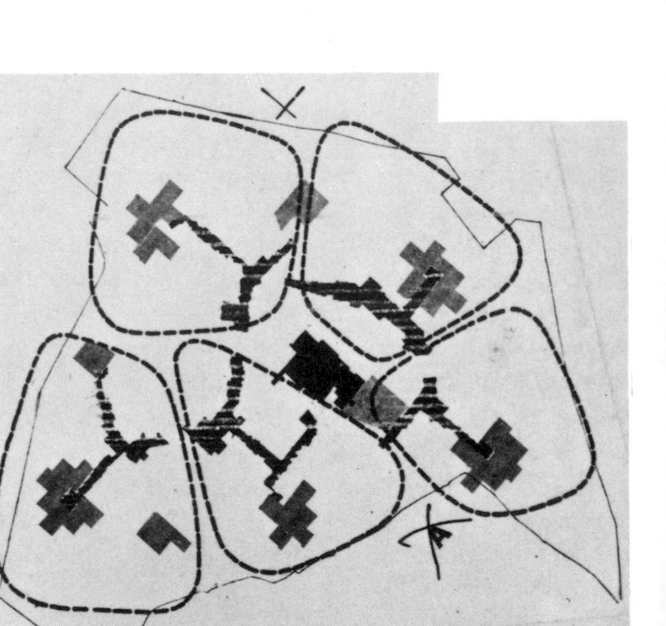

Fig. 6.13. Stages at which Toulouse le Mirail is to be built. Common facilities extend at ground level as housing proceeds.

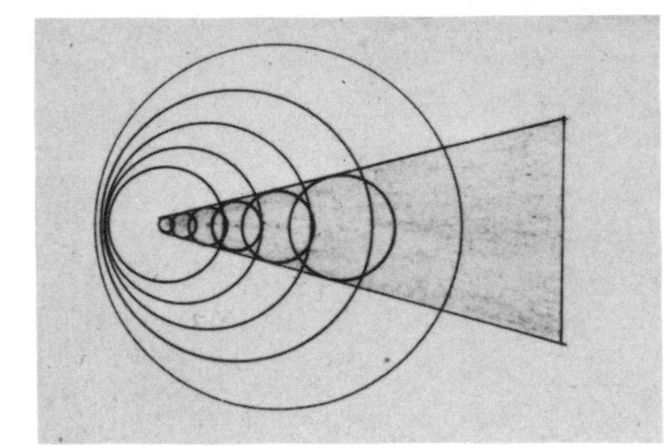

Fig. 6.14. Doxiadis principle of growth of centre of town from his study of Ekistics.

VI. 9. Examples

VI. 9.1. *Cumbernauld, nr. Glasgow, Scotland*

Site: A hill. 260 ft–480 ft (79 m–146 m) above sea level, exposed, rainfall annual 45 in. (1·1 m), 56° latitude.

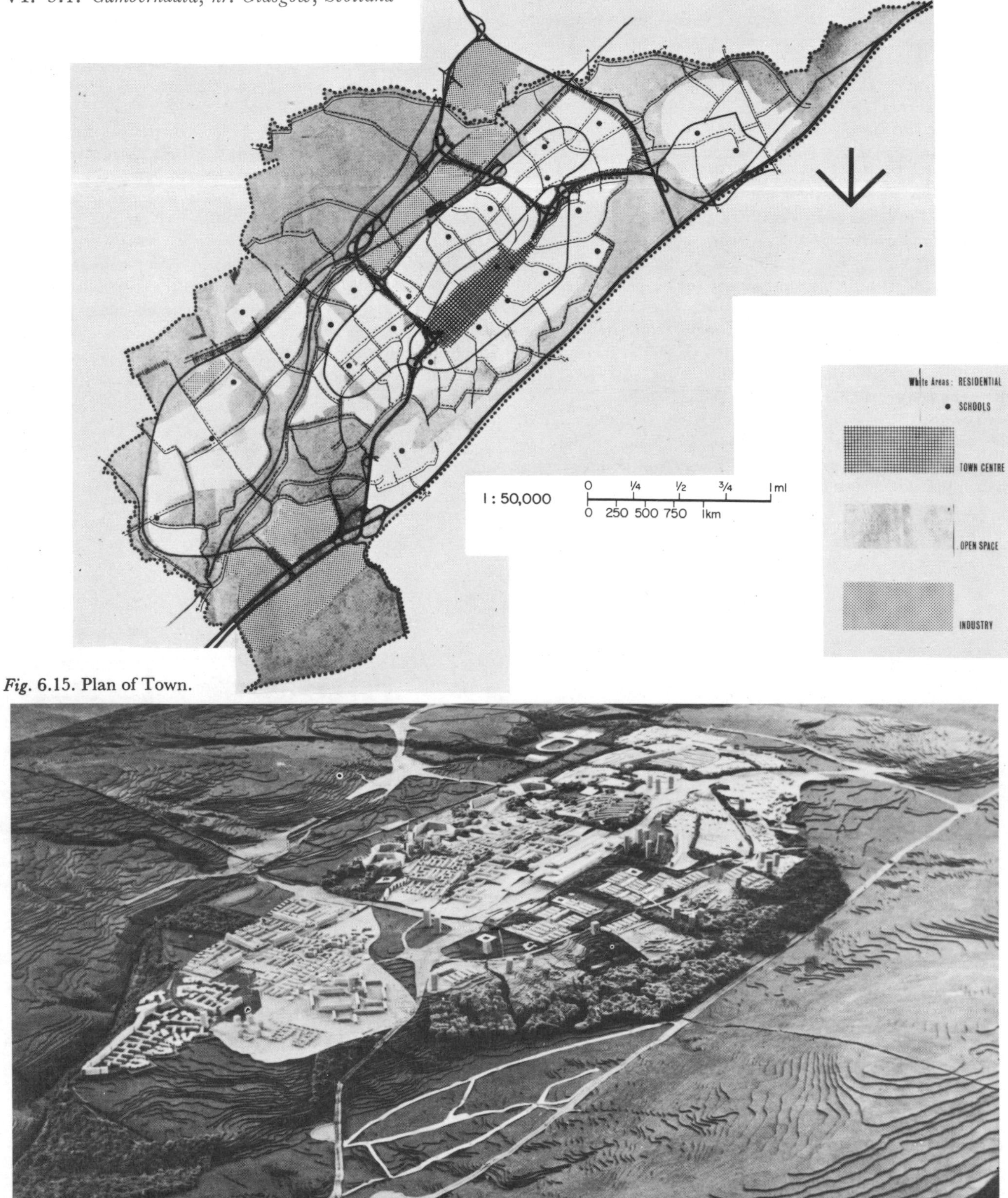

Fig. 6.15. Plan of Town.

Fig. 6.16. Aerial view of model of *esquisse* scheme for hilltop (1960) (50,000 inhabitants). Deckhousing (middle-back) replaced by alternate scheme.

KEY

Red	—Pedestrian Areas.
Black	—Motor Roads.
Dark Grey	—Car Storage.
Light Grey	—Dwellings.

Unless otherwise stated this key applies to each plan.

Plans have not been redrawn to retain character and variety. The red colour, denoting pedestrian areas, is therefore shown in a number of ways; either solid red, or edged with red, or main red paths leading through squares or greens, or public buildings and shops about a pedestrian area are solid red or edged in red.

Design: Chief Architect and Planner to the Development Corporation, Hugh Wilson, successor and late deputy architect D. R. Leaker, with Hugh Wilson as consultant since 1962, and new deputy architect, Derek Lyddon. Interprofessional Design Teams: A. G. McCulloch; J. A. Denton; A. K. Gibbs; G. P. Youngman; W. C. Thomson; J. R. B. Palmer; A. Moscardini; A. H. Bannerman; E. Browning; G. Copcutt; J. R. Hunter; G. Callaghan; W. C. Taylor; L. W. Buckthorp; D. H. Garside; A. S. Scott; B. J. Allan; W. Gillespie.

Origin: Designated a New Town under Act of Parliament in 1955, to relieve pressure on Glasgow. Completion by 1980.

Population: 50,000 on the main hill and 72,000 altogether.

Kind of Town: As other British New Towns, "balanced" to provide employment for the citizens, shopping and cultural activities as well as homes. The first of the British Mark II New Towns, with far greater density and a path system for pedestrians separate from the roads. The neighbourhood principle is not used. There is one main centre only for the 50,000 main hill inhabitants. But the 22,000 in their village extensions on the periphery have their own local social and shopping centre, similar to neighbourhoods.

Residential Accommodation: A great variety of maisonettes in the centre. Two-storey and three-storey housing; flats at varying heights. The proportion of houses will be lower than that in the earlier new towns.

Car storage for each family, partly with the dwelling, partly in separate groups, partly vertically separated—with pedestrian access above.

Densities vary from about 100 persons per acre (250 ppha), to about 60 (150 ppha), rising towards the centre, with some low density houses in scattered areas.

Centre: (Architect in charge, G. Copcutt.) The centre is to serve the population of 72,000.

It provides 9·4 acres (4 ha) of shopping 411,000 ft^2 (38,000 m^2) out of a total shopping provision for the whole new town of 14 acres, (5·5 ha) 617,000 ft^2, (57,000 m^2)

Office, educational, cultural, sports and public buildings are in the centre also.

Total area 66 acres, (26·5 ha).

Parking for 3000 to be provided with room for more.

The motor traffic and parking takes place entirely below the town centre, and access by lifts, escalators and stairs connects lower and upper levels. Sheep in adjoining meadows are an amenity, often mentioned by the author,(i) but here it may not be particularly effective so near to farms in a small tight urban development.

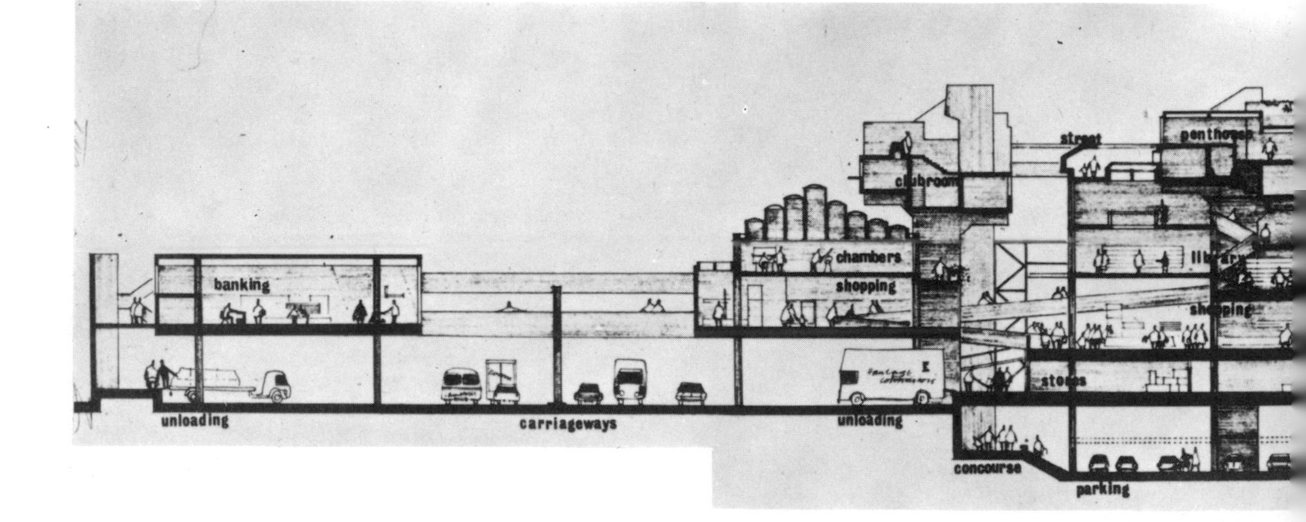

Fig. 6.17. Cross-section of town centre.

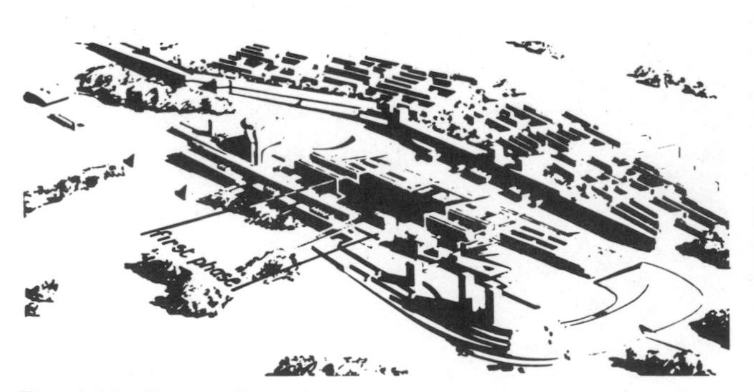

Fig. 6.18. Perspective of whole town centre showing first phase begun 1962.

References.
Cumbernauld Examples in Section VII. Copcutt G. Car Parking in Cumbernauld. *A.J.* 15 December 1960. Wilson H. Cumbernauld D. C. Preliminary Planning Proposals. Cumbernauld 1958. *A. B.N.* Cumbernauld Special Issue. 29 March 1961. Wilson H. Cumbernauld New Town Mark II. *A.J.* Shopping in Cumbernauld. *A.J.* 1 October 1959, 1 December 1960. Central Area Report, Part I, Cumbernauld Development Corporation, Architects Department 1960.

(i) *The Times*, 17 May 1961.

Public Transport: Public transport is mainly by bus also terminating below the town centre. The railway station, in the valley, is within fairly easy reach of the town centre.

Work: Industry is provided in two main locations, but some small areas of industry are mixed in with residential developments.

Walking Distances: The majority of the 50,000 citizens of the hilltop town are within convenient walking distances of the centre.(i) The satellite villages are on bus routes.

Road Network: The first time for a New Town that this was carefully calculated: 65% of cars were assumed on roads at peak hours, 0·7 cars per family. 1 car per family is now thought more correct.

Background Reports: The research work at the basis of this new town, both with regard to residential and central areas, is extensive and thorough. The documents issued by the development corporation are of outstanding value.

70% of families are projected as owning cars.

28% of the population will go to work by car.

44% by bus.

14% would walk.

The remainder would use the train or be passengers in cars.

T-junctions are changing to the Reichow type, acute angle flow from minor to major roads in the preferred direction.

Footbridges and underpasses have been supplied at an early stage to encourage the use of the eventual footpaths from the start, with considerable success.

General: Cumbernauld has had a very important effect: although it is too early of course to pronounce on the town as a whole, the high density areas planned so far seem very sympathetic, particularly in the exposed climate of the hilltop in Scotland. The paths are attractive, truly a linear playground for children, and in continual use already although much of their usefulness is still to come.(i) Rail communications are used only in a marginal way.

The Hook project by the London County Council and the extension scheme for Basingstoke show the important influence of Cumbernauld.

This is the first fully-fledged multi-level new town centre in the world which is under construction. Vertical traffic segregation in terrace housing is also scheduled for the Carbrain South area. (See p. 291 also pp. 41, 42, 43, 108, 112, 235, 238, 247, 248, 249, 254, 258, 280, 291 for other illustrations of Cumbernauld.)

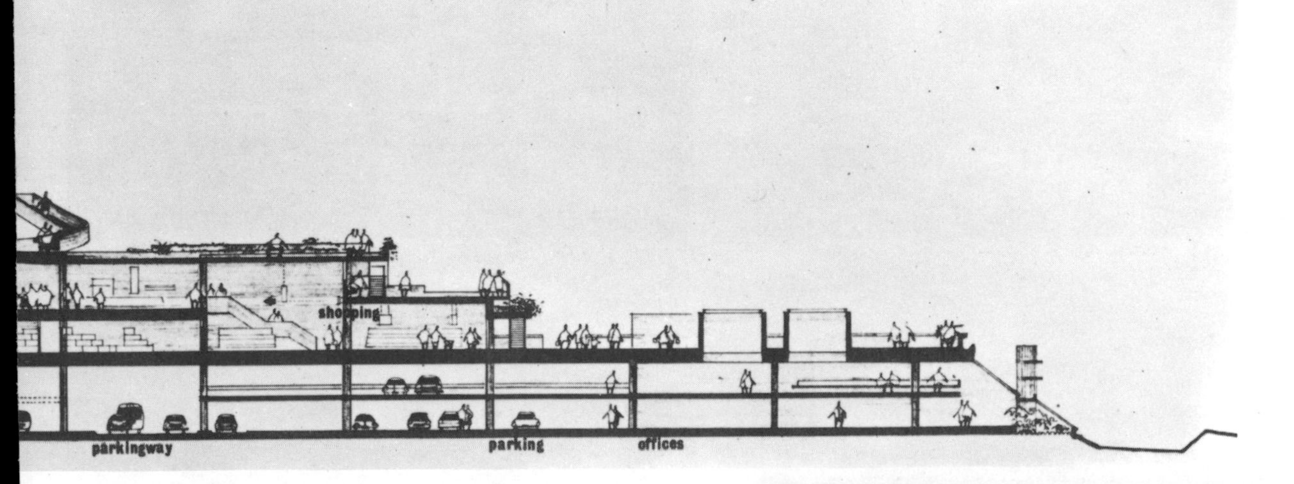

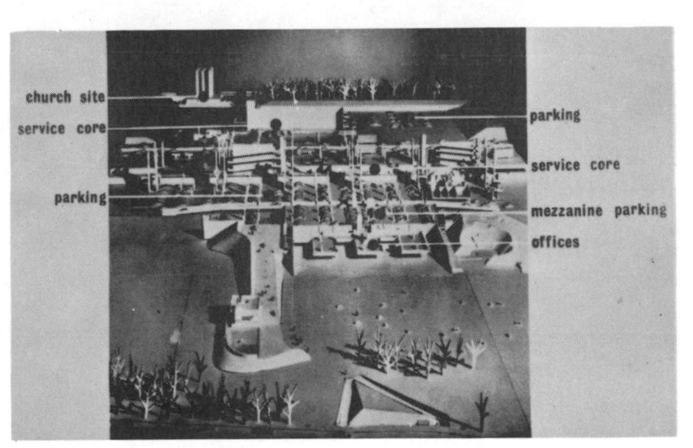

Fig. 6.19. View of lower level of town centre showing car parking. Note sheep by town centre. Underpass from residential area bottom right hand corner of photo.

(i) See comparative figures, p. 113.

Fig. 6.20. Aerial view of model of first stage of town centre.

(i) To walk about the town is decisively different from walking about in any other Mark I British New Town. The continuous path system separate from roads is a revelation, in spite of the weaknesses usual for any prototype.

Fig. 6.21. Aerial view of housing taken June 1962.

Fig. 6.22. Part of Main Path (Seafar 1).

Fig. 6.23. Children's play area in early low cost housing. Note predominance of hard surfaces where a wet climate makes play on grass for children impracticable, much of the time. (Kildrum 3).

Fig. 6.24. Typical path in early high density low cost terrace housing (Minehead 4).

Fig. 6.25. Main ring path by play area and "honey comb" flats.

Fig. 6.26. Typical garages for low cost housing.

VI. 9.2. *Hook, Hampshire, England*

Not to be built.

Site: Valley in pleasant rural area. Latitude 51°.

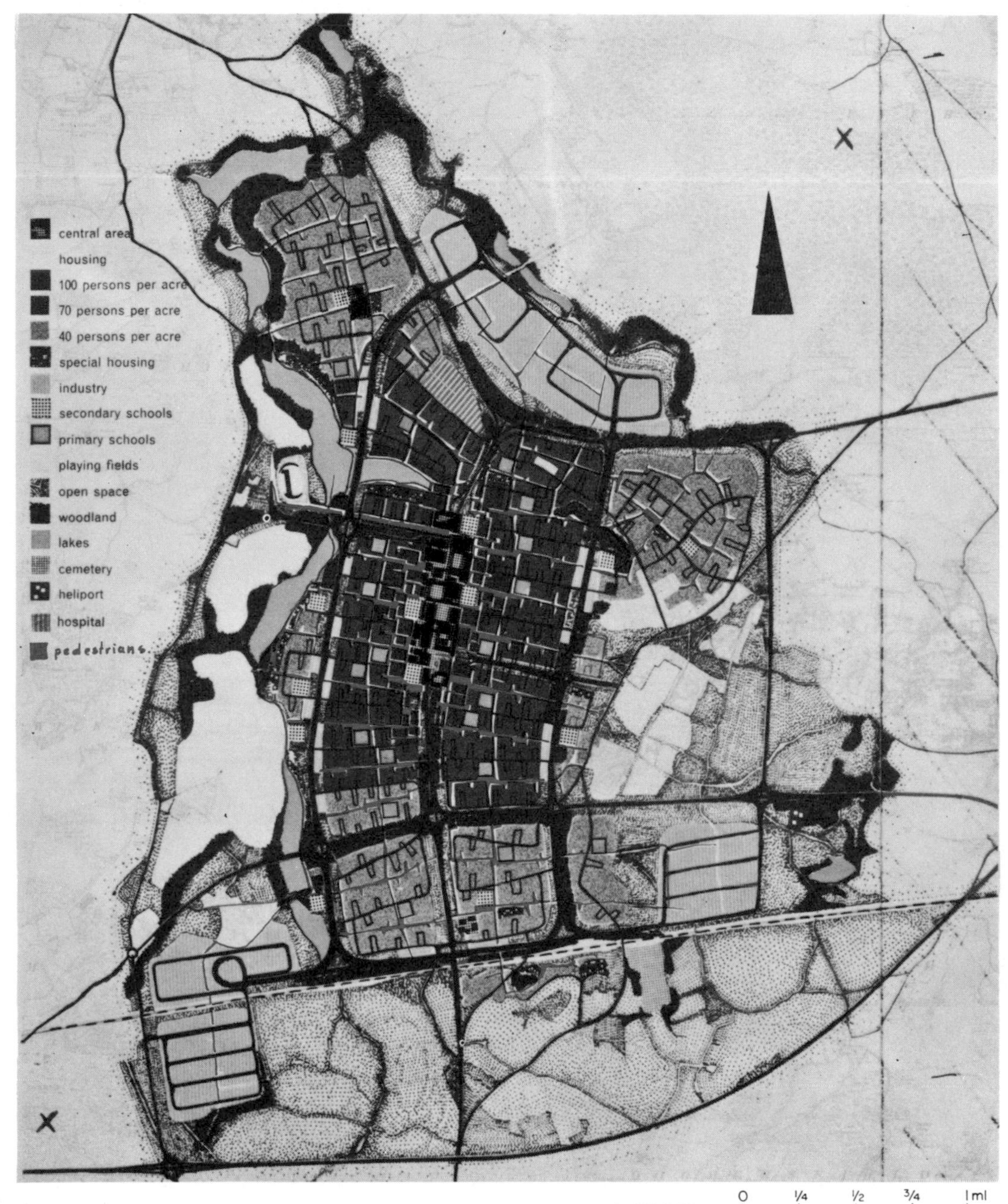

Fig. 6.27. Master Plan of town. 1 : 50,000 (scale: 0 ¼ ½ ¾ 1 ml; 0 250 500 750 1km)

Design: Architect of L.C.C. Hubert Bennett. Deputy architect, F. G. West. Head of Section, J. C. Craig. Senior Staff New Town Section, O. J. Cox and C. G. L. Shankland. Design Team. N. P. Allen; G. W. Ashworth; M. J. Ellison; Miss I. Foster; D. C. Hardy; G. I. Lacey; E. G. Lenderyou; H. C. Morris; R. C. W. Palairet; A. A. Percival; A. C. Perkins; M. N. Pickering; Miss M. E. Raffloer; W. J. Scadgell; G. E. Schoon.

Population: 100,000.

Reference.
New Town Design. A Study for a New Town of 100,000 *at Hook, Hampshire.* L.C.C., London 1961.

Origin: London County Council Plan for a New Town, to take some of its surplus.

Kind of Town: A town to provide work, shopping and cultural life as well as homes for its citizens. Like Cumbernauld, built to absorb a fully motorized population with a multi-level centre—vehicles on the lower level—and an entirely separate footpath system in all parts of the town. A considerable use of vertical traffic separation in high density housing was to be made at Hook, where it links with the centre. Density is from 100 ppa (250 ppha) downwards, bringing the majority of people within walking distance of the town centre.

Schools: Primary schools are within the residential area but secondary schools are either on the periphery, beyond the ring road, with their sports fields attached, or in the central area with their sports fields apart.

Walking Distances: The linear centre and high density housing makes walking to centres convenient, past school and work, local goals, corner shops and so on.

Roads: Ring collector roads are assumed to carry 3000 vehicles at peak hours and residential distributor roads joining them 500. Every other one of the distributor roads joins at a grade-separated junction. Generally, junctions are of a type that allows developments in stages from roundabouts to grade-separated flow.

General: The Hook project demonstrated that, given the need to bring thousands of cars into a centre, a valley is a good site for the centre because the vehicles can be hidden in it cheaply and easily (unlike Cumbernauld's excavation). This seems to have some general relevance to the selection of sites for future new towns and was foreshadowed in the use of a valley in the centre of Farsta for a service loop. The population studies as important in indicating the widely varying types of population structure that emerge.

Residential Accommodation: 70% of housing is arranged with "outdoor space" directly accessible, but only in some cases in the form of private gardens. The concept of linear playground was to be used as an extension of the indoor space. The plan provides for 48,000 residents of the inner town to be housed in a continuous system of residential areas of 70 ppa (175 ppha).

Centre: The central area is 100 acres (40 ha). It has as many varying functions as at Cumbernauld and the calculations for the shopping provisions have been based on the research done by G. Copcutt at Cumbernauld.

Parking for 8000 is provided.

Connection between lower level and decks 30 ft (6 m) above are by lifts, escalators and stairs.

Public Transport: Over 2500 people are estimated to be commuting from Hook daily. Over half by train and 700 of these to London. Bus and railway stations are placed at the junction of the spine road and railway line. This is at the southern fringe of the town centre.

Work: Three industrial areas are planned to reduce distances to work and spread the flow on roads at peak hours. There is also a liberal sprinkling of industries within the residential areas. 50% of the completed town's workers are seen as employed in local industry. (See also pp. 237, 247.)

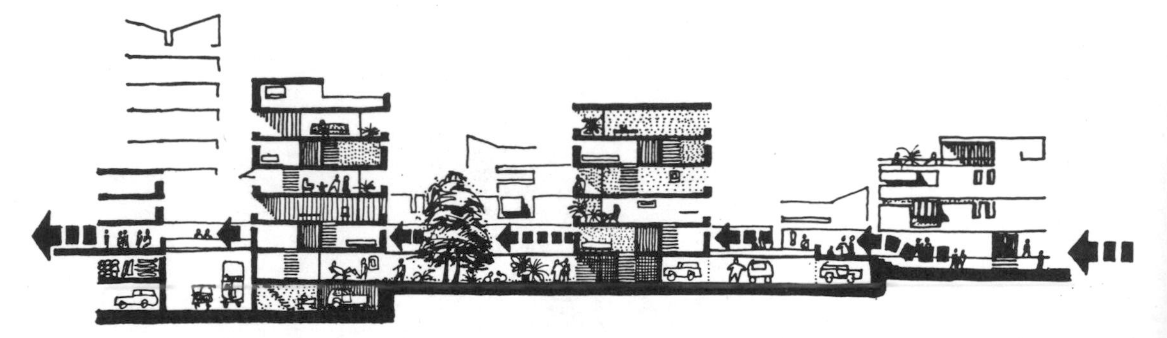

Fig. 6.28. Section through town centre and high density residential areas, showing pathways vertically segregated.

Fig. 6.29. View of town from a main access road.

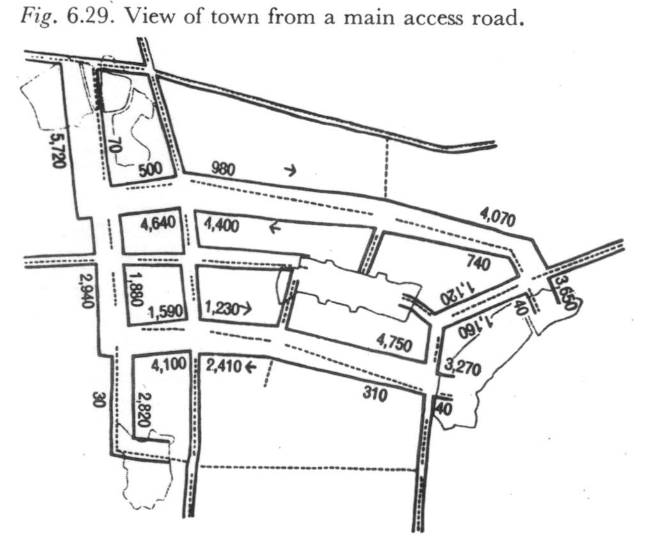

Fig. 6.30. Peak hour flow Hook. Directional proportions given in numbers and indicated by dotted line.

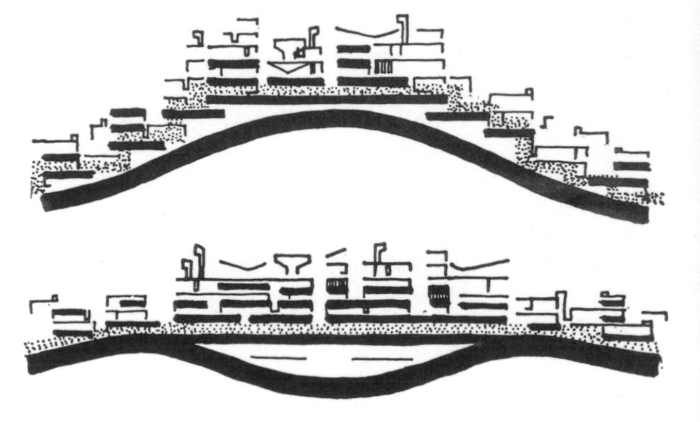

Fig. 6.31. Diagram showing pedestrian access to town centres on a hill (Cumbernauld) and "on" a valley (Hook).

Fig. 6.32. Residential area with linear playground, steps, ramps and railings designed as play-equipment.

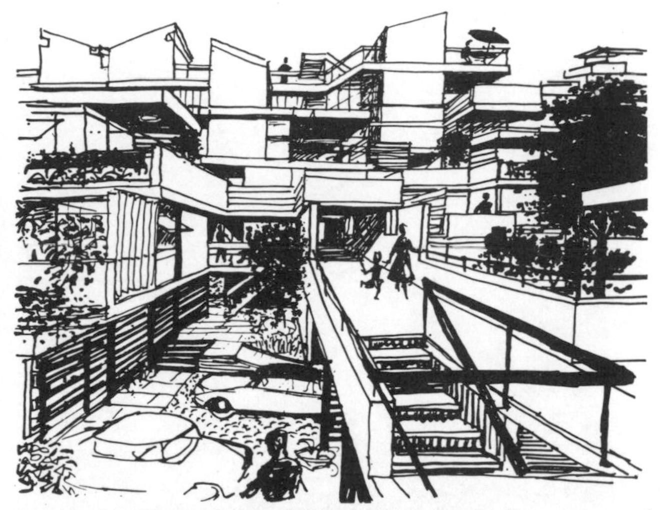

Fig. 6.33. Detail impression of what vertically segregated pedestrian walkways in high density housing, 100 persons per acre (250 ppha) might be like.

Fig. 6.34. Bus station below town centre.

Fig. 6.35. View of north end of central area, housing and lake.

V.I 9.3. *Erith, Kent, England*

Site: Fen country in the Thames Estuary. Local regulations prohibit habitable rooms less than 12 ft (4 m) above ground level as there is flooding in the area. Site adjoins older L.C.C. housing estate.

Design: Architect to the Council: Hubert Bennett. Deputy: F. G. West. Senior Staff: K. J. Campbell, John Craig, J. Whittle, G. C. Logie. Research and Design Team: E. Hollamby (Assistant Senior Architect), C. B. Mears, A. C. C. Jones, Miss A. F. Maclaughlin, Miss S. D. Graham, Mrs. N. A. Pears, Michael Ellison, E. D. Faulder, J. Markham, D. Davighi.

Origin: This development of "Housing for Erith" is a scheme by the L.C.C. in the county of Kent to provide for some of its population. In this area employment would be largely local so that commuting problems would not arise. Shortage of land within the L.C.C. boundaries make this site desirable.

Population: 25,000 persons.

Kind of Town: Although just called "housing" this development using local industry and probably a local enlarged shopping main centre has many attributes which make it more useful to include it as a "town". The principle of complete vertical traffic segregation, here enforced by the site conditions may prove to be of more general application if successful in this prototype.

"Villages" of 400 dwelling units each, containing a cross section of dwelling sizes as currently appropriate to London, are on individual platforms linked by pedestrian bridges. Six of these are grouped in every one of three clusters around a platform placed centrally to hold the primary school and local shopping facilities and social amenities. Two secondary schools are at the perimeter of the site.

Residential Accommodation: Each dwelling has one car storage place below the platform and there is some additional parking for visitors. Other service accommodation and some workshops and courtyards for play are also provided at ground level. The position of the 31-storey towers, with two and three room dwellings at the bridge connections, ensures that there is lift connection to the platforms for all those who need it. Four-storey, three-storey and two-storey dwellings form the "village". Hydroponic "gardening" is envisaged in the main. Allotments for gardeners are to be provided at ground level. The tall blocks are designed to enhance the fen-atmosphere of the flat landscape, and give a scale which can hold its own with the large industrial towers that there are in the area.

The net densities of the villages themselves are very high indeed, for the towers add to an already dense pattern. Scissor housing(i) might make the same densities possible without the tall blocks. Gross residential area is 48 ppa (120 ppha) net residential density 100 ppa (250 ppha). (The net density of one village is much higher of course.)

Communications: The road feeding the housing is to take peak flow based on the assumption that people all use cars. However, for local traffic it is obvious that all persons are within easy sheltered walking distance of goals. Two stations connecting Erith with London are alongside the site.

Prefabrication on the site is an important aspect of the scheme. It is likely that in British New Towns prefabrication will play a major role sooner or later.

Assessment: This important proposal may be implemented within a few years. It is in a preliminary stage only and variations to the road junctions, not rational for the function they fulfil, and the lack of connection by pedestrians with secondary schools and stations are not in any way deliberate omissions.

(i) Barr A. W. C. Housing in the 1960s. *J.R.I.B.A.* April 1962.

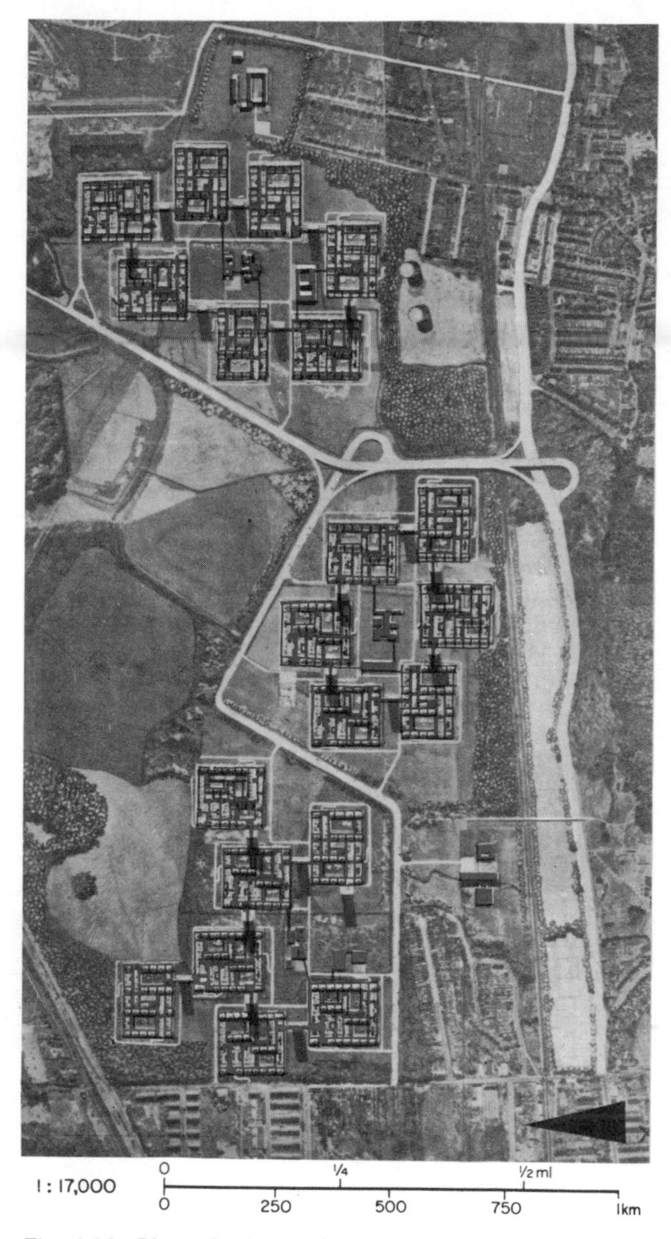

Fig. 6.36. Plan of whole scheme superimposed on aerial photograph.

Fig. 6.38. View of village from ground level court.

Fig. 6.39. View of village.

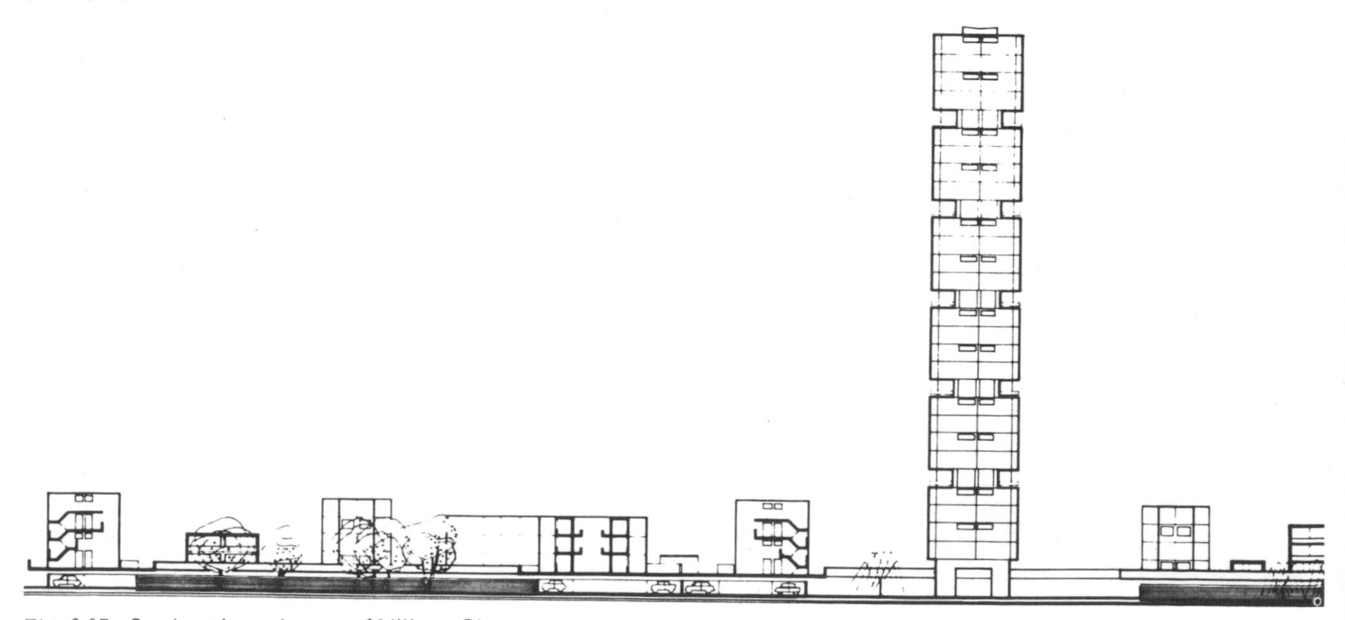

Fig. 6.37. Section through part of Village Cluster.

1:1800
0 50 100 150 200 ft
0 20 40 60m

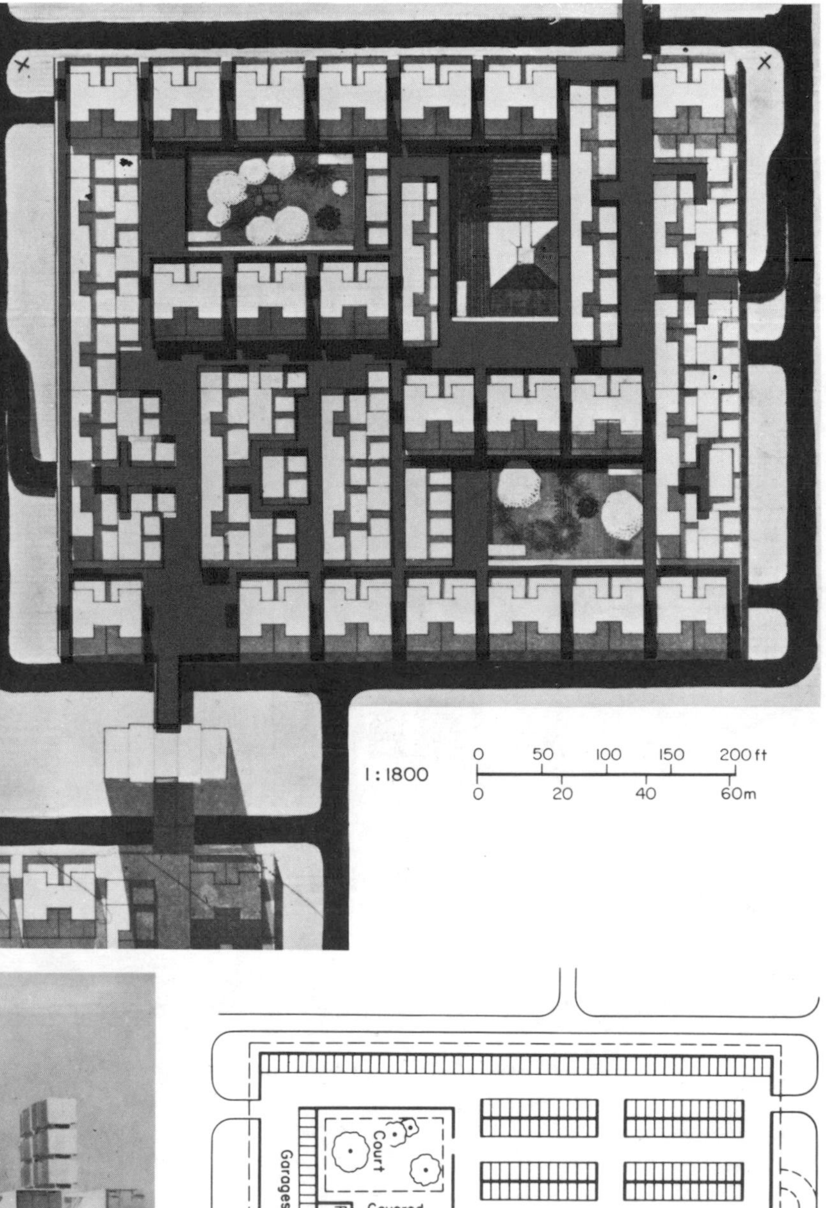

Fig. 6.40. Plan of one "village".

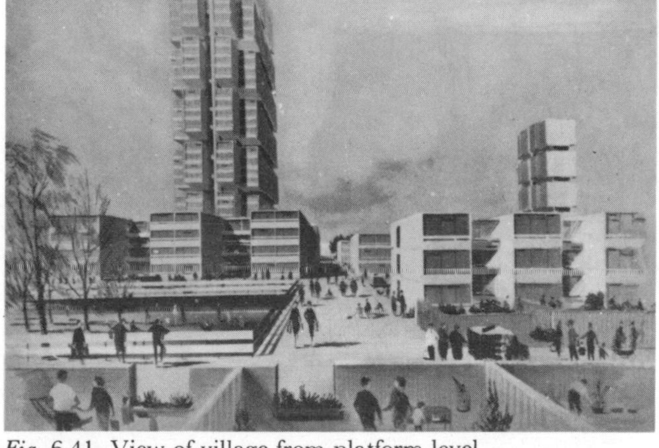

Fig. 6.41. View of village from platform level.

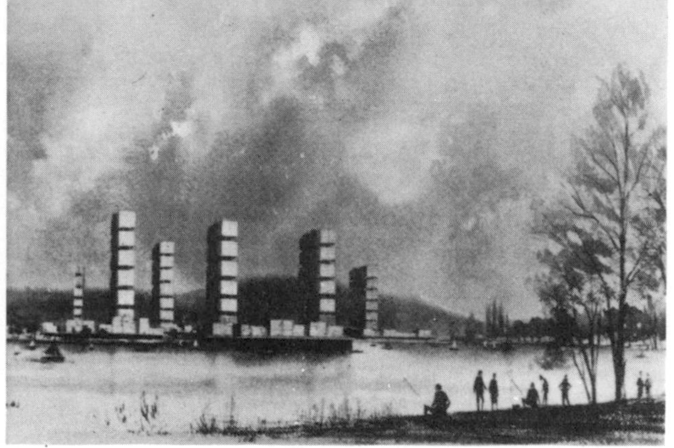

Fig. 6.42. Distant view of town.

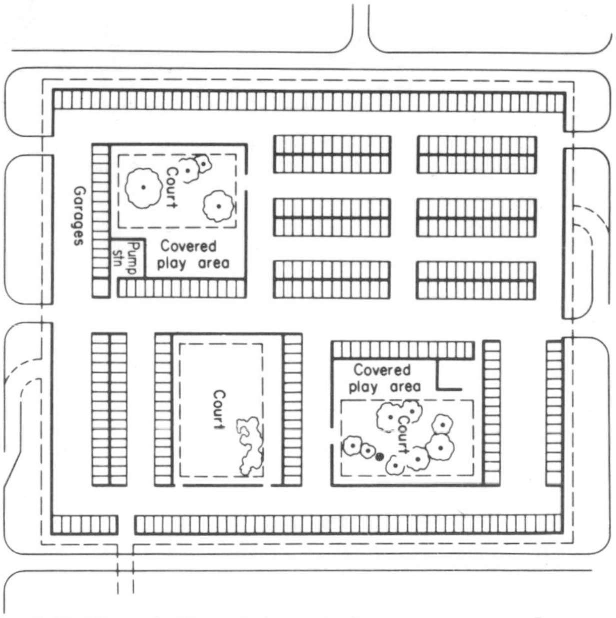

Court
Garages
Pump stn
Covered play area
Court
Covered play area
Court

Fig. 6.43. Plan of village below platforms.

VI. 9.4. *Toulouse le Mirail, Toulouse, France*

Site: On the other side of the bank of the Garonne from Toulouse itself, 43° latitude. Three miles (4·8 km) from the centre of Toulouse.

Fig. 6.44. Plan of town showing independent paths and road networks.

Design: G. Candilis, A. Josic, S. Woods, P. Dony, H. Piot, J. Francois.

Origin: First prize in competition for New Town for Toulouse by the city, 1962.

Population: 100,000.

Kind of Town: "*The plan derives from a linear association of activities to form a stem into which housing units could be plugged. The stem would contain commercial, social, cultural activities and traffic would be completely separated with automobile parking, service roads and courts being sunken so as to leave the pedestrian completely free in his movement. The dwelling along this stem would be of the 'deck house' type*[(i)] *with continuous horizontal ways through the buildings, linking their vertical circulations and various entry points. This would facilitate the localisation of automobile ways and parks in order to establish a harmony of pedestrian and mechanical circulations.*"[(ii)]

(i) Smithson A. & P. Golden Lane Competition, 1952.

(ii) From a letter to the author by S. Woods. January 1963.

Residential Accommodation: 17,284 flats in Deck Housing, 7-, 11-, 15-storey. 2752 3–4-storey flats. 2728 Terrace houses and other dwellings. Total 22,764.

This would be the first new town where deck-housing at a very high density would accommodate most of the people (75%). (Deck-housing was considered for parts of Cumbernauld and rejected on cost grounds before further consideration was given to it.)

Total of 28,000 car spaces: 7500 hard standing, 14,000 underground, 4000 multistorey, 2500 special uses.

Centre: Beyond the linear, continual growing centre there is a main centre with shopping, entertainment and cultural provisions, for the 100,000 and a hinterland of 125 miles (201 km) radius. Car parking for 4000 in the administrative centre.

Work: The light industry lies just outside the main collector road connected by paths and many offices and craft-workshops are below the deck development. Heavy industry, 3 miles (4·8 km) to the north and 1 mile (1·6 km) south-east.

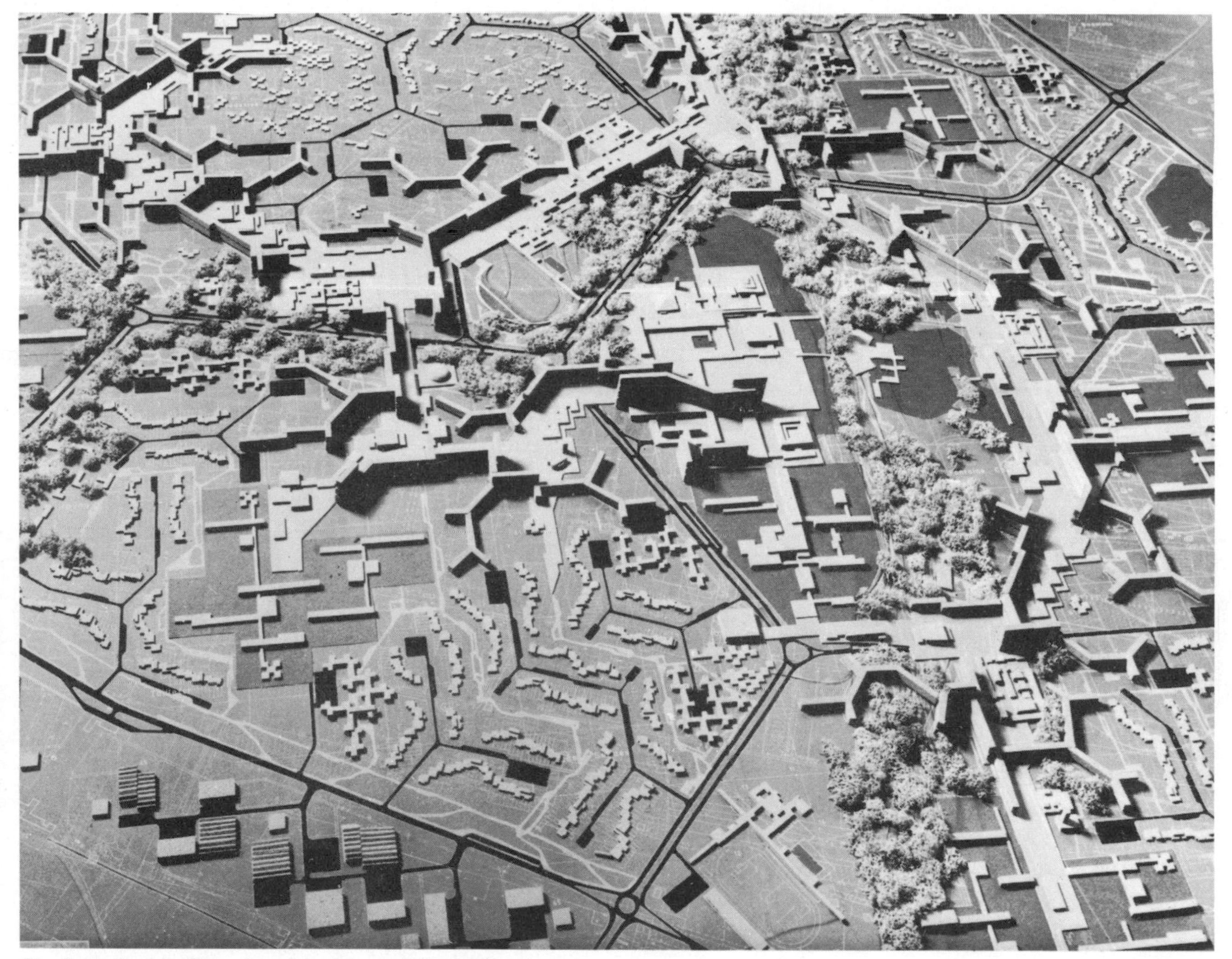

Fig. 6.45. Aerial view of model of town. (Photo. Yan.)

Public Transport: Monorail loop to Toulouse and to heavy industry proposed.

Walking Distance: Most areas are within 0·6 miles (1 km) of the centre.

Roads: Towards Toulouse the capacity of ringroad outlets and two dual carriage ways is 6880 vph, to the south ringroad two dual carriage ways, 4240 vph, each way. But the planners make it clear that the car is to be "un outil au service de l'homme"(i) and is not to enslave him. Most routes meet at economic junctions of three roads.

General: The idea of the scheme is to preserve the countryside as much as possible and to provide a clear distinction between urban and rural environment. The concept as a whole is designed to fit, by its strong character and contrast, into the historic surroundings of old castles and beautiful towns of the area.

A proper assessment of such a new idea awaits the building of the scheme. There is a determined plan here to limit the use of the car and to make it entirely subservient to pedestrian movement. It does seem as if the tall blocks might have been connected, if only by bridges, to a greater extent than they seem to be, making the three-dimensional system of routes complete and useful. There is also some doubt whether the peak hour flow from residential concentrations could be coped with by minor junctions. However, it is clear that in this town the planners have set out to minimize car use and restrict it largely to out of town journeys. The monorail or other connection to Toulouse must be built to make this realistic. This is the first deliberate attempt to plan for pedestrians while full demands for car space are not met, unlike Hook and Cumbernauld where both walking and driving are catered for fully in the central area.

See also diagram showing growth principle on page 116.

(i) *L'architecture d'aujourd'hui*, No. 88. May 1962.

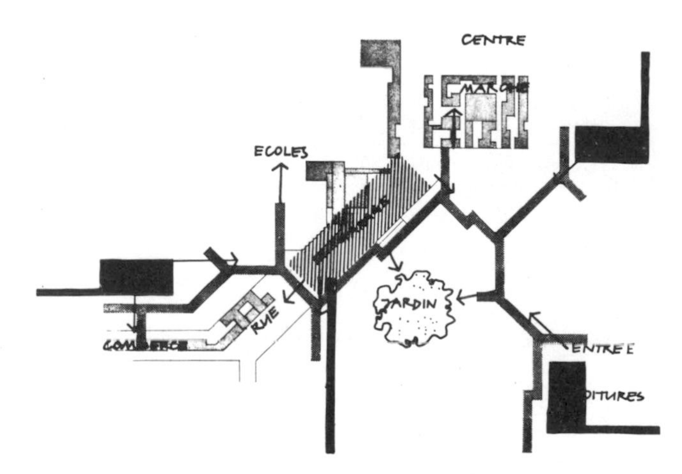

Fig. 6.46. Principles of arrangement at foot of deck housing.

Reference.
June 1963.

VI. 9.5. *Vällingby, Stockholm, Sweden*

Site: Flat, valleys hills, rocky and partly wooded. 60° latitude.

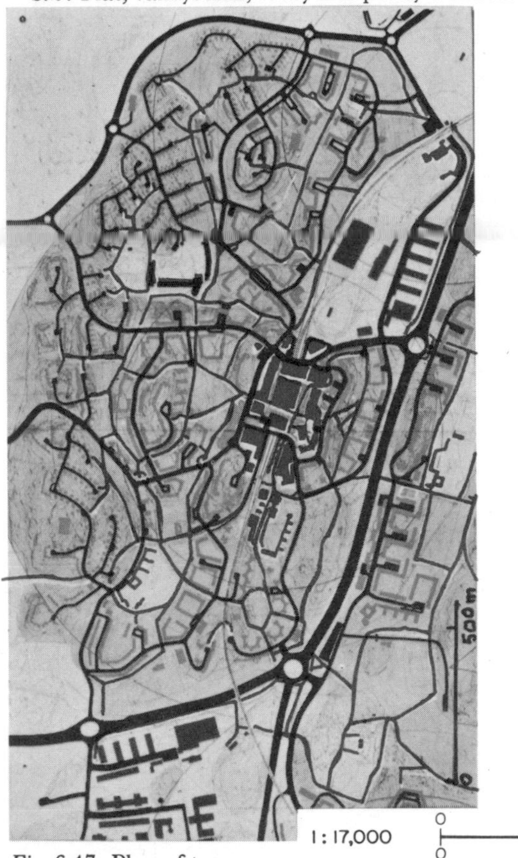

Fig. 6.47. Plan of town.

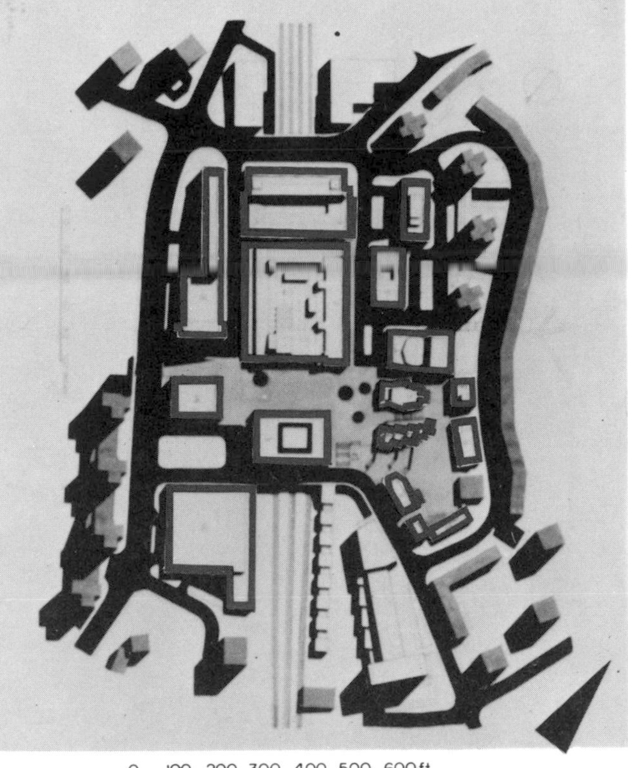

Fig. 6.50. Plan of town centre.

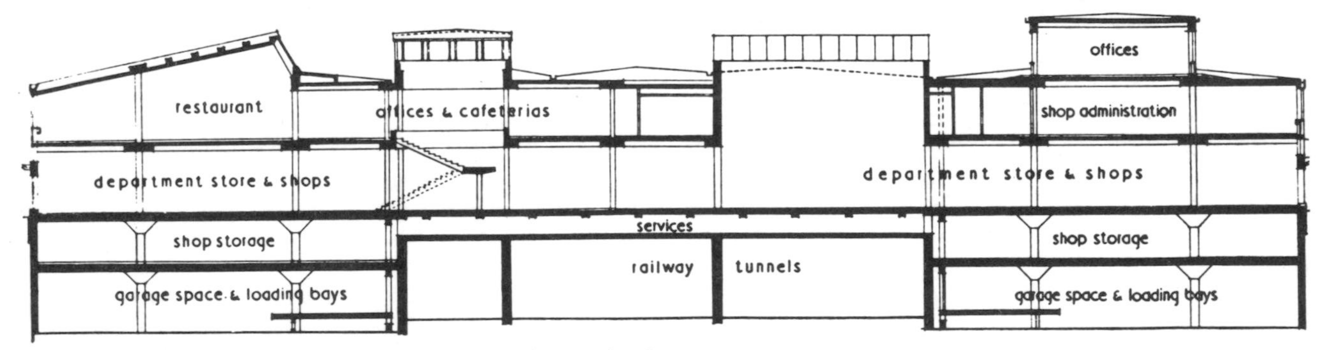

Fig. 6.48. Section through centre showing lower level loading bays.

Fig. 6.49. Access to centre down steps through a passage from residential area. Underpass in the background leads to residential areas on the other side.

Fig. 6.51. Aerial view of town centre showing railway passing under.

Design: Basic Plan, Sven Markelius, City of Stockholm Planning Commission.

Origin: Part of the creation of a chain of satellite towns for Stockholm. First plan 1952, completed 1960.

Population: The town itself 23,000, but a closely connected hinterland of five neighbourhoods and other residential areas.

Kind of Town: On the neighbourhood principle, though such centres, particularly within Vällingby proper, are kept quite small. Low density. Main centre connected with Centre of Stockholm by Rapid Transit Electric Railway, underground in the city. Partly a dormitory town. First European town with traffic segregation, with a separate footpath system leading to all neighbourhood centres and through the green parks of all neighbourhoods contained within superblocks. Ring roads are nearly always under or over-passed. The centre is to one side of the town close to the main expressway and above the rapid railway connection with Stockholm.

Residential and Walking Distances: Flats are built within 1500 ft (457 m) of the centre and beyond within 1800 ft (549 m) of the centre are all the houses.

Car storage was provided for every fifth inhabitant with housing.

Centre: To serve town and hinterland population of 100,000. 180,000 ft^2 (16,723 m^2) shopping area, 100 shops, 600 parking places, restaurants, theatre, churches, civic buildings, sports hall, swimming pool. The railway station directly beneath the centre. Three large and immensely enlivening fountains, naked children abound in summer.

Public Transport: Every five minutes trains leave for Stockholm and stops on the way. This is virtually the only public transport and explains, perhaps, why there is a far larger percentage of car users than was anticipated by providing 600 parking spaces in the centre (36·5% arrive on foot, 26·3% by car, 9·2% by cycle and 29% by trains). Of course, given more parking spaces the number of car users might be larger.

Work: Forty percent of workers are employed in industrial areas and in smaller works among residential development. But the town is not intended to provide all the work for its inhabitants.

General: This town, as the first in Europe, had an immense effect. Whatever its weaknesses, to be expected in a prototype, it is a revelation and remarkable contrast to the British New Towns, Mark I, emerging with a vastly inferior lay-out at the same time. The combination of public transport to the centre and the pathways make the town pleasurably accessible on foot. Walking along paths through one superblock after another gives the average visitor the impression that he has arrived at a park inhabited by specially civilized people. (Tulips bloom without protection on main pathways.) Provision for car storage is on the other side of dwellings only.

Winter sport is possible right outside dwellings. The immense success of this lay-out has led to the development of the idea in many countries. Markelius was in close relation with Clarence Stein when he evolved the plan. (See also pp. 43, 250, 256.)

Fig. 6.52. Railway platforms and low level service access to shops.

Fig. 6.53. Clear example of superblock adjacent to centre.

Fig. 6.54. Two Radburn culs-de-sac high class housing at Hässelby Strand.

Fig. 6.55. Pools in the town centre in use. Naked children find pleasure in hot weather.

References

GENTILI G. Satelite towns of Stockholm. *Urbanistica.* September 1958.

GUERIN E. Vällingby. *Architecture and Building.* December 1958.

VI. 9.6. *Farsta, Stockholm, Sweden*

Site: A beautiful, hilly, wooded, resort area, south of Stockholm. 60° latitude.

Fig. 6.66. Plan of town.

Fig. 6.67. Section through shopping centre, A.-A.

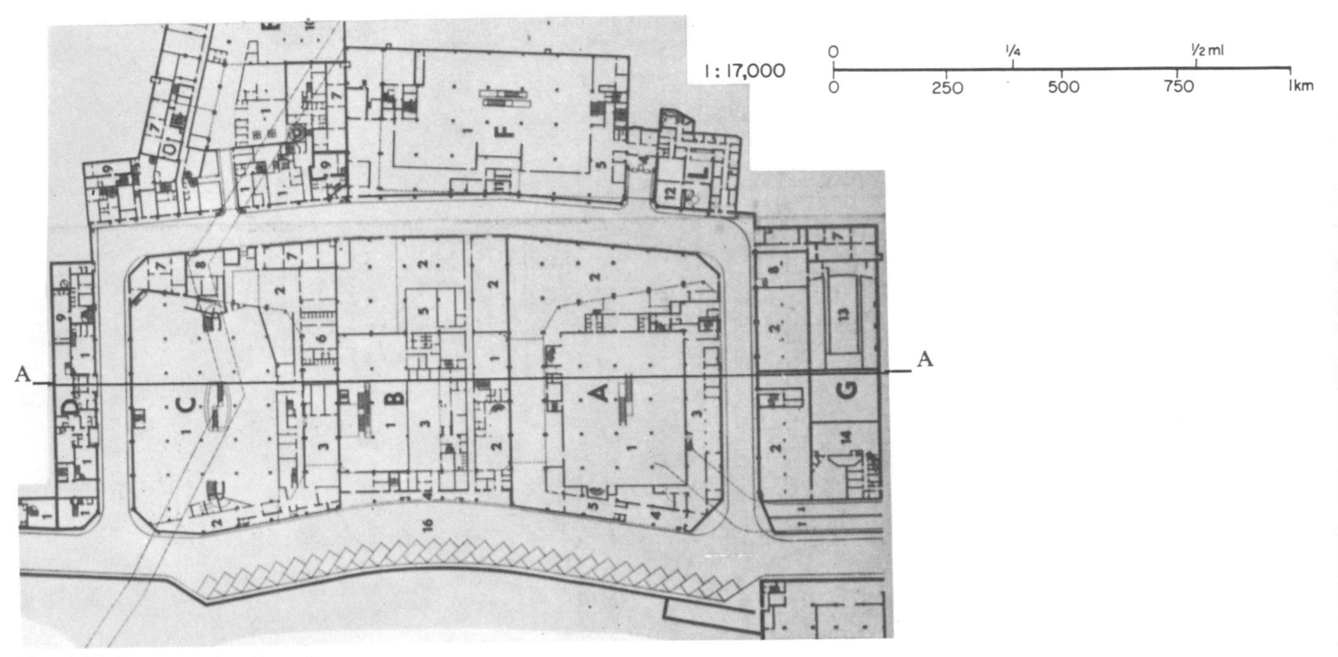

Fig. 6.68. Loop under shopping centre.

Design: City Architects' Department, Stockholm, lay-out only.

Origin: One of a number of Stockholm City Satellite Towns. Designed 1957, completed 1963.

Population: 35,000.

Kind of Town: A higher density town than Vällingby. Employment for half the working population is local. A centre for many surrounding districts. Vällingby experience has resulted in greater car facilities.

Centre: The centre is supplied almost entirely from underground, where orders can also be collected. A valley is utilized to good effect for this service loop of considerable dimensions. Parking is still on the surface and acts as a barrier to residential areas on two sides of the centre. Otherwise it is intimately related to the dwellings along the separate path system. It is the size of Vällingby. The immediate hinterland is only 65,000 (to Vällingby's 100,000), but attraction of many customers from other areas is expected. There are 200,000 people within a 15 minute car ride. Car parking for 2350 vehicles is provided. The centre is very much in the middle of the town. Restaurants and many public facilities supplement the 387,500 ft² (17,000 m²) of shops with 183,000 ft² (36,000 m²) of offices. This has been increased by about 10%.

Public Transport: Direct frequent electric (underground) train connection to Stockholm make journeys to the city very convenient. Buses, leaving from the centre, cover most areas of the town which is hilly for walking. There are also bus connections with other parts of Stockholm.

Residential Area: Flats 76% and 24% houses. Tree-covered landscape is left intact to a very large extent. Point blocks take 7·6% of the population and slab blocks the majority. Car storage: one car per family.

Work: Industrial areas are small and well within the town area.

General: The centre is lively on all days of the week, day and evening. Walking about the town is a great pleasure, but the point blocks and slabs are at times too overpowering and massive.

The calculation of customers certainly seems not to have been over optimistic. On Friday evenings, the top shopping time, all the car parks are full and the road sides are parked up also.

In winter the safe pedestrian environment makes an excellent background for winter sports. (See also pp. 42, 254)

Fig. 6.69. Pool in middle of shopping centre.

References

Gentili G. Satellite towns of Stockholm. *Urbanistica.* September 1958.

City of Stockholm. Farsta. Stockholm 1960.

Farsta Centrum. *Sartrick nr Arkitektur.* Stockholm 1961.

Fig. 6.70. Aerial view of centre. Top right hand corner shows the valley of Figs. 6. 73–77.

Fig. 6.71. Aerial view of centre, showing parking areas.

Fig. 6.72. Station at centre, coach can be seen in the centre between buildings at first floor level.

Fig. 6.73. Underpass in the housing area leading to centre.

Fig. 6.76. Service road and parking at high level.

Fig. 6.74. Parking at high level seen from the path system with play lot adjacent.

Fig. 6.77. Looking down the valley towards road bridge.

Fig. 6.75. Play ground below, cars above.

Fig. 6.78. Commercial building on a footpath, seen from an underpass.

VI. 9.7. *Cologne, New Town, West Germany*

Site: Flat land in the Rhine Valley, 51° latitude.

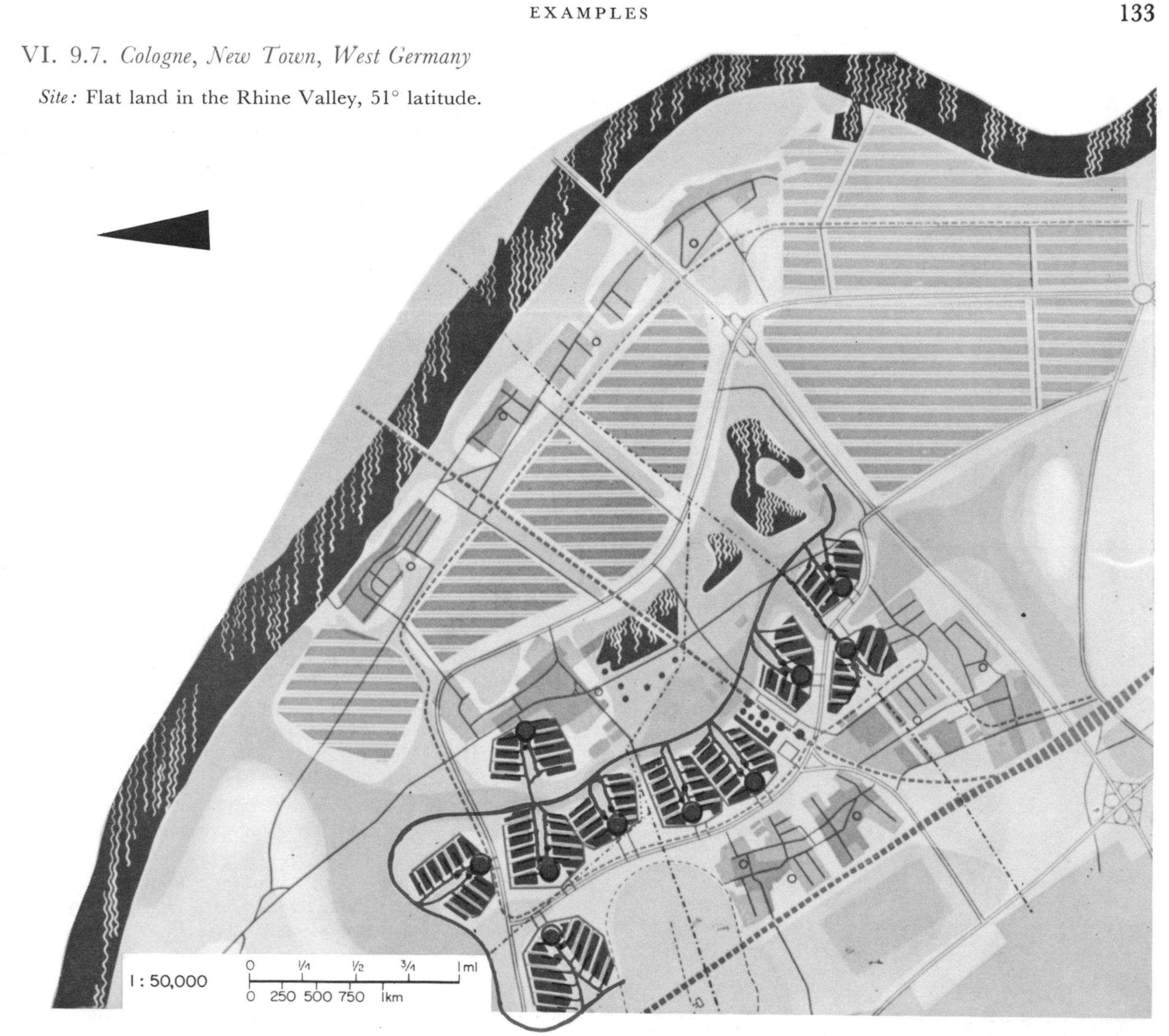

Fig. 6.79. Plan of town.

Design: City Architect Peck.

Origin: Expansion by the City of Cologne on a large remaining site. Design published 1958. Construction begun 1960.

Population: 100,000.

Kind of Town: To be regarded as a balanced unit with industry, shopping and cultural activities suitable for a town of that size. Specific relations with the mother town of Cologne and its amenities are borne in mind. Planned for walking, cycling and driving. Designed on the neighbourhood principle.

The town is laid out according to Reichow's principles of branching to some extent. The main stem is the expressway connecting to the centre of the town from two sides. On to this the neighbourhoods fit, like individual leaves. These, each a superblock of something like 6000 people, do not have a ring road but two main culs-de-sac into which merge, at an acute angle, the minor branching culs-de-sac. Joining two such neighbourhoods allows the double use of "ring" roads giving an extremely cheap road lay-out. There is only one outlet to the freeway from each neighbourhood. At this point is the neighbourhood's shopping centre with tram station and small industry. The footpath system leads more or less directly to this centre through a large green open space with schools and clinics in it.

Residential Accommodation: Gross density 15 dwellings per acre (37·5 dwellings per hectare) which is given to be 55 ppa (137·5 ppha) 50% garage provision, and all parking for visitors within the curtelage of the private dwellings. Garages within the house or as near as possible. Percentage of houses and flats not settled.

Centres: There are to be some shops along the "streets" in the residential areas. The neighbourhood centres are related to the public transport stops and junction of neighbourhood collector road with freeways leading to Cologne, etc. The main centre, where the heaviest population lies, has railway connection and road connection in three directions adjacent.

Public Transport: This is by tram along the nine neighbourhood centres and is supplemented by buses leading to the main industrial areas.

Work: In main industrial area and also near to each neighbourhood centre.

Walking Distances: Maximum distance to tram stations or centre at any neighbourhood is 0·6 of a mile (1 km) taken to be 12 min walk. Twenty minutes walk will take most people from home to industry. Cycle tracks are invitingly direct for journeys to work.

Reference

PECKS, *Eine Neue Stadt Köln*, Cologne 1958.

General: There seems a danger that with footpaths leading away from the centre and desire lines before turning a sharp angle back towards them, as in Reichow's road branching system, they may be neglected in favour of pavements on roads and shortcuts. The journeys to school are direct. The strategic placing of the neighbourhood centres and the branching system of the roads is an important experiment to be observed at work. There seems little evidence of road capacity and flow calculations which, for the superblock–freeway junctions, are likely to be high. There are no crossings, only three streams of traffic meet at any point, according to Reichow's principles.

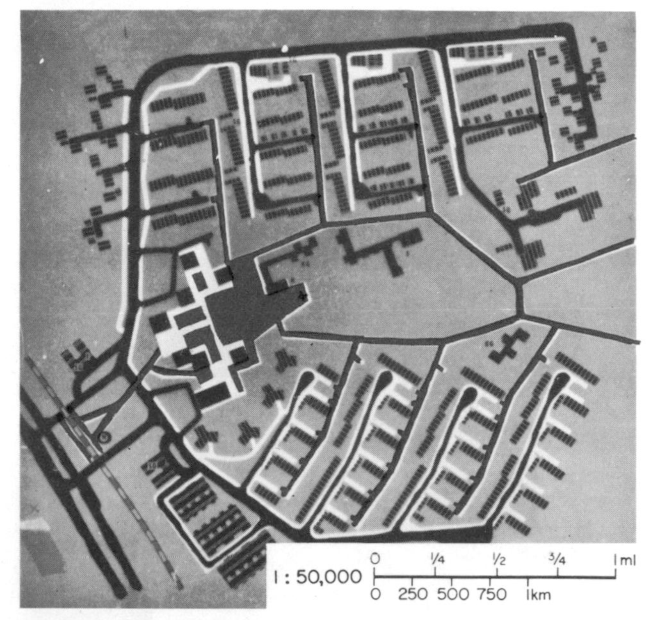

Fig. 6.80. Plan of neighbourhood.

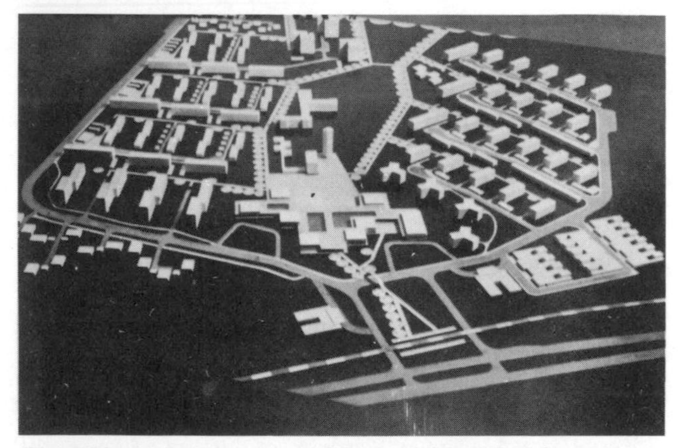

Fig. 6.81. Aerial view of model of neighbourhood.

Fig. 6.82. Overpass leading to town centre along side of road.

VI. 9.8. *Sabende, Guinea, Africa*

Site: A wooded valley among plateaux. 10° latitude.

Design: M. Ecochard, with Lagneau, Weill and Dimitrijevic.

Origin: To house the employees of Compagnie Pechiney, who mine bauxite in neighbouring areas.

Population: 20,000 to 40,000.

Kind of Town: On a neighbourhood principle of 5000 persons per neighbourhood, with schools, shops, mosque (in daily use). To allow for an uncertain growth pattern each neighbourhood is designed to be self-sufficient. One main centre serving the first 20,000 people would be augmented by another if the town grows to 40,000. Neighbourhood centres are linked by path-systems to the main centre. Each dwelling has direct access to the path system. Roads, culs-de-sac in the main, 250 ft (80 m,) long end in communal car storage with all dwellings within 150 ft (50 m,) maximum walking distance. Gross density of town, high for the area, is concentrated for economic and social reasons at 28 ppa (70 ppha).

Residential Accommodation: Densities vary from detached houses at 12 ppa (30 ppha) to terraces 72 ppa (180 ppha). The detailed house plans and lay-out is carefully considered to allow the proper development of the local community life in terms of small groups outdoors and indoors (40 to 50 dwellings). "*. . pedestrian paths and circulation have been designed with a wish to encourage contact.*"(i) There are some *unités d'habitation* in each neighbourhood centre.(ii)

Walking Distances: Maximum 1640 ft (500 m), to centre of neighbourhood.

Roads: A hierarchy of roads; a main east–west axis route with loops defining neighbourhoods, to centre of employment; rapid transit, ring roads, and slow culs-de-sac feeders to residential areas.

General: A careful study preceded the design and the inclusion of a growth principle, in terms of the neighbourhoods, is of interest. It is admirable indeed that this town, in West Africa, like Kitimat in Canada, also for a private company, should be designed in an enlightened manner.

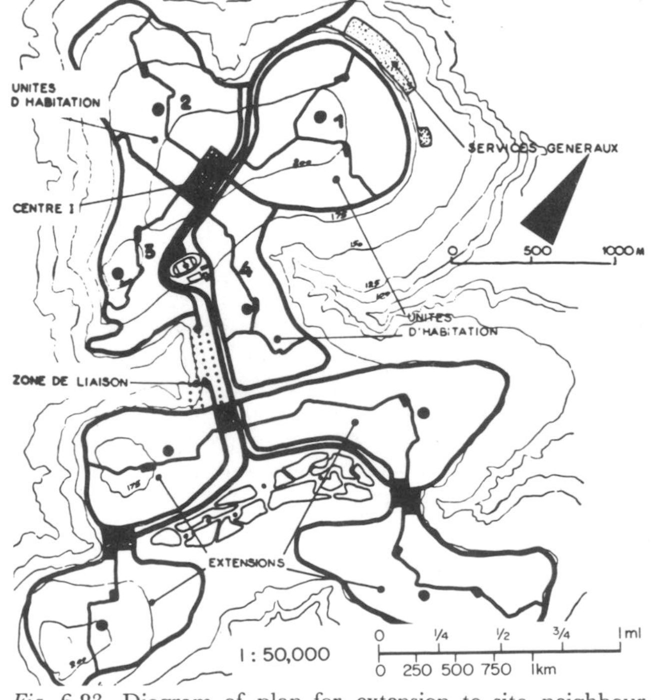

Fig. 6.83. Diagram of plan for extension to site neighbourhoods, shapes determined by valleys between plateaux.

(i) Letter from planner to author. 17 January 1963.
(ii) *L'architecture d'aujourd'hui*, No. 88 Paris 1960.

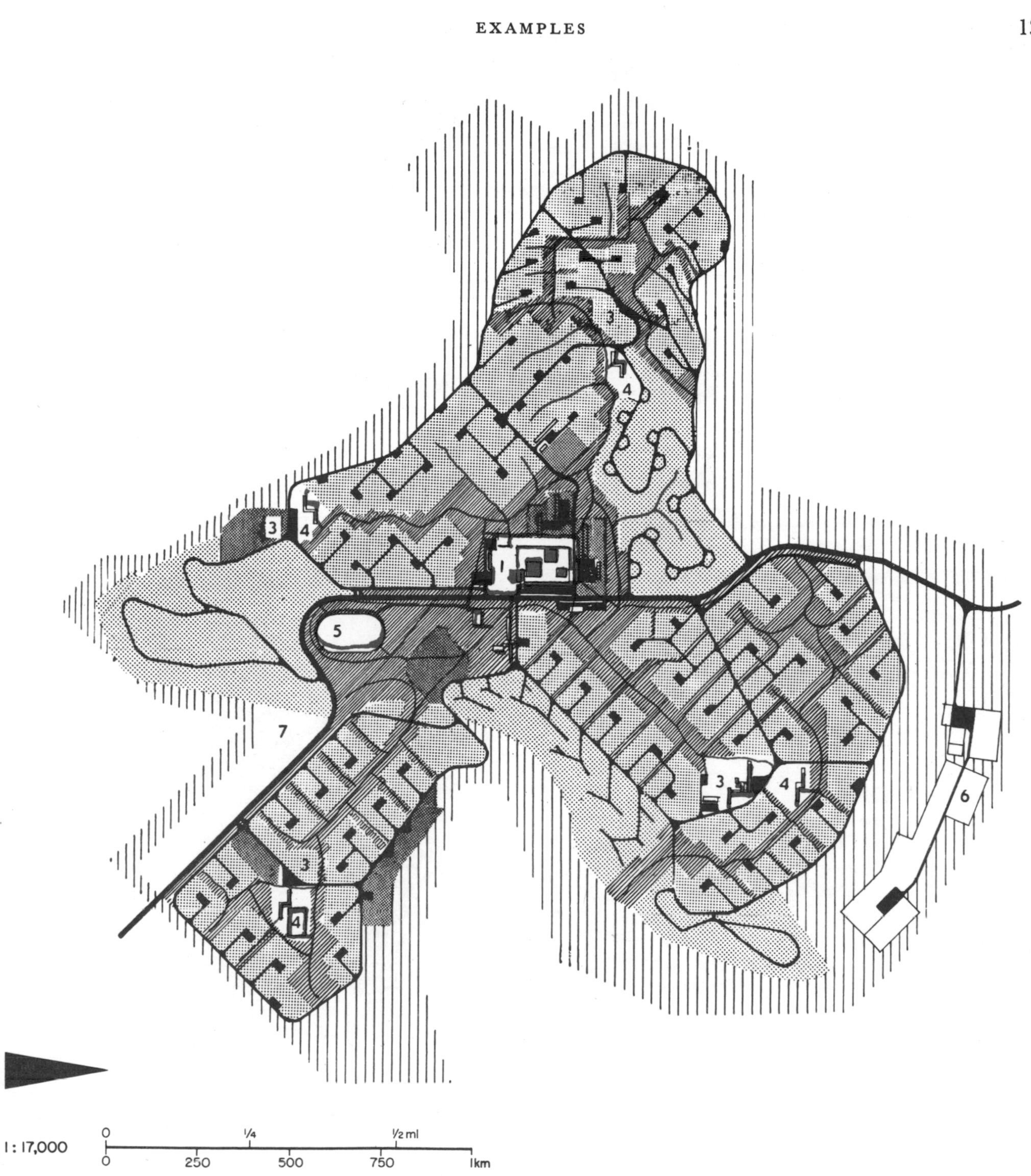

Fig. 6.84. Plan view, showing four neighbourhoods (20,000) and main paths and roads.

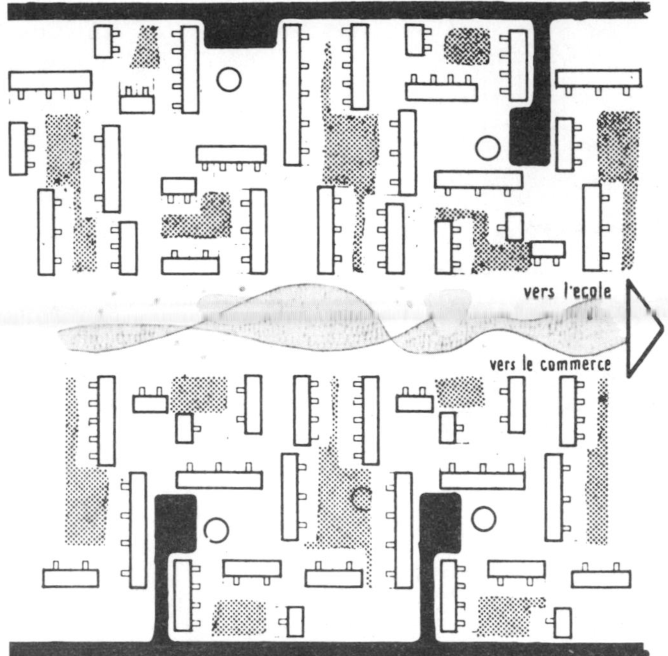

Fig. 6.85. Detail of terrace housing showing careful organization. From principal porch towards one pedestrian side with meeting place. Wives, for housework, meet on the other pedestrian side.

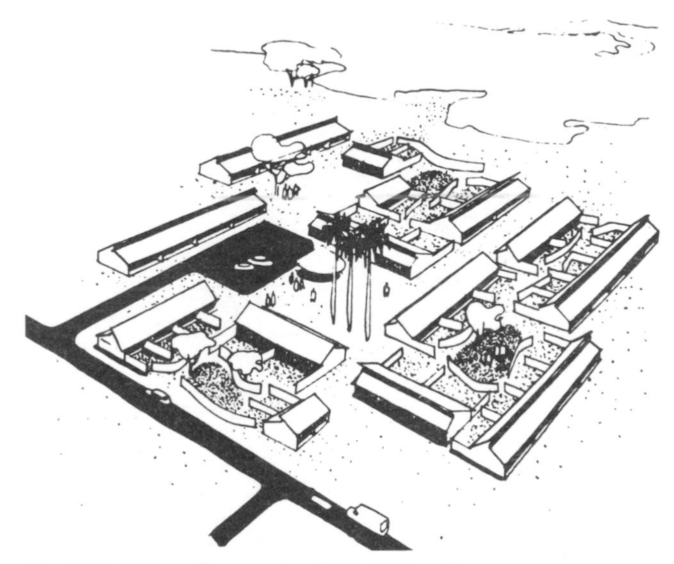

Fig. 6.86. View of one unit of 40–50 dwellings.

Fig. 6.87. View of town under construction.

VI. 9.9. *Sennestadt: New Town by Bielefeld, Germany*

Site: Undulating against the background of the Teutoburger Wald Hills. 52° latitude.

Architects: Dr.-Ing. H. B. Reichow, Fritz Eggeling.

Origin: Winner in 1954 competition; the first town in Germany with traffic separation. Designed for a population of 24,000 in 1956. Light industrial development only.

The density of this town has increased from 3·5 dwellings per acre (9·5 dwellings/ha) to 9 per acre (21·5 per ha) gross over the whole town. The reason for this is economic; the rising price of land. Also, 40% owner houses were to be added without reducing the number of rented flats; 5000 dwellings are now projected. Reichow's system of road branching and junctions has it first full expression: right-angled junctions are avoided to give safety and smooth maximum flow. Different coloured road lighting helps with orientation. The weakness of the road system lies in all traffic being channelled into the centre, even if bound outward. This is a very important experiment.

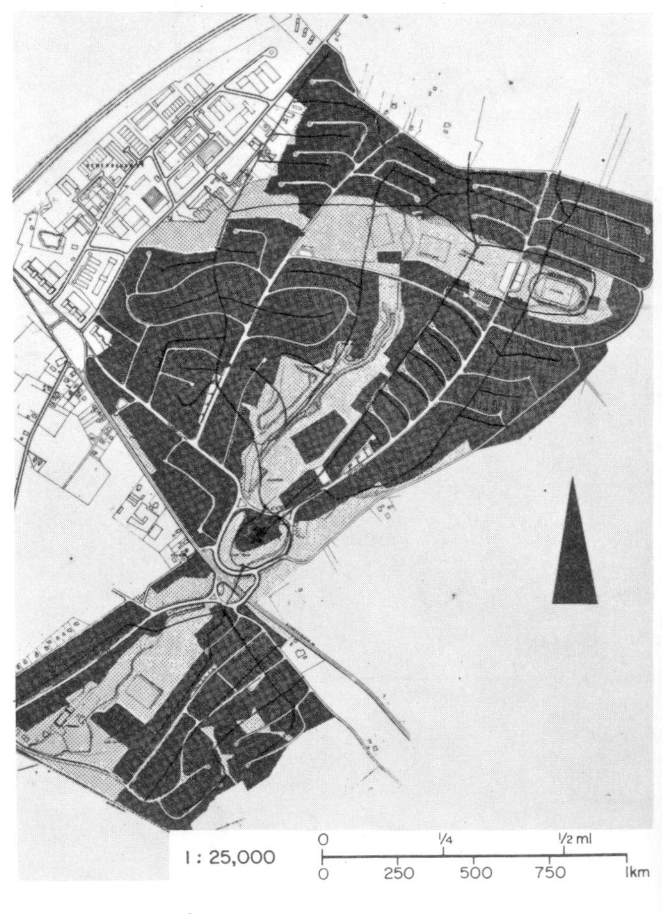

Fig. 6.88. Plan of town.

References

Reichow H. B. Bau der Sennestadt. *D.B.* June 1957.

Reichow H. B. Bau der Sennestadt. *D.B.* August 1959.

Reichow H. B. Bau der Sennestadt. *D.B.* December 1961.

Reichow H. B. *Die Autogerechte Stadt.* Ravensburg 1959.

Fig. 6.89. Aerial view of model.

Fig. 6.90. Green area between flats.

Fig. 6.91. Terrace housing with main footpath.

Fig. 6.92. Town centre.

VI. 9.10. *Neu-Winsen, near Hamburg, Germany*

Site: Flat country. 53° latitude.

Designer: Rolf Rosner, D. Haase.

Origin: A proposal by the designers to act as catalyst.

A population of 82,000 inhabitants, a balanced community. The net density is 90 ppa (225 ppha). There is complete traffic segregation in each of the 15 neighbourhoods. It is composed of 87% flats, 13% houses and has an industrial estate. The centre of the town has a pedestrian podium 14 ft (4·3 m) above car level.

Maximum distances to the centre or into the countryside are 20 min walking time. No worker need cycle for more than 20 min to reach his job.

The number of houses is proportionately small because nearby in the existing settlements there is a large preponderance of them.

Because the architects believe there are more important things in life, only 35% garage provision has been made.

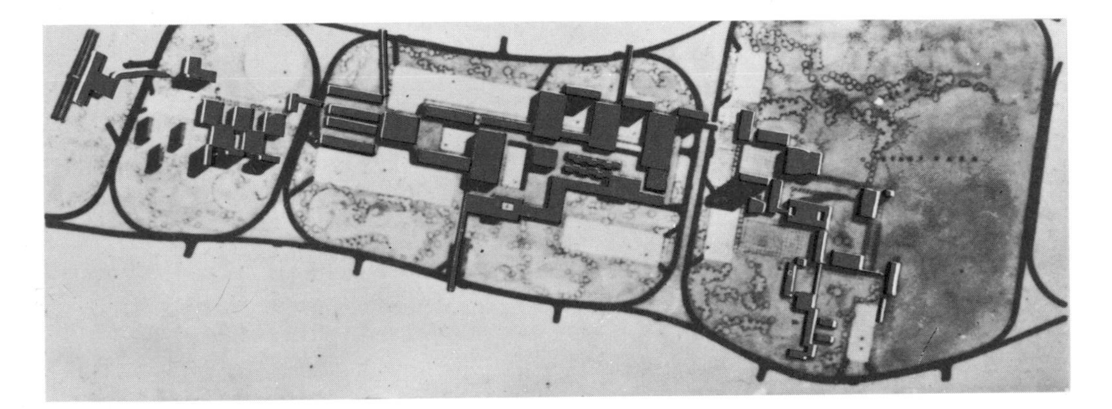

Fig. 6.93. Plan of town.

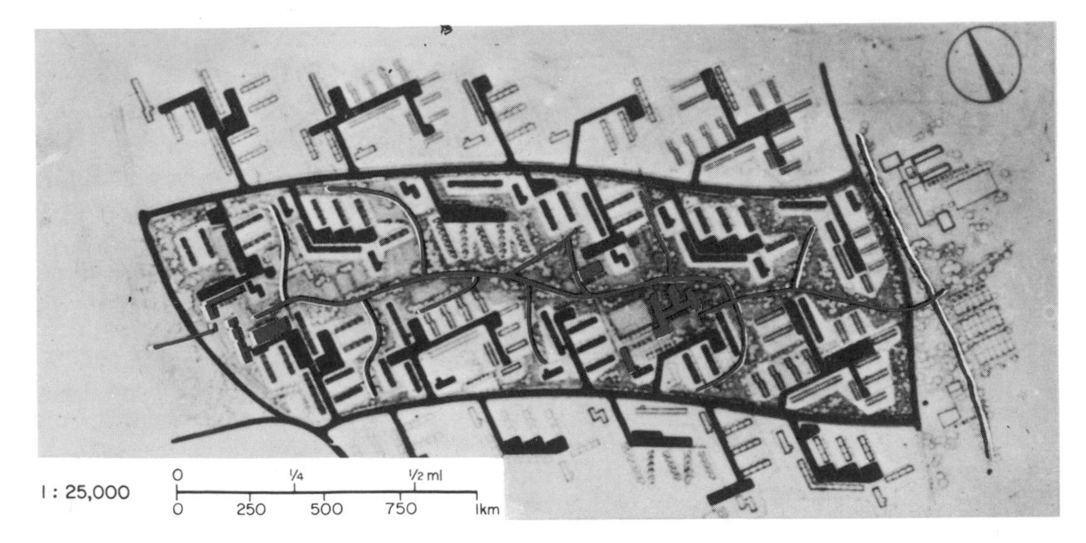

Fig. 6.94. Plan of central superblock and typical residential superblock.

References

ROSNER R., Hamburg, the Neu-Winsen Reconstruction. *J.T.P.I.* July-August 1960.

ROSNER R. Hamburg, the Neu-Winsen Reconstruction. *J.T.P.I.* June 1960.

VI. 9.11. *Noord-Kennemerland, Alkmaar, Holland*

Site: Flat. 52½° latitude.

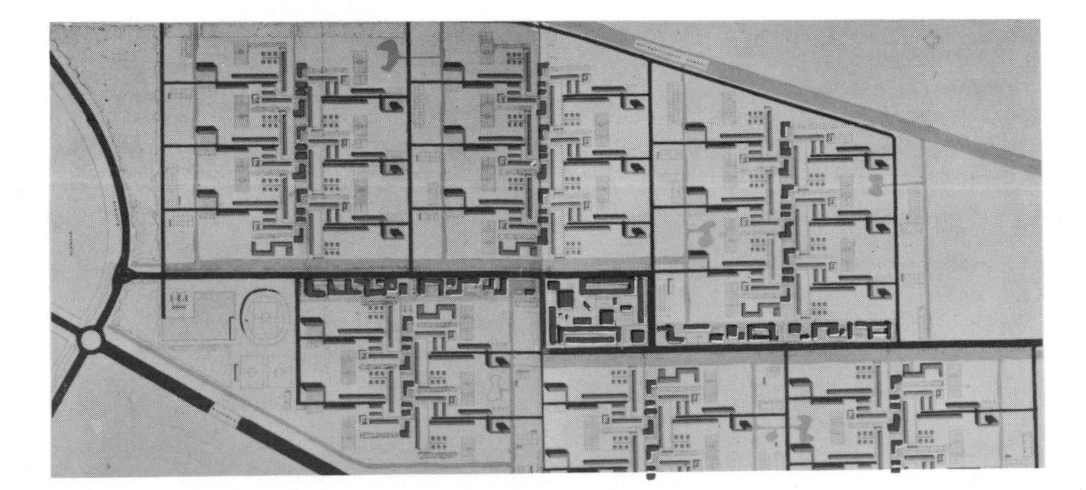

Fig. 6.95. Plan of town.

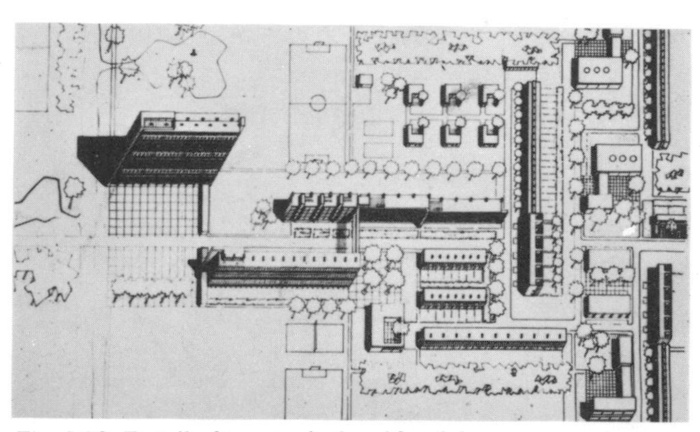

Fig. 6.96. Detail of one typical residential area.

Designer: Van den Broek and Bakema.

Origin: One of a number of Government projects to enlarge existing towns and build new ones, in the fifties.

Linear city concept worked out for 100,000 inhabitants, at a very high net density of 36 dwellings per acre, largely in flats (90 per hectare).

This example is typical of the strictly functional patterns that give a rational regularity on a flat site. Bakema pioneered planning in this way. Although the town as submitted to the authorities does not have traffic segregation it is nevertheless so designed that when this becomes obviously acceptable and so necessary that citizens will agree to it it can be used in this way which has been planned.

Road pattern with junctions of three two-way streams throughout is advanced for its time. It divides the area into six large superblocks with service cul-de-sac feeders. This clearly represents the backbone for a traffic-segregated city.

Fig. 6.97. Central area.

Reference
Alkmaar, *Bouw.* 18 April 1959.

VI. 9.12. *Chandigarh: Capital of the Punjab, India*

Site: 31° latitude.

Design: Maxwell Fry and Jane Drew, (original plan by Mayer and White) who called in Le Corbusier and Pierre Jeanneret.

Origin: Government scheme design, begun 1952, now under construction. The initial project is for 150,000 people; a development to 500,000 is envisaged.

The plan was based on an analysis of road functions thus:

"*V-1—Inter city, all purpose highway—no development.*
V-2—City Distributive Road, all purpose, tracks for fast, slow, cycle and walk—development on slow moving frontages.
V-3—Segregated fast moving traffic road—no sidewalks or frontages.
V-4—Slow moving mixed-traffic, main street shopping, business, parking.
V-5—Residential slow-moving distributive road minimum frontages.
V-6—House access road.
V-7—Recreational or cycle road in open space or parkway."

The right-angled crossings are not a rational way of contriving junctions for motor traffic and the "parks" crossing the mixed traffic (slow?) seem questionable. The mixed traffic (V-4) may not work for either cars or pedestrians.

The government buildings are famous for their architectural symbolism.

Fig. 6.99. View of mixed purpose road.

Fig. 6.100. New housing emerges.

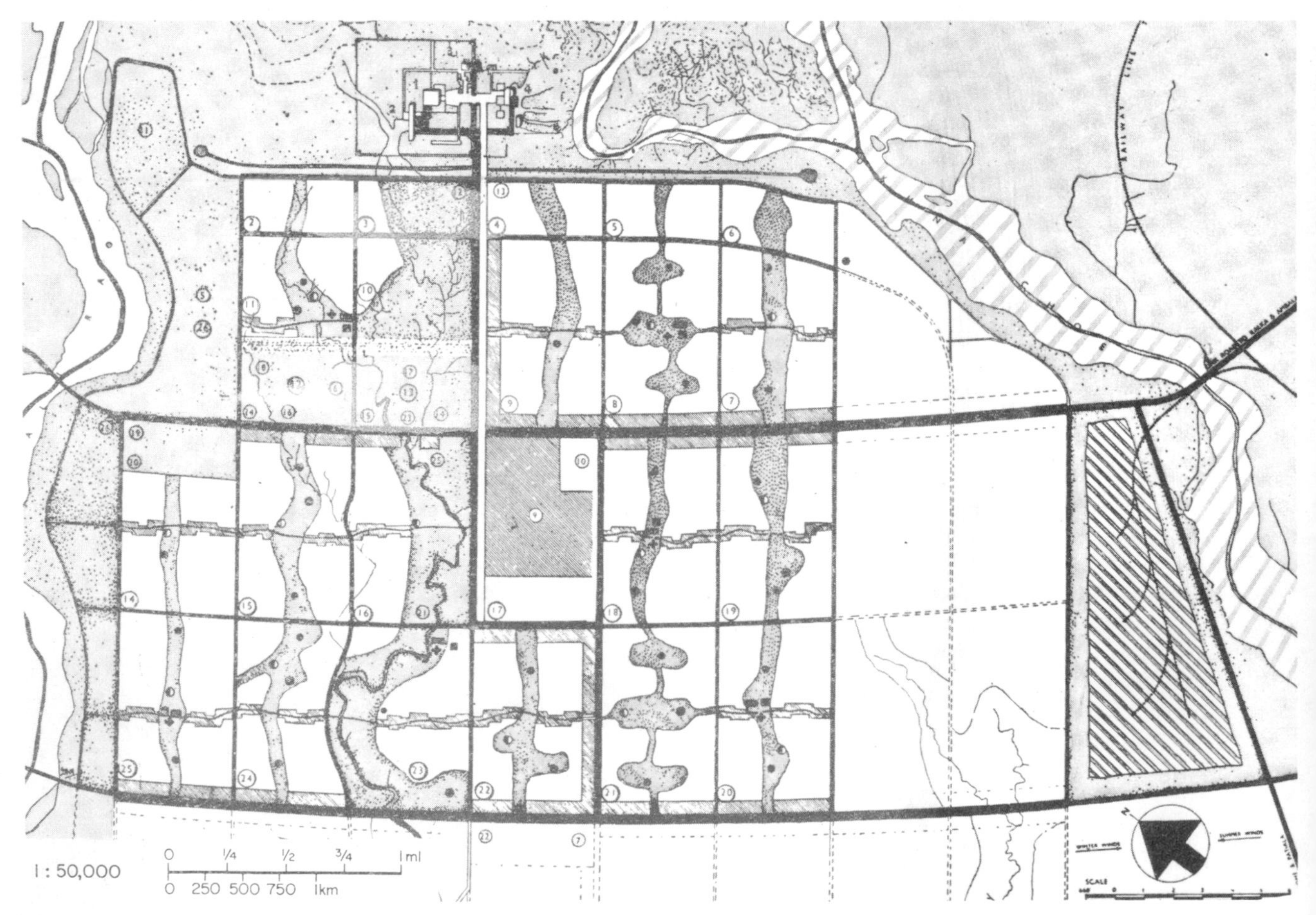

Fig. 6.98. Plan of city.

References

Fry M. Chandigarh. *A. Rec.* June 1955.
Fry M. Chandigarh. *R.I.B.A. Journal.* January 1955.
Hervf L. *L'architecture d'augourdhui,* No. 101. Paris 1962.

Fig. 6.101. Plan of same area.

Fig. 6.104. View of paths in residential area designed by Le Corbusier.

Fig. 6.105. School.

Fig. 6.102. Wide roads awaiting traffic beyond bicycles and cycle drawn carriages.

Fig. 6.106. Detailing for shade.

Fig. 6.103. Roundabouts as standard solution for four-way junctions.

Fig. 6.107. Secretariat and typical traffic.

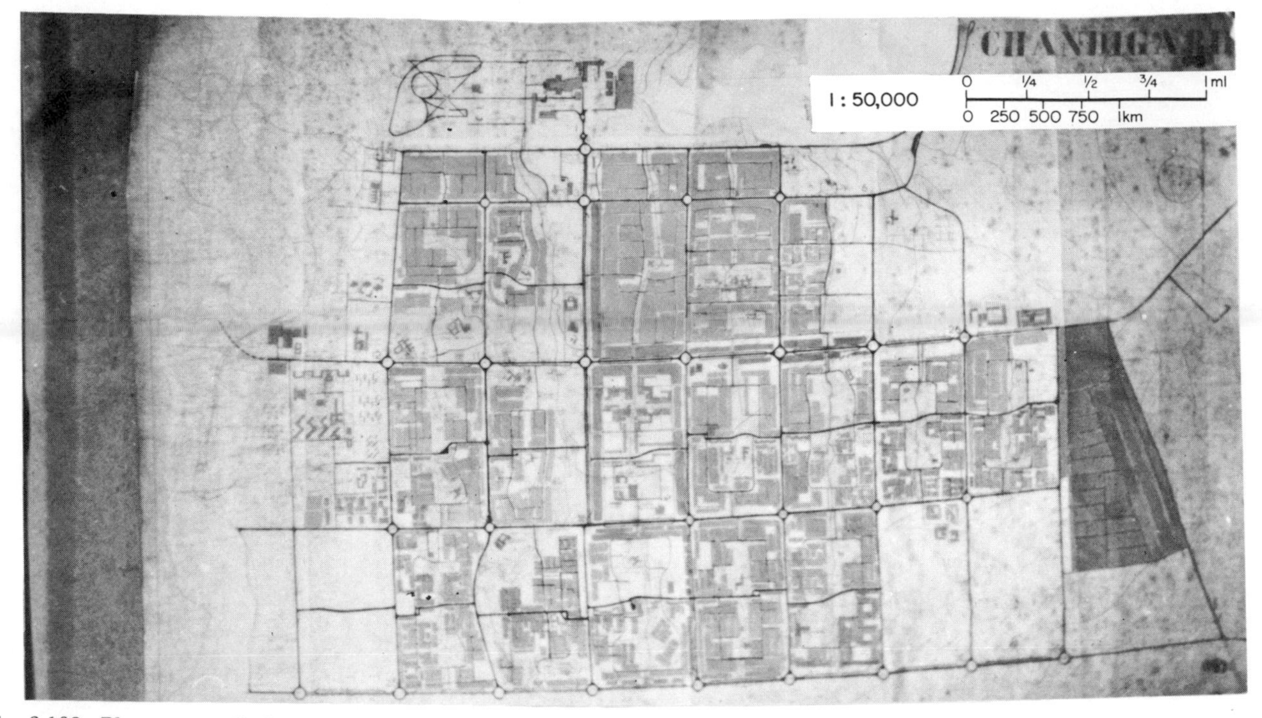

Fig. 6.108. Plan as supplied by Chandigarh Planning Office in 1963.

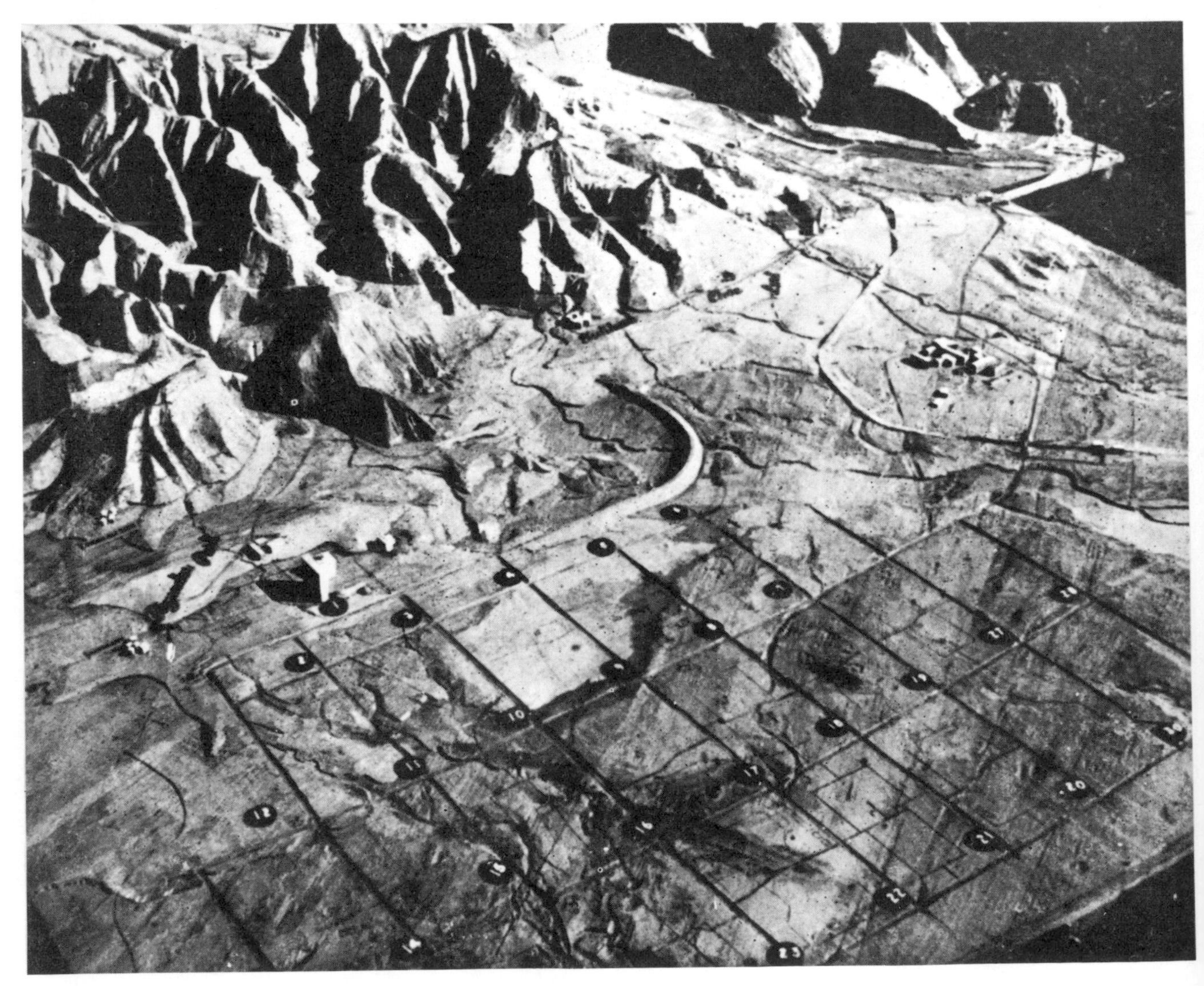

Fig. 6.109. Relief map photo from Time and Life showing the dramatic site of Chandigarh.

VI. 9.13. *Kitimat. New Town, British Columbia, Canada*

Site: Dense woodland. 54° latitude, Pacific coast.

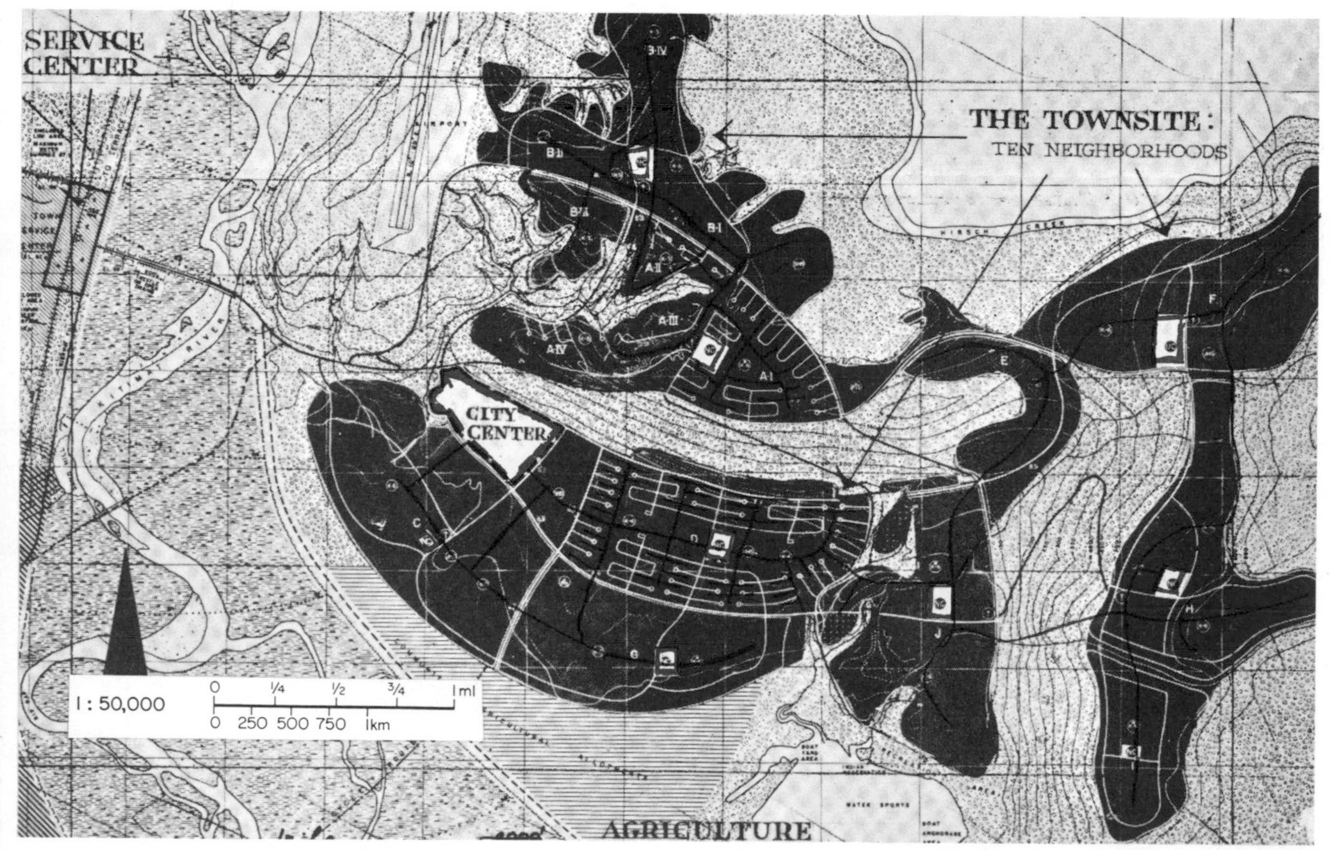

Fig. 6.110. Plan showing neighbourhoods.

Fig. 6.111. Main Centre model showing covered walkway connections and school in background.

Design: Co-ordinator and Director of Planning, Clarence Stein; Architects and Engineers of Master Plan, Mayer and White.

Origin: Sponsored by the Aluminium Company of Canada. Master plan completed 1952.

This town for 50,000 people is at present under construction. It is being built on the neighbourhood system, industry being the *raison d'être* of the new town. The Centre with shopping and entertainment is to bring permanent workers to the region. Covered links connect all parts of the centre and the school.

Density is very low, largely detached family houses and bungalows. Complete traffic segregation is aimed at. Underpasses are planned to link main footpaths. However, loops replace Stein's original concept of culs-de-sac and the plan of neighbourhood "A" shows that there seems to be a regrettable orientation towards and indeed on to the roads. However, it does seem promising that a private developer has decided to plan a city on principles of segregation of traffic and employed those famous for this. It is the first private "Radburn town" scheme in the world! To provide a path system even in very low density is a positive and far-sighted step.

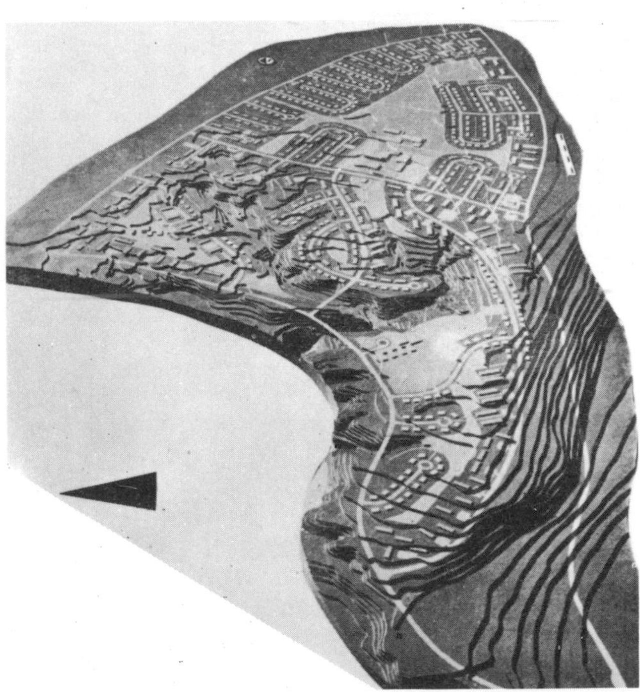

Fig. 6.112. Model of one neighbourhood showing contours with centre on the plateau.

References
McGuire Wild. Kitimat. *C.G.J.* November 1959.
Stein C. Kitimat. *A.F.*, July 1954.

VI. 9.14. *Tapiola, Garden City, Finland*

Site: Beautiful wooded area. 60° latitude. Near Helsinki on the Baltic Sea.

Design: Aarne Ervi won the competition for the town centre and designed other buildings. Several other architects were also engaged.

Origin: A variety of organizations pledged to produce good living conditions, particularly for the welfare of young people. Begun in 1952, completion date 1962.

Overall density 30 ppa (75 ppha).

The town is planned for some 15,000 people; the pedestrian centre, one of the earliest, is to cater for 30,000, including the hinterland, and has only 300 car parking spaces. It will be of interest to see whether this will work. Buildings on the beautiful site chosen are to take a secondary place to natural forms of trees, etc.

The objectives are set out as follows in a leaflet distributed to applicants for housing:

"*Present-day urban milieux, with their terrorizing traffic, persistent nervous tension, exhaust gases, soot and dust, are unfit for human life from all points of view including the biological. A better milieu must be created, and it is modern housing that holds the key to that. The most important target of present day town planning, therefore, is the creation of a socially and above all biologically suitable environment for man to live in. To this fundamental aim everything else must be subordinated. This vision was the starting point for planning.*"

"*The Housing Foundation will exert every effort to make Tapiola Garden City into a housing district protecting for man, his home life, rest and recreation. A good and safe environment must be provided for children to grow up in and for the young. Traffic may not dominate here; it must be kept in check and limited to the needs of the inhabitants. Tapiola is intended for residents who appreciate these viewpoints and are prepared to accept them.*"

In 1956 there were 4000 applications for 300 dwellings.

Fig. 6.113. Centre at Christmas.

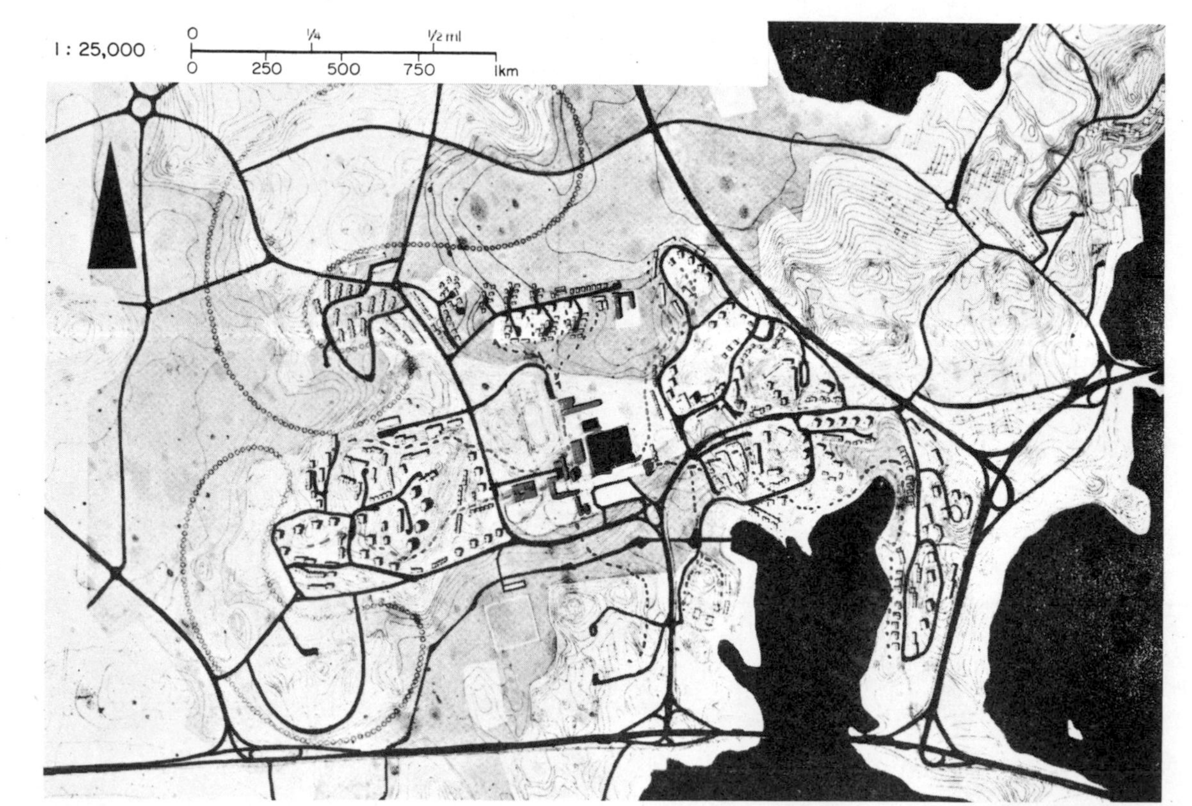

Fig. 6.114. Road and Path plan.

Fig. 6.115. Flats in the woodlands.

Fig. 6.117. Aerial view of terrace housing by Rewell and walk up tower blocks by Tavio.

Fig. 6.116. Centre of Ervi. Intimate, and free from traffic. The large central lake solves the problem of a large existing gravel pit.

Reference
Tapiola: Garden City. Finland 1959.

VI. 9.15. *Sputnik, Moscow, Russia*

Site: 25 miles (40 km) from Moscow on the motorway to Leningrad. Forests. 56° latitude.

Fig. 6.118. Town plan showing roads, footpaths, industrial areas, recreational and central areas.

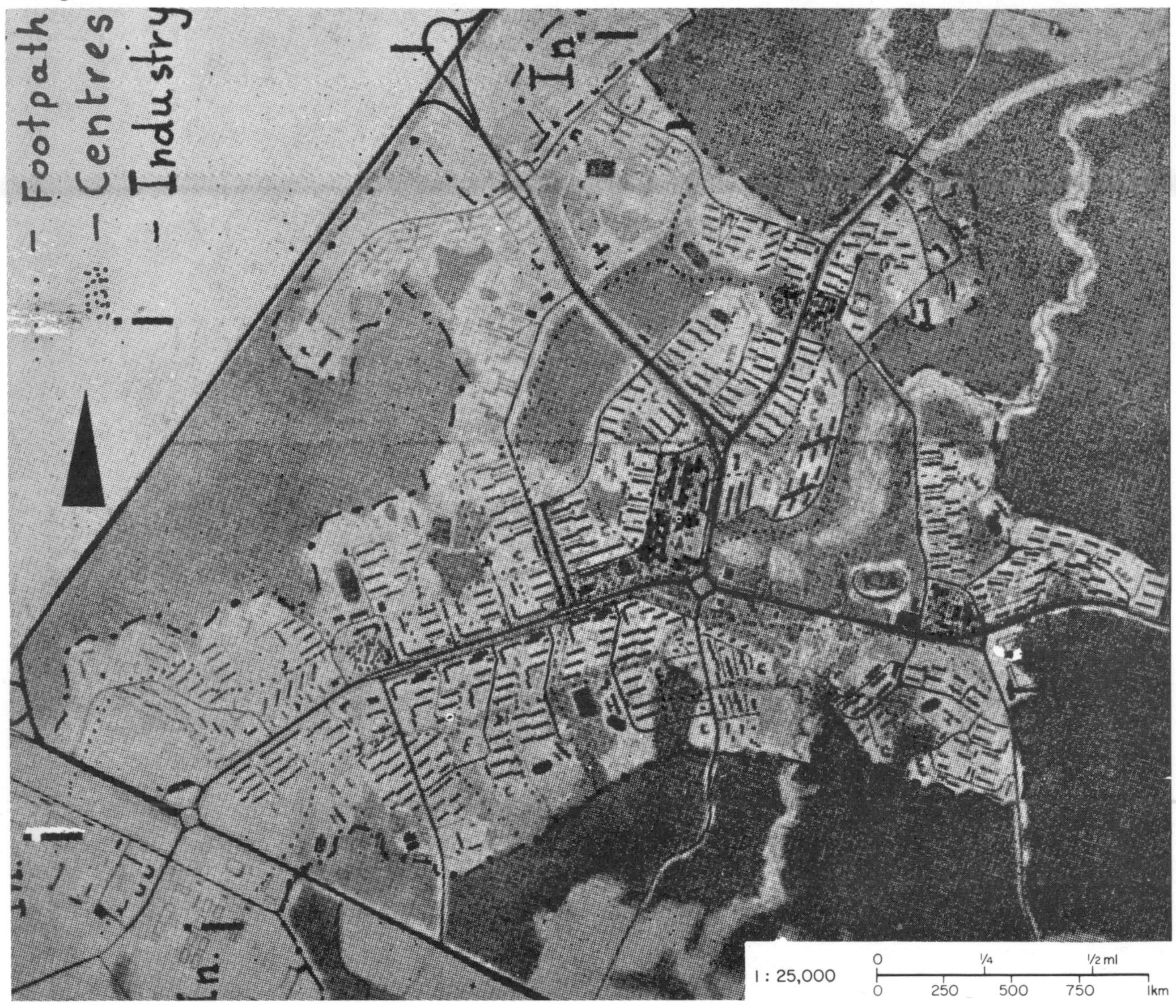

Design: Institute "Project for Moscow", Atelier 14. I. Rogine, Chief Architect.

Origin: First official Satellite Town for Moscow.

Population: 65,000 minimum to 80,000 maximum.

Kind of Town: A balanced community with three industrial zones; ten sections of 6500 inhabitants each; three sub-centres to serve these as well as one main centre. Main arterial roads connect residential areas with industrial zones. Road, mainly culs-de-sac, are within 200 ft (60 m) of dwellings. A footpath system connects recreational areas, schools, shopping and residential quarters. The markets are by the railway. The forest, in the shape of green fingers, enters the town separating the residential quarters. The open areas continue as communal orchards, 645 ft^2 (60 m^2) per family. The dwellings, none above four storeys, stand among gardens. There is to be district heating. There is a hospital and considerable recreational and cultural provision, with technical colleges. The total town area is approximately 1480 acres (600 ha) and of this 53% is built up, 17% is devoted to services, 13·5% to green space, and a similar percentage is taken up by roads.

Fig. 6.119. A perspective of the town showing residential buildings on either side and higher central area buildings in the middle.

Reference

L'architecture d'aujourd'hui, no. 88. Paris, March 1960.

VI. 9.16. *Cambridge Village, Cambridge, England*

Site: Bar House Farm, Huntingdon Road, 5 miles (8 km) from the centre of Cambridge. A valley with banks rising 60 ft (18 m).

Design: Covell, Matthews and Partners.

Origin: Private development by Holland and Hannen and Cubitts Ltd., in response to the County Planning Authorities encouragement for development of New Villages to the North of Cambridge, respecting its own development plans to limit the increase of the town itself.

Population: 3500 rising to 4000 through natural increase.

Kind of Village: One third of the population are to work in new local industry and commerce in the village so that this is not merely a commuter settlement. About fifty shops and adequate educational and social entertainment facilities are planned, to contrast with the old villages surrounding Cambridge. Thus the centre is likely to attract such villagers. The whole of the development is one superblock in the purest Radburn manner, with parkland in the centre and about twenty culs-de-sac serving clusters of some 100 houses directly, or by footpath access. Walking distances are short enough to make local use of cars unnecessary. Car space for each dwelling is to be provided. Density will be twice the normal for a village: private gardens very much smaller, in many cases only the size of the dwelling itself. 300 houses are to be completed by 1964 and the development completed within ten years. The village and open space beyond the ring road is to be owned by developers and a community trust and administered by a residents committee.

Reference
SENIOR D. Cambridgeshire Development, Village Scheme. *A.J.* 24 April 1963.

Fig. 6.120. Sketch for house type.

Fig. 6.122. Preliminary Housing Areas.

Fig. 6.123. Sketch for house types.

Fig. 6.121. Master Plan.

VI. 9.17. *Reston, New Town, Fairfax County, Virginia, U.S.A.*

Site: Rolling pastures and woodlands, lakes and brooks. 250 ft (76 m)–450 ft (137 m) above sea level. Annual rainfall 42 in. 18 miles (45 km) from Washington. 39° Lat.

Design: Planners, Whittlesey & Conklin. Architects for first phase; Geddes, Brecher, Qualls & Cunningham, Charles M. Goodman Associates, Satterlee & Smith, Whittlesey & Conklin.

Origin: Owner and developer, Simon Enterprises; the first city built as part of a presidentially approved regional development plan. A new zoning statute gives freedom of location, setting only a maximum of 11 persons per acre.

Population: 75,000 by 1980.

Kind of Town: A balanced community like the British New Towns, with industry shopping and recreation facilities.

"*The plan of Reston thus assumes the continued use of the automobile, but we have as a goal reducing the complete dependence upon the automobile that is characteristic of suburbia.*"

"*The home itself, in terms of transportation, is a pedestrian island. We meet our friends and have family life as pedestrians. The school is a pedestrian island. Most forms of recreation are pedestrian islands: the golf course, the baseball field, the gymnasium and the tennis court. And shopping, as presently developed, occurs in a pedestrian island.*"

"*Now, all these pedestrian islands which are the terminals of our vehicular transportation routes are all the good things of life. All of our social life, our recreation, and our education occur when we are outside of vehicles, during that portion of our life in which we are within pedestrian islands. So it seems quite logical that, in planning for a modern community, we should try to maximize the opportunity of being a pedestrian, for all the people of the community, we should enlarge these pedestrian islands as much as possible. An ideal diagram would combine them and make a single pedestrian island, with all the activities of our life surrounding Home, School, Work, Recreation and Shopping. But this is not physically possible. The land area requirements of each of these elements make a combination incompatible. But combinations of two or more are possible.*"(i)

(U.S. suburbs show 30% of trips are to work, 20% shopping, 10% school, 22% recreation, 18% others.) These figures produced by the planners back their decision to concentrate on walking connections between home and recreation.

There are seven villages, each with its own shopping centre in close contact with the high density "sinews" of the plan, and connected with their own recreational facilities. There is also a main centre to have a regional function beyond the town. "The villages are linked by walks, bicycle and bridle paths, while a network of roads is carefully separated." The Washington Centre of Metropolitan Studies is making a full-scale study on Reston's first 3 years.

(i) From *The Reston Story*. Brochure, 1962.

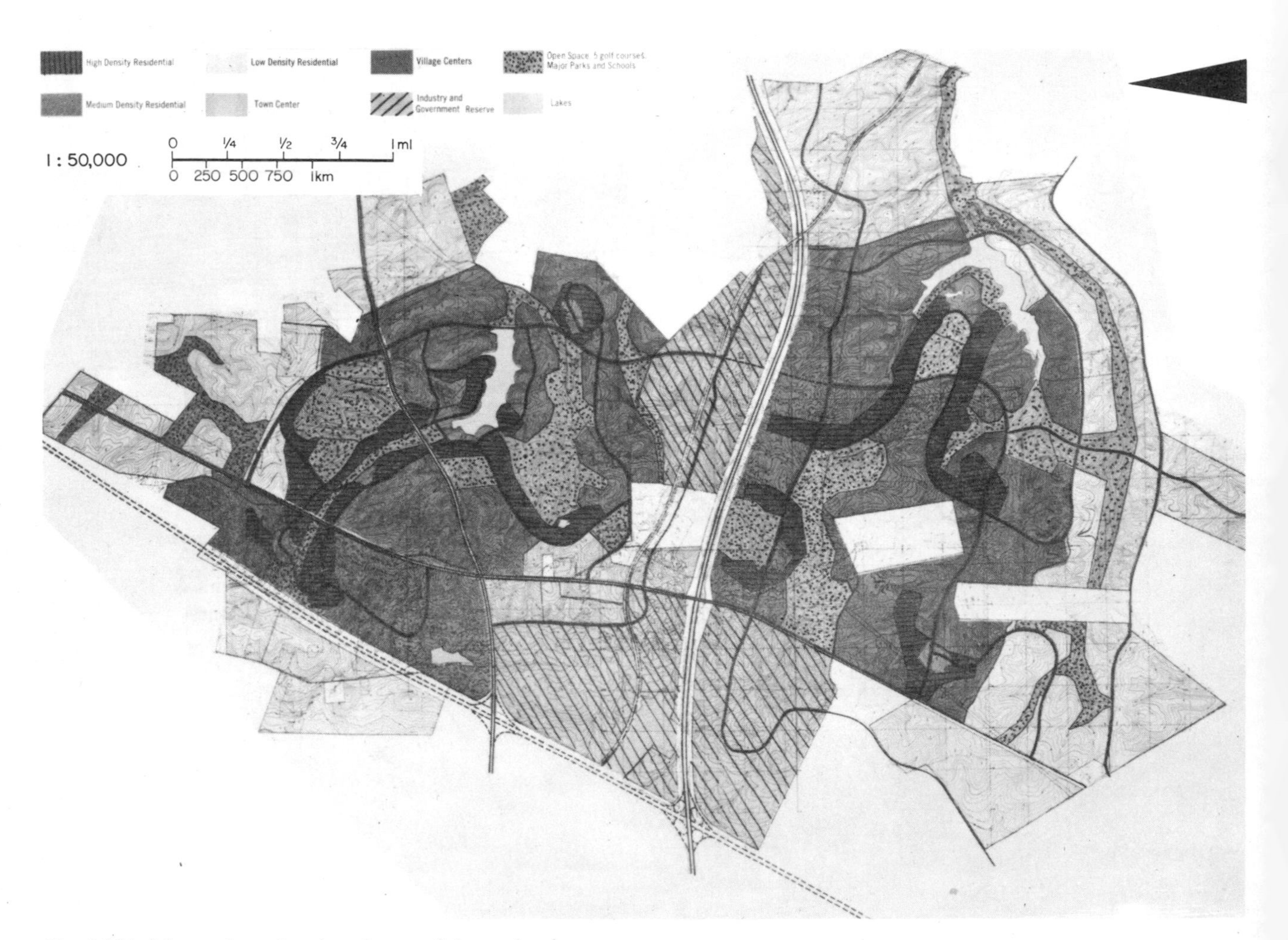

Fig. 6.124. Masterplan, showing sinews of dense development with main paths access along these, not shown on the plan. Existing freeway cuts Reston but no access to it is as yet allowed.

Residential Areas: The first residential areas designed show clusters of dwellings with immediate access to recreational areas. The design of these clusters is a great contrast to the endless strings of widely spaced dwellings spawned across the States. Densities increase up to 60 ppa (150 ppha), and there is a great variety of dwellings and buildings, from tall flats to terrace houses. Beyond this it is intended that a large number of privately designed owner occupied houses should contribute to the overall design of Reston.

Robert Simon, the father of the originator of the town, worked with Clarence Stein on Radburn.

One particular idea planned for Reston is residential clusters around a particular hobby such as horse riding or yachting. The popularity of riding, on the increase, has led the planners to position bridle paths right through the town so that shopping on horse back is possible. Hitching posts will be provided in the village centre.(i)

"*The Reston Plan is based on two convictions: people should be able to do things they enjoy near where they live; many Americans want the stability of belonging to one community for a lifetime. They are tired of rootlessness.*"

Fig. 6.126. Model of first village centre, by Whittlesey & Conklin, with housing in point block, terraces and detached dwellings.

Fig. 6.127. The same centre from the lake.

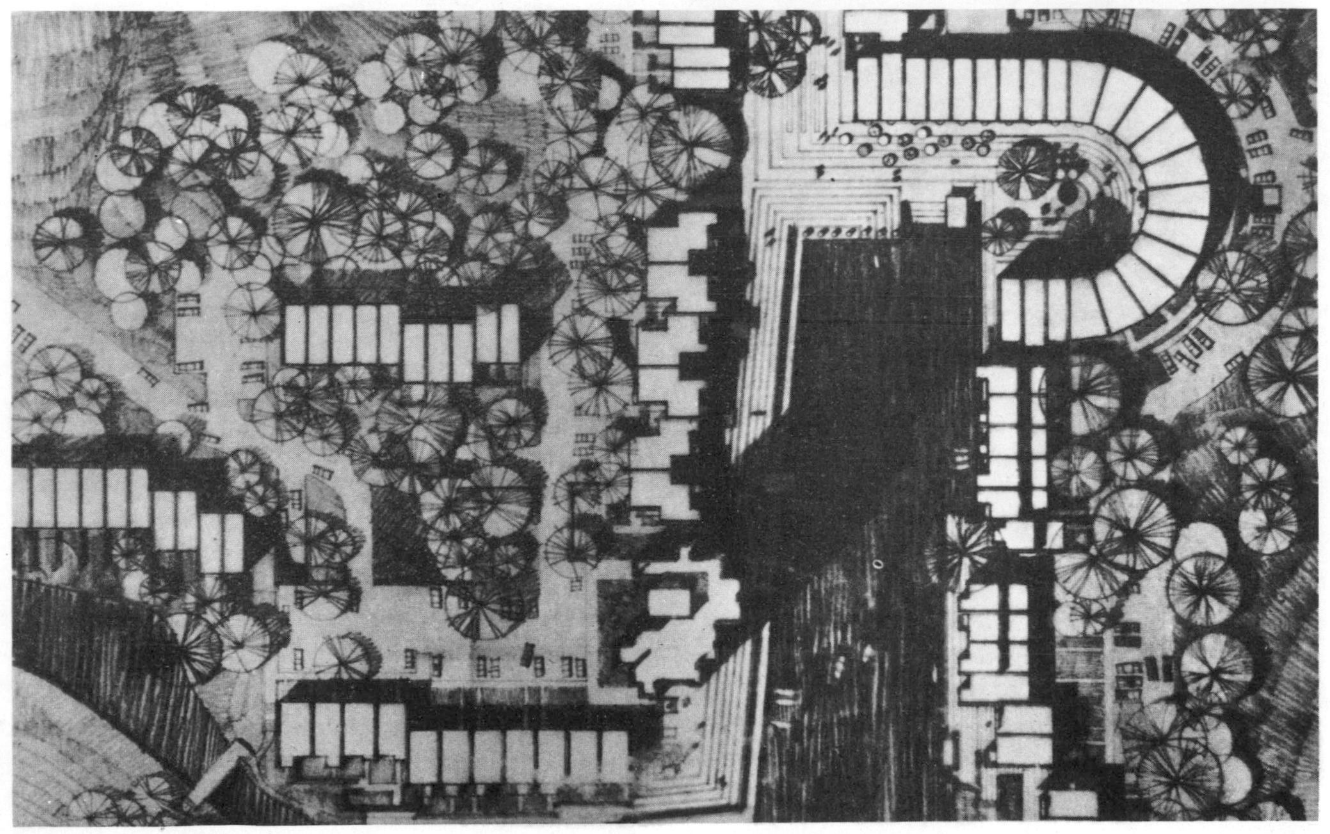

Fig. 6.125. Plan of first village centre.

(i) Riding Schools etc. in the path environment were supported by the author in his address of the ILA Symposium 1961.

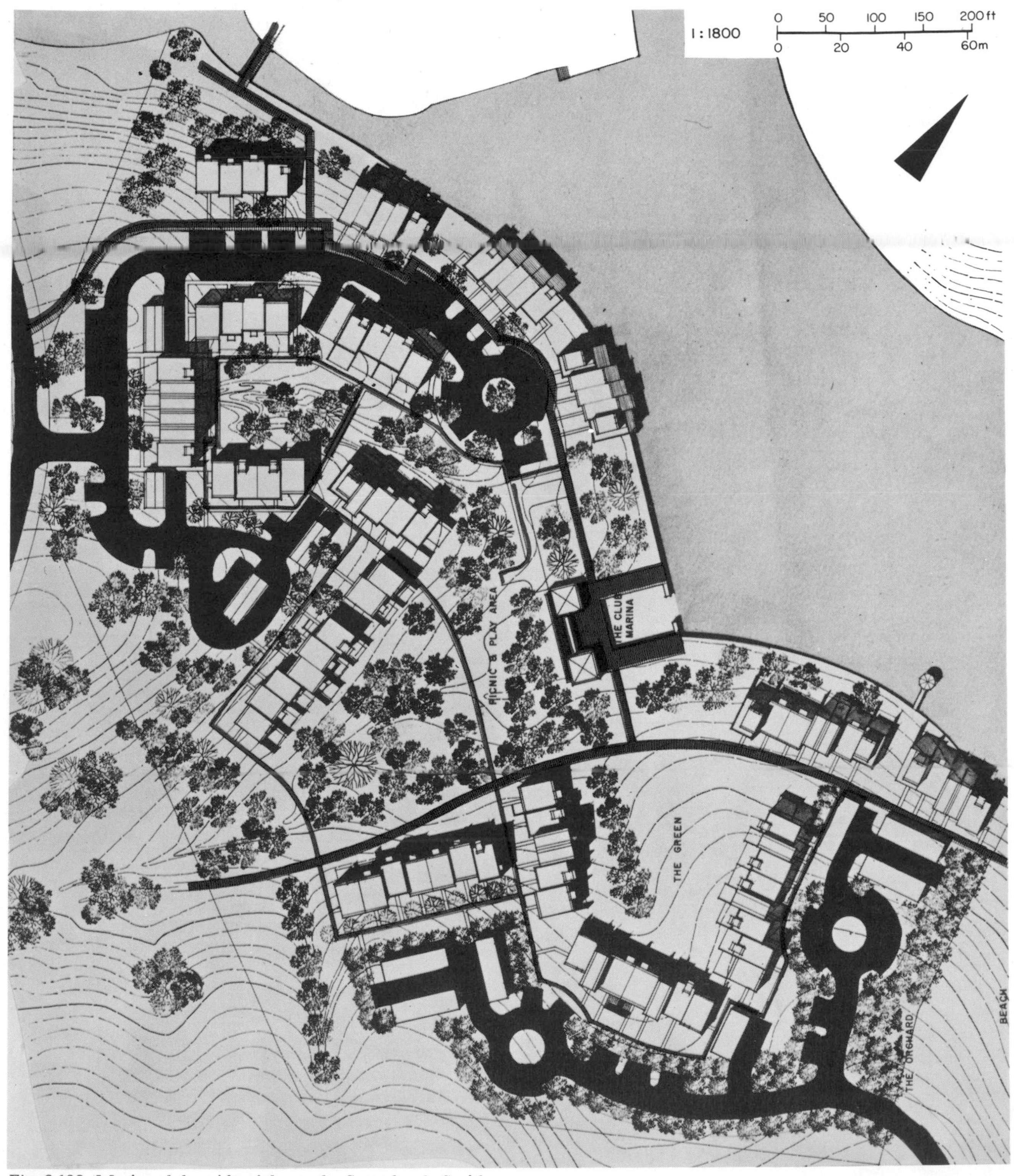

Fig. 6.128. Marina club residential area by Satterlee & Smith. Plan.

Fig. 6.129. Marina club area, view from lake.

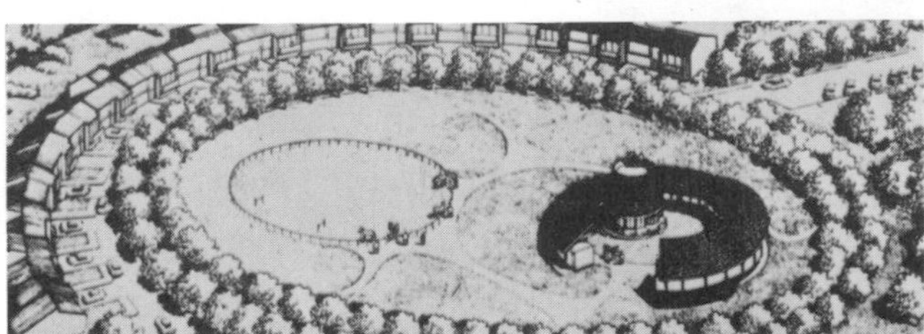

Fig. 6.130. Community stable cluster, by Geddes, Brecher Qualls and Cunningham, for those keen on riding.

VII. Urban Renewal

VII. 1. General Policy 152
- VII. 1.1. The Case of New York. 153
- VII. 1.2. The Case of San Francisco. 154
- VII. 1.3. The Case of Detroit. 154

VII. 2. Design for Urban Renewal 155
- VII. 2.1. Basic Principles. 155
- VII. 2.2. Horizontal Traffic Segregation. 157
  - VII. 2.2.1. Segregation by time. 158
- VII. 2.3. Vertical Segregation, Pedestrians Above. 158
- VII. 2.4. Vertical Segregation, Vehicles Above Ground Level. 159
- VII. 2.5. Vertical Segregation, Vehicles at Ground Level, Pedestrians Below. 159

VII. 3. Urban Roads 159
- VII. 3.1. Improving Existing Road Systems. 159
- VII. 3.2. Introduction of Motorways or Freeways. 160
- VII. 3.3. Motorways in the Townscape. 161

VII. 4. Parking 162
- VII. 4.1. Parking Estimates. 162
- VII. 4.2. Other Factors Influencing Parking. 163
- VII. 4.3. Kerbside Parking. 164
- VII. 4.4. Parking Meters. 164

VII. 5. Economic Considerations, General 165
- VII. 5.1. Cost of Full Motorization. 165
- VII. 5.2. Comparative Costs of Commutor Journeys. 166
- VII. 5.3. Cost of Traffic Improvements. 166
- VII. 5.4. Cost of Congestion. 167
- VII. 5.5. Economics of Pedestrian Shopping Streets. 167
- VII. 5.6. Comparative Values. 168

VII. 6. Examples 170
- VII. 6.1. Coventry, Traffic Plan. 170
  - VII. 6.1.1. Coventry, Shopping Precinct. 172
- VII. 6.2. Stockholm, Traffic Plan. 173
  - VII. 6.2.1. Stockholm, Sergelgaten. 174
- VII. 6.3. Philadelphia, Traffic Plan. 176
  - VII. 6.3.1. Market East Section. 177
- VII. 6.4. Liverpool, Traffic Plan. 180
  - VII. 6.4.1. Ravenseft Area. 182
  - VII. 6.4.2. Civic Centre. 184
  - VII. 6.4.3. Paradise Street Area. 184
- VII. 6.5. Newcastle, Traffic Plan. 186
  - VII. 6.5.1. Two Redevelopment Areas. 187
- VII. 6.6. Hong-Kong, Central Area. 188
- VII. 6.7. London, Traffic Plan. 189
  - VII. 6.7.1. Trafalgar Square Area Redevelopment. 191
- VII. 6.8. Fort Worth, Central Area Reconstruction. 192
- VII. 6.9. Nottingham, Central Business Districts. 194
- VII. 6.10. Banbury, Central Area. 197
- VII. 6.11. Swindon, Central Area. 198
- VII. 6.12. Västerås, Sweden, Traffic Plan 199
- VII. 6.13. Bishop's Stortford, Hertfordshire, England. 201
- VII. 6.14. Sutton Coldfield, England. 202
- VII. 6.15. High Wycombe, England. 205
- VII. 6.16. Beeston, Nottinghamshire, England. 206
- VII. 6.17. Horsholm, Denmark. 208
- VII. 6.18. Chester, Central Development. 208
- VII. 6.19. Andover, Hampshire, England. 210
- VII. 6.20. Lijnbaan, Rotterdam, Holland. 212
- VII. 6.21. South Bank Development, London, England. 213
- VII. 6.22. Stevenage, England. 214
- VII. 6.23. Stevenage, Shephall Centre, England. 215
- VII. 6.24. Treppenstrasse, Kassel, Germany. 215
- VII. 6.25. Pedestrian Shopping Street, Essen, Germany. 215
- VII. 6.26. Greyfriars Redevelopment, Ipswich, England. 216
- VII. 6.27. Sea-front site, Brighton, England. 216
- VII. 6.28. Strøget, Copenhagen, Denmark. 216
- VII. 6.29. Ellor Street Comprehensive Development, Salford. 218
- VII. 6.30. Hammersmith Central Development, London, England. 219
- VII. 6.31. Piccadilly Circus, London, England. 220
- VII. 6.32. Knightsbridge Green, London, England. 220
- VII. 6.33. Basildon New Town, Main Centre, England. 221

VII. Urban Renewal

VII. 1. General Policy

Until recently there was no doubt that improving our town centres meant the widening of streets. The Ring and Spoke policy was generally accepted as inevitable and good. The vehicle was the criterion, though radials and rings are not efficient for this either. It has become obvious that the car's demands are insatiable. It devours any city that invites it unreservedly. Although more money (multi-level use) can accommodate more cars, each and every private vehicle of the ever-increasing number cannot act as a few were able to years ago. It was not a valid precedent but rather the motorist's honeymoon. A city does not live by private cars alone.

Today a different criterion applies. The town centre can become more convenient and prosperous by creating a harmonious pedestrian environment served by excellent public transport, while providing efficient parking for many private cars. The number will vary with the character of each city and usually cannot be enough for all the motorists of the future. The inconveniences involved, inevitable for some, in the use of public transport, or parking at some walking distance from their goal, can be more than adequately counterbalanced by the delight in the compact, car-free, care-free centre, with easy access to a multitude of shops and other accommodation. The U.S. suburban shopping centres offer pleasant pedestrian shopping only after a longish unpleasant walk from some point in the sea of parking surrounding each one and are popular even so. This can be improved upon considerably in town centres.

There is no need for the city centres to die or erode. On the contrary, the freeing of the core from vehicular domination spells a new era of more civilized city life and an important means whereby town centres can become a good residential environment. Imagine the evening: from the dance halls and the cinema, the theatre and the concert hall, the lecture room and the prayer meeting, the restaurant, the pub and coffee bar, the clubs, hotels and libraries, the museums, the church, the political meeting—the people emerge. The hilarious, heated, jovial, serious, pensive, reverent, inspired, passionate, restless or romantic mood is not immediately reduced to the same traffic-jarred common denominator, the stinking, noisy, dangerous road. One may linger, laugh and love, sway, swoon, stand and stare, delve, deliberate, ponder, postulate and prattle, in a multitude of urban spaces, each appropriate in character and detail. People must be allowed to live in the centre. It is the right place for many, it can be the right place for more.

Fig. 7.1. Seven Corners, Virginia.

Fig. 7.2. Northland Detroit.

Fig. 7.3. Bötcherstrasse, Bremen, one of the first streets to be limited to pedestrians in Europe in recent years.

Thus, it requires a value judgement to decide how many cars shall be allowed in the centre of any particular town. The question is "How many private vehicles can this or that city take?", not "How many will wish to drive their vehicles in?" The answer will determine the extent of roads and parking. These are closely related. Dallas has empty parking lots but its road speeds are among the lowest in the U.S.A., indicating where their lack lies. In Europe, of course, roads are not adequate, but most cities have such a dearth of off-street parking that not even the influx of cars possible along its streets is anywhere satisfied and hardly ever rationally considered. In New York a recent application for construction of thousands more underground parking spaces in Manhattan was turned down on the basis of the sort of value judgement mentioned: it was pointed out that this would tempt more cars into the already overcrowded streets, and make life even worse for the pedestrians.(i) Such judgements should rely on a proper understanding and feeling of what it means to have say, ten thousand cars on the road, or parking single or multi-storey and on profound knowledge of the city involved, allowing a full use of latent existing facilities.

A decision on the degree of motorization that a particular environment can stand has become basic to the art and science of urban renewal and is closely linked with the prevailing economic policies, which may themselves be irrational and antisocial. A Ministry of Planning might help in supporting priorities and induce action by grant allocation.

Fig. 7.4. Main shopping centre, Basildon New Town, Essex, England. Office block and shops. (See also p. 221.)

VII. 1.1. *The Case of New York*

New York is doubtless a special case. But as its space restriction is near to the problem of many European capitals it is worth while learning from an instructive study by Leiper.(ii)

Midtown, the core of New York City, employs about a million people and serves as a centre to a hinterland of about 16 million. In an automobile orientated country New York Midtown has a mere 25,000 off-street parking spaces which must be by far the worst U.S. provision. In spite of this, during the decade 1948–1958, retail sales rose by 16% whereas the five next most important central areas in the U.S.A. lost 4% of trade during that time. This seems to be in spite of four large department stores closing down within that area during the same time period, as reported by Bendtsen.(iii)

In discussing the future of New York City, Leiper writes:

"*In January,* 1960, *the city's Department of Traffic made public a study recommending the construction of* 16 *municipal parking garages providing more than* 10,000 *additional off-street parking spaces to be located generally within the Midtown core area. This* $57 *million program would represent an increase of nearly* 40% *over the existing off-street parking supply in this area. The proponents of the new garages claimed that rate structures designed to encourage short-term parking would attract substantial numbers of new shoppers and business travelers to the Midtown area, and that traffic congestion on Midtown streets would be alleviated by resulting reductions in street traffic and parking.*"

"*In March,* 1961, *the New York City Planning Commission released a report disapproving the Traffic Department's application to construct the first three garages in the proposed program. The Planning Commission's report concluded in effect that, rather than attracting primarily new travelers, additional parking garages might divert many commuters, shoppers and business travelers presently using mass transit or less convenient parking facilities. The commission also concluded that more automobiles might be attracted to the Midtown area, thus adding to the traffic congestion in the core. In exercising its zoning powers to restrict the development of off-street parking the Commission expressed its concern that benefits for a few motorists would be outweighed by detrimental effects on the city as a whole.*"

And the tentative conclusion he reaches is this:

"*Evidence could be presented to suggest that the vital Midtown area faces an awesome but inevitable task of future rebuilding to accommodate many more cars. Perhaps a better case could be made, however, for the theory that the decision-makers now rebuilding the Midtown area have already predetermined that the car will continue to play a back-seat role here.*"

(i) Dealt with fully later in this chapter.

(ii) LEIPER J. McM. The Role of the Automobile in Midtown Manhattan. *T.Q.* Saugatuck. April 1962.

(iii) BENDTSEN P. H. *Town and Traffic.* Copenhagen 1961.

"An ultimate solution to the problems of circulation in Midtown Manhattan will probably combine features of the different alternatives we have discussed. The facts of life, however, point strongly to the conclusion that the New York Region has no real alternative but to put maximum emphasis on improving its mass transportation systems. Even then it may be overwhelmed."

(Manhattan's special problem lies with the very poor public transport, about to be decisively overloaded, it is supposed, by 50,000 extra workers in the PANAM building, about to be completed right above Central Station.)

VII. 1.2. *The Case of San Francisco*

San Francisco is only 350 miles (563 km) from the only city with a four level fly-over, Los Angeles, also with the only freeway to relieve a freeway in its centre(!). Yet in San Francisco freeways can be seen sticking into the air, unfinished, signs of a population not accepting the motor-car at any cost, as in the other city.

In 1959 the legislative body of San Francisco disapproved freeway proposals as shown on the plan which had been officially adopted. The Board said,

... *"the demolition of homes, the destruction of residential areas, the forced uprooting and relocation of individuals, families and business enterprises; and ... the property taken from the tax rolls will shrink the already restricted taxable area of San Francisco ..."*

but the matter seems far from settled:

... *"The recent approval by the voters of a bond issue for a rapid transit system will have its effect in that certain peak hour conditions, particularly on the Bay Bridge, may be alleviated. However, the freeway system as originally planned was predicted on the need for a separate transit system, although many people may have voted for rapid transit on the basis that it would obviate the need for more freeways."* (i)

Fig. 7.5. Not to be completed freeway "ruins" sticking in the air at San Francisco.

(i) Letter to the author from J. W. KEILTY, Senior City Planner, San Francisco. December 1962.

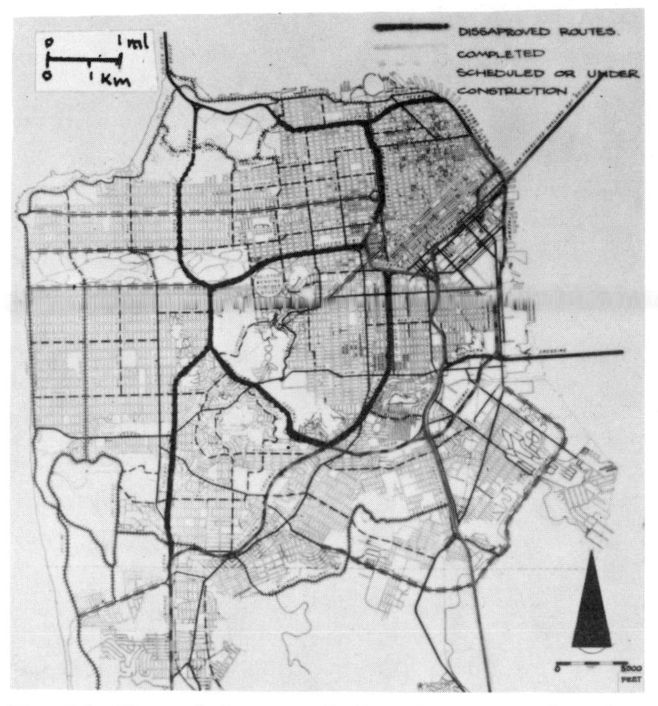

Fig. 7.6. Plan of freeway disallowed by San Francisco legislative body.

VII. 1.3. *The Case of Detroit*

Yet further evidence of the need for substantial public transport comes from Detroit where more people seem to use private cars than in any comparable town. During peak hours in 1960, only something like 40% of commutors entered the centre of Detroit by public transport. Yet Alger Malo, Director of Traffic in that city, writes,

"With all of the streets leaving the District now used to capacity during the peak hours, street space would have to be nearly doubled in order to accommodate the same number of people without mass transit." (i)

Even this sounds like a very conservative estimate.

Fig. 7.7 Los Angeles, Harbors freeway in cutting.

(i) MALO A. The Relation of Mass Transportation to Total Transportation in Detroit. *T.Q.* Saugatuck. April 1961.

VII. 2. Design for Urban Renewal

VII. 2.1. *Basic Principles*

The basic principles to make possible the change from our present-day chaotic townscape to a better one are fourfold:

a) a growing pedestrian-only environment.
b) an improved public transport service with convenient termini linked directly with parking complexes.
c) an optimum provision of parking and road space, efficiently connected with shops and public transport—including the park-and-ride principle.
d) the careful use of vacant space and multilevel use of central space.

These principles can be implemented in a number of ways, and they can be usefully combined in endless permutations. Small towns, of which there are many thousands—as well as the large ones—need careful planning. A larger percentage of small town inhabitants can and want to travel to the centre by private car—particularly in low-density areas. The larger the city, the greater the problem of accommodating the car population. In the latter, particularly, where suburban railway lines penetrate to the heart, the provision of parking at outlying stations can be extremely useful. This policy is in use in Stockholm, St. Louis, Philadelphia and San Francisco, among others. It allows the efficient use of the car for the journey to the station. Subsequently, however, the train will take the worker to his goal more quickly, and as nearly to his place of work as the car might in many instances, and as cheaply, bearing in mind the inevitably rising costs of parking as the demand increases.

Quite close to the centre there is, for example in Nottingham, the opportunity to accommodate a great many cars in a strategic position, so that the commuter can conveniently change to a rail shuttle service, using existing lines in tunnels, right into the city centre, with opportunities for new development at a number of vantage points. It is a very general point that railway stations can be adapted to provide space for the commuter's cars and connections for his last lap to work.(i)

It will become a general principle for city centres to be guarded from cars by an outer ring of "defences"—the car parks which conveniently transfer the driver to train or bus or other means of rapid transit. Then there will be the inner ring of defences in the form of car-parks directly off the inner ring road, or motor way, or tangentials, or the equivalent, which allows the driver to find easily the available empty spaces. These must, as must also the public transport stops, be closely associated with the pedestrian precincts and the various pedestrian levels below and above ground.

The creation of precincts, and the linking of these, must proceed in each city according to a three-dimensional plan which indicates the circulation (pedestrian and vehicular), the scale and the approximate use of any area to be redeveloped, as a whole or piecemeal. Beyond a purposeful network of paths within the centre, the development plan must extend to show how the educational, industrial and residential areas bordering on to the centre will be connected by paths to redevelopment in the centre. The town-planner who has hitherto knitted with one strand only, namely the roads, must learn to knit with two, the paths and the roads. This in itself is a fundamental growth principle. Bremen has gone a long way towards such a plan (thanks to Professor Wortmann), using the much used existing park-paths in the centre of the city, and connecting these with newly created pedestrian ways.

*Fig.*7.8. Part of the wall-walk of York, showing a section where it passes over a main road.

Fig. 7.9. Bremen's footpath plans, by Prof. Wortman.

(i) See examples, Section VII, pp. 173, 194.

The town expansion of Basingstoke will create a path network connecting existing areas as well as the new ones with the centre, as also at Andover.

It should be statutorily required for each authority to produce a footpath survey and proposals, just as is now the case with roads in many countries. Each city has many interesting pedestrian ways and many more are sorely needed to make a comprehensive system. The long walk-ways on York's city wall are an outstanding example.

It is remarkable that the most up-to-date literature on this subject is the joint publication by the Ministry of Transport and the Ministry of Housing of Great Britain. To quote from this important, well-illustrated official 22-page Bulletin;

"*The traditional town-centre with its variety of uses has developed over the centuries as a market place and a meeting place. Essentially it was a place for people on foot, a place where you could move about freely and safely. Some of these features remain in our town centres: narrow pedestrian ways have survived, unexpected open spaces have been retained, and historic buildings preserved. Redevelopment should not ignore these qualities; they strengthen the sense of continuity between the past and the present without which a town becomes anonymous and dull.*"

"*Circulation: traffic movements in the town centre are interdependent with function and layout. The basic objectives must be worked out in terms of pedestrian and vehicular movement. Traffic alone should not dictate the planning of the town centre, but most of the functions and activities of the centre depend on the movement of vehicles and people, and can be analysed in those terms. A satisfactory solution to traffic problems is fundamental to the success of town centre renewal. The primary objective must be to sort out conflicting types of traffic and provide adequately for each—in particular to separate pedestrian and vehicles so that both can move freely and safely.*"(i)

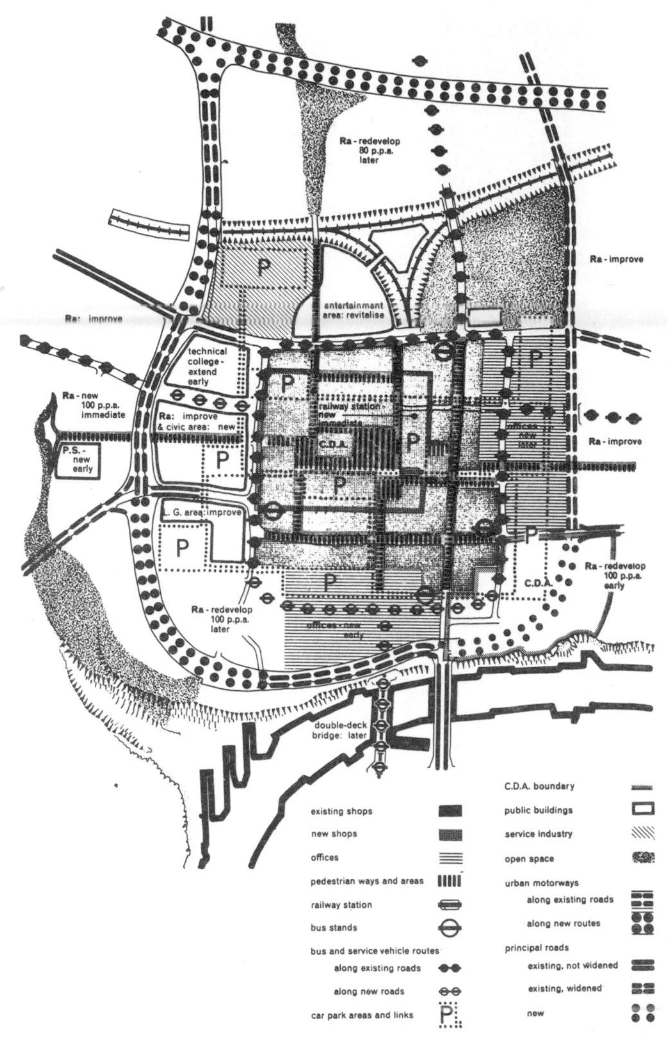

Fig. 7.10. Ministries of Transport and Housing Bulletin suggestion for town centre map, indicating the kind of development plans each city ought to work out to solve its long-term problem, step by step.

Traffic separation in central areas may be a combination of any of the following systems. The chief criterion for design is the convenient, pleasant, continuity of the pedestrian's routes and spaces.

Fig. 7.11. The section showing the "safe" pedestrian route from Stockholm's station across the town centre suggests that here "convenience and pleasantness" are not sufficiently considered, not even with escalators!(i)

(i) Ministry of Housing and Local Government, Ministry of Transport Planning Bulletin 1. *Town Centres, Approach to Renewal.* H.M.S.O. 1962.

(i) EDBLEM, STRÖMDAHL, WESTERMAN. Moten ny Miljö. *Arkitekten Stockholm* 8, 1962.

VII. 2.2. *Horizontal Traffic Segregation*

In areas where there are economically well established and architecturally worthwhile buildings, a street, or street network, can be closed to vehicles, giving pedestrians the freedom of the whole width of the streets. This is particularly suitable where this width is not great. In small market towns, many of which have in fact only one main street, which widens perhaps into a market place, this is usually the right solution. Often, a connection with the open space around cathedral, castle or the town park is possible.

Fig. 7.12. Söge Strasse, Bremen, closed to traffic: note new lighting brackets.

Fig. 7.13. Nottingham, not even a one-way street!

Fig. 7.14. and 7.15. Lincoln Road, Miami—a wide street changed into a mall.

It means that all along the street the areas of highest value, in every way, can be left intact. Modifications take place all around the back. Here a ring road may be constructed or an existing set of roads used for one-way traffic right round the central strip. Car parking, single or multilevel, can find space among the lower cost property at the backs. The short access routes from car parks to main pedestrian street or precincts become commercially desirable areas. Single or multiple pedestrian crosses are formed in this way. The possibilities for a better environment in the regained, or indeed newly found, pedestrian spaces are numerous. Bus stations must be safely and conveniently connected with the pedestrian areas.

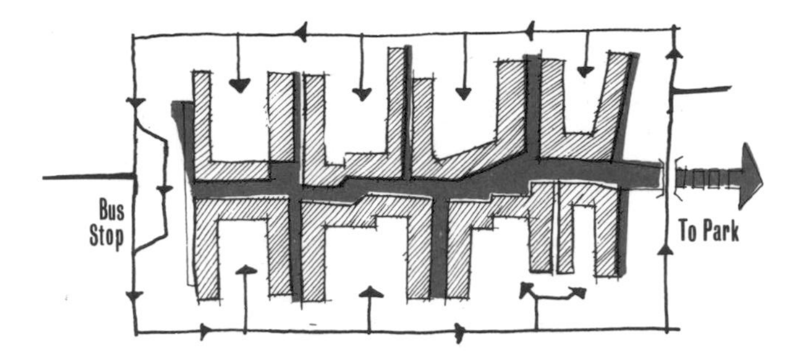

Fig. 7.16. Diagrammatic shape of horizontally segregated shopping centre, particularly applicable to smaller old market towns on level sites. Where slopes exist, parking and service at lower levels become possible.

VII. 2.2.1. *Segregation by time.* Closing a street off to traffic during certain hours each day can give a great deal of benefit in retail trade and so real estate value, virtually without any capital outlay. Delivery is during the convenient, usually early and late, hours of the day. In such a pedestrian haven in Gothenburg, created out of one of the main shopping streets and originated by the police, trade increased.

Subsequently there was a request by shopkeepers in adjoining streets for the same arrangement. It is particularly suitable, to the grid-iron street net of Gothenburg, and many other towns, to close, say, every other parallel road. This system has been used in a number of German towns. With good public relations it is very easy, compared with other types of traffic segregation, to make such an experiment for a time.(i)

VII. 2.3. *Vertical Segregation, Pedestrians Above*

In larger towns, where it is difficult to cut off traffic from some streets, where precincts can only occur at intervals, there is a problem, particularly where increased pedestrian flow from precincts crosses vehicular routes. In such places multi-storey car parks have their proper place outside the immediate central area to keep traffic at bay. The connection at basement, first floor or even second floor level, of these car parks with the inner shopping areas, can be contrived via low level shopping arcades, pedestrian bridges, decks and platforms which link with shops and may give protection from weather right into the precincts. Vertical connections and long horizontal stretches can have mechanical aid from moving pavements and escalators. The urban scene can be enlivened by such multiple traffic streams. To gain additional shopping floors, with direct access from the car parks, is an advantage to customer, shopkeeper and property owner. It is now well understood that unless people are introduced naturally and easily to additional levels of shopping they will not normally use them. The exceptions are the smaller, special shops, philatelists, herbalists, second hand book shops, etc., who have their own special lunch-time and evening clientele.

(i) See also Strøget, Copenhagen.

Fig. 7.17. Upper Deck, Sheffield, England, under construction. This illustrates the slope leading naturally to the upper level which connects with the volume of the new multi-storey market for small traders (City Architect, Womerley; Architect, City Centre proposals and of Market, A. Derbyshire. (See also p. 286.)

Where the centre of a city slopes, change in level opens up many cheap possibilities for vertical segregation. Pedestrian decks have to be much higher, to clear double decker buses, than raised roads. The latter have merely to clear the heads of people. The structure for the former although higher is slighter and cheaper. High halls of department stores might, with this idea, become part of the usable town centre volume, penetrated by the pedestrian on his routes.(i) Such connections apply similarly to public transport termini.

(i) See Philadelphia, p. 176.

VII. 2.4. *Vertical Segregation, Vehicles Above Ground Level*

The danger with this solution, in all but the large towns, is the size of elevated motorways which tend to dwarf and destroy the scale of the town. They are much wider, very much wider than railway viaducts, and cannot stride elegantly across a town as the latter do. The multiple level connections and the space these occupy are forbidding: 200 ft (61 m) with sliproads as against 35 ft (10·7 m). Where levels make this sort of solution desirable, buildings should be designed with the road structure in mind. Views from the motorway become an important consideration. Services and storage can take their place below such high level roads. (See Liverpool and High Wycombe.)

The carbon monoxide fumes, which descend, since they are heavier than air, must be taken care of. In complex, grade separated junctions a whole multi-storey composition of roads is involved, as much as 70 ft high. (See Section V.)

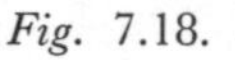

Fig. 7.18.

VII. 2.5. *Vertical Segregation, Vehicles at Ground Level, Pedestrians Below*

This solution, used to date in large towns (Philadelphia, Stockholm, Vienna, Montreal, London, Piccadilly), brings with it the strict limits beyond which a mole-like existence, however glamorous, becomes depressing to human beings.

After visiting the Opernring, Vienna, the City of Stockholm decided to cover the large oval area to the south of Sergelgatan, previously open to the sky though on a lower level. This step seems regrettable to many.

Fig. 7.19.

VII. 3. Urban Roads

VII. 3.1. *Improving Existing Road Systems*

Bendtsen[i] believes that in spite of increases in ownership, the speed on our crowded urban roads will stay constant, at least in the large cities. He says that as soon as improvement has made possible better movement for the cars on the road, others are attracted to this and reduce the speed to the average tolerated, which Bendtsen takes as about 6 mph (10 km/hr) (1961 average speed in Glasgow was 8 mph (13 km/hr); Newcastle 9 mph, (14·5 km/hr); Edinburgh 13 mph (21 km/hr); Worcester 12 mph, (19 km/hr). He states, on the basis of American statistics, which show that traffic runs more orderly and safely now than in 1920 despite the enormous increase in cars since then, that the minor road improvements can increase flow considerably.

Recently a peak-hour travel time study showed that average speed had risen from 24 mph in 1957 to 30 mph in 1962 in Metropolitan Los Angeles. Before this can be accredited as a gain it would be interesting to know how much extra distance of travel the new routes represent. It is only a gain in time which is worthwhile, in travelling from *A* to *B*, not just speed on a longer route.

Factors which increase flow in roads and which can often be easily applied to existing town centres include:

- Reduction of junctions.
- Reduction of traffic stream crossings.
- Rational and obvious precedence at junctions.
- Creation of curves to allow proper town speeds.
- Taking pedestrians out of the road network.
- Segregation of varying types of motor traffic.
- One-way traffic with "progressive" traffic lights changing with flow and rational arrangement of right and left turns, with space for those waiting to turn, to reduce conflict of traffic streams.
- Taking out unnecessary forests of signs and, where possible, traffic lights. Painting of lanes.
- Taking out street parking.

But if this increase in flow is towards an area which itself cannot cope, then improvement is meaningless and very wasteful, i.e. Hyde Park Corner Improvement, London 1962. It must be taken as a basic law that road improvement cannot be considered rationally in isolation from other factors. Whether it is a matter of road widening, one-way traffic or other furthering of flow, it must be ascertained that it does not have adverse effects on the town as a whole. This may be so, for improved road speeds attract further traffic; the supply is endless

(i) Bendtsen P. H. *Town and Traffic.* Copenhagen 1961.

and likely to remain so, to fill any road in any city centre.

Increase in flow has been recorded for example in the following cases:

The effect of parking prohibition on speeds ranges from 27% increase to 16% in odd/even date parking, the latter applicable to London.

Allenby Road in Tel Aviv increased peak-hour capacity from 1103 to 2120 vehicles and speed from 3 mph (5 km/hr) to 9 mph (15 km/hr) after the following measures:

installation of traffic light
provision of pedestrian crossroad
lane marking
no left turns
no kerb-side parking.

In Philadelphia, abolition of kerbside parking reduced accidents by 10%. The increased speeds along certain roads in London, such as Gower Street, and the greater flow, have had the good effect of making it easier to drive and resulted in less accidents,(i) but have had the bad effect of making it more difficult to cross a road in an area where there is much walking by a large pedestrian population. It has cut down the use of buses unable to come to the more convenient place of the previous stops. This is a serious drawback of one-way traffic. *A* to *B* might be as long in the new scheme as the statistics refer to speed and flow, not to extra distance or relative times taken over a journey, particularly local ones. In the Piccadilly/Pall Mall scheme the disadvantages far outweigh the advantages.(ii) University College Buildings alongside Gower Street are more noisy and the environment generally more hostile to man. Accident figures given by the Borough Engineer and Surveyor(iii) show a staggering increase of over 300% in Gower Street, but it is suggested that Hampstead road figures should be taken into account also. The figures for the three streets are:

TABLE VII.1.

	1960	1961	1962
Tottenham Court Road	68	46	76
Gower Street	8	14	27
Hampstead Road	65	—	48
	141		149

(i) Only in the first year.

(ii) CHURCHILL J. D. C. Effect of Traffic Engineering Measures on London's Buses. *T.E. & C.* January 1963.

(iii) Letter to author 29 March 1963.

At the same time a newspaper report quotes that all other roads in St. Pancras of similar importance have reduced accident figures. In this case, unlike other claims, it can not be said that the one-way system has led to greater safety. In residential areas the simple re-routing of major traffic arteries into so called "ring-roads" or relief roads can have equally dubious effects, although statistics may show them to be successful.

H. A. Barnes, who has introduced electronic computors in 8000 intersections at Baltimore, cutting accidents by 46% and increasing speeds from 6 mph (9·7 km/hr) to 18 mph (29 km/hr), still maintains that there has to be a limit on car usage. His long-term solution was a three-lane bus system "travelling on motorways (cantilevered over backyards at $3 million a mile)"—to be more rapid than private cars.(i) To take the pedestrian out of the motor-road (no zebras) is also likely to speed up traffic as well as make it safer. But there are no figures for such an experiment although almost all other factors seem to have been studied.

An interesting idea has been developed for Holborn, London, by Cook, Borough Architect, where 4000 pedestrians cross in 15 peak minutes every day. Highways ramped 1:15 allow pedestrians to cross below.

Dr. Smeed(ii) has also shown that parked cars and other obstructions affect road capacity considerably. Whereas car capacity rises, the slower the speed, in London, according to work by the British Road Research Laboratories(iii) it is also true that wider streets and particularly motorways are the most efficient way of using each square foot of road to shift vehicles at about 35 mph (56 km/hr).

VII. 3.2. *Introduction of Motorways or Freeways*

The introduction of motorways has complicated effects. One view is that "an area's overall freeway needs are relatively independent of the degree of concentration down town."(iv) The same book says:

"*Planning Implications.—Non-work trips for shopping business, social and recreational and other purposes have been shown to increase in frequency as income and car ownership levels rise. Since automobile ownership in urban areas will probably continue to increase at a greater rate than population, between now and* 1980, *non-work trips will represent a larger share of all travel as the ratio of cars to population increases.*

(i) *A.J.* 26 November 1961.

(ii) SMEED R. J. *The Traffic Problems in Towns.* Manchester 1961.

(iii) WARDROP J. G. *Traffic Capacity of Town Streets.* R.R.L. London 1954.

(iv) Smith and Associates, *Future Highways and Urban Growth.* New Haven 1961.

Most non-work trips are made during midday and evening hours, rather than at peak hours. Thus, peak-hour travel will become a smaller proportion of 24-hour travel. The implication is that new highways built to accommodate peak-hour travel will receive relatively more over-all use in future years than they do at present."(i)

A more detailed analysis of the effects on the other city streets has been made by Downs. His main conclusions are:

"1. *Peak-hour traffic congestion on any expressway linking a central business district and outlying areas will almost always rise to surpass the optimal capacity of the expressway.*"

"2. *Therefore, in relatively large metropolitan areas, it is impossible to build expressways wide enough to carry rush-hour traffic at the speed and congestion levels normally considered optimal for such roads. The forces of traffic equilibrium will inevitably produce enough overcrowding to drive the actual average speed during peak hours to a level below the optimal speed.*"

"3. *Commuters driving on expressways should resign themselves to encountering heavy traffic congestion every day, even though they may spend less time commuting than they did before using expressways.*"

"4. *Since urban expressways cannot be designed large enough to eliminate rush-hour traffic congestion, other design goals must be employed to decide how large a capacity each such expressway should have. The only goal regarding commuting which is feasible is reduction of the average amount of time spent by commuters per trip. Low congestion and optimal speeds during non-peak hours probably constitute much more practical goals for the design of urban expressways than any goal connected with rush-hour traffic movements.*"

"5. *Under certain conditions, the opening of an expressway may make rush-hour traffic congestion worse than before the expressway was built. This outcome can occur only in cities in which a high proportion of commuter traffic is carried by segregated track transportation facilities before the expressway is opened.*"

"6. *Thus any program of expressway planning and construction must be integrated with similar programs concerning other forms of transit in the area if it is not to cause unforeseen and possible deleterious effects upon the level of automobile traffic congestion therein. In particular, marked improvement of roads without any improvement in segregated track transit may cause automobile traffic congestion to get worse instead of better. Since the U.S. has already launched a massive road improvement program, which includes construction of many urban expressways, continued failure to undertake an analogous program for other forms of urban commuter transit may result in a generally higher level of rush-hour automobile traffic congestion in those cities which now have extensive segregated track transit facilities serving commuters. Therefore such possibilities as the charging of direct tolls on expressways and the expenditure of auto toll collections or gasoline tax revenues on segregated track transit should be thoroughly explored as means of developing a balanced urban transportation system.*"(i)

(i) Downs A. Peak Hour Expressway Congestion. *T.Q.* Saugatuck July 1962.

VII. 3.3. *Motorways in the Townscape*

It is most difficult to manage the very wide road in general townscape. They may become something positive or they can be hidden in enormous cuttings as in Chicago, Detroit, Los Angeles. Their efficiency and cost, i.e. three million pounds per mile, makes them a rational choice in many situations. Their correct placing and planning is crucial as they are so big and costly.

The earth work can be minimized, as in Los Angeles, by alternately cutting below the cross traffic and rising above it. It can be exciting to view the city from roof-top height (equivalent to the fourth floor) from the motorways. But raised roads cut a city visually into parts and it is important that this should be meaningful. Whether sunken or raised, or level, the positive design of wide roads in an urban environment is an almost virgin art.(i)

In old town centres and in those old residential areas which have to be pierced, the art is even more taxing, the technical skill required even greater. Where the motorway is carefully designed it can be an asset visually and socially. The Hammersmith fly-over in London adds to its environment. By taking traffic off adjoining roads, a motorway on stilts, far from cutting off an area, may allow passage below, and also easier crossing of the old relieved street system nearby.

Peripheral motorways without frontages determine that social and economic values grow within precincts away from them. Ring-roads which, with crass illogicality, allow shopping development along them, for example in Birmingham and Nottingham, suck the life out of precincts into the dangerous, stinking, noisy environment, contradicting all rational planning principles. Such wide and busy roads really do cut the town into pieces. When underpasses are liberally provided, the cost difference between the "ring-road" and motorway shrinks. Reasons for such planning are invariably short term and short sighted eagerness by the Local Authority to make money by selling sites bordering on "ring-roads" for shopping. This is a matter for legislation to put right.

(i) See Liverpool, p. 180.

Fig. 7. 20. Motorway junction in Chicago.

VII. 4. Parking

VII. 4.1. *Parking Estimates*

Basically five kinds of parking take place in a city centre.

a) Shoppers and visitors for short-term parking. The percentage of cars parked for less than one hour varies inversely with the size of the town from 88% of the cars for towns of 5000–10,000 to 44% in cities over a million,[(i)] in the U.S.A. This is from the economic point of view the most lucrative and from a functional point of view the most worthwhile provision.
b) Those who work in the city requiring half day or whole day parking, and who arrive first and somehow have to be kept from the spaces for short-term parking.
c) The parking of service and delivery vehicles for varying periods, which should be provided with the premises.
d) Residents' parking. This may be insignificant or it may be large if a concentration of flats is placed above the centre in tall blocks.
e) Evening entertainment parking, which can use a) and b).

There seem some basic points about parking estimates which have hitherto been ignored. First there is the decisive distinction between towns built to take full motorization (which can only apply to moderately sized new towns) and other towns. Only in the first instance is it of any use to calculate what is likely to be wanted ideally. Given existing towns, and particularly big ones, we now know that all parking provided will be filled, if it is of the right sort and in the right position. Thus, the provision of parking becomes a matter for careful assessment of how many people are best provided for, by which kind of parking, in what position, for how long, and at how much per hour.

This is not a simple matter of counting parked cars and multiplying by three to give the Ministry recommended figure for 1980. To list more exactly the interacting decisions that have to be made:

a) How much total parking space (and associated road) can be provided without ruining the town (in the centre) at what levels?
b) What percentage of this shall be for commuters and for shoppers?
c) Where therefore are the strategic places for the parking?
d) How does this fit in with existing road capacities and proposals?
e) What is the rational programme of provision according to growth?
f) What price is right for parking at various points in the town? This last affects the kind of structure (levels) that can be built economically and who will build and run the parks.

The general principle of keeping car parks near or right next to main routes must be repeated, for this keeps down the need for traffic and the roads to carry it. The law-endorsed provision of parking within each site is obviously nonsense although individual contribution to communal strategic garages is a rational way of providing. The flexibility and full use of parking through radio communication or some similar advice informing the traveller where there is space, has hardly been considered, but is obviously worthwhile. Intercommunication between car parks is important. In Coventry[(i)] this takes place on a separate ring route.

The enforcement of the planner's intentions, with regard to who should park where and for how long, is a very difficult problem.

When the shopper arrives in town, the extra parking spaces provided have often been filled by workers in the city, as there seems to be an endless waiting list of such workers. Mere provision of parking places will hardly help shoppers. The provision of short-term parking with parking meters is not to be relied on. A survey in London has shown that nearly half the available meter space is used illegally for long term parking![(ii)] Parking offences number many thousand per day in every large city, beyond those noticed by the law.

In Boston, Mass., whole rows of cars can be seen right by a "Towing Area" sign. So much for the law, even when it threatens quite specifically.

Professor Bendtsen[(iii)] concludes that the construction of motorways has not helped retail trade; but the construction of underground parking right in the shopping centre of San Francisco made the real estate values soar.

(i) Burrage and Mogren. *Parking. Saugatuck.* 1957.

(i) See example, p.170.
(ii) Kirby C.P. *Utilization of Parking Meter Spaces.* L.C.C. 1963.
(iii) Bendtsen P. H. *Town and Traffic*, Copenhagen 1961.

TABLE VII.2. *Development in retail trade in central area and city as a whole*, 1948–54

Place range 0·5–13 million	Cars parking/ 1000 inhabitants between 10 a.m. and 6 p.m.	Retail trade Central area	Retail trade Whole town
Cleveland	47	+ 6%	+31%
St. Louis	50	−10	+32
Seattle	71	+ 5	+38
Louisville	83	+15	+41
Portland	92	− 3	+20
Dallas	110	− 2	+59

The fact that Dallas, which has the best parking facilities, shows a loss of trade in its central area, which contrasts dramatically with the gain in the town as a whole, makes sense only when one reads in another table that peak traffic speed at Dallas is one of the slowest among cities of similar size, namely 3 mph (4·8 km/hr) for public transport and 5·6 mph (9 km/hr) for private cars. This example demonstrates that neither better access roads nor better parking facilities alone, without the other, can solve the problem. Road capacity and parking capacity must balance.

In large towns extra parking places do not necessarily increase trade. But, in a survey of 16 Californian towns of 10,000–140,000 inhabitants, the number of parking places per citizen did make an important difference to trade.

VII. 4.2. *Other Factors Influencing Parking*

The level of car parks can follow three basic principles. They can be on the ground, overhead, or underground. There are many variations and combinations possible, but the following factors should be borne in mind:

Car fumes are heavier than air, and fall. Wherever the car is driven it takes noise with it and its weight. Where tall blocks exist, the roofs of the lower buildings are vital, both visually and from the point of noise generation (reflection, etc.). Long-term parking can afford a longer journey to the car than shopper's parking. The latter ought to be in the most direct contact with shops possible, vertically or horizontally. The number of cars parked at one particular outlet (particularly long-term parking) should be determined by the road capacity, with regard to rush-hour needs. The existing flow on such roads and whether crossing traffic is permitted, will be vital factors.

Fig. 7.21. Part of the centre of Minneapolis. Federal aid enables local authorities in the U.S. to pull down blighted areas quickly. Rebuilding is much more difficult and complex. In the meanwhile every available foot is filled with parking, four deep! First in, last out. Photo taken by author from Foshey Tower, 1962.

Both limited parking for services and service access to shops are always required. They should be in direct contact with the shops, either horizontally or vertically, by running belt, trolley service, lifts, etc. Where horizontal (from the back) contact takes place there is the special problem of avoiding the crossing of the service route by those who have parked their car on the ground at the perimeter parks. Visually it also means special care with yards and backs. These areas are open to view from high buildings in the centre. At Stevenage (see example in Section VII)[i] careful design of the backs with yards for most shops represents a feasible solution. Entrances through to some of the main shops, direct from the parking areas, are valid where the service backs are replaced by service below ground. At Farsta, the wife, her shopping done, waits at the exit, facing the parking. The goods are in a little trolley, with room for a small child. The husband drives up and unloading takes place straight into the car at the door of the shop. The empty trolley is left. Orders can also be collected by car in the tunnel loop.

From an aesthetic and economic point of view very much could be gained if there was no "front" and "back" but shopping façade all the way round. It is by multi-level service that the Fort Worth super-block manages to be so expansive and provide so much frontage in a relatively small area.

Under or over-passes are essential in connecting the inner town and car parks from outside the ring road to the core itself. These ways through are of particular importance where a considerable flow of shoppers and workers can be encouraged to enter.

Garages must be strategically placed. But hitherto only the parkers' view has been emphasized. It is equally important that the dead form of a garage does not sterilize the frontage of an area. This applies to residential and commercial areas. In Nottingham this mistake is about to be made. A delightful site, previously occupied by a church, in a visually strategic place, is to be filled by a garage, four-storey parking with no other kind of accommodation at all. This is a typical example of planning for one function instead of the town as a whole in an integrated fashion, typical of cities without plans and proper planning departments.

VII. 4.3. *Kerbside Parking*

This kind of parking becomes invalid as soon as the efficient use of road space is required. Street side parking contributes to accidents to a considerable extent. Since comparatively very few cars can stop even along the front of a long shop, the abolition of kerbside parking is not the commercial disaster it is often said to be. Most parking—particularly in Europe—is still kerbside parking, in many towns almost exclusively. Double depth kerbside parking was rampant in many places and still must be in many towns. Where kerbside parking is allowed, in any area, the speed of cars ought to be drastically reduced to, say, 15 mph (24 km/hr). "Clearways" should become general. In Stockholm's main streets even taxis are not allowed to pick up passengers.

(i) See p. 214.

Day[i] protests rightly against the many different parking regulations and enforcements. Particularly alarming is the information that in Britain the police now have the right to tow away (£2 fine) any vehicle an officer thinks obstructs the highway. There is no right of appeal and no justification by the police is required.

VII. 4.4. *Parking Meters*

Parking meters installed in London are judged to have improved traffic flow. Half are misused by long-term parking even though the financial returns are far less than expected because of administrative costs, largely wardens guarding against misuse.[ii] It is degrading that they should take up such a large amount of pavement. Pedestrians on the inner side of roads ringing London Square have to squeeze past meters which have destroyed the usefulness of these considerable miles of pavement.

Day recommends a far more finely graded cost to make meter use continual and allow extended parking in places. He also describes how to feed meters successfully beyond the legal time.(!)

In Leicester, wardens alone are reported to be most successful in enforcing short-term parking.

Fig. 7.22. Traffic jams are nothing new in London!

(i) Day A. *Roads*. London 1963.

(ii) Kirby P. *L.C.C. Report* 1963.

VII. 5. Economic Considerations, General

The economics of redeveloping central areas are complex. The following basic relevant factors can be isolated.

The values in any central areas depend largely on retail trade and related real estate values. To bring retail trade into the centre, ease of access is essential and shopping should be convenient, safe and enjoyable.(i) Ease of access is achieved through good roads for smooth travel and well placed parking, and, even more important, really efficient public transport which can be continually improved. Pedestrian shopping networks give safety and pleasure in shopping.

It is impossible to provide all the roads and all the parking ideally wished for and still retain the space for shops. In other words, when 49,000 parking places and two thirds of all the area is given over to cars in Los Angeles, the real estate values drop. The roads have attracted more cars and the traffic jams are similar but much bigger, involving many more people and ever longer distances. Shops and life have moved out of town although there is much building in the centre at present by the city itself.

Where city councils care for their city and planning officers produce the schemes, property owned in the centre by the city itself can, of course, be used in ways more rational and civilized than those based on monetary value pure and simple. Indeed, this is the only possible way of properly reconstructing cities, of bringing people back to live in the centre and providing against the unbalanced overpopulation which acutely afflicts London, New York and many other towns in the world.

In the redevelopment of centres, private capital is essential in the economy of many countries. As this goes hand in hand with the danger of profiteering at the expense of the community, problems of administration and public relations are involved. It is difficult to have full communication with the public with plans so as to have the support and response necessary for democratic planning and at the same time prevent the ruthless speculator from gaining knowledge of property which will gain value rapidly. The way out, which has been taken by a number of authorities, is to state plainly that the decision on what will be developed privately and what will be developed by the authority itself is not determined. This does not solve the problem in some instances.

(i) "*The economic losses in London had become apparent as early as* 1635: '*The greatest number of hackney coaches of late time seen and kept in London, Westminster and their suburbs, and the general and promiscuous use of coaches there was not only a great disturbance to his Majesty . . . and others of place and degree in their passage through the streets; but the streets themselves were so pestered, and the pavements so broken up, that the common passage is thereby hindered and more dangerous; and the prices of hay and provender, and other provisions of stable, thereby made exceedingly dear*'."
From: BRUNNER, C., Roads and Traffic Economic Aspects, *Traffic Engineering and Control*, July 1960. A proclamation in London in 1635, quoted by R. M. C. Anderson in *Roads of England*.

VII. 5.1. *Cost of Full Motorization*

Bendtsen clearly shows, in the case of Copenhagen, that European capitals must in the future restrict private car use.(i) Copenhagen would need 16 four-lane motorways to bring in and out its fully motorized population (500 vehicles per 100 residents. 20% town centre use assumed). The junction of these with the ring distributor road would take up an area of 17 × 660 × 660 yd (15·5 × 600 × 600 m) or a total of a circle slightly over 1 mile (1·61 km) in radius. Beyond that 80,000 parking places would be required. The costs estimate of this is given (see Fig. below).

16 motorways	125 million pounds
Cost of junctions	25 million pounds
80,000 parking places	100 million pounds
	250 million pounds

Bendtsen's costs might be reduced by multiple use of land. But he shows that the undesirable is also too expensive. The conclusions of the calculations apply more or less to any city of similar size.

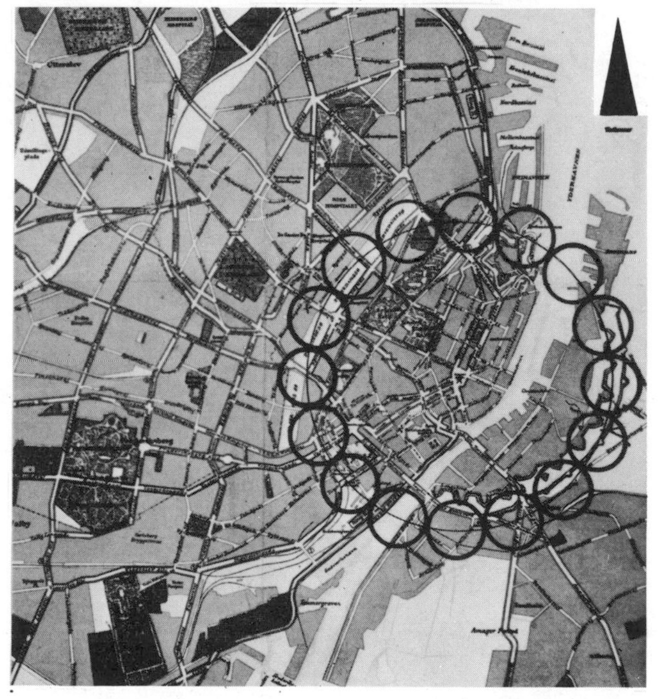

Fig. 7.23. According to Bendtsen a full motorization of Copenhagen would mean radials (16) and 16 junctions of a size such as could only be accommodated in the position shown on the plan, taking up the areas shown. (Diagram by Paul Ritter.) If parking and roads are now added, it will be seen that little of the centre of Copenhagen could remain.

(i) *Traffic Congestion Costs Money*. Plan, Stockholm 1953.

VII. 5.2. *Comparative Costs of Commuter Journeys*

Bendtsen shows an interesting economic comparison between the cost of private car journeys and public transport. Both in Copenhagen and in New York the public ride came out about twice as expensive, which for short distances seems quite remarkable, but is confined to immediate petrol and similar expenses only. If it were not for the dollar parking fee, private car travel would be cheaper in the U.S.A. If we consider that the private car does go faster (which has also been shown by *Fortune* magazine by sending out drivers from central areas of many cities at rush-hour and giving them 30 min to reach their maximum distance), and that to this time saved must be added time saved through not having to walk to the public vehicle and from it, there are many cases where personal economics may be in favour of the private car. But purely economic considerations are not always the decisive ones.

Day,(i) Reader in Economics at London University has written that: (see also writings of Beesley, M.)(ii)

"*One of the biggest troubles with road policy both here and abroad is that far too little attempt is made to apply these basic common-sense rules of equity. On the one hand, there are many new roads which ought to be built but are not—even though their benefits would far outweigh their costs, nothing gets done. On the other hand, there are far too many cases where the benefit to an individual arising from his use of a car or lorry is far outweighed by the costs he imposes on the rest of the community—the outstanding example, being the commuter who drives in and out of a city at rush hour, at a cost to himself in petrol and oil no greater than the bus or train fare and in the meantime parks free of charge all day on congested streets. The costs he imposes on the rest of the community are far greater than the cost he himself has to pay.*"

He suggests that the best solution would be to charge for each mile driven and that this is entirely practicable, if politically a hot potato.

Passive electronic "blocks" on vehicles would reflect messages to electronic roadside scanners. For a few pounds each individual could have one of these with its own code. The information would be passed from the scanner to a central computor. Given a proper spacing and a charge per mile a bill would be sent monthly to the driver for use of road. The idea is flexible. Central London might cost 4*d.* per mile, outskirts 1*d.* per mile and at night roads could be free.

Such proposals must be given immediate serious consideration. To introduce them when a more or less congested town has been taken for granted is too late. Rational habits can spring from early introduction. The Ministry's report(i) indicates such policies:

"*In general, the urban problem will have to be tackled partly by the expenditure of larger amounts of money than hitherto and partly by regulation of traffic, especially of the use of the private car. The Group cannot yet make recommendations about the appropriate scale of investment nor suggest the ways in which effective traffic regulation should be achieved. But the most efficient allocation of available road space, as a generally accepted objective, might be approached by administrative action, by the use of the pricing mechanism or by a combination of both. Administrative action might take the form of parking prohibitions, the exclusion of certain types of traffic from certain streets at particular times, or subvention of classes of traffic, such as public service vehicles. Under the heading of the pricing mechanism one may include parking charges, which are already widely applied to kerbside and off-street parking in London and to some extent elsewhere, and charges for the use of road space assessed on moving vehicles by the use of meters, electronic scanners, stickers, etc. A good deal of experience is being obtained, mostly in London, and the Group think that as much experiment as possible should be made. But further research into the economics of urban traffic and the possibilities of applying any suggested solutions is needed.*"

VII. 5.3. *Cost of Traffic Improvements*

Buchanan(ii) states that the traffic flow improvement is proportionate to money spent on any given area. This is an over-simplification leaving the "bargain" of inventiveness out of consideration.

In an overpass experiment in the centre of Birmingham, a steel structure designed to last ten years is costing only £80,000 whereas the estimate for the structure in permanent form was £500,000. The erection time of the single carriage fly-over is one week-end. The experiment is likely to be an important precedent. Original thought may bring many more economic and satisfactory solutions.

Structural ingenuity is helped by the use of cheap peripheral land at the back of existing and thriving property.

The Road Research Laboratory has suggested a number of ways in which it can be assessed whether a road improvement scheme is worth the money spent on it. These invariably leave factors out of account and are only an approximate guide(iii).

(i) DAY A. *Roads*. London 1963. See also CHURCHILL J.D.C., R.S.H. 1963 (for criticism see T. E. & C. October 1963).

(ii) Beesley M., Roads, Town Planning Institute Summer School, Cambridge, 1963.

(i) Min. of T. *The Transport Needs of Gt. Britain in the Next Twenty Years*. H.M.S.O. 1963.

(ii) BUCHANAN C. Town and Traffic. *R.I.B.A. Journal*. London August 1962.

(iii) WARDROP J. G. *The Capacity of Roads*. R.R.L., London 1954.

Liverpool's 5 mile inner ring motorway is to cost £23,000,000 for construction, including interchanges, and land costs a further £8,500,000, just over £6,000,000 per mile. The road is raised 20 ft above ground level and is to have six lanes.

VII. 5.4. *Cost of Congestion*

At this point a further factor arises, with various complex facets: the cost of traffic congestion. Such is the undeveloped state of Traffic Engineering that "congestion", the term universally used to gauge capacity of roads, is defined in very different terms by different authorities. Wardrop and Duff(i) list the following:

a) mean speed of traffic;
b) mean difference in speed between successive vehicles;
c) standard deviation of speeds;
d) actual speeds in relation to desire speeds;
e) actual numbers of overtakings in relation to desired numbers;
f) proportion of vehicles impeded, based on the distribution of time-intervals between successive vehicles.

The following table shows the annual cost of traffic congestion in Manhattan. It is taken from the report of the Citizens Traffic Safety Board, 1952. (Neither the basis of calculations nor the "normal" traffic speed is given.)

Extra	
in wages	200 million dollars
in petrol	40
in insurance	15
in repairs	40
in depreciation of vehicles	90
taxi fares	30
reduced retail	50
entertainment loss	15
frustrated building	50
Total	530 million dollars

Bendtsen suggests that the reduction on the retail turnover, judging by the increase of other business, may be 350 million dollars underrated. Thus this cost would provide any city with a handsome underground railway system every year! And it must be borne in mind this is Manhattan only; losses are also incurred in the remaining districts of New York. Many other differing calculations for other places have been made. All the figures are astronomic and point to the need for thought and action. The Road Research Laboratory in their work have concentrated on average speed–flow relation which, in towns, is useful as it takes into account journey time, important to the urban driver. They take 15 mph as the accepted minimum speed between controlled intersections.

Official calculations in Britain are limited to losses in working hours, at about 15*s*. per vehicle hour, a "puritan" approach which is not realistic. This leads to an underestimate of the money worth spending on improvements—£500,000,000 for 1958 is given as the total cost of traffic congestion in Britain by Wardrop.(i)

One important principle emerges. The degree of congestion is twice the increase of vehicle numbers. The more cars on the road, the worse the effect and cost of each extra one.

VII. 5.5. *Economics of Pedestrian Shopping Streets*

Shopkeepers in particular still seem to believe that parking and window shopping while in the car are important sales attractions. Contrary to this view it was discovered that sales in the centre of Philadelphia rose 3%–8% after prohibition of kerbside parking at some 500 strategic places out of 23,500, admittedly a small proportion. The *Engineering News Record*(ii) in an article on Pedestrian Shopping Malls says in heavy type: "*The mall is the quickest, least costly and most dramatic way to attract business.*" Further, the Danish Institute for Shopping Centre Research strongly backs pedestrian shopping areas. This remarkable organization is financed by all retail traders in Denmark, co-operative and private. Recently the British Retailers Association has expressed a similar opinion.

Very recently the Multiple Shops Federation in a booklet, *The Planning of Shopping Centres*, concedes that the Federation has "been driven to the conclusion that the only satisfactory basis on which to plan a new shopping centre or rebuild on old one is the principle of the traffic free precinct".

It is clear that once parking is prohibited in a street and the motorist cannot "stop and buy", as in the shopkeepers' popular concept, he will not gain from passing traffic. In contrast, the pedestrian mall will attract the customer, who, because on foot already, may more easily stop and buy. And if a *non-parking* shopping road is turned into a pedestrian area it does not stop parking occurring

(i) Wardrop J. G. and J. T. Duff. Factors affecting road capacity. International Study Week in Traffic Engineering. October 1956. Stresa, Italy.

(i) Wardrop J. G. and D. J. Reynolds. Economic Losses due to Traffic Congestion. R.R.L. Nice 1960. 5th International Study Week in Traffic Engineering.

(ii) *Engineering News Record.* 17 October 1959.

wherever it occurred before. The logic of this and other such arguments is entirely left out of consideration. The perfection demanded of a remedy before it is admitted as worthwhile, may be out of all proportion to an existing mess. It is important to differentiate clearly between the advantages gained by an improvement and the perfections which, although desirable, may not be obtainable.

An example is quoted in the *Engineering News Record* of a shopping mall in Boston which came to grief after it has been in "highly successful operation" for some time, because it was found that *one* department store had no supply line.[i] The traffic commissioner had to decide between hurting commerce by barring the delivery trucks or sacrificing the mall. The mall went and the narrow streets reopened to traffic! This is quite irrational, as the one store could have been supplied from within the mall, along the route left for fire engines (as in the Kalver straat, Amsterdam and others).

In another American town an old scheme for making the main road of a small town pedestrian was tried out, without any proper public relations merely by closing the street to traffic one fine day, for a few weeks "experiment". This was, as may have been foreseen, disastrous for retail trade and the "experiment" had to be abandoned almost immediately. The sad thing is the conclusion: the idea was faulted—not the lack of proper communications with the citizens. Such "experiments" may have ulterior motives!

In Lincoln Road Mall at Miami trade increased appreciably, as at Gothenburg, on the exclusion of the car.

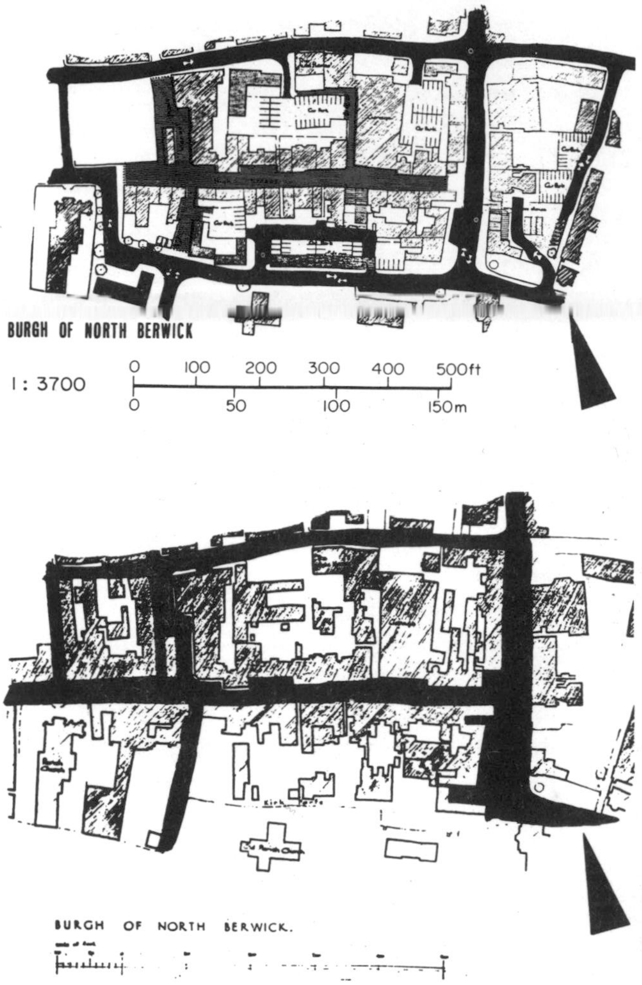

VII. 5.6. *Comparative Values*[ii]

North Berwick with its congested High Street, typical of thousands of European towns, has given rise to a most interesting economic comparison. Two schemes have been proposed, one for widening the street to ease congestion and the other for creating an easy route around the high street, making this the spine of a 600 ft (183 m) long pedestrian precinct, with good service access from the back end and considerable increase of nearby parking. The second scheme, as the report of the East Lothian County Council shows, proves to be by far the more economical. This finding and documentation is an extremely important precedent.

Fig. 7.24. Plans of existing High Street and Pedestrian Mall proposals, with 132 parking spaces.

"*In consultation with the District Valuer the spot values were determined for all the properties, which includes the normal compensations for disturbance. The demolition costs were worked out by estimating the cubic content of the buildings.*

	a) *The High Street Widening as originally proposed.*	*b*) *Kirk Ports Widening as proposed in this report.*
Property	£57,750	£22,500
Demolition	2,500	1,900
	£60,250	£24,400
Less residual site value	5,250	5,250
	£55,000	£19,150

(i) *Engineering News Record.* 17 October 1959.

(ii) Supplement to Report on Redevelopment of High Street Area, North Berwick, by County Planning Officer, East Lothian County Council, 1960.

(iii) TINDALL J. *Town Planning Review.* North Berwick, April 1962.

"*While this effectively compares the cost of widening the two traffic routes, the proposal to make the High Street a pedestrian shopping street can only be carried out if car parking and back access is provided for the north side off Forth Street. The equivalent figures here, which include for the acquisition and demolition of the lemonade factory and Ben Sayers factory, are as follows:*"

Acquisition	£16,500
Demolition	1,900
	£18,400
Less residual site value	1,500
	£16,900

"*The other costs which have been estimated are the excavating, bottoming and tarmacing of the car park and back access on both sides of the High Street,* £2,700, *and the re-paving of the High Street, Market Street and Law Road* £1,750."

"*There is considerable amount of recoupment that can be obtained through the redevelopment of the new frontage that became available. This is estimated above at* £6,500, *but could be more. The proprietors of the shops also stand to gain by the provision of better back access and this can be taken into account when fixing the value of any ground acquired from them.*"

"*These partial costs indicate not only the economic comparison between the two schemes but also show that the figures involved are not very great for the benefit of this imaginative and forward looking proposal.*" (See also p. 200.)

Fig. 7.25. View from hill in Elysian Park, Los Angeles, near Centre of City. Golden State Freeway running left to right crosses Pasadena Freeway running from right bottom corner. Note 3 major routes run side by side in that direction. Freeways do not replace the existing road system.

VII. 6. Examples

VII. 6.1. *Coventry, Traffic Plan*

Designer: Donald Gibson, City Architect 1941–1955; A. Ling, present City Architect. Planning for traffic segregation continuously developing since first precinct suggestion (see Section 10.38).

Population: 300,000 approximately.

Road System: A fairly tight inner motorway circle, ½ mile diameter approximately, leads off directly to an independent system of connected car parks, mainly at roof level. 7500 parking spaces presently recommended, with possible further increase to 10,000. An inner circulatory road is mainly for bus traffic and leads to other feeder roads.

There are enough parking facilities and roads for 20% of the commuting population to use private cars. Public transport in the form of buses is envisaged for the rest.

The existing, very extensive, pedestrian network of precincts, etc., the finest in the world, recently created, is planned to grow as the development of the centre continues. Path and cycle tracks are planned to connect residential areas with town centre.

Fig. 7.27. Pedestrian and cycle way system, showing links with residential areas, bus route and stops.

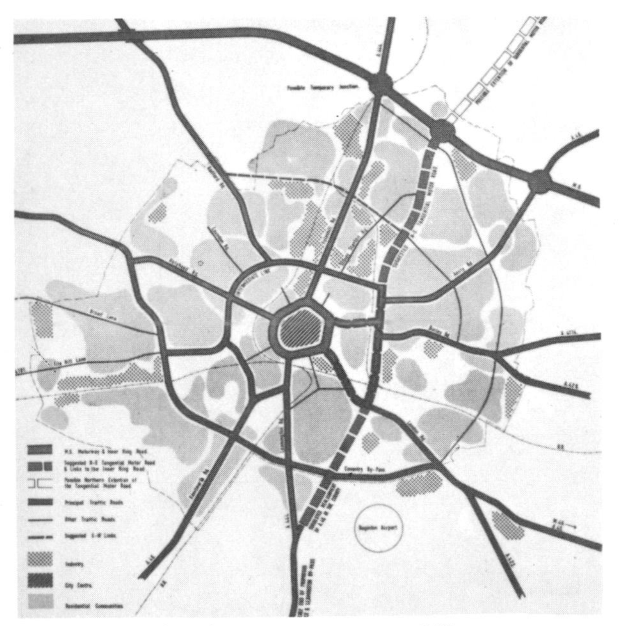

Fig. 7.26. Road and motorway system of Coventry.

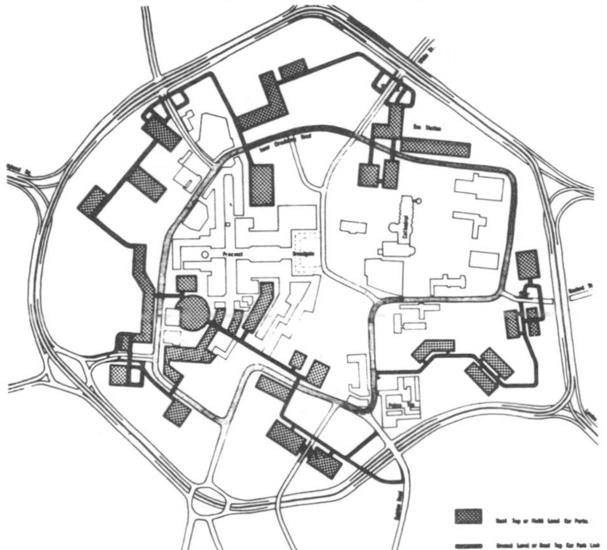

Fig. 7.28. Road system and parking and bus stops.

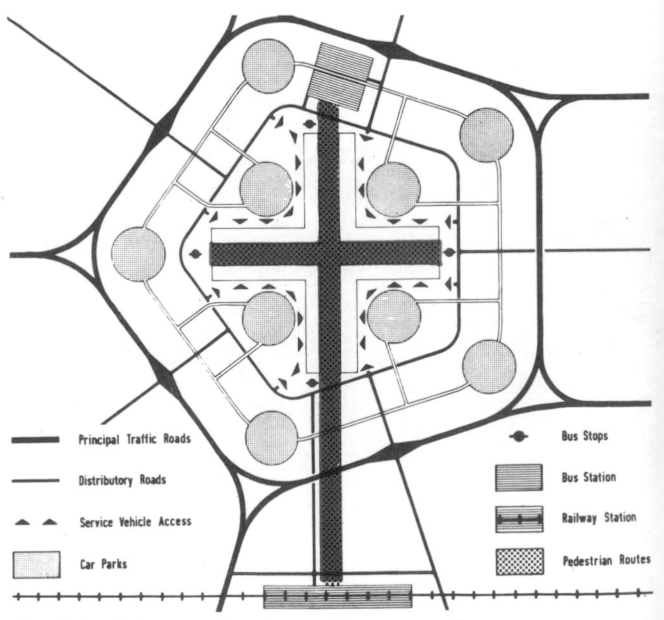

Fig. 7.29. Diagram showing traffic segregation.

References.

Coventry Development and Redevelopment, City of Coventry 1961.

Coventry Rebuilds, *A.D.*, London.

A. J. Special number on Coventry, 11 June 1962.

A. & B. N. Special number on Coventry, 11 June 1962.

The City Road Pattern, City of Coventry.

KEY Red —Pedestrian Areas.
Black —Motor Roads.
Dark Grey —Car Storage.
Grey —Dwellings.

Unless otherwise stated this key applies to each plan.

Plans have not been redrawn to retain character and variety. The red colour, denoting pedestrian areas, is therefore shown in a number of ways; either solid red, or edged with red, or main red paths leading through squares or greens, or public buildings and shops about a pedestrian area are solid red or edged in red.

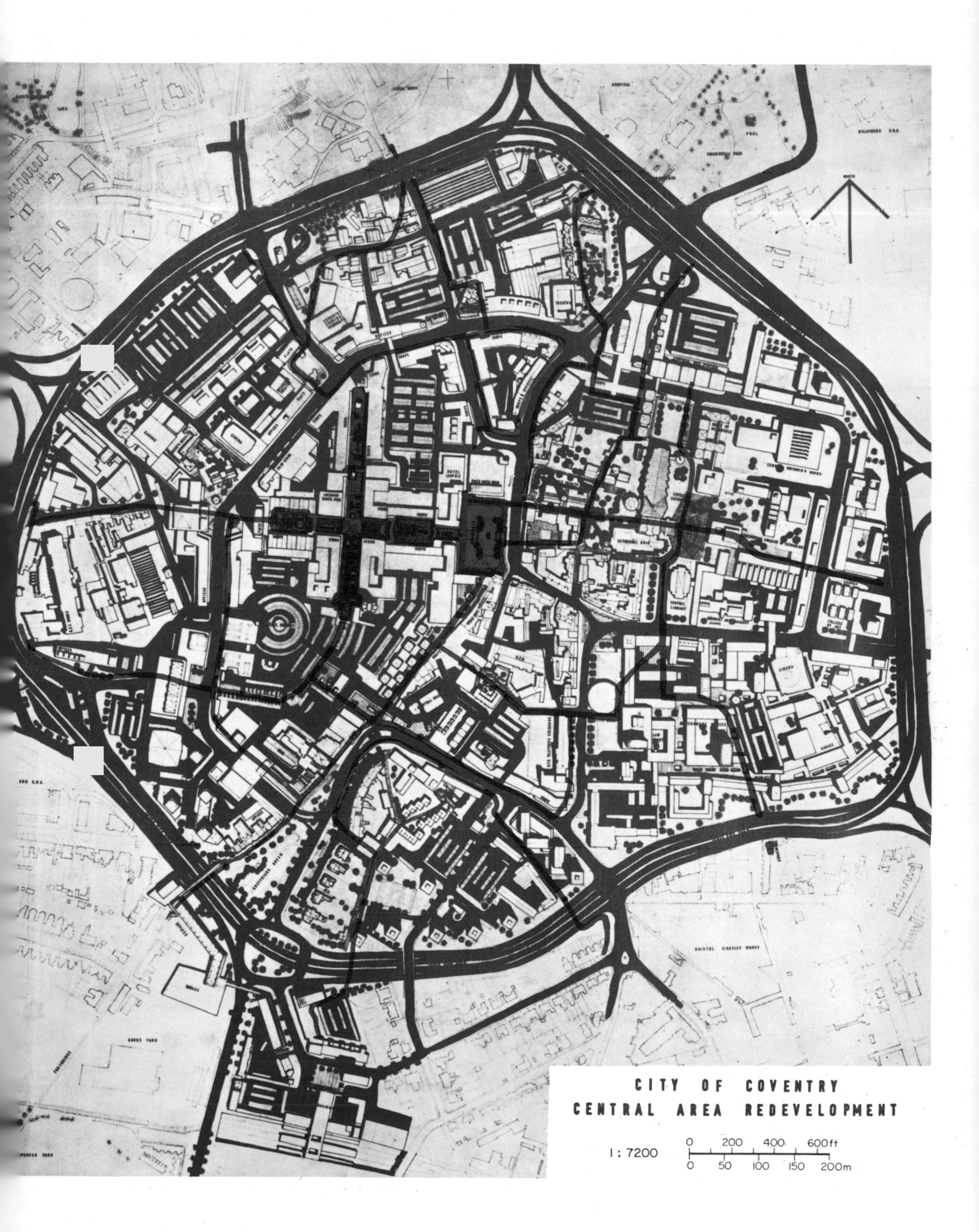

Fig. 7.30. Central area development plan.

VII. 6.1.1. *Coventry, Shopping Precinct.*

Designed: Donald Gibson, City Architect 1941–55; Arthur Ling, City Architect from 1955.

This is the first wholly pedestrian centre in Europe, a historic achievement against much opposition. 7500 parking lots are to be provided.

The shopping is at two levels. The variety and extent of provision are truly impressive. The emergence of this centre is a matter of good fortune. The long member of the cross was very nearly made into a motor road. After six years the original idea of the wholly pedestrian centre was regained by the new City Architect. The cultural and civic areas including the new cathedral are to be linked by underpass to the shopping centre giving immense scope for the pedestrian. (See also p. 43.)

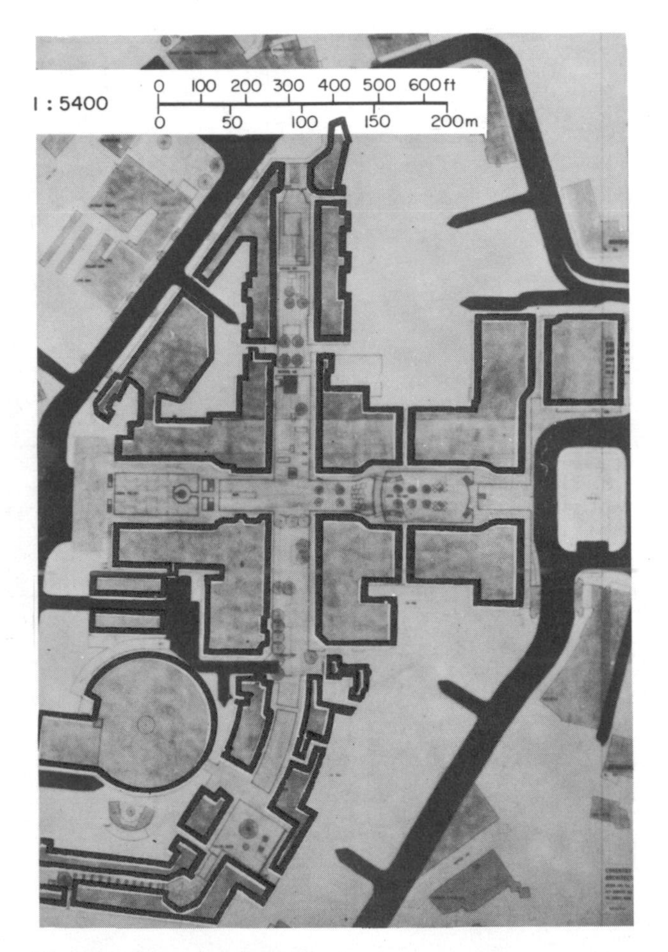

Fig. 7.31. Plan of main precinct.

Fig. 7.32. Model 1958, showing parking areas at roof level.

Fig. 7.33. View towards cathedral.

Fig. 7.34. View in the same direction from behind restaurant.

Fig. 7.35. Part of the extensive City Arcade.

Fig. 7.36. The Market Way—Note Lijnbaan type covered link.

VII. 6.2. *Stockholm, Sweden, Traffic Plan*

Design: Master Planning Commission of Stockholm. Submitted, February 1960.

Population: One million approximately.

This plan, designed to meet traffic demands in 1990, introduces motorways for Stockholm and a system of main arterial routes and sub-arterial routes. The motorways form a ring and a cross, two arms of the latter forming tangents to the central area. Within this area 45,000 off-street parking spaces are to be provided.

Stockholm has as its main traffic artery the Underground Railway which has been in existence since 1950. Underground trains arrive and depart at 47 stations along the 40 km (25 miles) of track. Today the Underground Railway accounts for almost 50% of the public passenger transport and more than 20% of the total traffic of the Swedish capital. Almost half a million passengers travel by Underground every day. Only by Underground is it possible to travel quickly across a capital whose roads are blocked by motor-cars.

The importance of the Underground has grown much more rapidly than was anticipated at first. The decision taken by the Stockholm City Council in 1941 to build an underground railway proved to be not only wise but far-sighted.

Only relatively few buses are envisaged. However, if the 75% of people commuting are to be accommodated, then bus traffic would take up one of the lanes on the motorway and this would reduce its capacity decisively. An extra lane for public transport would then be required.

Many "park-and-ride stations" are provided to ease the private car pressure on the city centre.(i) It is estimated that 1970 traffic will be about four times that of 1953.

Fig. 7.38. Underground station, well designed and clean.

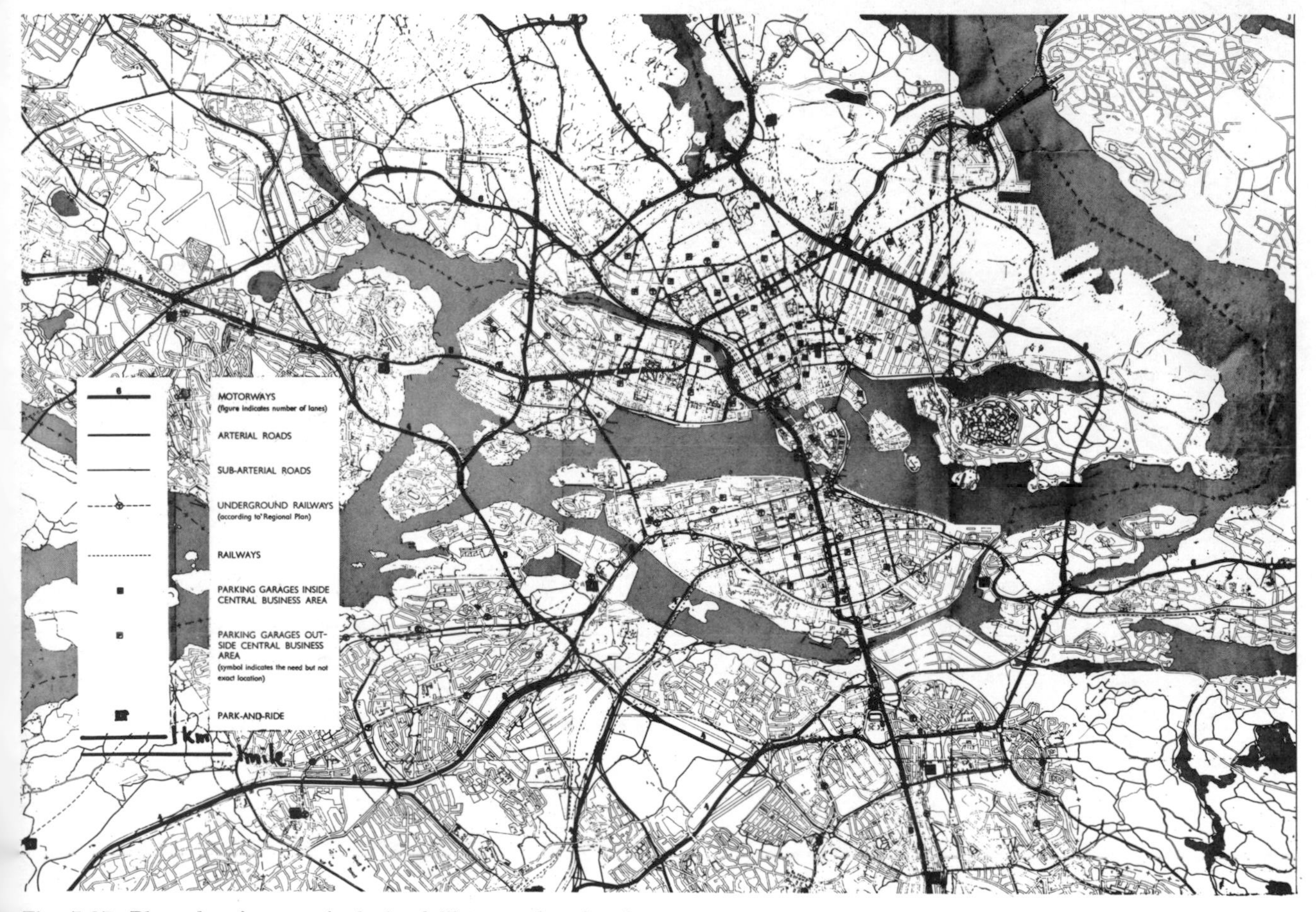

Fig. 7.37. Plan showing particularly deliberate planning for "park-and-ride" stations along underground in the outskirts.

(i) A traffic route plan for Stockholm, 1960.
MARTERN B. Trafiken och det Framtida City. Stockholm 1958.
City of Stockholm; City of Stockholm. Stockholm 1959.
HOLMGREN P. *Stockholm Today and Tomorrow*, Stockholm. 1 November 1960.

VII. 6.2.1. *Stockholm, Sergelgatan.* Planned since 1946 when first proposals were submitted and developed.

This is a comprehensive redevelopment in the commercial centre of Stockholm, providing a new hub, with underground station, parking, motor traffic and pedestrian provisions. It contains five multi-storey office blocks, multi-level pedestrian terraces over large areas with the Sergelgatan as a specifically pedestrian boulevard. There is garaging for 700 cars, with service facilities below this area. Lifts lead directly into shops. Road tunnels and garaging under the area became practicable when construction of the underground railway met sand.

It is an early attempt to restructure the centre of a metropolis gaining efficiency for the motor and regaining intimacy for man, giving a lead to Europe's capitals. It seems a pity that present plans do not include more pedestrian areas in adjoining parts of the city.

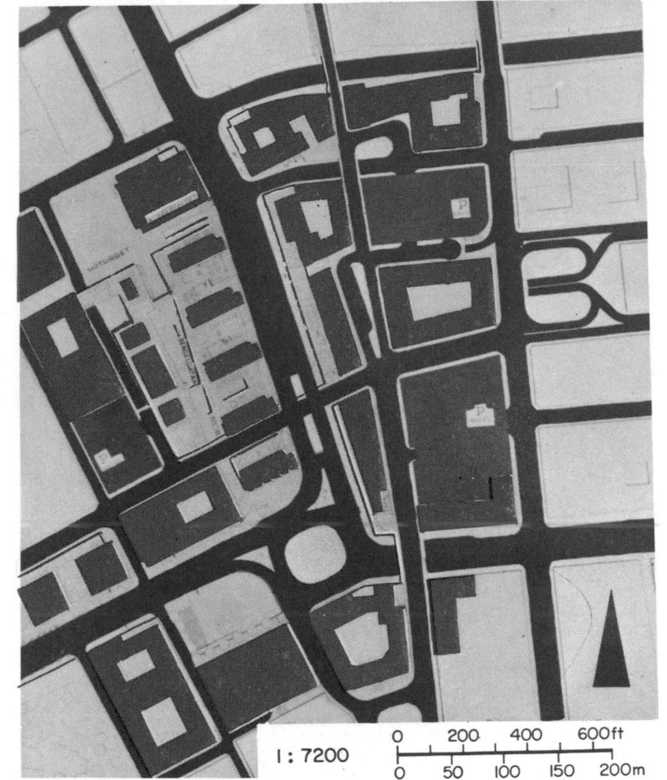

Fig. 7.39. Plan of Sergelgatan area.

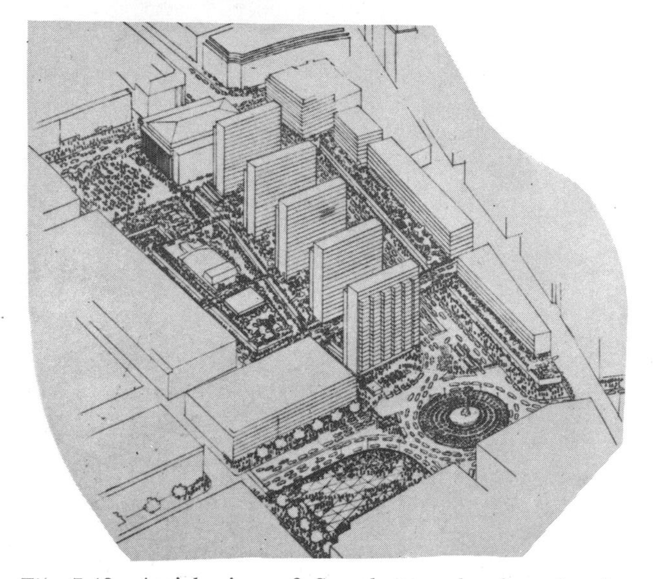

Fig. 7.40. Aerial view of Sergelgatan showing the lower concourse in the foreground.

Fig. 7.41. Aerial view of model showing the whole central area with the five tower blocks and Sergelgatan in background centre.

Fig. 7.42. Section through office block with shops in podium and basement, roof gardens and garages with lift connection two and three floors below ground level.

References

Bor W. *Stockholm Central Redevelopment.* Stockholm; New City Centre. *A. J.* 15 August 1962. *A. D.*, May 1963.

Fig. 7.43. View from a tall office block onto Sergelgatan.

Fig. 7.44. View of model of Sergels Torg, lower concourse.

Fig. 7.45. Day view of life in Sergelgatan precinct.

Fig. 7.46. Night view.

Fig. 7.47. View of Sergelgatan.

Fig. 7.48. Perspective drawing of Sergels Torg, under construction.

VII. 6.3. *Philadelphia, U.S.A., Traffic Plan*

Design: City Planning Commission, Chairman, Dean Perkins; Director of Design, Edmund Bacon.

Population: 2,500,000 approximately.

Origin: In the late forties, the Greater Philadelphia movement was formed and in 1952 Government reorganization started the continuing comprehensive redevelopment of central areas which has now been partly carried out.

Nature of Scheme: Starting point was a limiting of the invasion of cars by providing better public transport. The Urban Transportation Board, and later the Passenger Service Improvement Corporation, co-operated closely with the Planning Commission. The rebuilding will provide public transport with stations really centrally situated, pedestrian networks at three levels and integrated parking in the central area for 46,000 cars. This is the number the central area is gauged to be able to take. A pedestrian shopping mall, Chestnut Street, with only slow public trolley transport running along it, is planned to go right along the Central Area rectangle.

Beyond the central area described above there are 45 other comprehensive redevelopment projects within the city of Philadelphia.

Philadelphia Planning Commission has the finest public relations with its citizens of perhaps any city in the world. Free pictures, information and an immense, superb exhibition of ingenious models, showing present areas and, reversing each section, proposals. These are permanently available in the city commercial museum regularly visited by schools.

General: Philadephia, combining public transport and traffic proposals in a meaningful way with city planning, has led the way for the rest of the United States and indeed the world (see p. 101).

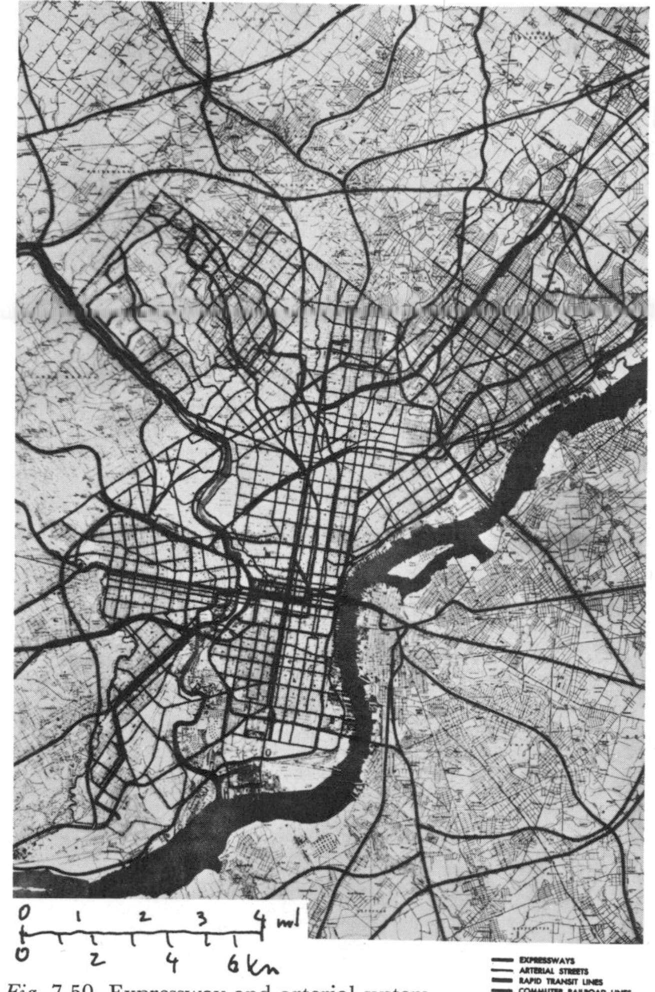

Fig. 7.50. Expressway and arterial system.

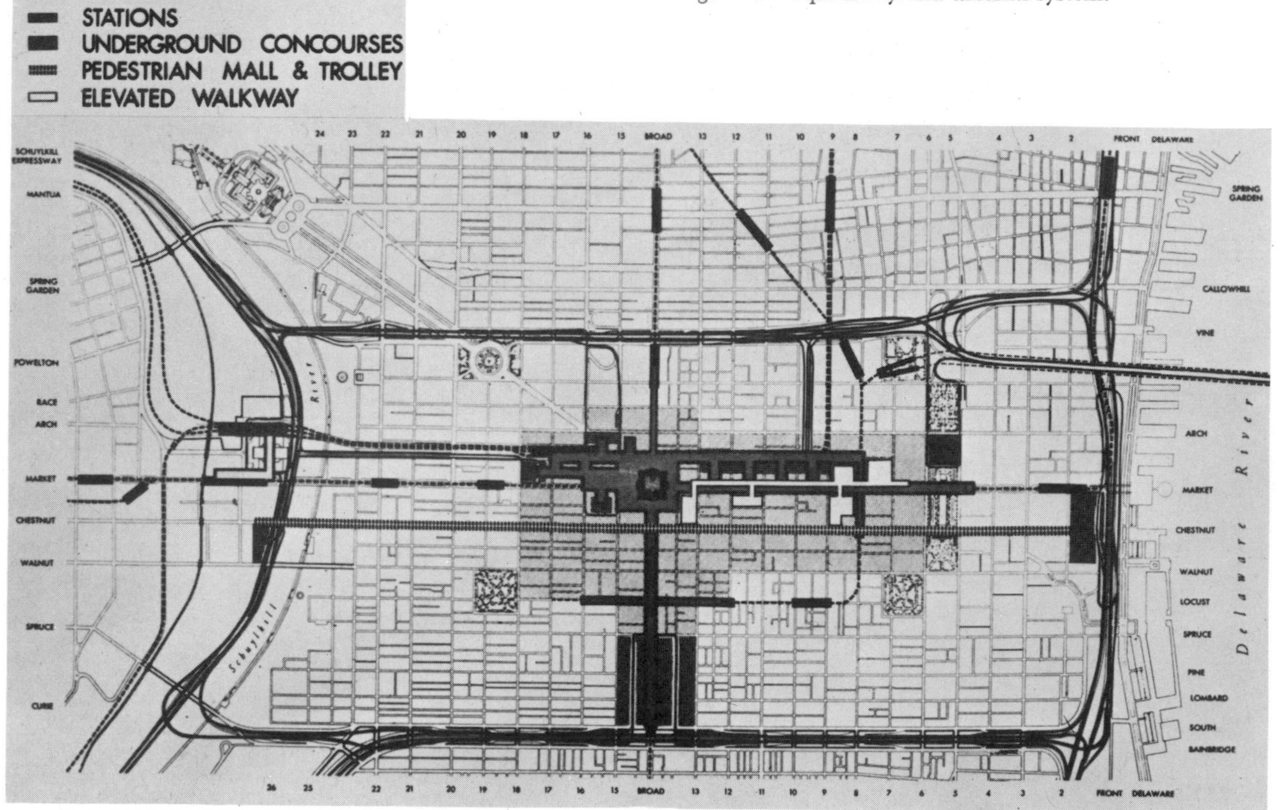

Fig. 7.49. Philadelphia Central area showing public transport motorways, largely below main street level, pedestrian network and the mall of Chestnut Street.

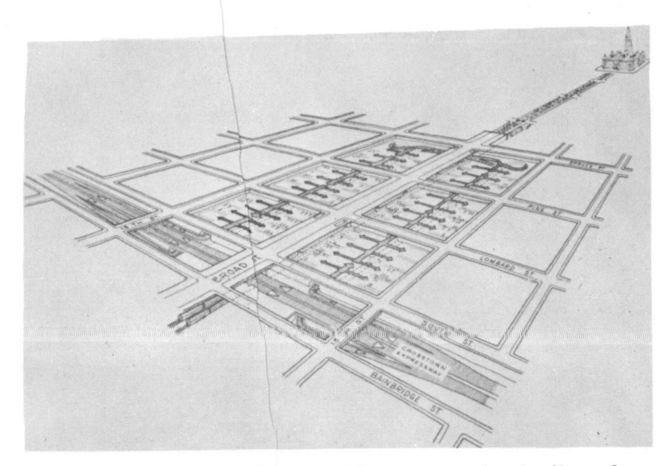

Fig. 7.51. Underground car parking on two levels direct from motorway ring, at a depressed level, Broad Street.

Fig. 7.52. Vine Street Expressway, part of the motorway ring and the Central Area.

Fig. 7.53. An important aspect of the development.

References

Philadelphia City Planning Commission, *Annual Report* Philadelphia 1957.

BACON E. Philadelphia. *A. Rec.* May 1961.

Philadelphia, *J. A. I. P.* August 1960.

Comprehensive Plan for the City of Philadelphia. Philadelphia Planning Commission 1960.

VII. 6.3.1. *Market East Section.*

City Centre: The extensive multi-level planning scheme is to be carried out by private developers, each observing the basic needs for vehicular and pedestrian routes as planned. 320 acres (129·5 ha) plan based on the following five principles:

"1. *An efficiently concentrated core to house the dynamic centre of private commerce and public services for a region of six million people.*

2. *A modern, balanced transportation system (auto, bus and rail) penetrating close to the heart, with easy and pleasant transfer from vehicle to foot; and the exclusion of traffic not destined for the core.*

3. *Major parking facilities and bus terminals directly connected with the expressway system to reduce congestion on local streets.*

4. *Pedestrian movements in the core separated from automobile traffic insofar as practicable by upper and lower level concourses, and the removal of automobiles from the City's primary retail street.*

5. *Parks and open space to provide an appropriate and dignified setting for public buildings and historic shrines, squares for public gatherings and celebrations, quiet spaces for relaxation and places for enjoying the bustle and excitement of the crowd.*"

Fig. 7.54. Market East, lower level concourse, used by restaurant.

Fig. 7.55. Lower level concourse used as skating rink.

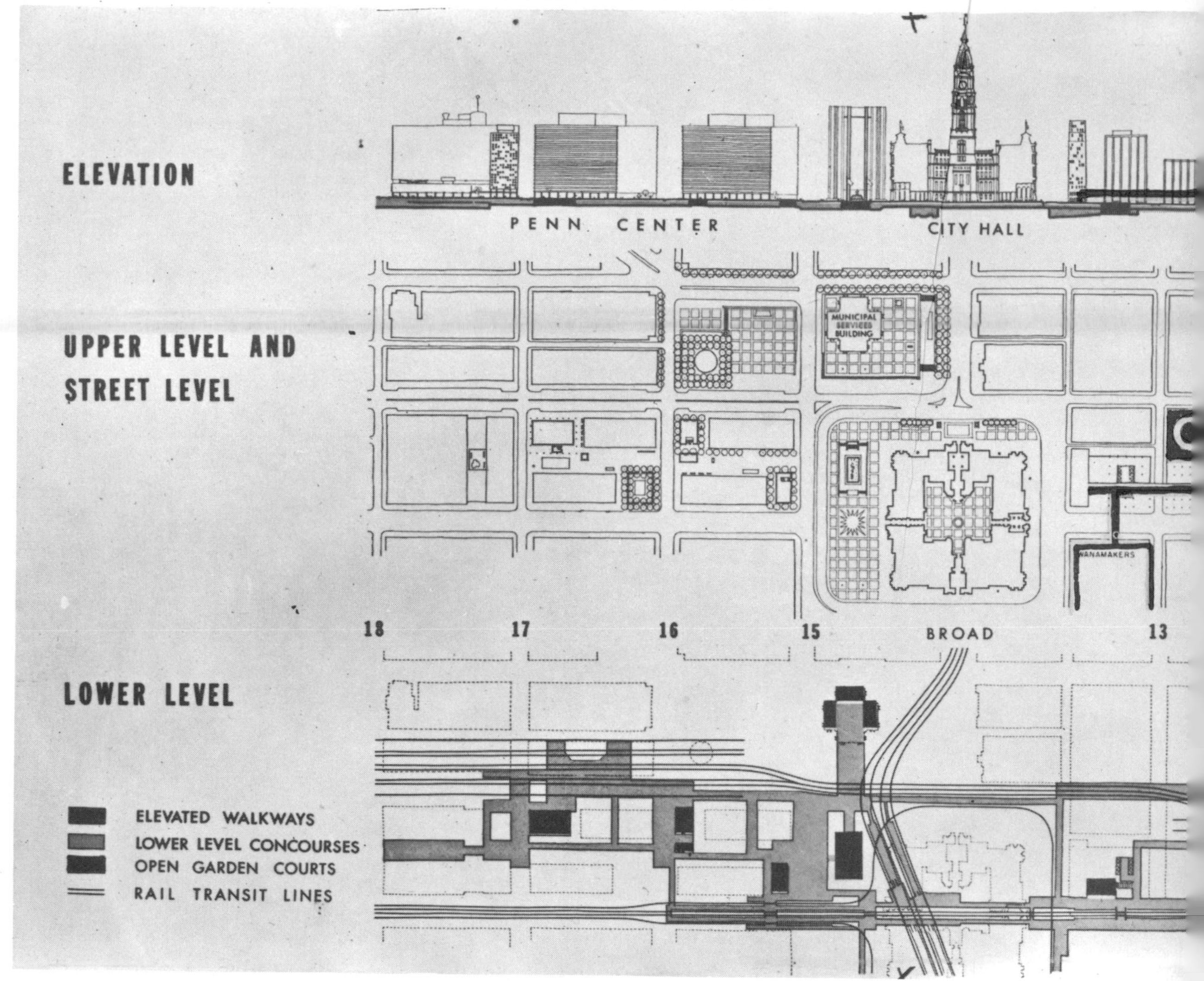

Fig. 7.56. Central Development. Market East Section with section through centre. Upper level and street level plans. Note upper level bridges connecting new buildings across Market Street right into eight-storey central hall of Wannamaker Department Store and two others. Lower level concourses plan.

Fig. 7.57. Perspective section through one of the ten lower open garden courts relieving the very long underground shopping concourses giving daylight and views out; multi-level traffic is clearly shown.

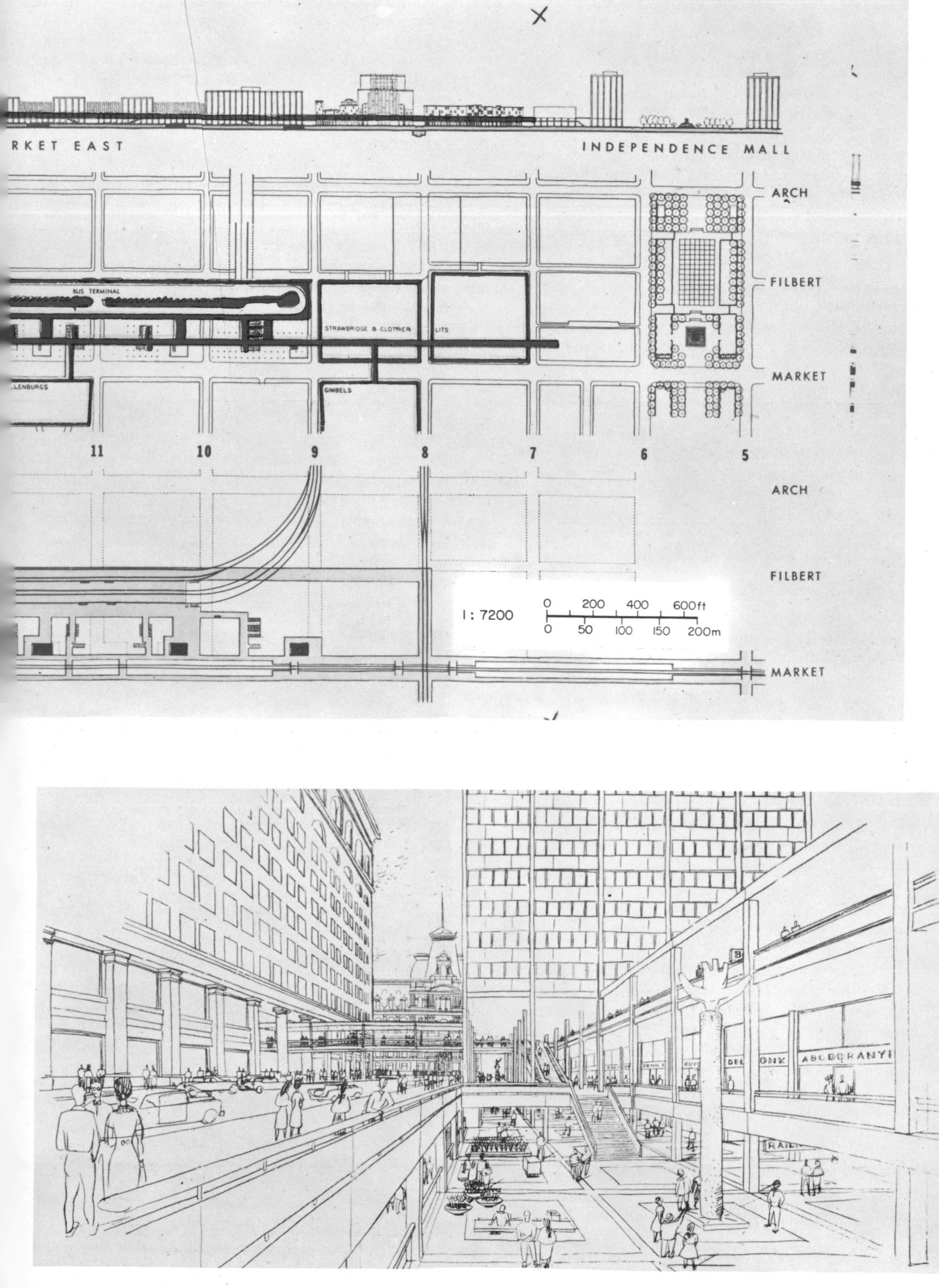

Fig. 7.58. Multi-level pedestrian circulation, with bridge connection to existing department store, and thence to station.

VII. 6.4. *Liverpool, England, Traffic Plan*

Design: Graeme Shankland Associates. Graeme Shankland, D. M. Gregory-Jones, A. Proudlove, D. C. Robinson, D. A. Searle, G. Ashworth, A. Meats, E. Schoon, V. E. Wilshaw, I. Masser, R. Harrington, E. Simonds.

Population: Approximately 800,000.

Origin: Planning Consultant Appointed by the City and County Borough of Liverpool, January 1962, to make three-dimensional proposals for the redevelopment of the central area.

Nature of Scheme: Excerpts from "General Planning Principles" of Consultant:

"*The central area should be planned to enable it to continue and to develop as the main regional centre for Merseyside and its hinterland for the foreseeable future.*"

"*The extent of the central area for these purposes should be sufficient for the redevelopment and expansion of these uses to take place without resulting in the congestion of building development on individual sites and with adequate provision for car parking and space around buildings.*"

"*All new building development road, car parks, open spaces etc. should be designed to make an effective and sympathetic architectural contribution to the appearance and function of their immediate surroundings in addition to being well designed in relation to their own specific purposes.*"

"*Particular attention should be given to certain areas of special architectural importance, including those containing historic buildings.*"

"*Proposals for the siting and shape of developments and the spaces around them, and the overall design of the new parts of the central area, should be designed with attention to important views out of the centre, within it and views of it from outside.*"

"*The communication system, by rail and road, public and private transport, should be considered as a whole in relation to the quantity, extent and functions of the various elements in the central area. In this connection:*"

"*i. The road and parking plan together with the public transport and pedestrian networks will be based on estimates and usage 20 years and more ahead.*

"*ii. Notwithstanding the projected further rise in car ownership, public transport by rail and bus should be sustained and improved. At no time in the future can it be envisaged that Central Liverpool will not have to rely primarily on public transport for its continued health and growth and the maintenance and improvement of the public transport system should be a major planning objective at all times.*

"*iii. Pedestrian–vehicular segregation should be planned where this is necessary to (a) provide safer routes for pedestrians in directions they wish to follow, which are not necessarily the same directions in which vehicles need to travel, (b) keep pedestrians off important roads thus increasing their vehicle capacity, (c) provide special traffic-free areas for meeting, shopping and recreation and areas free of vehicle noise and fumes where functions of adjacent buildings make this desirable.*

For this purpose, provision should be made in redevelopment schemes for upper pedestrian level circulation over streets and across the roofs of buildings and through them as necessary to effect a comprehensive network of upper level movement where needed.

"*iv. Bus routes should be planned in conjunction with the operating concerns to penetrate the Central Area so that bus stops will be conveniently near to all major destinations.*

"*v. Wherever possible, new roads should be planned neither to sever established communities nor functional zones.*

"*vi. New car parks will be planned to be as near as possible within easy walking distance of the destinations of their occupants. In general, new car parks should be planned horizontally on several levels with their roofs being covered and available in whole or part for pedestrian movement and open space.*

"*vii. The parking and stopping of all vehicles on main arteries and main distributor roads in the central area should eventually be prohibited in order to use the road space to its maximum capacity.*"

"*The programming of redevelopment will be conditioned by the state of existing property, the priority requirements of major works, and the economic viability of the central area as a whole.*"

"*Consideration in programming must also be given to the unexpired terms of the City's freeholds and the extent to which the council is prepared to exercise compulsory powers.*"

"*The value of the City Council holding so large an amount of land in the central area cannot be overestimated both from a short term and long term planning point of view.*"(i)

The inner motorway is designed for 50 mph (80·5 km/hr), six lanes, two slip ways, total length 3½ miles (5·6 km), total width average (including slipways) 200 ft (61 m), most of its length elevated 20 ft (6·1 m). It will cost £32 million. It is designed to take 0·3 cars per person and a traffic flow 4·75 times the volume of 1959. This is to satisfy what is considered the ultimate car ownership in Merseyside. Parking for 36,000 cars is considered the maximum. Increase, it is gauged, would destroy the function of the central area for it would mean not

Fig. 7.59. First motorway loop superimposed on existing street plan with three proposed comprehensive development areas. 1, Ravenseft; 2, Civic Centre; 3, Paradise Street.

(i) City and County Borough of Liverpool. *Planning Consultants Reports No. 1–8.* Liverpool 1962–3.

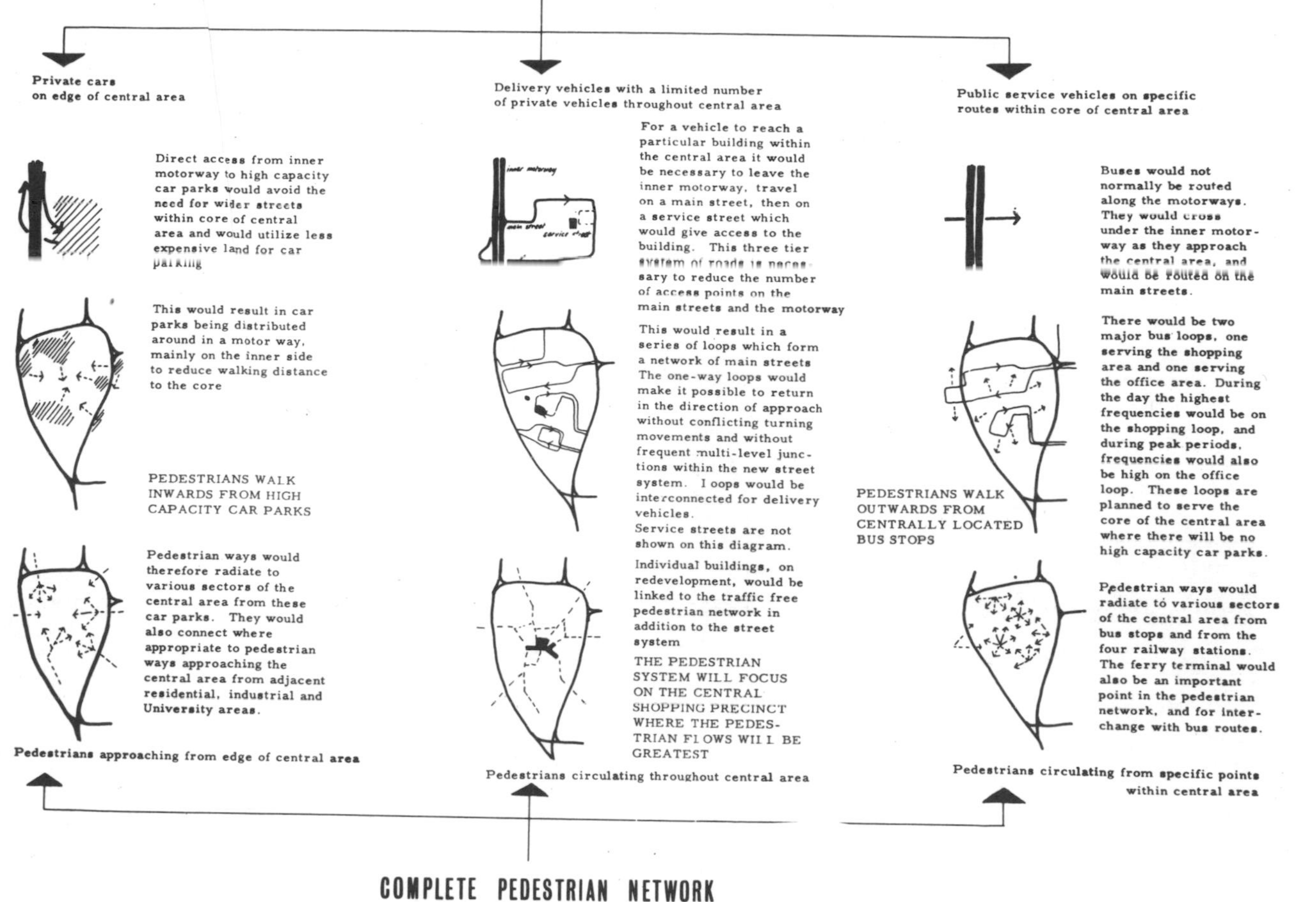

Fig. 7.60. Road system and pedestrian network from consultant's report analysis.

Fig. 7.61. Inner motorway system first loop around the centre (second will be around the University area) connected to the Mersey Tunnel in the centre, showing main streets, bus routes (broken lines) and parking (grey areas), with pedestrian network.

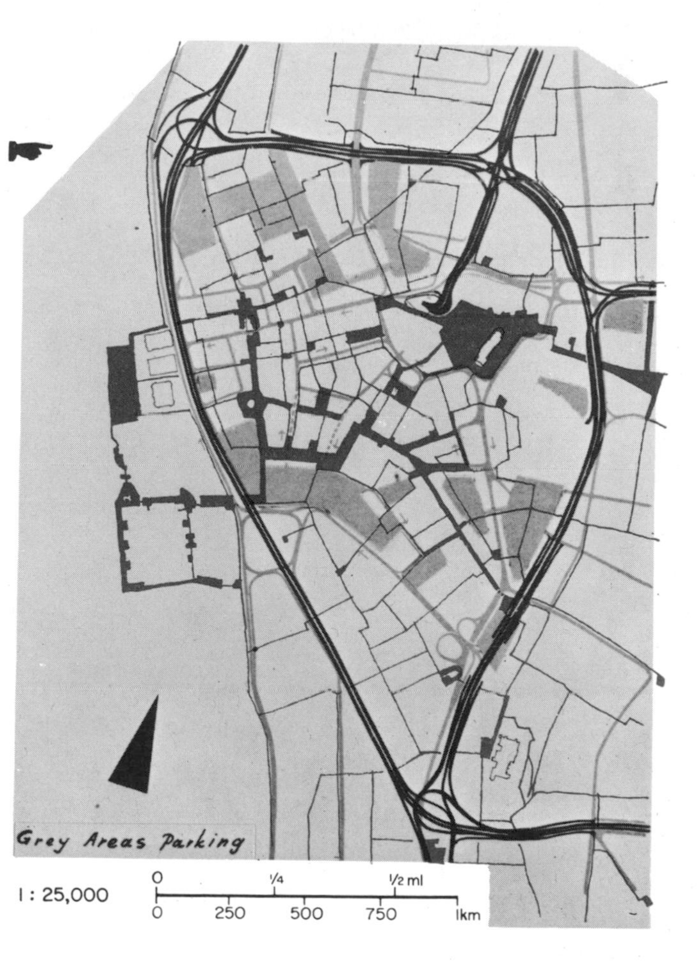

only the increase of the inner motorway by one lane and so 33% capacity but also increase of radials, junction as well as the parking itself. This means that 75% of commuters use public transport, trains and bus. Maximum speeds on main roads 20 mph (32 km/hr). Some roads reserved for buses and service roads third tier after main roads. Dimensions of central area within loop is ¾ mile (1·2 km) and 1¼ mile (2 km).

Fig. 7.62. Motorway on modelled embankment. The embankment has been taken up above the level of the carriage way and the slope eased into the open space.

Fig. 7.63. Warehousing and service industries under motorway.

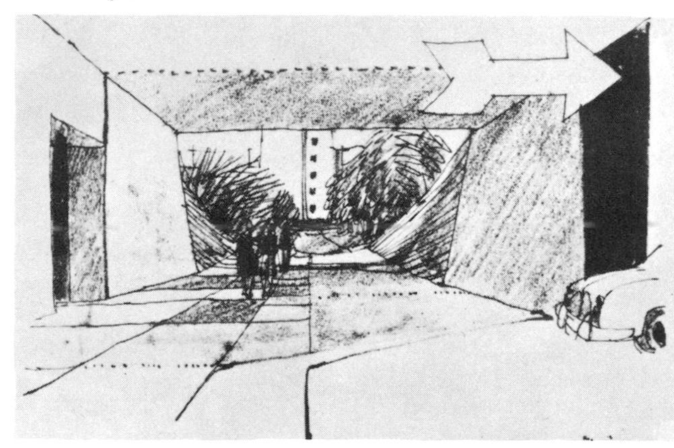

Fig. 7.64. Pedestrian underpass to motorway.

Fig. 7.65. The space contained within junction at Duke Street. Designed to link adjacent sectors of the city for pedestrians.

VII. 6.4.1. *Ravenseft area.*

Area 41·3 acres (16·4 ha).

> "*. . the first major operation in converting the City's Centras area into the largest continuous pedestrian precinct in Europe, outside Venice.*"(i)

Access by foot, bus, train and private car are carefully assessed and connected..A network of walks links stations shops and concourse at basement level. Service is by cul-de-sac, from above and below shop level. Emergency vehicular access to precincts is included. The corridor street is changed into a succession of spaces of varying size and character.

First stage of Liverpool Central Redevelopment.

Fig. 7.66. "*A detail of the Ranelagh Centre . . .*"

Fig. 7.67. Bird's eye view including J. A. Roberts' scheme for St. John's precinct, 40 acres for Ravenseft Ltd. including retail market, 100 shops served from basement, 8-storey hotel, multi-storey garage for 800 cars, roof garden, beacon tower 460 ft high (140 m), restaurant at 350 ft (106 m) level.

(i) City and County Borough of Liverpool. *Planning Consultant Report No. 3,* Liverpool 1962.

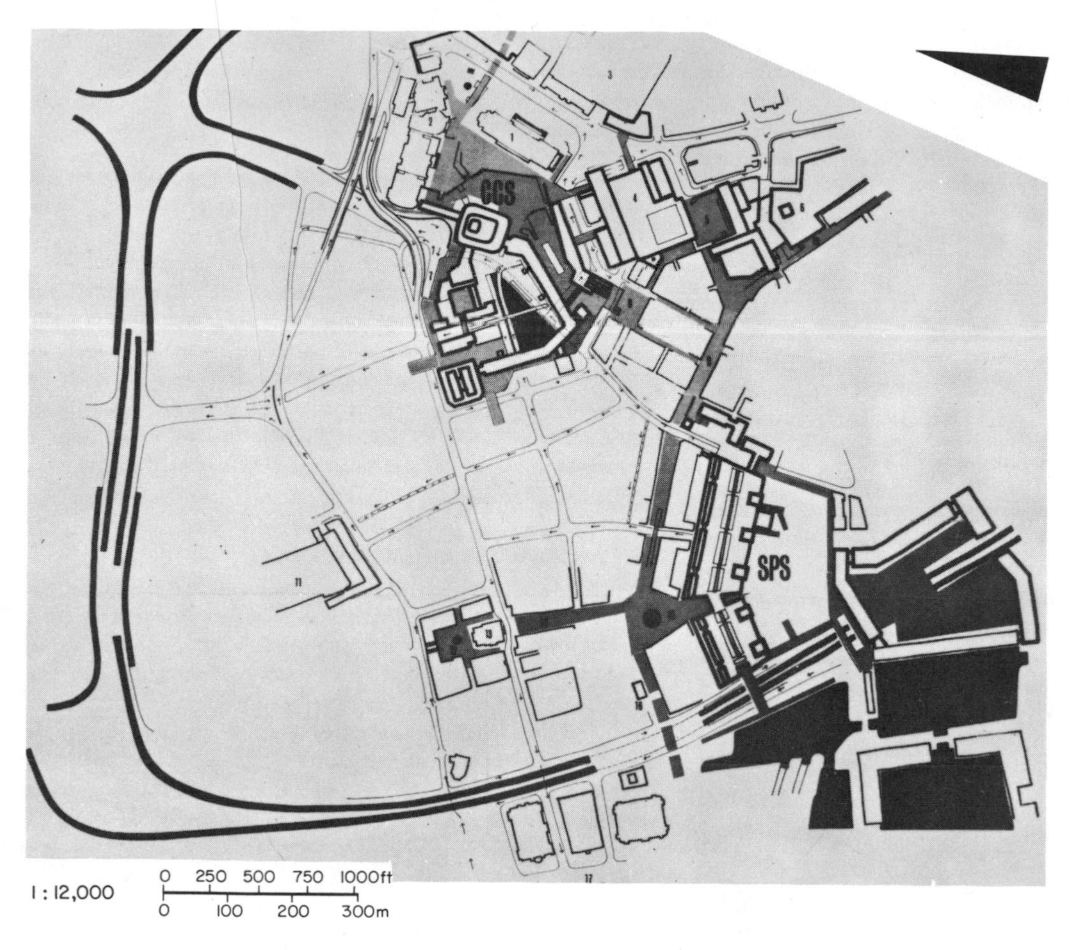

Fig. 7.68. Location and traffic plan of the Civic Centre and Paradise Street proposals.

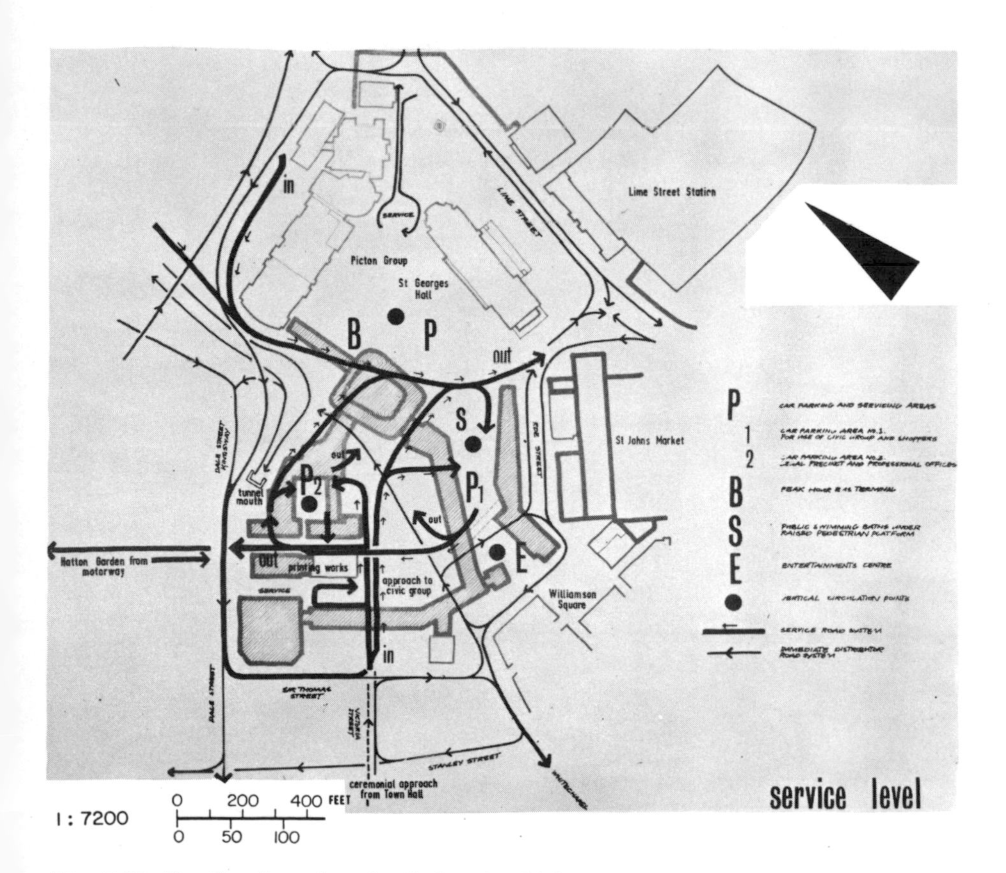

Fig. 7.69. Detail of service circulation in Civic Centre.

VII. 6.4.2. *Civic and Social Centre.* 31 *acres* (12·5 *hectares*). *Location on Fig.* 7.59: Changing the site already designated for the Civic Centre to the more logical, exciting and difficult site adjoining the fine set of Civic Buildings in the traditional centre of the city, a remarkable proposal is possible, thoroughly studied in its multi-level traffic arrangements. A splendid pedestrian environment could be provided and the private car is well provided for (1600 parking places). The aesthetic dilemma of the Mersey Tunnel entrance is solved at last. Some of the points made in the consultant's report, prepared in conjunction with the new City Planning Officer, Walter Bor, are:

"*A Civic Centre is the efficient government of the City to consist of a group of buildings devoted to the political and administrative functions of governing the City.*"

"*The construction of a Civic Centre should be used positively to enhance the quality of the City both visually and as a place in which to live, work, shop, and play, and be so designed to incorporate where suitable the livelier uses of the central area including, among others, cultural and entertainment facilities, shopping, an hotel, flats, a swimming pool, licensed premises, restaurants and cafés.*"

"*The City's Courts together with related professional offices are centralised into the form of a legal precinct, self-contained but closely associated with the Civic Centre.*"

Fig. 7.70. View past monumental column across St. John's precinct towards proposed law courts and Liver Building beyond.

Fig. 7.71. View of St. John's Precinct from St. George's Plateau; St. George's Hall on the right.

VII. 6.4.3. *The Paradise Street–Strand Area. Location on Fig.* 7.59: This area can be developed into a residential community, bus station and other accommodation because the Civic Centre has been relocated.

"*The total area of property, including all roads within the site boundary is* 19·7 *acres* (8 *hectares*) *of which approximately* ¾ *is in the freehold ownership of the City Council.*"

"*The opportunities presented by this site are exceptional. Containing a very large proportion of cleared land, it is strategically placed at the meeting point of the City's business and shopping districts and is easily accessible from both. The site is immediately accessible from the proposed Inner Motorway. James Street station is also within easy walking distance, so that if the proposal to locate the City's Bus Station in this area were to be acted upon, access to the site by train, bus and private car would be uniformly excellent.*"

"*The site is sufficiently close to the river and docks to enjoy a magnificent outlook at high level.*"

"*Any redevelopment must form, once again, a significant termination to Castle Street. It must also stand up to the massive scale and dynamic visual character that the adjoining motorway will bring to this area, and to the waterfront development. Redevelopment on this site must be designed with due regard to views of the city from the Mersey, and of the Mersey from the City.*"

"*This new residential community backs all the good reasons for providing dwellings in centres.*"

"*Building uses provided for, and which are allowed for in these proposals, are as follows. (Provision for the various uses will be described in detail later.)*"

"*a*) *bus station directly connected to main pedestrian shopping centre via an arcade.*
b) *shopping centre, containing* 60 *shops,* 2 *stores,* 1 *supermarket, kiosk shops, licensed premises, restaurants etc.*
c) *entertainment centre.*
d) *car parking for* 2500 *cars straight from motorway.*
e) *housing containing* 600–800 *dwellings.*
f) *park, approximately* 6½ *acres* (2·4 *hectares*).
g) *hostel.*
h) *hotel.*
i) *warehousing, light industry, etc.*
j) *Telegraph House replacement.*"(i)

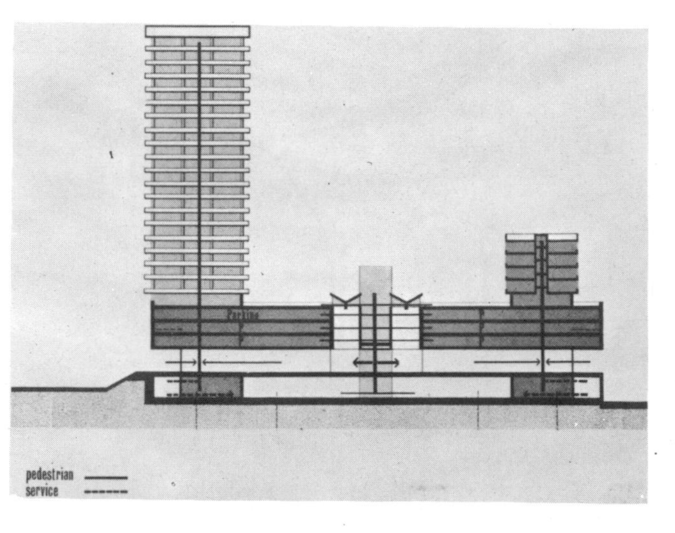

Fig. 7.72. Section through Paradise Street scheme.

(i) City and County Borough of Liverpool. *Planning Consultants Report No.* 8, Liverpool 1963.

Fig. 7.73. Aerial view of model showing Ravenseft Development on the left, Civic Centre in the middle and Paradise Street area at the rear.

Fig. 7.74. Aerial perspective from Dale Street.

VII. 6.5. *Newcastle-upon-Tyne, England, Traffic Plan*

Design: City Planning Officer, Wilfred Burns.

Population: 300,000.

Origin: Wilfred Burns was appointed Chief Planning Officer with status equal to the City Architect and Engineer to prepare a plan for Newcastle. This was the first plan for a large British city incorporating a motorway system, comprehensive footpath network and proper integration of parking and comprehensive redevelopment areas 1961 (with perhaps the exception of Coventry, Burns' previous place of work).

Great care is taken not to damage the good existing building and areas of value and prosperity. A new bridge across the Tyne is needed for one of the two motorways to form the important north–south route incorporating the Great North Road. An east–west loop runs under the centre.

The arterial main roads go under the motorway to reach the town centre. The steep banks down the Tyne make an exciting multi-level network of all transportation routes possible. All the main shopping streets and places are turned into pedestrian streets and precincts and service is from several levels below, utilizing the slope of the site.

Unlike other schemes a great proportion of garages seem to lie on the arterial inner ring and service roads and are not directly accessible from the urban motorway.

Fig. 7.75a. Model of Pilgrim Street Scheme.

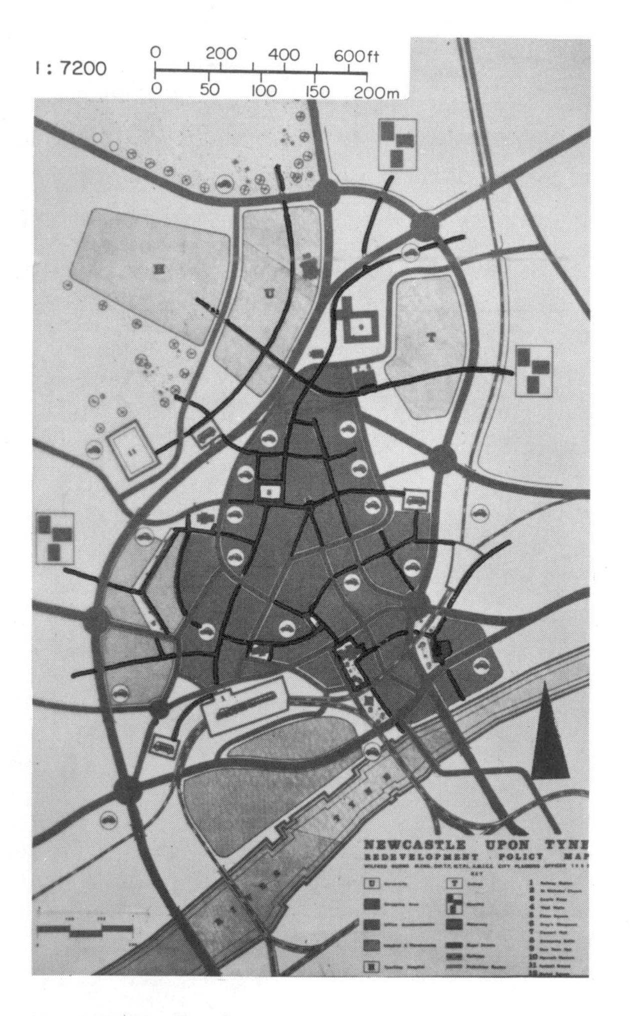

Fig. 7.75. Traffic plan

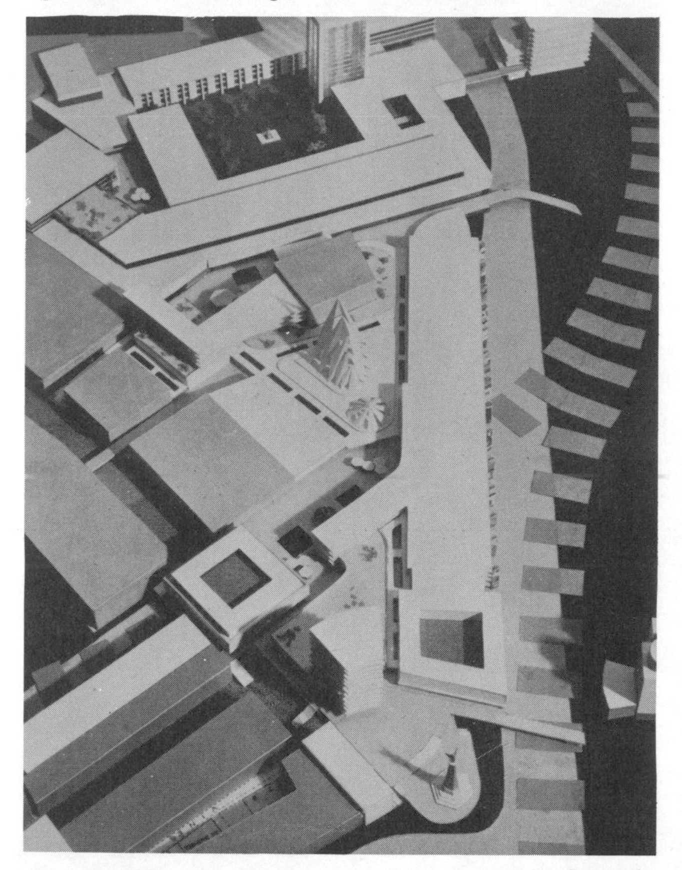

Fig. 7.76. View from the north showing motorway from the north in its junction with east–west loop coming from under the centre.

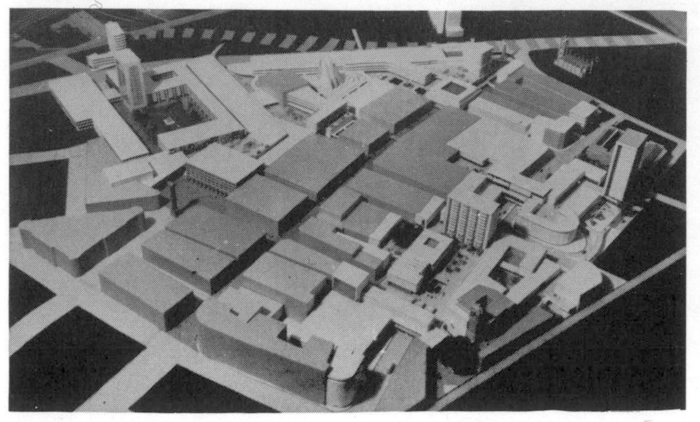

Fig. 7.77. View of the two areas. New building light. Northumberland Street shops remain.

Reference

NAIRN I. New Newcastle, *A. J.* 20 April 1960.

VII. 6.5.1. *Central area, overall redevelopment. The first two areas combined:* The first proposals turn one of the busiest shopping streets in Britain into a civilized pedestrian area without disturbing its valuable frontages, providing at the same time two and a half times the present parking spaces in the area with convenient access via bridges and upper decks to the pedestrian areas.

The second redevelopment provides for civic buildings connected in a meaningful and effective way, by a pedestrian network at deck level and served by efficient service roads from ground level.

A large number of flats are introduced into the town centre on the basis that the proposals make it a desirable place to live in.

The path networks proposed are part of the comprehensive system linking not only all parts of the central area and its parking but also the colleges and housing beyond the motorway ring, by underpasses and bridges.

The existing Eldon Square and other famous places are set in a new and finer townscape.

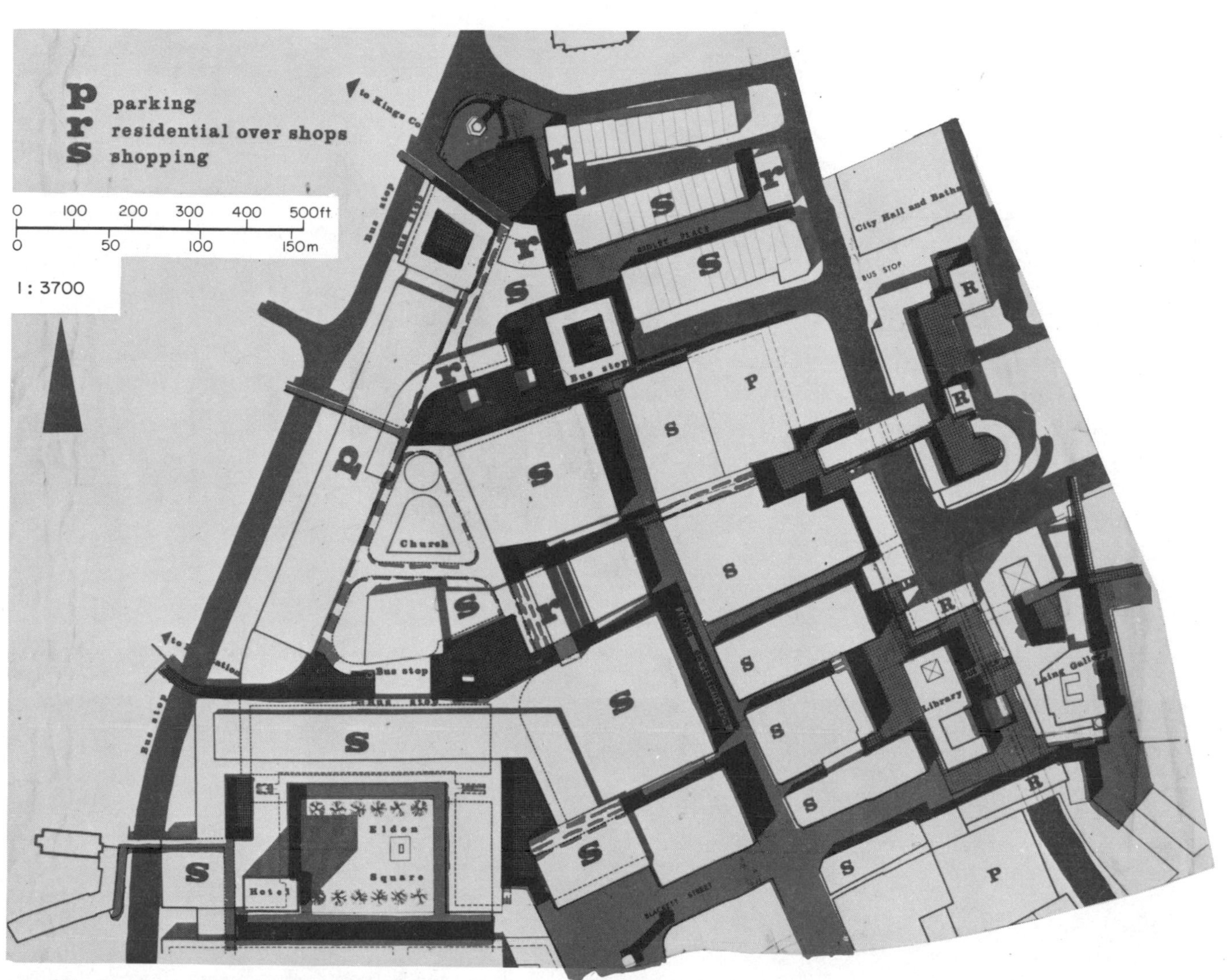

Fig. 7.78. Proposals of the first two areas combined.

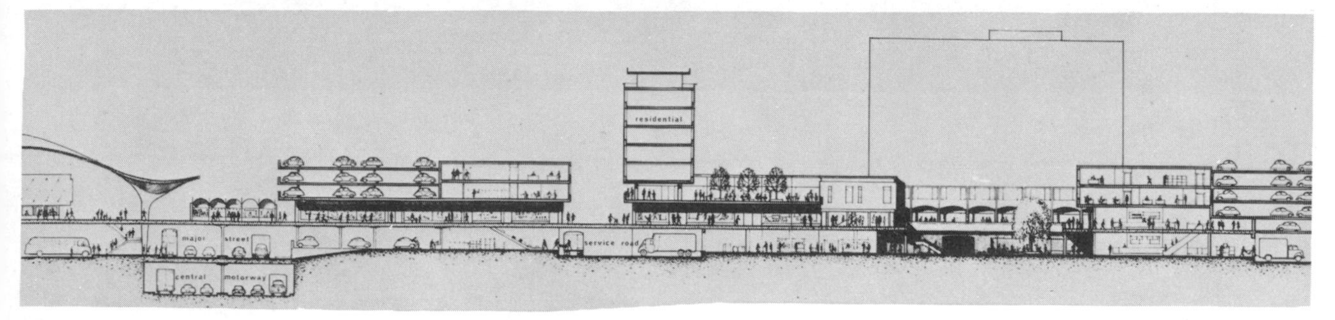

Fig. 7.79. Cross section showing use of different levels.

Reference
Central Area Report 3 and 4, October and November 1962.

VII. 6.6. *City of Victoria, Hong Kong, Central Area*

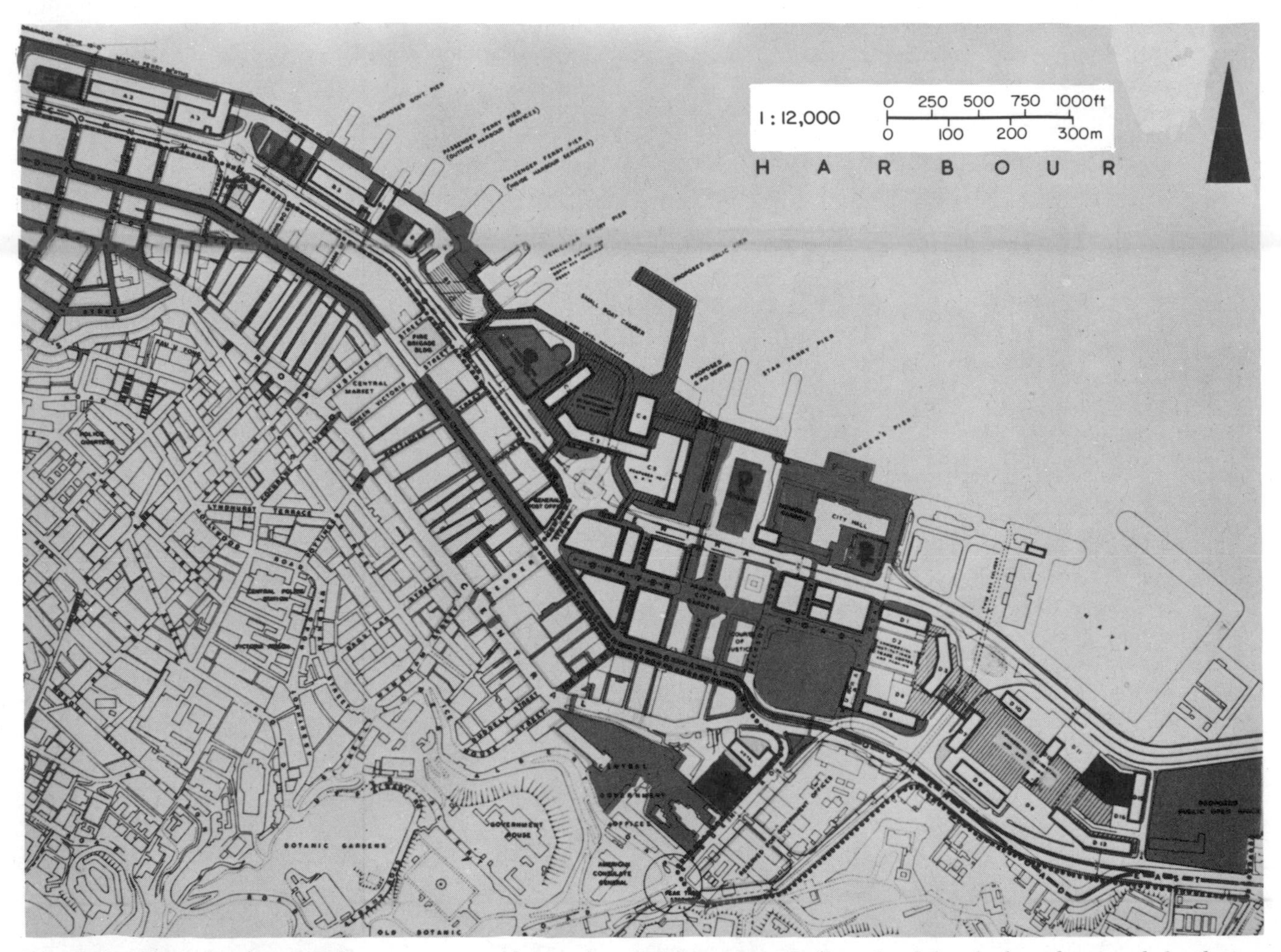

Fig. 7.80. Plan showing multi-storey car parking and pedestrian areas. Podium level hatched and ground level grey.

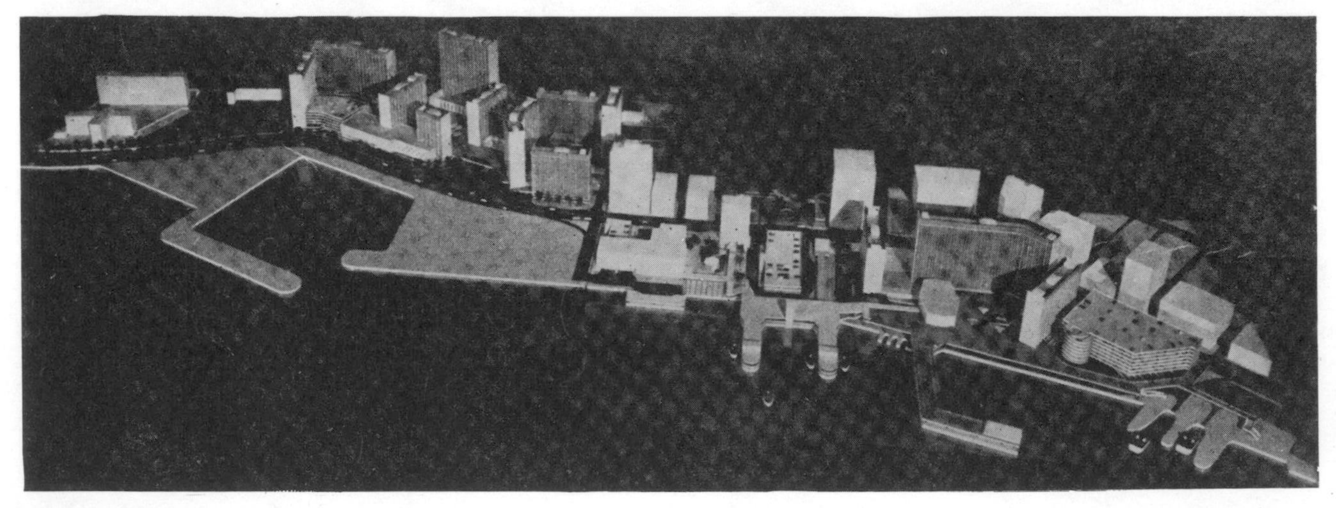

Fig. 7.81. Proposals, aerial view of model.

Fig. 7.82. Charter Road "precinct" as existing

Design: Office of Director of Public Works.

Population: 450,000.

Nature of Scheme: It is estimated that the reclaimed area will cover land purchasing costs. 76 acres (31 ha) of reclaimed and government acquired land are dealt with comprehensively. The report says of the area:

"*Its function is essentially one of personal contact between shopper and salesman, wholesaler and retailer, professional man and client, banker and business man, actor and audience, civil servant and public.*"

"*A western city inhabited largely by orientals; a city of sky-scrapers fighting each other for light and air, it contains the unique characteristics of an oriental city; a 'happy' mixture of noise and confusion, movement and crowdedness; rickshaw vying with Cadillac, hawker stall with emporium, shopper with clerk, pedestrian with motorist; each intent on his own affairs.*"

"*The Board has been concerned not to destroy what is good in these characteristics but at the same time has wished to ensure that crowdedness does not deteriorate into congestion, that confusion does not become chaos. Its first thoughts therefore, have been to separate motorist and pedestrian and, to a lesser degree, business man and tourist, shopper and clerk. This has led to a system of pedestrian precincts. Only thus can the prosperity of a compact business or shopping centre be maintained.*"(i)

The precincts planned are linked by bridges and underpasses wherever possible and the underpasses are shopping arcades. Traffic roads are to be cleared of pedestrians and parked cars entirely and in those areas where vehicular traffic and pedestrian traffic remain mixed, clear precedence of pedestrian or vehicle will be established, depending on the prior claim.

"*Conflict between pedestrian and motorist is ever present where the two must intermix and it is usual for the former to have to give way to the latter. However the Board has endeavoured to provide facilities for the pedestrian in certain areas to off-set the severe restrictions which must be placed on him in others. The two areas immediately east and west of Statue Square have been designated areas of pedestrian predominance, that is, areas in which vehicles are permitted but in which it is considered they should at all times give way to pedestrians. This suggestion is a somewhat novel one and needs working out in detail prior to trial.*"(i)

Among the new features are multi-storey car parking for 8000, which is to be charged for on a carefully graded scale, rising towards the centre of the centre; vertical segregation of a pedestrian podium which runs at a height of 19 ft along much of the river front; a replacement of Queen Victoria's statue in its rightful place "in the heart of the city". Public transport trams which move 15,000,000 people a month extremely cheaply, and buses moving seven million, are to be improved.

Fig. 7.83. Charter Road precinct as proposed.

(i) City of Victoria, *Central Area Redevelopment.* Report by the Director of Public Works, Hong Kong, August 1961.

VII. 6.7. *London, Traffic Plan*

Design: W. K. Smigielski.

Population: Ten million with a very large hinterland.

Origin: Second Prize Winner in the "New Ways for London" Competition of the Roads Campaign Council, 1959.

It is illustrated because it showed special regard for the needs of pedestrians as well as vehicles. Competitors were asked to make overall proposals for the traffic of London and to select an area for detailed development.

Nature of Scheme: Smigielski shows a sensitive knowledge of London. His scheme is based on three main factors: a) the "soft belly" of the South Bank where less traffic approaches and so suggest a horseshoe rather than a ring road for traffic; b) the tangential road system which fits London's needs, makes routes short and convenient and prevents congestion in the centre; c) the "historic route" roughly parallel to the river which, from Buckingham Palace to The Tower, touches on most of the precious formal heritage of London. This last is conceived as a largely pedestrian and vehicular traffic-free route with main motorways in tunnels where they cross. The urban motorways in this scheme are sensitively placed to keep existing areas which seem to have affinity and character in one whole, as precincts, in Alker Trip's sense. They have no through traffic, and form many pedestrian areas and with one mile overall dimensions. On their edges, garaging for 40,000 cars is accessible directly from the motorways so that cars are kept off the local road system and areas are right for walking. The railway termini are roofed with parking for several thousand cars each and linked by motorways.

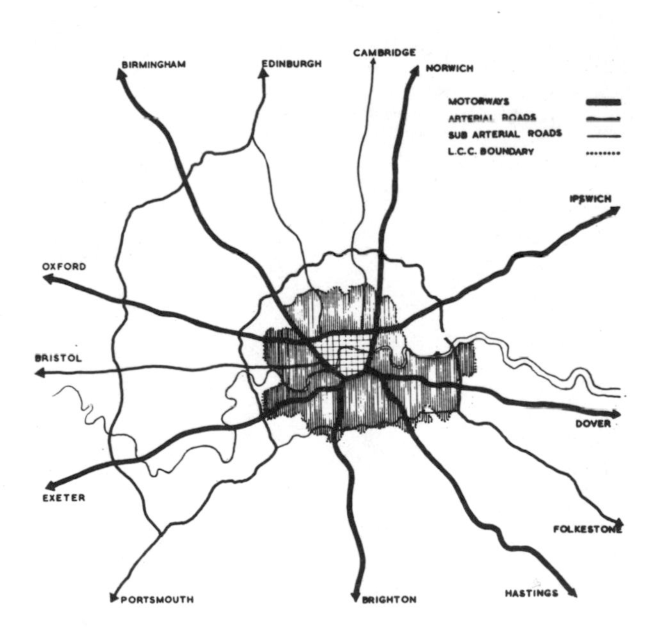

Fig. 7.84. Proposals in the national setting.

(i) Smigielski M. K. Traffic Plan for London. *J.T.P.I.* London, April 1960.

London Roads Competition, Results and Appraisals. *A. J.* 24 December 1959.

New Ways for London, The Five Award Winning Schemes, The Road Campaigns Council, London 1960.

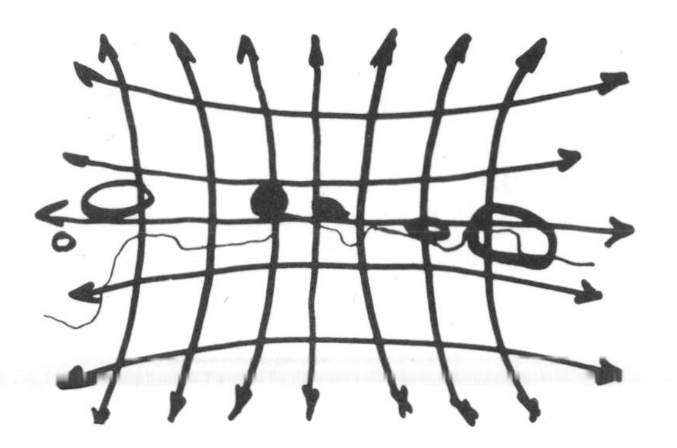

Fig. 7.85. Tangential, functional road network for London.

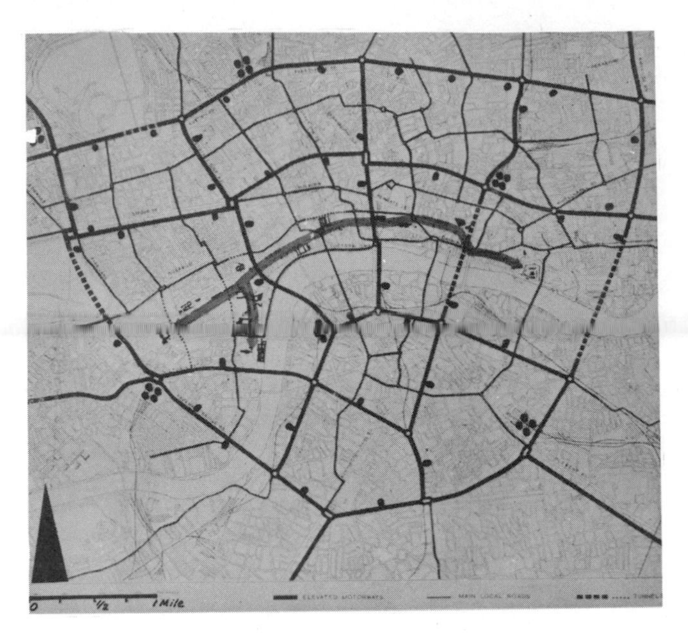

Fig. 7.86. Road proposals for Central Area, showing large car parks over six railway stations (multiple dots) and car parks directly accessible from motorways (single dots). These define the precincts. The historic route from Buckingham Palace and Westminster to the Tower is shown running parallel with the river.

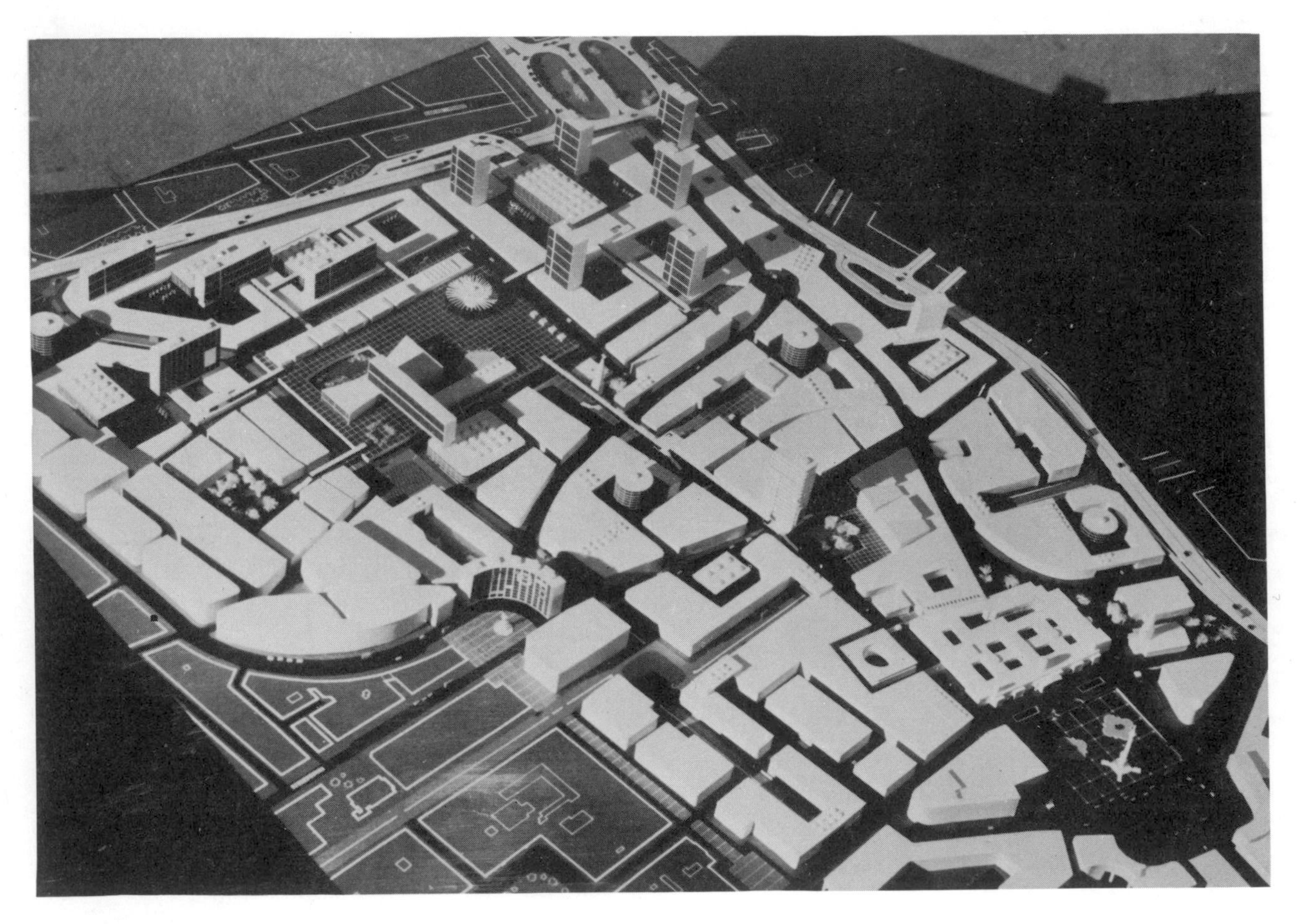

Fig. 7.87. Birds-eye view of model, Piccadilly Circus in the centre foreground. Trafalgar Square is no longer severed from the National Gallery and other surrounding buildings by motor traffic, taken now at a lower level.

VII. 6.7.1. *Trafalgar Square Area.* The area detailed by Smigielski, from Trafalgar Square to Leicester Square, shows a multi-level pedestrian environment well served by roads and cars. The peace and beauty of such car-free areas give great promise. Moving pavements help the pedestrians to overcome any tiring distances.

10,000 parking places are provided in this area. The main pedestrian level is 25 ft above ground on the entertainment level, and another at ground level, with shopping arcades linking the area commercially with the surrounding, celebrated shopping streets. Most existing streets are maintained without widening and service the area adequately.

The outstanding merit of the scheme is the recognition of the "historic route" which would give London a fine spine to orientate the visitor, a main pedestrian network, in a linear centre.(i)

Fig. 7.89. Looking past St. Martin's towards Nelson's column in Trafalgar Square. The foreground is at present a seething mass of traffic.

Fig. 7.90. Plan of area detailed including Piccadilly Circus, Leicester Square and Trafalgar Square, with existing street network not widened but relieved by motorway shown on two sides and extensive new pedestrian environment.

(i) New Ways for London. R.C.C. London, 1960.

VII. 6.8. *Fort Worth, Texas, U.S.A. Central Area Reconstruction*(i)

Design: Victor Gruen & Associates. E. Contini, B. H. Southland, R. Simpson, B. Zwicker, N. Crawford, B. Rawdon, D. Solon, K. Norwood, W. Greub.

Origin: Planned by the designers as a catalyst to produce a Greater Fort Worth.

Planned in 1955 and the report, lavish and beautiful, with magnificent and realistic photographs of the model, was published in 1956. The multi-level solution of the large pedestrian centre had a great influence. It was very widely illustrated and published.

Nature of Scheme: The longest side of the central area is about 1 mile. The centre is planned for a population of over a million: 9·2 million ft^2 (·85 million m^2) of retail space, 2·3 million ft^2 (·21 million m^2) of office space, a million ft^2 (92,903 m^2) of wholesale space, 3200 hotel rooms, half a million ft^2 (46, 451 m^2) cultural and entertainment, and a little more than that for administrative accommodation of all kinds. There is parking for 60,000 private cars in six multi-storey car parks and underground delivery space for 2600 lorries per day adjoining the inner ring motorway. Public transport is integrated in an efficient manner, playing a vital part. As service is from below ground all sides of buildings are free for commercial development and open to pedestrians only. "Pedestrains" take people at very slow speeds along the main pedestrian routes, quietly and safely. Moving pavements and escalators are anticipated and the guiding of cars to free parking places is to be perfected with electronic means.

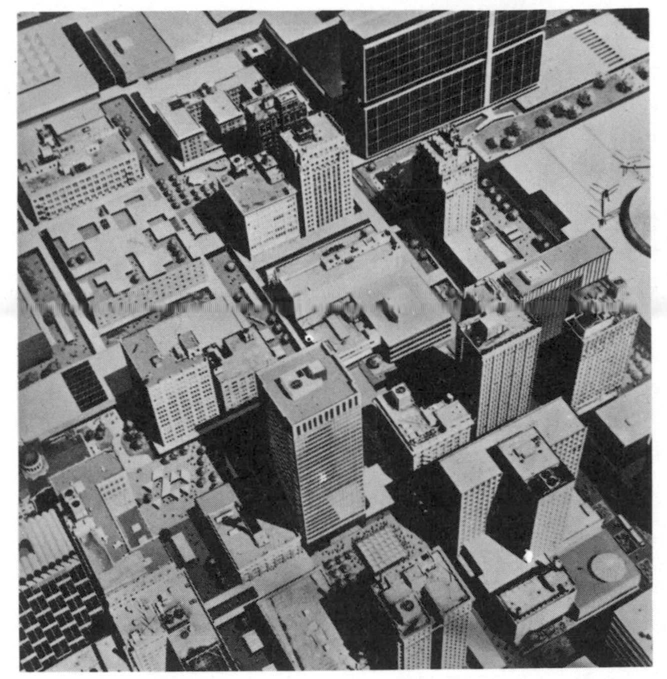

Fig. 7.92. Detail aerial view of part of scheme.

Fig. 7.93. Perspective of one of the pedestrian areas.

Fig. 7.91. Aerial view of model of the whole area, railway station in foreground, showing helicopter ports on top of garages, 10,000 cars each.

(i) GRUEN V. *The Greater Fort Worth.* Fort Worth 1956.

DEVELOPED LAND USE

TABLE VII.3

	Av. floors	Ground area	EMPLOYEE			PUBLIC		SUMMARY	
			No. employ	Parking basis	No. car spaces	Parking basis	No. car spaces	Total no. car spaces	Total parking area
RETAIL AND SERVICES	3	73·5 acres (29·9 ha)	19,200	50% by car 1·5/car	6400	4/1000 ft² (4/92·9 m²)	38,400	44,800	308·5 acres (123·4 ha)
Setbacks, private open		18·4 acres (7·3 ha)							
WHOLESALE	2	11·0 acres (4·5 ha)	960	60% by car 1·5/car	385	3/1000 ft² (3/92·9 m²)	2880	3265	22·5 acres (9·0 ha)
Setbacks, private open		2·8 acres (1·1 ha)							
OFFICES, BUSINESS, FINANCIAL PROFESSIONAL	10	5·3 acres (2·1 ha)	15,300	60% by car 1·5/car	6120	1/1000 ft² (1/92·9 m²)	2320	8440	58·1 acres (23·5 ha)
Setbacks, private open		1·3 acres (0·5 ha)							
HOTELS AND APARTMENT HOTELS	10	5·8 acres (2·3 ha)	3200	30% by car 2/car	480	1 car /3 rms	1100	1580	10·9 acres (4·5 ha)
Setbacks, private, open		0·9 acres (0·4 ha)							
CIVIC AND INSTITUTIONAL	2	7·1 acres (2·9 ha)	2050	50% by car 1·5/car	685	1/1000 ft² (1/92·9 m²)	620	1305	9·0 acres (3·6 ha)
Setbacks, private open		1·8 acres (0·7 ha)							
CULTURAL AND ENTERTAINMENT	1½	6·8 acres (2·8 ha)	225	50% by car 1·5/car	75	1/1000 ft² (1/92·9 m²)	450	525	3·6 acres (1·5 ha)
Setbacks, private open		1·7 acres (0·7 ha)							
MISC. BUS TERMINAL AND AUTO SERVICE	1	4·6 acres (1·9 ha)							
Setbacks, private open		1·1 acres (0·5 ha)							
PARKING TERMINALS		47·8 acres (19·1 ha)							
TOTALS		190·0 acres (76·7 ha)	40,935		14,145		45,770	49,915*	412·6 acres (165·0 ha)

(* adequate only with highly efficient rapid transit system.)

SUMMARY Total ground area required in acres and hectares.

TOTAL AREA—developed land use	190·0 acres	(76·7 ha).
TOTAL AREA—public malls, walks, parks	75·7 acres	(30·3 ha).
GRAND TOTAL—within perimeter road	265·7 acres	(107·0 ha).

VII. 6.9. *Nottingham, Central Area, and Traffic Plan*

Design: Paul Ritter. Sketch Designs for areas of the A.D. 2060 model were by Paul Ritter, C. Fox, F. Trout, D. Green, P. Cook, D. Brindle, T. Gwilliam, J. Shimwell, J. Knowles, C. Millard, Eric Lee, H. D. Hogg and G. A. Bowler.

Origin: The City of Nottingham, without a separate Planning Department, was implementing, piecemeal, highly dubious proposals. The new plan was created to show the population and councillors that integrated planning is vital and what its criteria should be.

Large models of the centre of Nottingham, as in 1830, 1960, and as it might be in 2060, made vivid that change takes place whether we plan or not, and that, as the scale of renewal grows, so does the need for planning.

Population: 300,000 approximately.

Basis for Proposals: A. General Information from 1952 City Council Report.

B. Transport: the main arrival points for public and private transport are at the north, south and south-west of the city.

C. Pedestrian Flow: determined by the termini and the shape of the centre in its south–north valley, rising to the north.

D. Townscape: scale in valley set by Dome and St. Peter's, and on the rock dominating the entrance to the valley by St. Mary's, the Unitarian Chapel Spire and the Castle.

E. Geology: sandstone rock riddled with caves invites car parking and service access provision below ground.

F. Micro-climate: the area below the rock at Broad Marsh is not ventilated. The square is not adequately sheltered from winds and the large amount of successful arcade invites an extension of the system.

G. Traffic: shopping streets are a vicious mixture of pedestrian and motor traffic, particularly Wheeler Gate, Pelham Street, Clumber Street, Long Row and the ring road to the north of the city.

The proposals have three central aims:

i) To provide a city where people have peace from vehicles in places of human scale.

ii) To provide adequately for private, public and service vehicles.

iii) To make the most of the city economically through redevelopment and the improvement of site use and values.

Fig. 7.94. Plan of the city with outline traffic proposals. The extensive areas of housing from various periods, all with traffic segregated path access, are estimated to have a total population of nearly twenty thousand. The central area is serviced from below ground and by a high level road placed inconspicuously among the buildings, leaving all shopping streets and squares free from traffic. (For path system in interwar housing, see IX.2.6.)

General Traffic Plan: Two networks, one for pedestrians and one for motor traffic are recommended. The pedestrian network to be formed out of most of the existing shopping streets in the centre. New path ways from the dense housing areas, ready for comprehensive renewal and within walking distance of the centre, are to be directly connected via under and over passes.

Within the low density housing areas path networks to be created as recommended for Aspley (see Section IX.2.6).

Tangential Urban motorways replace the old concept of the ring road. These motorways bring traffic as near to the centre and relate directly to parking facilities. There is no ring of traffic round a city. This idea has been shown to be quite false.

On the western side of the city, towards the Park, there is no need for a motorway as on the other three sides. The Park and the Castle, the Rope Walk, etc., should be integrated with the city centre in a number of ways. Helpful in this respect should be the Victorian Tunnel, leading into Derby Road, the latter freed from traffic, and escalators are planned to overcome the steepest points at other places.

Lifts from Castle Boulevard car-parking, within the rock, up to town-level would be very convenient.

Bus stations are connected by underpasses and escalators and moving pavements to the town centre.

Commuter traffic from the estates will be brought into the city by bus or private car, either via the motorways, or the radial main roads passing under or over the motorway where necessary. The Radford Boulevard, Lenton Boulevard, Gregory Boulevard ring route should ultimately become a motorway also.

A park-and-ride system can be developed in Nottingham. The marshalling yards and railway lines are ideally placed to allow public transport via underground rail to bring several thousand people into a car-free city centre, via the existing tunnel under the city centre: Termini at Wilford Bridge Marshalling Yards, with parking for the traffic from Clifton Bridge; at high level London Road Station; and at Hucknell Road, Mansfield Road Junction, Park and Ride provisions are also planned at Midland Station, high level. In each case access is excellent and parking space plentiful and the siting ideal for multi-storey parking. The motorway would connect directly to the London Road Park and Mansfield Road. Three underground stations in the city centre are recommended.

Areas for Redevelopment:

A. Area between Huntingdon Street bus station, Central market and the Market Square. An underpass to cope with Parliament Street traffic is the key to design of this important and convenient route, to raise the property values, and make full use of a basement level for shopping.

B. Similarly the installation of an escalator to cope with the steep incline to the top of the rocky crescent, delineating the southerly side of the centre facing the river, would directly connect the centre with Broad Marsh Bus station and create very valuable and useful sites between. (See p. 157.)

C. Most lucrative and important is the restructuring of the Market Square itself. Its sterling qualities are weakened not only by roundabout and traffic but an ugly weak slope and funnelling-off at one end. The latter, though ancient, is nevertheless open to improvement. To create a number of small spaces with very valuable new shopping would also give a finer form to the square itself. This should be creatively and usefully developed as a market at a lower level, without disturbing the public functions it now serves. Queen Victoria's Statue could come back to its original place.

Parking: 12,000–15,000 is estimated to be the number of vehicles that Nottingham as a city and its improved road system might cope with without destroying itself.

Fig. 7.95. Nottingham 1830; virtually the whole of Nottingham. Plan view showing minute scale and view of Market Square.

Fig. 7.96. Nottingham, 1960. Central Area. Plan view of model.

Fig. 7.97. Nottingham, 2060. Central area multilevel development, vehicular traffic systematically segregated from shopping, town shape improved, town scale in the centre intact.

Fig. 7.98. This is virtually the whole of the area on the models. The ugly tapering shape of the square is seen in the foreground. On the extreme right the two branches of the railway line are seen entering the tunnel below the centre. In the very right top corner the space for parking and the London Road High Level station is visible. On the other side of the railway are the little terrace houses, overshadowed by the rock, the area proposed for a park and entertainment centre. Parliament Street, severing the centre from Huntingdon Street bus station and the Central Market, runs along the left edge. The great outdated factories and warehouses of the lace market are to the right beyond the Council House. (Aerial photo.)

Fig. 7.99. 1830 aerial view into square, showing market. St. Mary's Tower dominates.

Reference
Some Pedestrian Planning Proposals. *The Builder,* 8 June 1962.

Fig. 7.100. Nottingham, 2060. The view into the market square shows the reshaping of the square itself, the new intimate spaces and the upper level shopping running into it and around it. Tall blocks are kept out of the valley of the town centre. The raised road in the foreground allows pedestrian flow underneath to the castle and its open space much used by the city workers. To the right of the Council House the high-level, "service-only" road is served by cranes and gantry running along it. In the background the high class residential development of the Lace Market includes a 500 ft high structural frame designed for the insertion of rented, prefabricated elements connected by circulation. The open effect of the structure is intentional so that the smaller but solid mass of the medieval St. Mary's may still dominate the hill top. (C. Millard.)

(The models were built under the direction of C. Millard, by C. Dalton, C. Massey, R. Short, Sylvia Cooper, E. Lee and Paul Ritter, with many minor assistants.)

VII. 6.10. *Banbury, Central Area Redevelopment*

Design: Civic Trust, Chief Architect, Noel Tweddell, with R. Worskett and V. Rose.

A population of 21,000, which is expected to grow to 48,000, but the central area will serve 70,000 people.

Local objections to the Borough Council's proposals to widen main shopping streets led to the Council asking the Civic Trust to act as consultants, in association with Hillier, Parker, May & Bowden, acting as economic advisers. Several schemes were prepared and assessed on their merits, including the financial return possible on development of new shops within the area to be redeveloped.

"*The cost of carrying out the whole of the selected scheme (Including Phase I) would amount to an annual loan charge of about £31,000 (a 1/8d. rate) over a twenty year period.*"

"*We are advised by the Borough Engineer and Surveyor that his estimates for civil engineering works are as follows and these have been taken into account in our valuation:—*

Paving Pedestrian Areas	£27,500
Service Roads	£19,500
Car Parks	£36,500
Service Areas	£11,500
New Road	£36,000
Bus Station	£40,000
Demolition	£22,000
Alterations to service mains	£28,000
	£221,000" (1)

A ring road brings buses and parking for 500 cars (and possibly more) within immediate reach of the central pedestrian network. Service is by loops and culs-de-sac. The consultants stress the importance of architectural merit to keep the character and scale of Banbury as a market town. They advise strong control under the Planning Acts. Phases of construction are recommended.

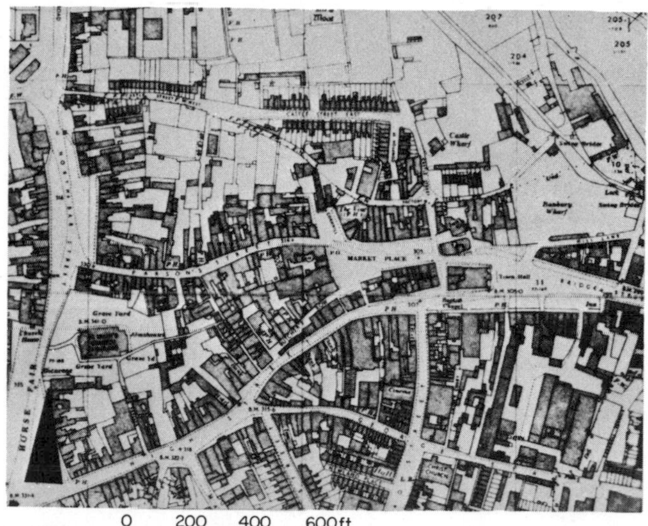

Fig. 7.102. Plan of Banbury as existing.

Figs. 7.103 and 7.104. Old and new view of Baptist Chapel.

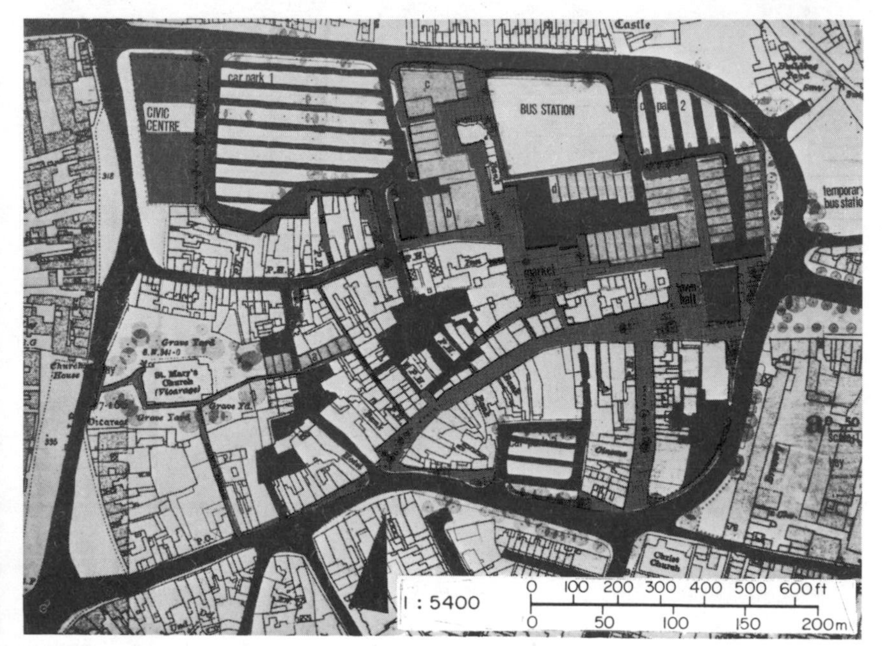

Fig. 7.101. Plan of Banbury as finally proposed by Civic Trust and accepted in principle by Borough Council.

(1) Civic Trust, Final Report 3D, Banbury Central Area, 1963. London.

Reference

Civic Trust Scheme for Banbury Central Area. Official Architecture and Planning, London, November 1962.

VII. 6.11. *Swindon, Central Area Redevelopment*

Planning Officer, L. R. Robertson.

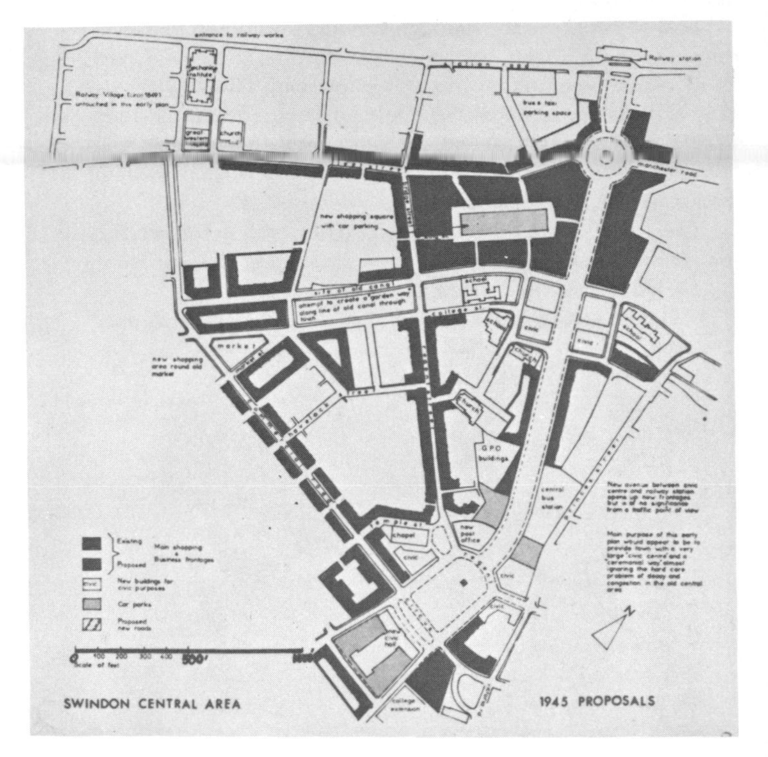

Fig. 7.105. Plan as proposed in 1945. No provisions for pedestrians, street widening in a manner typical of most local authority plans of that time.

Fig. 7.106. Plan as proposed in 1960, and substantially passed in 1962. In 1958 a single pedestrian street had been proposed.

This scheme was one of the first with traffic segregation for a small town in Britain. Its size had grown from 69,000 to 92,000 and the target population is 110,000. The plan was worked out with admirable co-operation between the tiny Borough planning office, the Wiltshire County Planning Office and the Ministry of Housing. It is being built as a joint enterprise by private developers investing in areas bought and leased by the Borough for Comprehensive Redevelopment under the architectural supervision of F. Gibberd as consultant.

Car parking for 3000 is to be provided. There is an outer ring road and quite close to it runs an inner ring which feeds the service loops and culs-de-sac.

An underpass from the county bus-station under the main ring road to the main pedestrian precinct is planned.(i)

Fig. 7.107. Aerial view of Central Area. (Tresco Reproductions Ltd.)

Fig. 7.108. First stage in pedestrian precinct system now under construction.

(i) GINSBURG L. B. Swindon, A Breakthrough in Central Area Redevelopment. *Official Architecture and Planning*, London, November 1962.

Borough Surveyor, Swindon Central Development, Swindon 1960.

VII 6.12. *Västerås, Sweden, Traffic Plan*

A Traffic Plan, for the central area.

Design: Orrje & Co., Consultant Engineers. Proposals asked for in 1958 by the city council.

This is a town of 80,000 inhabitants which expects to grow to 120,000 by 1990, the date for the final proposals as shown. Interim proposals are worked out, particularly for 1970. Proposed in 1958 it must rank as one of the first traffic segregated schemes for a small town.

The scheme aims for a pedestrian only network, a main T restricted to pedestrians and bus routes, along the most important shopping streets, and separation of two main cycle tracks where, near large factories, a great flow of bicycles takes placed at rush hours.

The plan carefully assesses parking needs, both for bicycles and cars, the fostering of public transport, and provision of parking, in a thorough report. Parking garages of 500 are to coincide with the abolition of 250 kerb-side parking spaces. By 1990 one car to $2\frac{1}{2}$ persons is anticipated. Bus routes are to be very much improved to attract the public by their convenience and comfort.

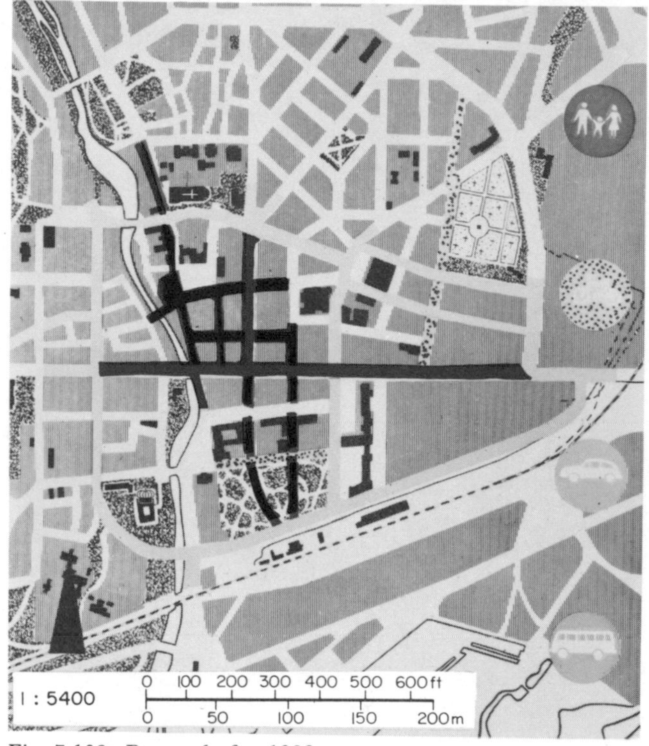

Fig. 7.109. Proposals for 1990.

Fig. 7.110. Diagram showing traffic accidents in 1959. Most of the cross-roads responsible for the accident concentrations would be replaced with safe T junctions by the proposals.

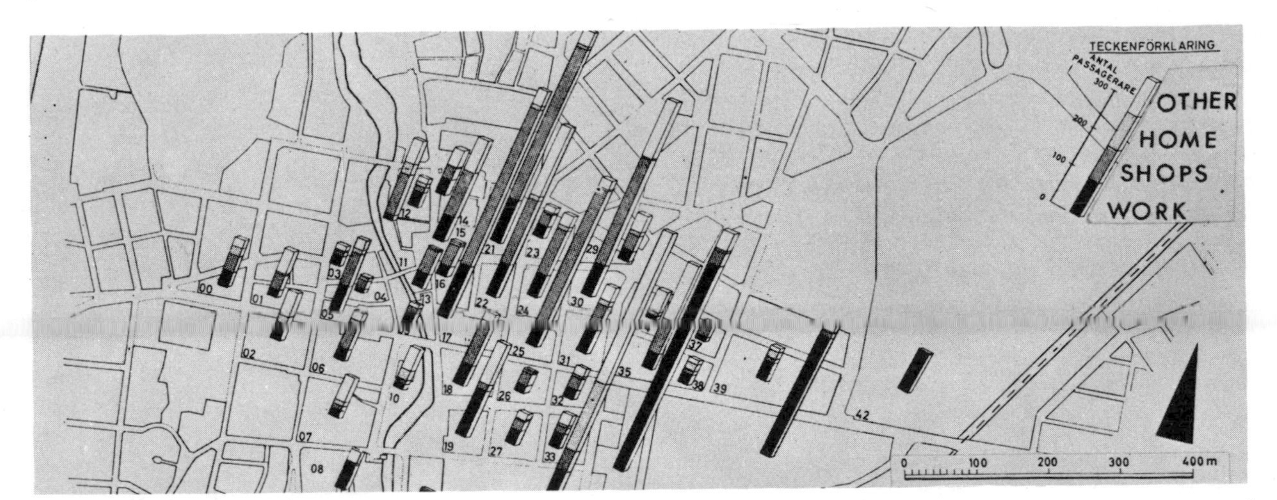

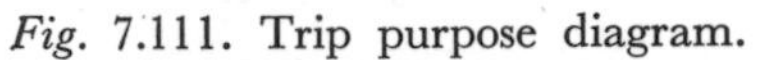

Fig. 7.111. Trip purpose diagram.

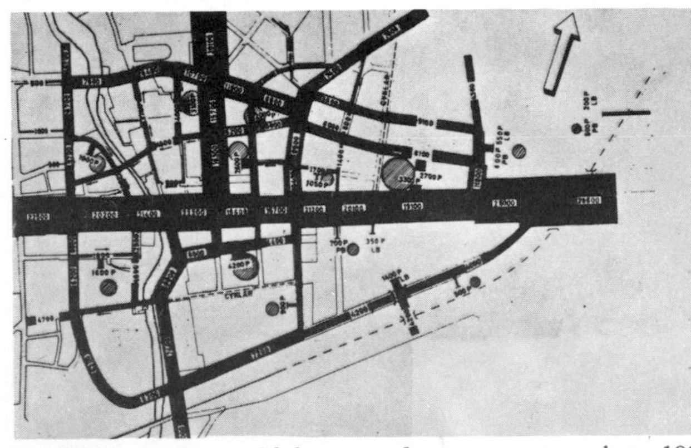

Fig. 7.112. As it would have to be to accommodate 1990 traffic flow.

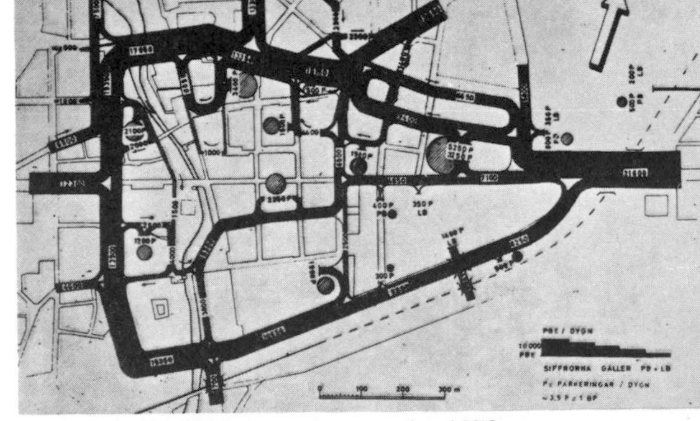

Fig. 7.115. Traffic flow as forecast for 1970.

Fig. 7.113. The main street as it was in 1905.

Fig. 7.116. As it is today.

Fig. 7.114. As it will be according to recommendations.

Fig. 7.117. Traffic flow as it would be if the main street were widened, 1970.

Reference
Samordnad Trafikplan Västerås, Utbedringar och forslag 1958–1960. Orrj & Co. Västerås 1961 (English summary and captions).

VII. 6.13. *Bishop's Stortford, Hertfordshire*(i)

Design: County Planning Officer, E. H. Doubleday, with G. W. Cowley, C. W. Smith and G. S. Probert.

Origin: Draft Plan 1962.

Population: 19,000 in the town itself and 50,000 in the hinterland.

Nature of Scheme: To allow the proper commercial and entertainment use of the town centre with ancient well established markets. The difficulties of pedestrian and vehicular traffic in the present town were established by survey and the three main stages of the draft plan are carefully worked out.

The entire shopping area and historical centre of the existing town is planned to become a system of pedestrian precincts. A ring road with secondary loops and culs-de-sac services the central area. An overpass links the latter over the ring road to the castle and grounds along the widened waterways. Eventual parking for 2000 cars, 1300 of these in two multi-storey car parks.

Fig. 7.120. Land use plan 1961.

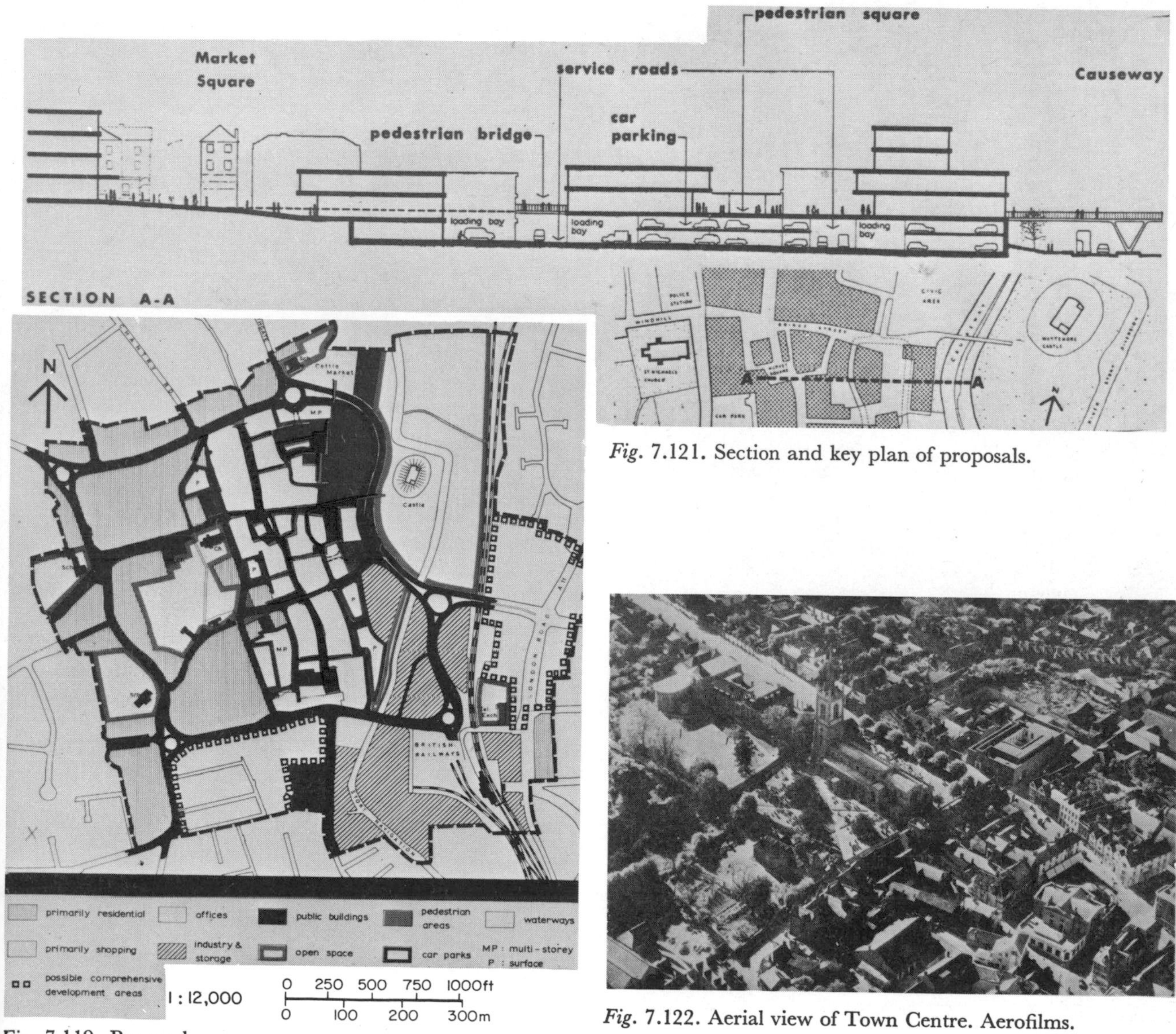

Fig. 7.121. Section and key plan of proposals.

Fig. 7.119. Proposals.

Fig. 7.122. Aerial view of Town Centre. Aerofilms.

(i) Bishop's Stortford Central Area Draft Plan, 1962. Hertfordshire County Council.

Figs. 7.123 and 7.124. View down High Street existing and proposed.

Figs. 7.125 and 7.126. View of heavy traffic in centre and proposals for the area.

VII. 6.14. *Sutton Coldfield, Central Area Redevelopment*

Design. Max Lock and Partners.

Origin: The above were appointed consultants to the Borough Council and produced report plans and model June 1962.

Population: 72,000, plus increase of 50% in ten years. The town is a very wealthy residential area for Birmingham.

Nature of Scheme: Some of the basic points which are in the proposals are summarized at the outset.

"*i. The major part of the scheme is in single ownership and the leases lapse in 1968, thus providing an unparalleled opportunity for comprehensive redevelopment.*

ii. The hilly topography of the site favours an economical and unusually concentrated type of development on two or more levels, facilitating the separation of pedestrian from wheeled traffic.

iii. The existing arrangement of roads provides a convenient simple circulatory system separating the traffic having business in the centre from that which has not.

iv. Around the shopping centre there is enough empty land to permit a well distributed group of car parks each with immediate access to the shops they will serve.

v. The contours of the site permit pedestrian foot bridges across the roads extending the safety walks beyond the centre to the north and south.

vi. The density of existing building on the site is low; so that all persons who are uprooted by the scheme, whether shopkeepers or residents—can be re-accommodated in new premises without being deprived of shop or home for a single day.

vii. The replanning of the town's existing shopping centre is not complicated by the presence of historic scheduled buildings protected from demolition. It can therefore be redeveloped comprehensively in well planned stages."(i)

A hexagonal ring road system of one-way traffic serves with culs-de-sac at three levels the shops added to the existing and pleasant shopping parade on the top of a ridge covering the whole area. This, along with the whole centre, becomes pedestrian. Car parking for 1750 is provided within the ring road, and the proposals double the shopping area to 430,000 ft^2 (40,000 m^2) and office accommodation is increased to 160,000 ft^2 (15,000 m^2).

The lower level service allows for blocks of shops to have frontages on three or four sides. Small intimate walkways link pedestrian precincts and the play of children while parents are shopping is borne in mind. Convenience is gained by proposals for covered links with train and bus station and between covered arcades serving all shops.

Residential areas bordering the centre are to be connected by pedestrian bridges with the centre, continuous safe walkways from home to shops. The architecture is to keep the small scale and intimacy now found in the town and favoured by the residents, many of whom contributed advice to the consultants during good public relations while the plan was under preparation.(ii)

(i) The Royal Town of Sutton Coldfield. Redevelopment of the Central Area. Consultant's Report to the Borough Council. June 1962.

(ii) Sutton Coldfield, *A.J.* London, 17 October 1962.

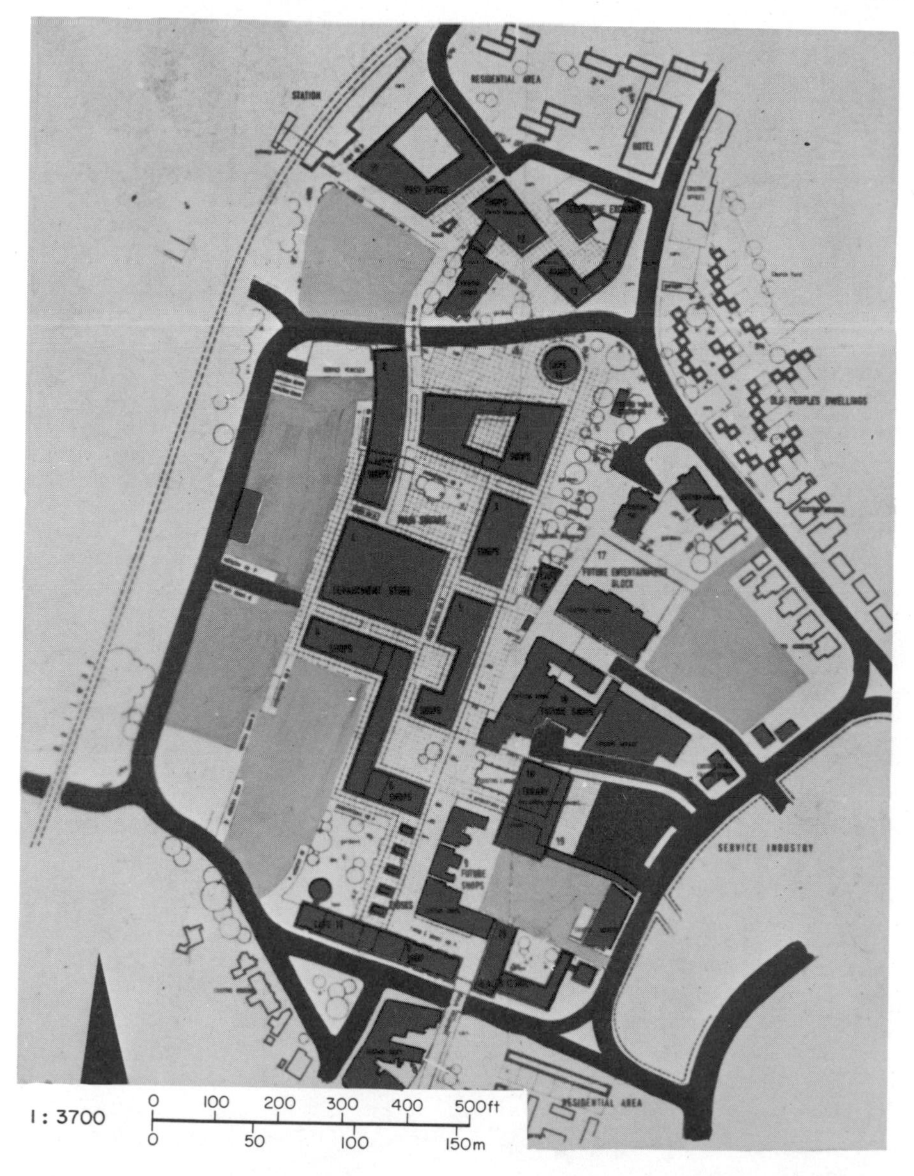

Fig. 7.127. Overall plan of proposals.

Fig. 7.128. Model of the town centre.

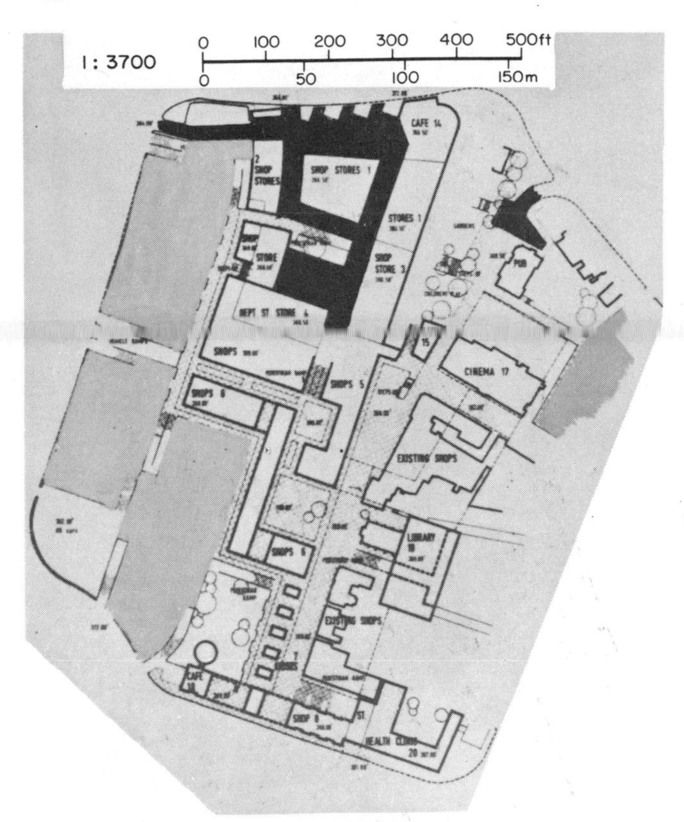

Fig. 7.129. Plan between levels 383 ft and 372 ft.

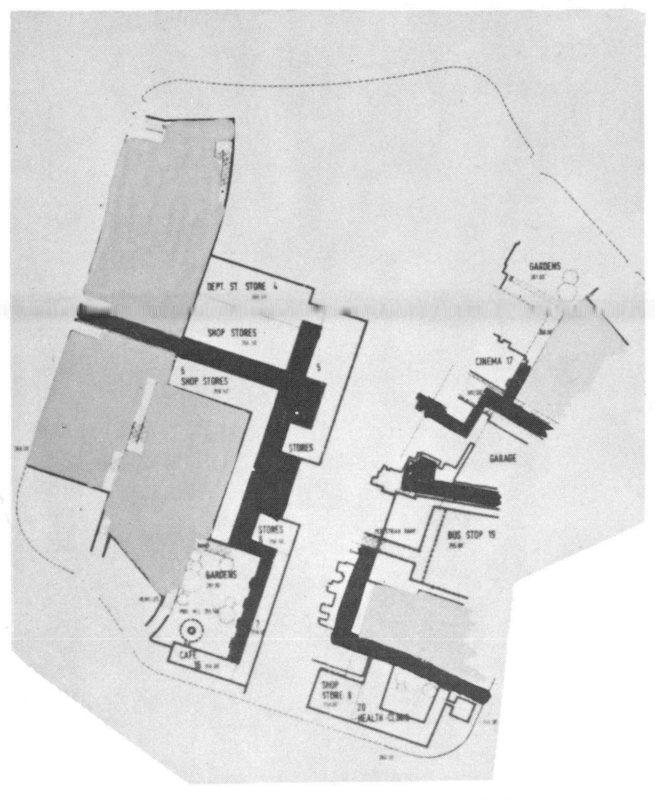

Fig. 7.131. Plan between levels 361 ft and 351 ft.

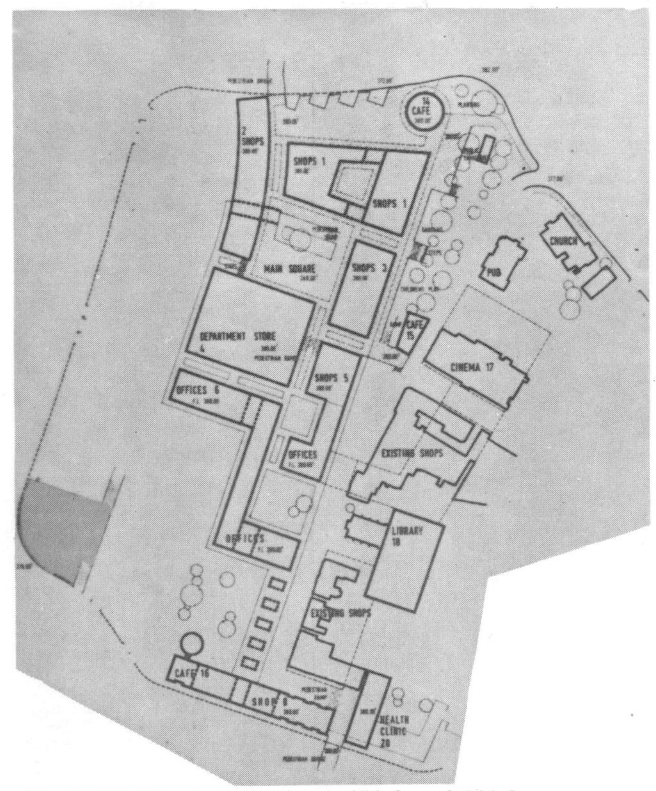

Fig. 7.130. Plan between levels 372 ft and 361 ft.

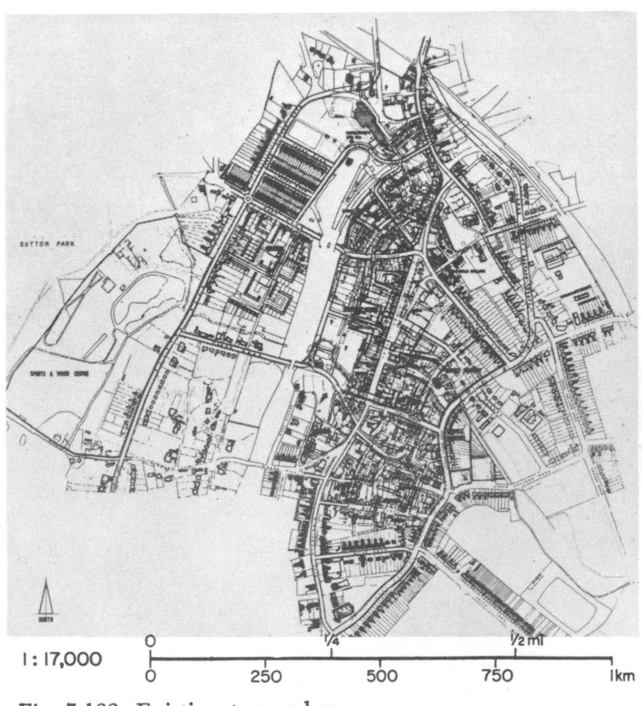

Fig. 7.132. Existing town plan.

VII. 6.15. *High Wycombe: Central Redevelopment*

Design: County Architect, F. B. Pooley.

Origin: Proposals by County Council still under consideration by Minister of Housing (1963), over a year after a compulsory purchase order in 1961.

Population: 40,000 with a considerable hinterland.

Nature of Scheme: The scheme includes proposals by the local authority of this town to make its centre efficient for vehicles and for pedestrians; 600 ft (183 m) of a 44 ft (13·4 m) wide relief road limited to vehicles is raised above extensive pedestrian areas. Shopping is to be linked with bus station and gardens on the lower level. The river Wye runs through the shopping precinct. The upper road level is to be integrated with building so that the viaduct does not sever the architectural unity of the town. Car parking is provided for 2600 and service to shops is from the viaduct at first floor level.

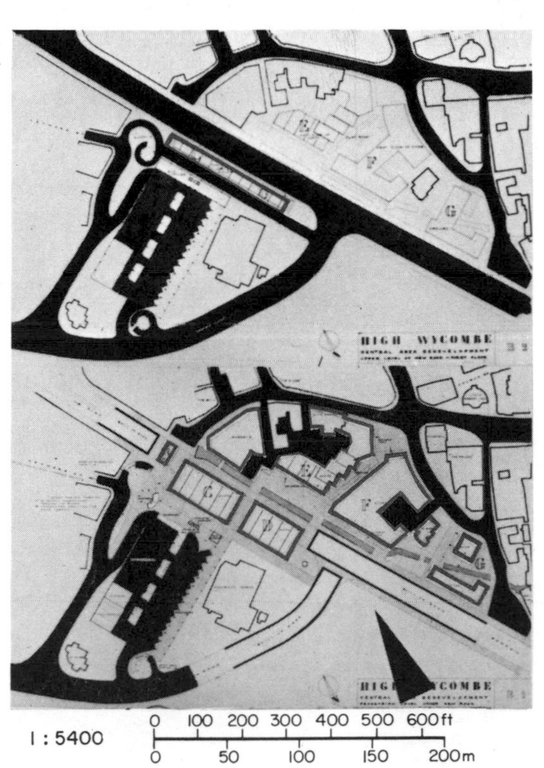

Fig. 7.136. Proposed plan, ground level and upper level.

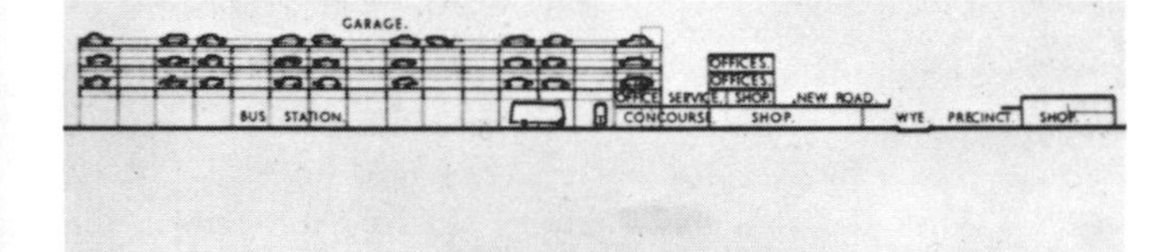

Fig. 7.133. Longitudinal section through parking road and precinct.

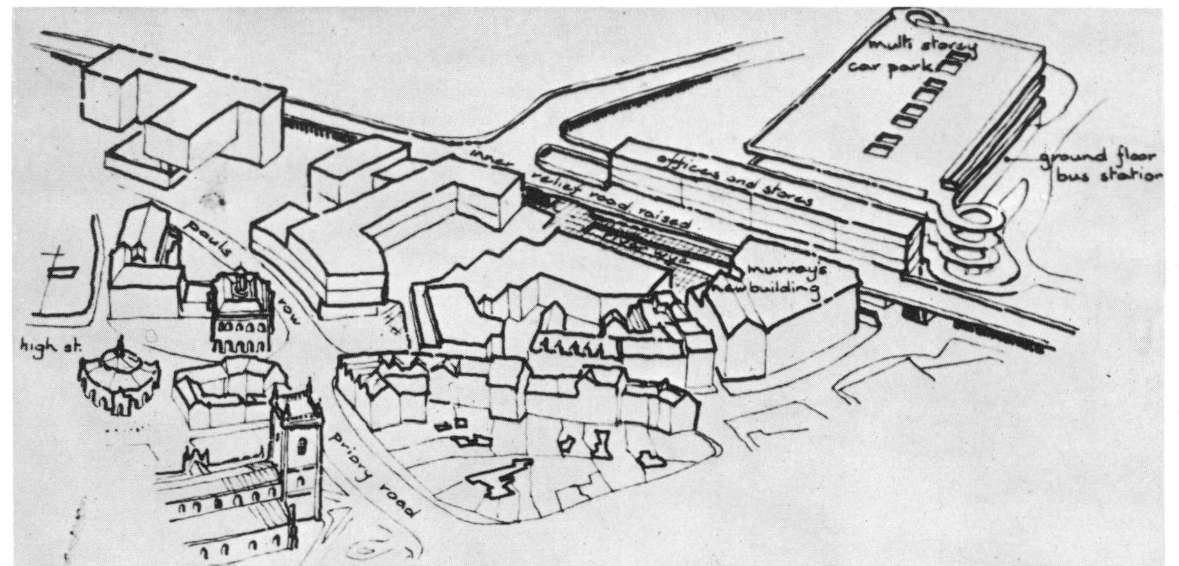

Fig. 7.134. Aerial perspective sketch.

Fig. 7.135. View towards precinct and church.

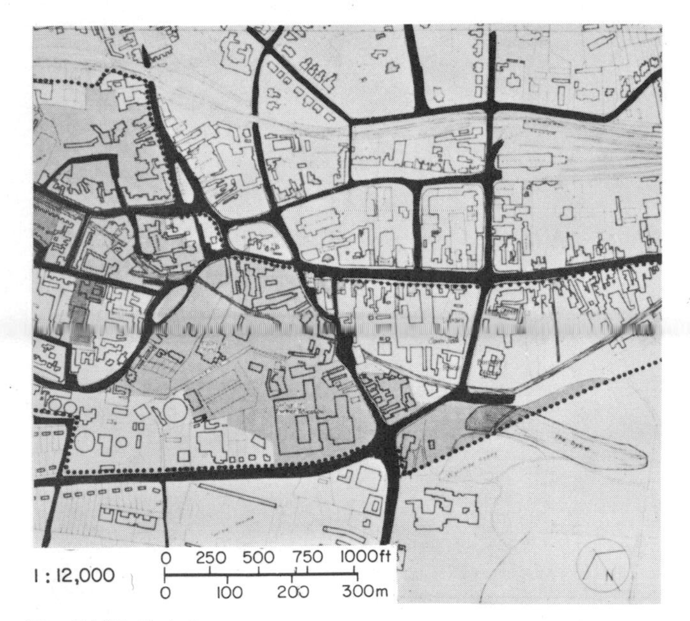

Fig. 7.137. Existing street pattern.

Fig. 7.138. Another view of the church and shopping precinct from the new road.

Reference
Pooley F. B. Central Development of High Wycombe. *H.R.* 1963.

VII. 6.16. *Beeston, Central Area Redevelopment*

Design: Harry Chadwick and Derek Oram.

Origin: A private proposal, stimulated by the apparent lack of comprehensive planning at the time (1961).

Population: About 56,000 including Stapleford (the other centre included in the urban district area).

Nature of Scheme: To transform the congested High Road, central cross-roads and mainly derelict side-streets into an attractive urban area and shopping district for the inhabitants and the students of nearby Nottingham University, with a complete ring road system and service culs-de-sac and loops. Most of the existing roads are transformed into pedestrian ways, thus involving as little disruption as possible whilst the redevelopment proposals are in course of implementation. Inhabitants would still be familiar with the new routes, albeit pedestrian rather than vehicular.

Bus routes are devised to travel all round the ring road, with main stops cutting into the heart of the scheme, thus making it possible for the traveller to alight and board at different points without re-tracing footsteps in the central area when shopping or on business.

Where, in some instances, the footpath system crosses service roads, the walker has been given precedence. It is suggested that the footpaths continue across the road without any change in level, with the roadway ramping up to it on either side, treated in some rough surface material to make it quite plain that here is a hazard, and the motorist must slow down and stop if anybody is about to cross.

Initial parking for over 1000 vehicles is allowed for at ground level, with accommodation allowed for a multi-storey car park at a later date. Car parks are situated as near to the shops and other buildings as possible, at the heads of culs-de-sac, with an indicator system provided at points of entry. Two large open spaces are provided in the area (extensions of a park and churchyard already existing) linked into the commercial areas by main footpaths.

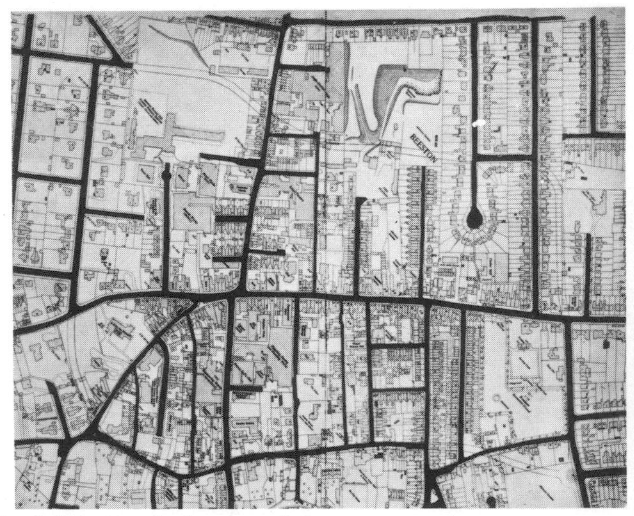

Fig. 7.139. Road plan as existing.

Reference
Chadwick H. and Oram D. *Beeston Centre Redevelopment.* May 1962. Unpublished Report.

Figs. 7.140 and 7.141. Model views showing the string of varied spaces into which High Walk has been transformed in the centre with parking each side.

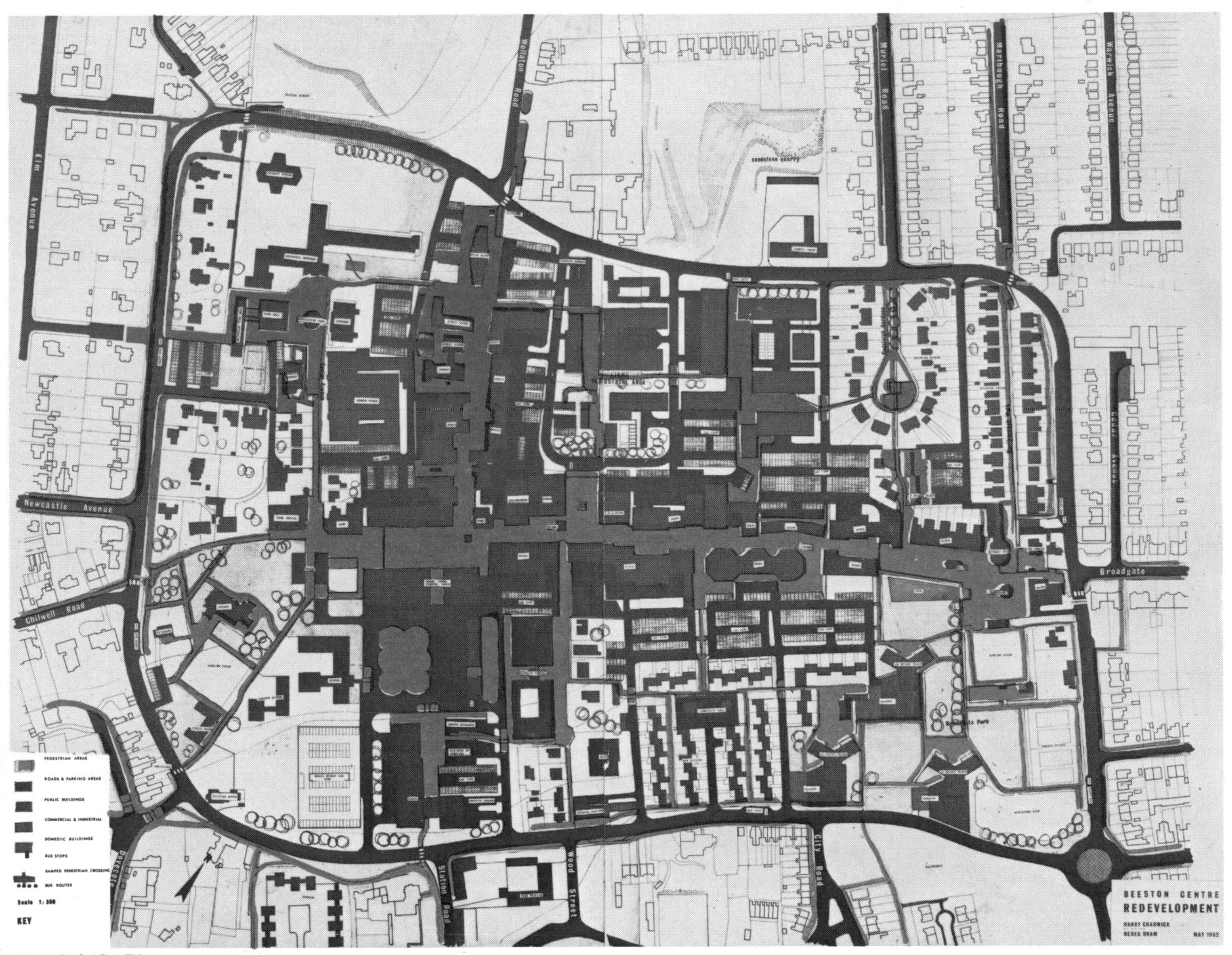

Fig. 7.142. Plan of proposals.

VII. 6.17. *Horsholm, Denmark*

Design: H. H. Skaarup in collaboration with others.

Origin: Scheme developed for the town to show the effectiveness of making a main street pedestrian in a small town, in 1960.

Nature of Scheme: Horsholm was planned in the eighteenth century. The castle of Hirschholm was to stand on the axis of a wide approach road. This scheme was not fulfilled and the main street and shopping street of the small town developed parallel to the original axis.

Skaarup's scheme proposes to create the original axis with the vehicular road which is to relieve the main shopping street, presently crowded with vehicles, which is to become a pedestrian area. The new bus stop is to be at one end of the precinct and, one hopes, connected by underpass to the precinct itself across the relief road.

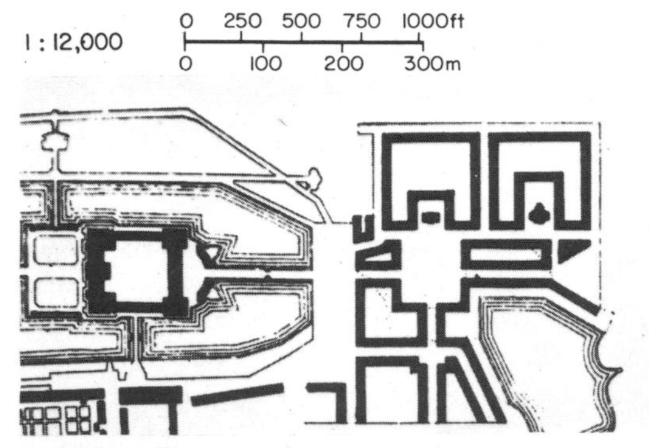

Fig. 7.143. Plan of original 1737 plan by Thura.

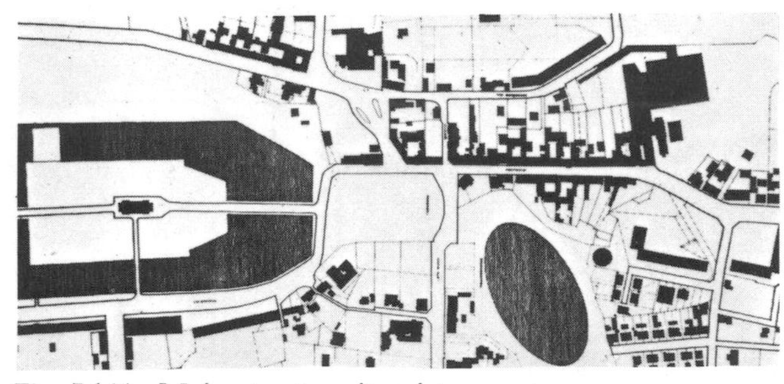

Fig. 7.144. Main street as it exists.

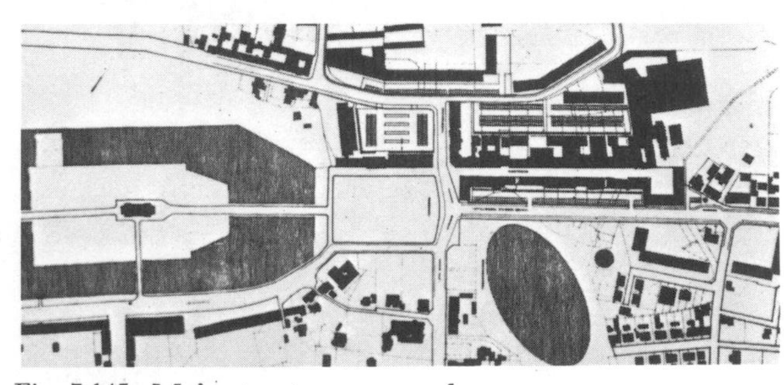

Fig. 7.145. Main street as proposed.

References

By Plan, Special number No. 5. Copenhagen 1960.

Bredsdorff, Rasmussen, Skaarup, *Horsholm*.

VII. 6.18. *Chester City, Central Development*

Design: Sir Percy Thomas & Son.

Origin: Grosvenor-Laing, Private Development.

Population: 60,000 approximately.

Nature of Scheme: $3\frac{1}{2}$ acres are to be developed at a cost of about 3 million pounds, to provide 60 shops, kiosks, a supermarket, department store, offices, ballroom and banqueting suite and multi-storey car park. The unique character of Chester—with its shopping "rows" of medieval origin, in many of the central streets, and at a level a few feet above the streets running parallel to them—is to be developed. Directly related to "The Rows" pedestrian shopping piazzas and decks are proposed in the new area. These connect also with yet another exciting pedestrian system, the city wall walk, which entirely encircles the centre of Chester. Completed, this scheme will add to and connect protected shopping over nine acres.

Service access takes up much of the ground level. On two roofs there is space for 100 cars, in addition to the 400 spaces in the multi-storey park.

The City Council has announced a competition for further redevelopment of a section of The Rows. Chester's Planning Consultant, G. Grenfell Baines, is to be the assessor. (See also p. 45.)

Fig. 7.146. Old drawing of part of The Rows.

Fig. 7.147. The new piazza from the old St. Michael's Row.

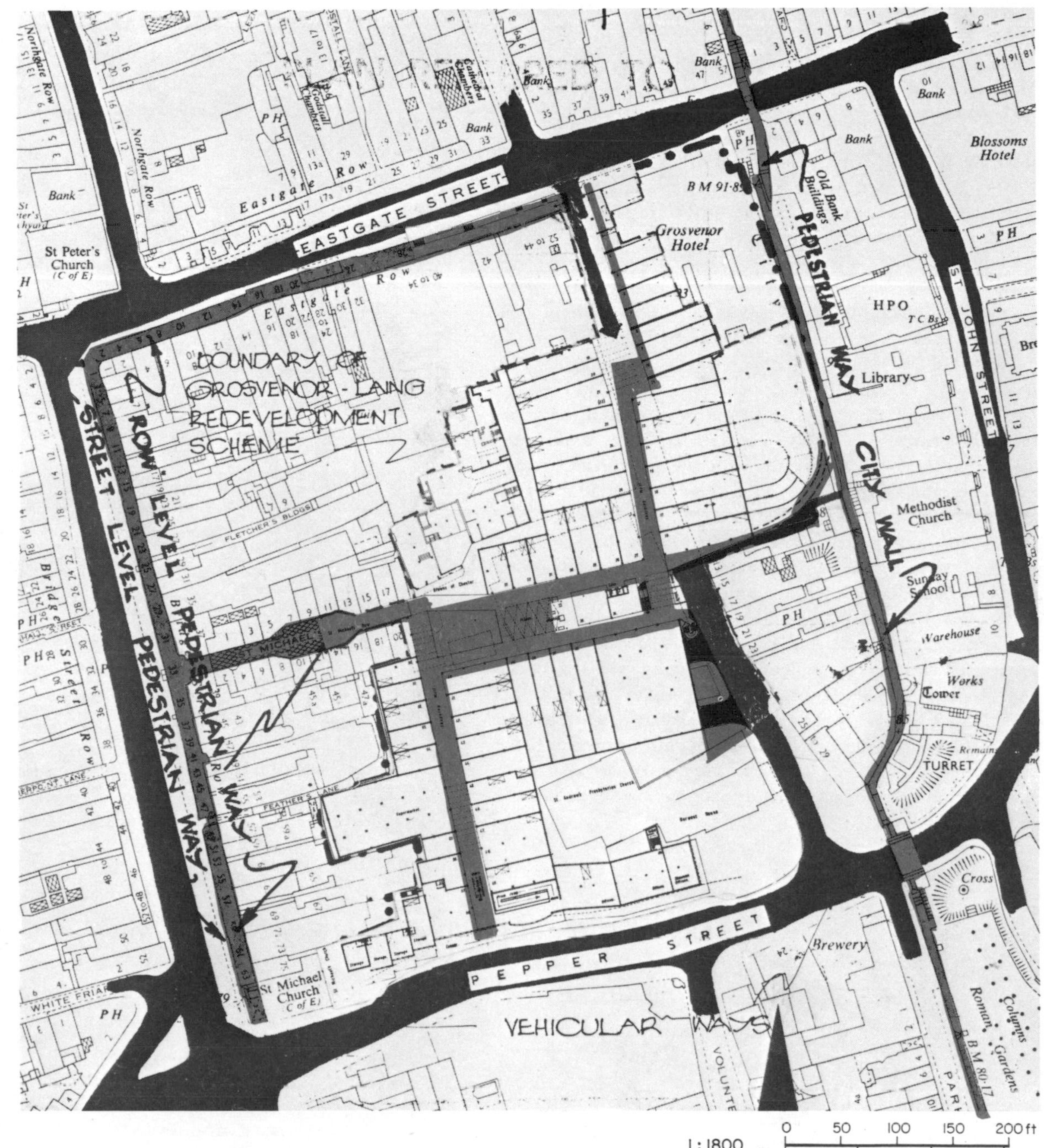

Fig. 7.148. Plan,(i) showing the new pedestrian complex connecting with the City Wall Walk on the right and The Rows top and left.

Fig. 7.149. View of The Rows.

Fig. 7.150. Banqueting Hall and multi-level car park from the City Wall Walk.

(i) Plan published by kind co-operation of the City Engineer and Surveyor, Chester: Mr. Higgins.

VII. 6.19. *Andover, Town Development*

Design: Town Development Division of the L.C.C.

Origin: A major town expansion under the 1952 Town Development Act.

Andover, together with Basingstoke and Tadley, were proposed by Hampshire to be extended in place of the new town of Hook as proposed by the L.C.C. (see example).

Population: Andover is to be enlarged from 17,000 people to 47,000 in the next twenty years.

Nature of Scheme: A main bypass to the south of the town is combined with a minor one ringing the north-east and west of the town and giving the main access to the new residential and industrial areas and the distributor roads. The main junction connects the distributor loop which rings the town centre. The latter has the existing shops service from the rear and the new ones from below, through a use of the slope of the site. Parking for 4000 vehicles is in low parking structures and bridge connections link these directly with the pedestrian precinct pattern of the town centre itself. The residential areas are connected with the town centre by a comprehensive path system and underpasses where the distributor road is met. A technical college and secondary school are planned for the town centre and all schools are connected to the path network of the town.

The pedestrian deck is 15 ft (4·6 m) above ground level. Service vehicles can thus unload in the storage areas and penetrate beneath the deck to service the inner "drums". The section here is taken across the main spine with residential dwellings over and through the larger of the drums which has an inner court. St. Mary's Church still dominates. But many existing trees in the heart of the development are ignored.

Fig. 7.153. Aerial view of town, by Howard's of Andover.

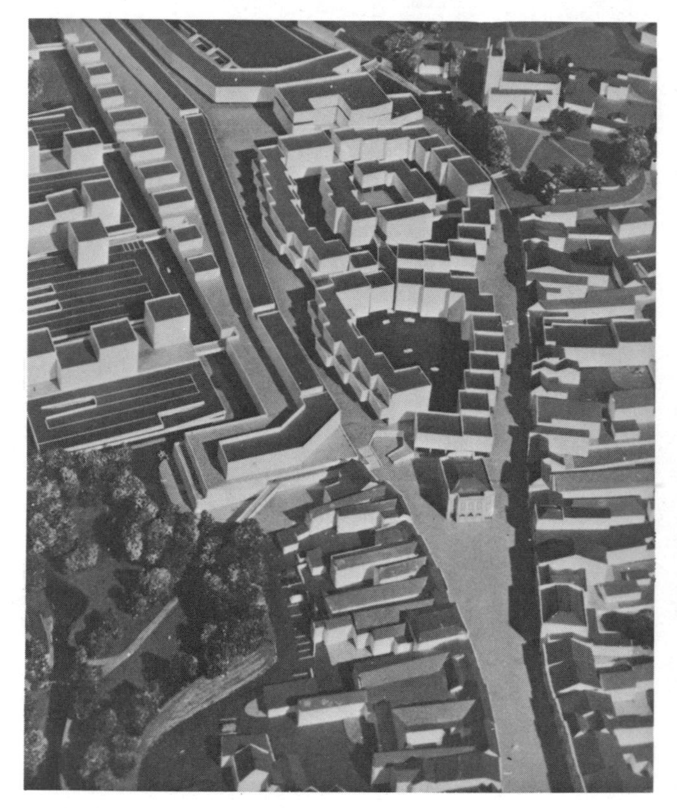

Fig. 7.154. Aerial view of model of proposals.

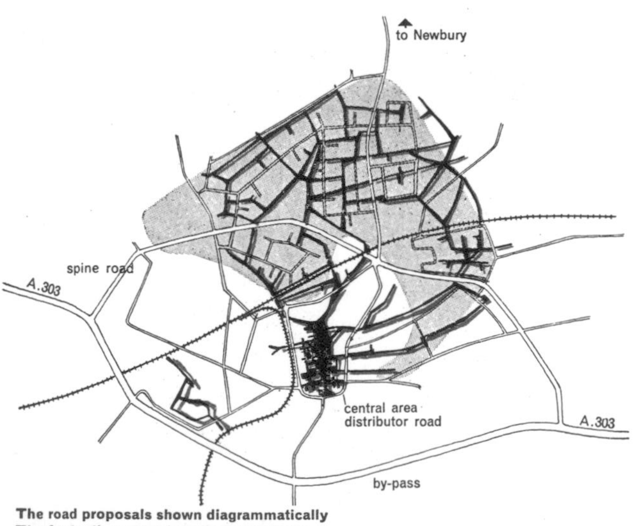

Fig. 7.151. Footpath system connecting centre and residential area.

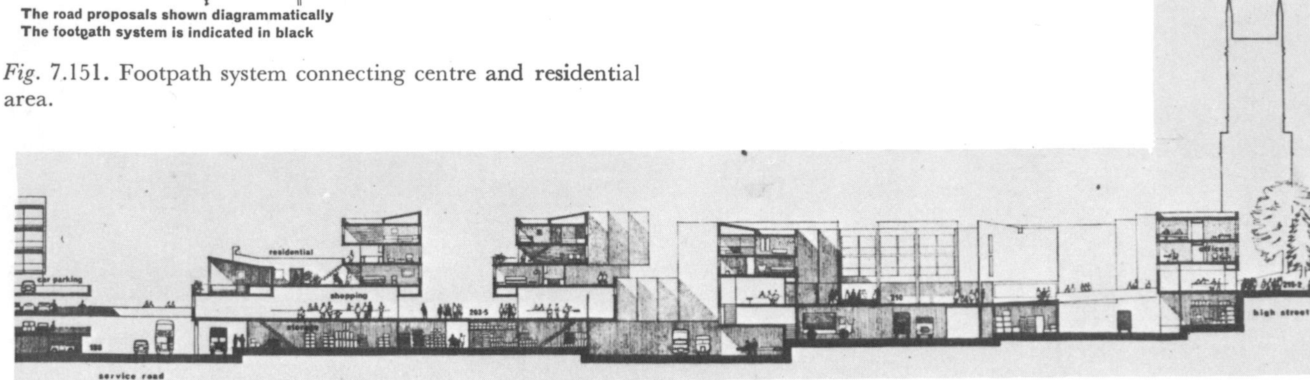

Fig. 7.152. Cross-section of proposal.

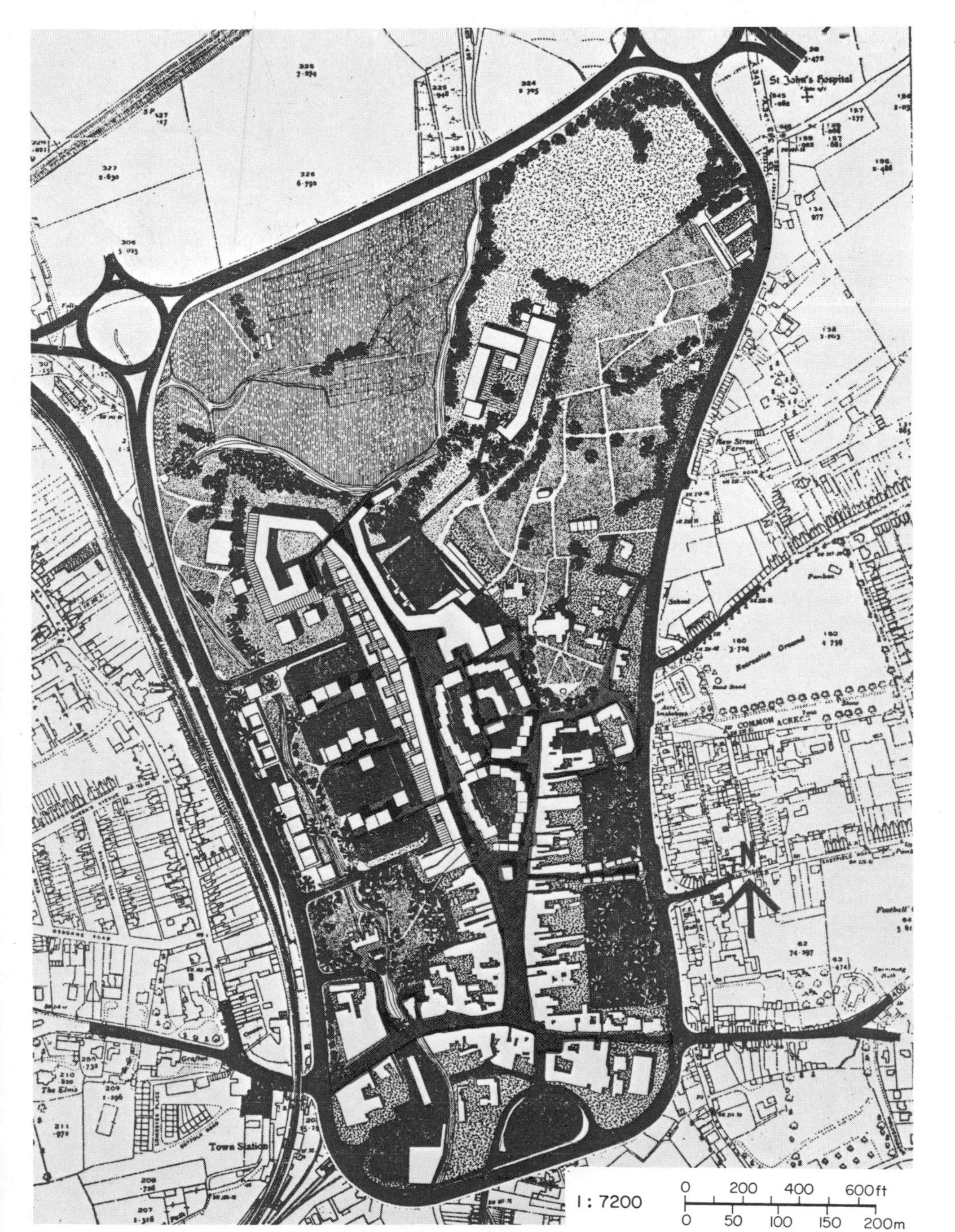

Fig. 7.155. Road and foot-path networks.

Fig. 7.156. The new pedestrian shopping deck seen from existing High Street.

Fig. 7.157. The public buildings area closely linked with St. Mary's.

References

Bendixson T. M. P. Expansion at Andover. *A.J.* 7 November 1962.
Bendixson T. M. P. Expanded Towns. *A.J.* 14 November 1962.
Craig J. Town Expansion at Andover. *J.T.P.I.* December 1962.
Andover Town Development. Town Development Division of the Architect's Department of the L.C.C. London 1962.

VII. 6.20. *Lijnbaan, Rotterdam, Holland*

Architect: van den Broek and Bakema.

Origin: Reconstruction of Rotterdam was planned after 1945. It had suffered almost complete destruction through bombing. The Lijnbaan is a great achievement for the architects who had to persuade a multitude of shop keepers to accept the idea of a pedestrian mall and of architectural unity for the individual shops in it.

Nature of Scheme: The Lijnbaan is a great pedestrian precinct in the form of a cross. There are 68 shops in the Lijnbaan itself, with kerbside parking for 200 cars and parking lots for 250. Three roads crossing the pedestrian precinct represent a regrettable compromise. One hopes that they will be shut when traffic increases. The architecture and detailing is to an overall prefabrication system. The canopies joining the two sides of the Lijnbaan are a very important innovation and have been successfully incorporated at Stevenage and other places.

The inclusion of high density dwellings at the back of the Lijnbaan gives life to the central area when shops are not open and sets an excellent precedent. The quiet precinct contrasts magnificently with the very wide vehicular boulevards surrounding the Lijnbaan, which, treated as shopping streets, are a bad environment for people.

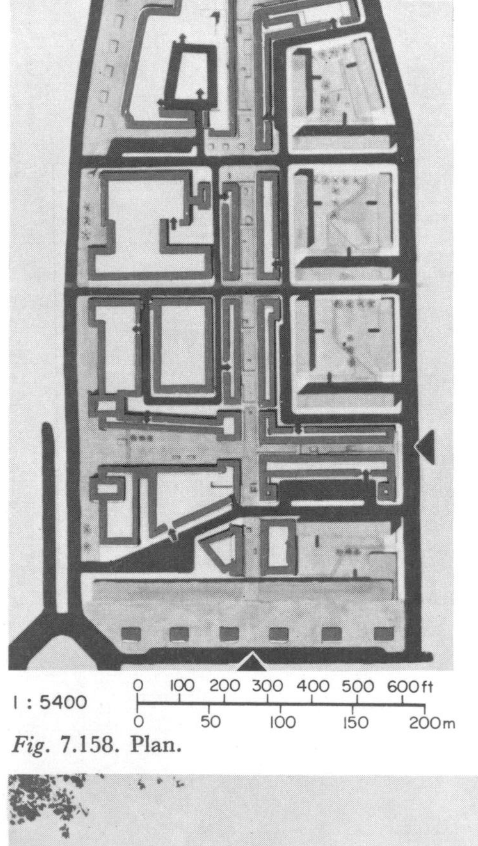

Fig. 7.158. Plan.

Fig. 7.159. Entrance short axis looking towards Townhall Tower which has a similar effect as the cathedral spire in Coventry's precinct. Canopies provide shelter from weather for shoppers.

Fig. 7.160. Flower beds, trees, kiosks give the pedestrian area variety and colour.

Fig. 7.161. High density flats at the back of the Lijnbaan.

Fig. 7.162. View from under canopy.

Fig. 7.163. Aerial view of long axis.

References

Marshall J. P. Rotterdam. *A.J.* 27 October 1955.

Rosner R. Rotterdam. *A.J.* 7 April 1955.

Berbiers J. L. *Rotterdam, the Reconstruction of the Inner City. A.D.* August 1954

VII. 6.21. *South Bank Development, London*

Design: H. Bennett, Architect to the Council. F. G. West, G. Horsfall, J. G. Cairns, N. Engleback, J. W. Chalk, M. J. Attenborough, R. J. Herron, R. C. F. Pocock, R. L. Martin, A. Waterhouse, J. M. Carss, D. Crompton, D. J. Curry, T. E. Kennedy, J. A. Roberts.

Origin: A development by the L.C.C. combining the New Front of the Festival Hall, a second building with two smaller concert halls and a sculpture exhibition gallery with a riverside walk and providing a multilevel pedestrian environment.

Nature of Scheme: "*The council on* 28*th March* 1961 *approved the scheme for the South Bank which will give a new appearance to the Thames side between Hungerford Bridge and Waterloo Bridge. The Royal Festival Hall will have a new front; downstream from it there will be a separate building comprising a smaller concert hall seating* 1,100 *and recital room seating* 400; *and further back from the river and with access from Waterloo Bridge, there will be an exhibition gallery on two levels with open-air sculpture courts cantilevered out from the main structure.*"

"*Upper level terraces will connect the three buildings and give access to Waterloo Bridge, Hungerford footbridge and Waterloo station free from vehicular traffic. These terraces will add to the open space available to the public and form a continuous pedestrian promenade.*"

"*Terraces at still higher level (over the lower level of the exhibition gallery and the foyer of the small concert hall) will provide a secondary access from Waterloo Bridge to the South Bank, and spacious observation and seating areas away from the more busy lower levels.*"

"*Vehicular access to the buildings and car parks will be provided by two loop-roads stemming from Belvedere Road, the upstream loop serving Royal Festival Hall, and passing under Hungerford Bridge and returning on the upstream side of the railway viaduct, and the downstream loop passing between, and serving the new concert hall and exhibition gallery and the repositioned entrance to the National Film Theatre, passing under Waterloo Bridge and returning to Belvedere Road on the downstream side of the bridge.*"

Parking for 170 *cars under the exhibition gallery, a further* 550 *on an adjoining site.*

"*The riverside gardens will be reshaped and there will be a continuous promenade through an avenue of trees between the County Hall and Waterloo Bridge.*"

"*The darker finishes of the new buildings (reconstructed Cornish granite slabs) and their sculptured outlines will contrast with the background of the Shell offices (*160 *feet high) and the Royal Festival Hall, in Portland stone.*"

"*Estimated to cost £*3,704,000 *in all.*"(1)

(i) L.C.C. Report, *South Bank from Royal Festival Hall to Waterloo Bridge*. 23 March 1961.

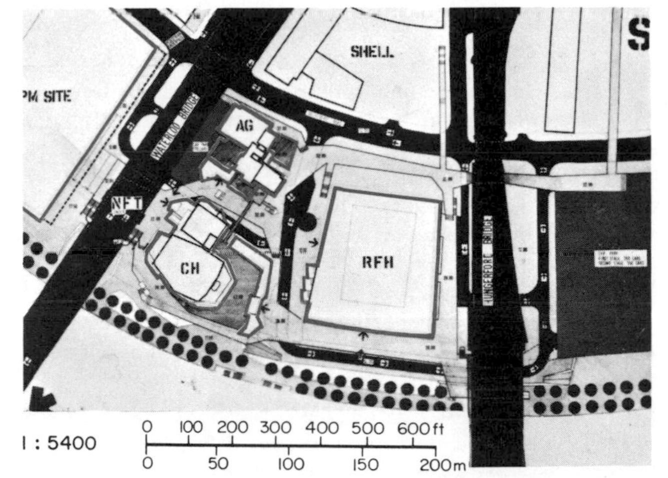

Fig. 7.164. Plan of whole development showing pedestrian access at high level to the adjoining areas including Waterloo Station.

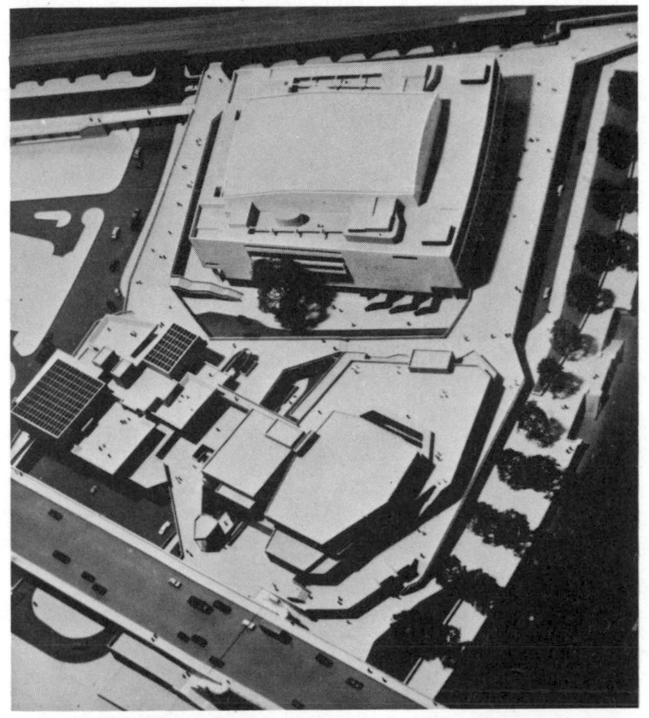

Fig. 7.165. Aerial view of model.

Fig. 7.166. View of pedestrian deck level on model.

References
South Bank, Next Stage. *A. J.* 30 March 1961.
L.C.C. South Bank Development. 23 March 1961.

VII. 6.22. *Stevenage, Town Centre*

Design: Chief Architect, Leonard Vincent.

Origin: When the economic advisors persuaded the Development Corporation against a pedestrian centre, the population pressed for it successfully. A first scheme was prepared by Clarence Stein, Professor Gordon Stephenson and Holiday, then Chief Architect. Donald Reay took a vital part in design and in arousing the interest and anger of the population when the idea was threatened by the Development Corporation. The present scheme is a different one. It is the best of the many New Town Mark I pedestrian shopping centres in Britain, but like the others sadly separated from unrelated housing. In that sense it is a matter of "Urban Renewal".

This scheme involves 150,700 ft² (14,000 m²) of shop area and 109 shops. 780 parking lots are provided, ultimate parking for 4400 vehicles. A weekly market is held on parking area reserved for this purpose. More parking is needed at the same time. Ultimately the population will be 80,000.

This centre is designed around existing trees, the slope of the site being sensitively incorporated. There is fine detailing and many handsome materials. The proportions of the square are based on an old square in a nearby Market Town. The influence of the Lijnbaan and Coventry centre can be seen.

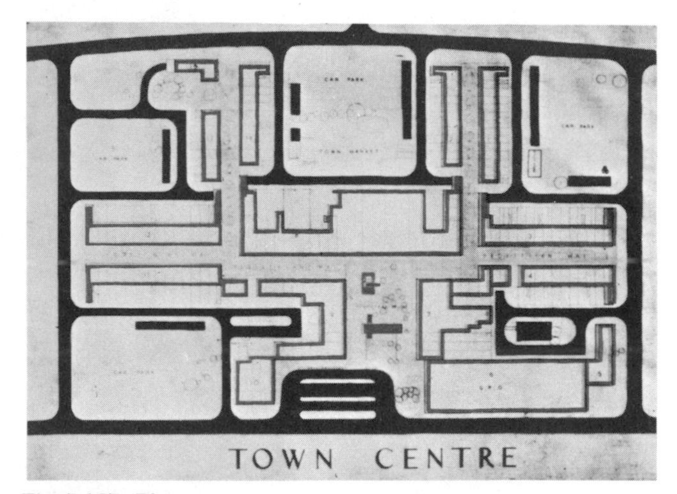

Fig. 7.167. Plan.

Fig. 7.168. Cycle parking in the main square.

Fig. 7.169. Main square at night.

Fig. 7.170. Fountain and platform accommodating public lavatories, in main square.

Fig. 7.171. One of the narrow subsidiary shopping streets showing the Rotterdam-like covered links.

Fig. 7.172. The service side of shops are designed carefully and effectively.

VII. 6.23. *Stevenage: Shephall Centre*

Design: Chief Architect, Leonard Vincent.

Nature of Scheme: This is a neighbourhood shopping centre with 27 shops, workshops, old people's dwellings and a site for a public house.

The shops are very popular with traders, the demand being greater than the supply. The use of round, continental type, advertisement columns is successful. Trees have been planted since the photograph shown was taken. The design of the service side of the shops is admirable.

The neighbourhood centre is typical of many in the new towns of Britain and Sweden. However, it suffers the grave disadvantage that, unlike the Swedish ones and like most British ones, no major public transport station is grouped with it, or is near it.

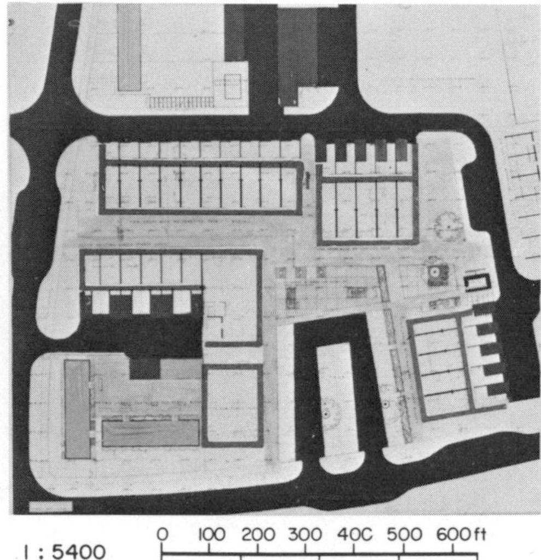

Fig. 7.173. Plan.

Fig. 7.174. View of main square of shopping area.

VII. 6.24. *Treppenstrasse, Kassel, Germany*

Design: Planning and Architectural Office of the City of Kassel.

This street was planned after war damage had destroyed the area. It is built in two parts: length, 394 ft (120 m) and 230 ft (70 m); width, 65 ft (20 m) and 33 ft (10 m).

The pedestrian street is a fine link between the railway station and the centre of the town. This pedestrian street represents the first part of a great scheme of creating a pedestrian network in the centre of Kassel. Some of it has been changed from normal traffic roads and service vehicles are admitted at certain hours only.

Reference

Der Wiederaufbau. Bremen December 1958.

Fig. 7.175.

VII. 6.25. *Pedestrian Shopping Street, Essen, Germany*

Typical of many German streets changed from mixed use to purely pedestrian traffic. Access for fire engines is assured. In some cases service traffic is allowed to enter at certain hours.

Reference

Der Wiederaufbau. Bremen December 1958.

Fig. 7.176.

VII. 6.26. *Greyfriars Redevelopment, Ipswich*

Design: Skipper & Corless.

Project by: Property Investment Company Consolidation Ltd., to serve a town to have 120,000 inhabitants with 70,000 in the present catchment area. First stage under construction.

"*Parking for* 857 *cars, plus* 164 *for tenants.*

"*Almost* 130,000 ft^2 (12,077 m^2) *of shopping space in a precinct arranged on two levels,* 75 *market stalls and an agricultural show-room of* 40,000 ft^2 (3716 m^2)."

"*A point block with* 63 *flats, bedsitter type, with two floors of office suites.*"

"*A slab block with about* 40,000 ft^2 (3716 m^2) *of lettable office space, restaurant on the top floor, an auction room suite, an exhibition hall and licensed premises.*"

"*Provision for a bank and a small cinema which will be usable for other purposes as well.*"

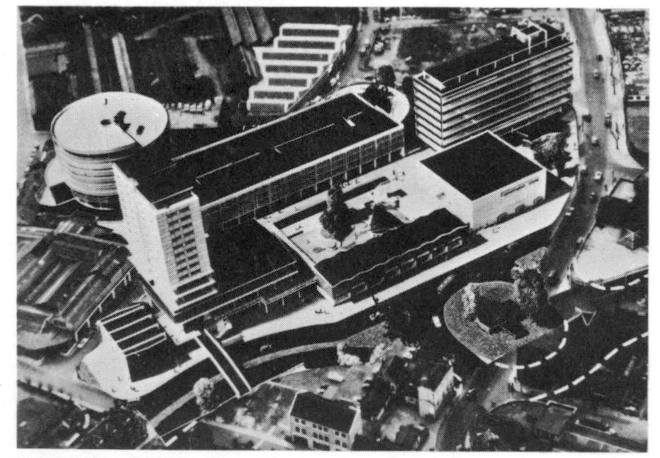

Fig. 7.177. Aerial view of stage one model.

Reference
Ipswich, Greyfriars Redevelopment. Skipper & Corless. Norwich 1962.

VII. 6.27. *Sea-Front Site, Brighton*

Design: Russel Diplock Associates, development by Taylor Woodrow Group.

Population: 156,000.

A comprehensive redevelopment area, the scheme is the one selected from three submitted to Brighton Corporation. It includes flats, shops, offices, cafés, restaurants, public house, five star hotel, entertainment centre, departmental store and ancillaries. 2065 parking spaces are provided with an optional extra of 700, if required.

Fig. 7.178.

Reference
A. & B.N. London, 28 September 1960.

VII. 6.28. *Strøget, Shopping Street, Copenhagen, Denmark*

Design: The Mayor and Council of the City of Copenhagen.

Origin: Against the advice of planners, police and economic experts the lay men in the government of Copenhagen decided to try out this scheme for a trial period of three months from 17 November 1962. In the middle of January some shops and newspapers asked for the trial period to include the tourist season. In March the scheme had proved a success, the warnings of the professionals had proved largely false and the trial period was extended indefinitely. It was felt that if the scheme was a success in winter then the tourist season should show its advantages vividly. Service traffic is allowed from 4.00 a.m. to 11.00 a.m.

The scheme has been studied in a number of ways which have provided very useful figures on closing a traffic road in a congested town. The shopkeepers themselves are only divided in that some want to be able to have servicing traffic until 14.00 claiming that they have lost some trade. The majority claim that they have gained trade.

The traffic pattern changed as follows: a bus route has been moved out and there has been a smooth adjustment, without loss of passengers. The trams are operating well but a new stop is asked for along the shopping street, where one of the roads crosses it. At such places, where radial roads cross this very long street, traffic lights are to be installed to allow safe crossings.

The streets which were thought to take the traffic stopped by closing the road showed 6546 extra cars (from 6.00 to 20.00) As Strøget had 8617 cars in such a count over two thousand cars seem to have disappeared. In the rush hour, 16.00–17.00, the percentage drop is even greater. 306 cars extra in other streets, compared with 810 in Strøget before it was closed. (62% of the rush hour traffic has vanished.)

Fig. 7.179. Plan of the Strøget, Copenhagen's main shopping thoroughfare made up of several streets. Total length three-quarters of a mile (1·2 km).

Police helped a great deal and said the problem was not as great as they had feared.

In March 1963 the City Council decreed that it shall be permitted to erect, each day after 11 a.m., cafés in the street, flower tubs, parasols, and that it shall be allowed to use the street as a meeting place. Great traditional Danish children's processions (Beech Day, in May), are planned to take place there. However it is not allowed to set up temporary sales kiosks or barrows in the area without very special permission which, it is thought, would probably not be granted.

Fig. 7.180. On the day of opening: pedestrians take possession, a band plays.

VII. 6.29. *Ellor Street Comprehensive Redevelopment Area, Salford*

Design: Sir Robert Mathew and Johnson-Marshall.

Design Team: Bigwood, Buszynski, Duncan, R. Stewart, A. Stewart. City Engineer, McWilliam. City Architect, Earle. Chief Planning Assistant, Whittle. Landscape, F. Clark. Valuation, Eve & Co.

Population: 175,000 approximately.

Site: 89 acres (36 ha), in the centre of Salford. The aim is to create within the Greater Manchester Region an attractive centre, much required to regenerate the industrial North of England. Civic and recreational buildings will have car parking for 600. 400,000 ft² (37,000 m²) of shopping, and 33,000 ft² of offices (3000 m²), space for 2070 cars.

There are to be 2450 dwellings at 185 persons per acre (462·5 per ha) in eight storey slab in the main and 145 persons per acre (362·5 persons per ha) in four storey and eight storey interlocking blocks. 100% garage provision in residential areas. A 5 acre (2 ha) primary school forms part of the development.

Balancing revenue and capital outlay, the design of the centre was aimed at yielding a return of 10% on capital cost.

Vehicular access is at ground level. The residential area is the first of six to connect to this regional centre by a greenway system.

A broad pedestrian way runs through the whole site to the shopping centre with pubs, smaller shops and community facilities along side.

Early realization of the landscaping assured by purchase of a tree nursery by the city.

22-storey blocks are used in places where point blocks are considered an important part of the design concept, as at the south-east, where the A6 motorway borders the site.

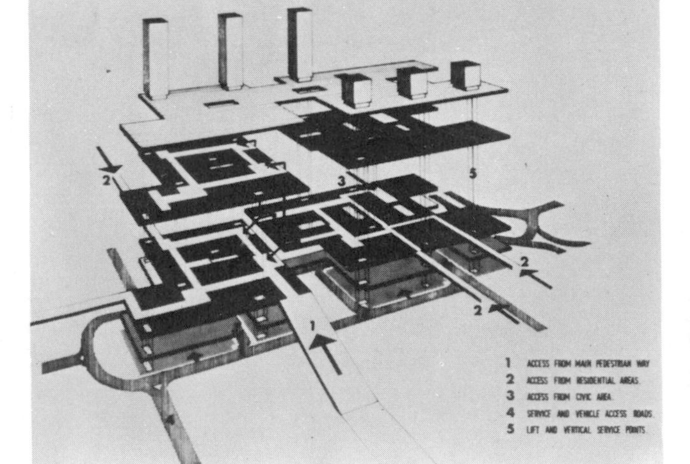

Fig. 7.181. Access and circulation diagram, showing vehicular traffic taking up whole of ground level.

Fig. 7.182. Final model of the shopping and civic area

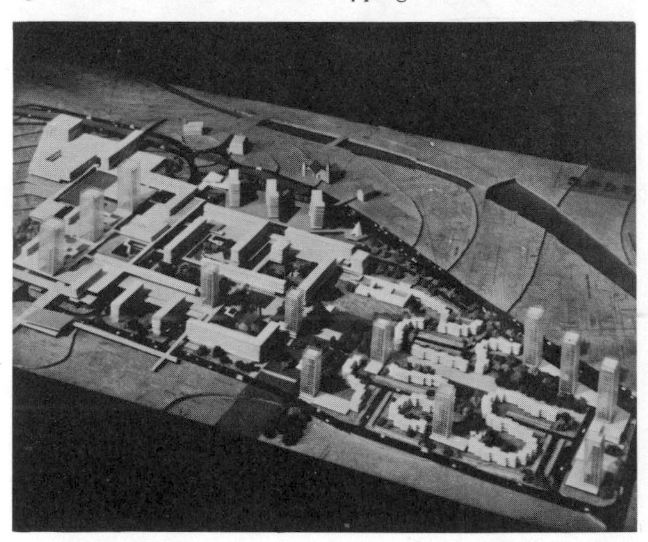

Fig. 7.183. Preliminary model of the whole area showing centre on the left and residential area on the right.

Fig. 7.184. Perspective of main pedestrian promenade linking residential area to centre.

VII. 6.30. *Hammersmith Central Development, London*

Design: Gollins, Melvin, Ward & Partners.

Approximately 14 acres (5.6 ha). Bounded on the south by the Hammersmith flyover.

Although the building of the flyover relieved traffic congestion in the area as far as through east–west flow was concerned, it has not solved the problems of heavy traffic build-up in the two main shopping streets.

The scheme re-aligns certain sections of the roads and creates a one-way system of circulation. Off-street car-parking and warehousing is under a pedestrian deck to be built over the whole site, 20 ft (6 m) above street level.

Contained in the podium are two main concourses interlinking the two underground systems and new bus termini. With the exception of these two places, and the perimeter access points, pedestrians never come down to street level.

At pedestrian deck level it is proposed to provide an under-cover shopping centre, market and recreational facilities, such as a "Palais de Danse", cinema, bowling centre, restaurants, public houses, etc.

Public buildings include office accommodation, crèches, library, fire station, etc., together with an hotel and banqueting halls.

Residential accommodation is to be provided in the form of a cluster of seven tower blocks, varying between eighteen and thirty storeys above podium level.

Fig. 7.186. View of model showing pedestrian deck built over the whole site.

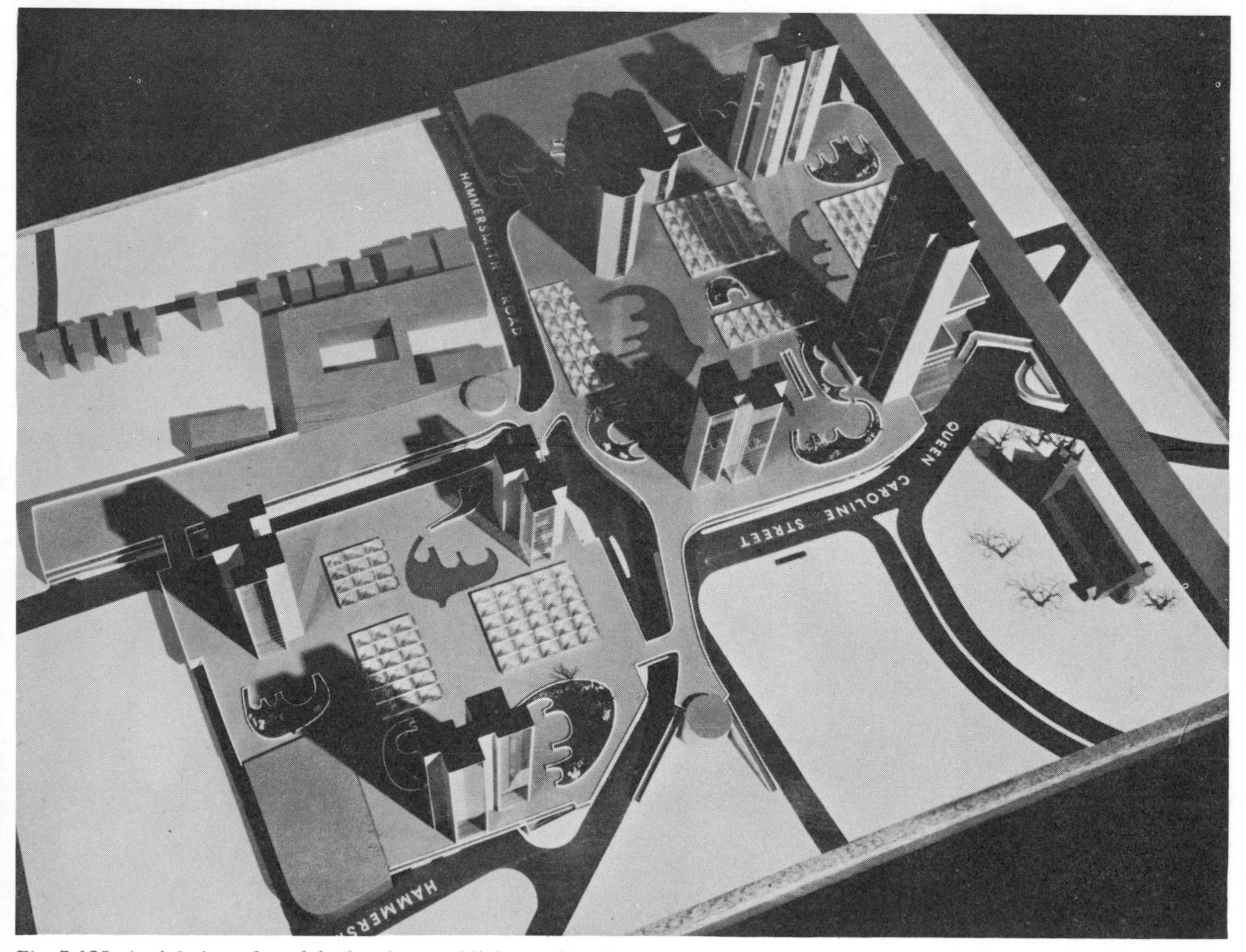

Fig. 7.185. Aerial view of model, showing roof lights and roof treatment for pedestrian deck which is fully enclosed, air conditioned, and continuous across streets.

VII. 6.31. *Piccadilly Circus, London*

Design: Sir William Holford & Partners.

A comprehensive redevelopment scheme for the L.C.C. The preliminary scheme submitted for comments to developers includes restaurants, cafés, three-dimensional sky signs, a piazza at street level. Both platform at higher level with access to restaurants, and links to underground concourse, are to be extended when further development takes place.

Although traffic segregation is only partial this is an immense advance on the previous proposal to develop the Monico site by a private company, turned down by the Minister after the historic enquiry.[1] Piccadilly Circus remains basically a roundabout for traffic, but its diameter is much enlarged.

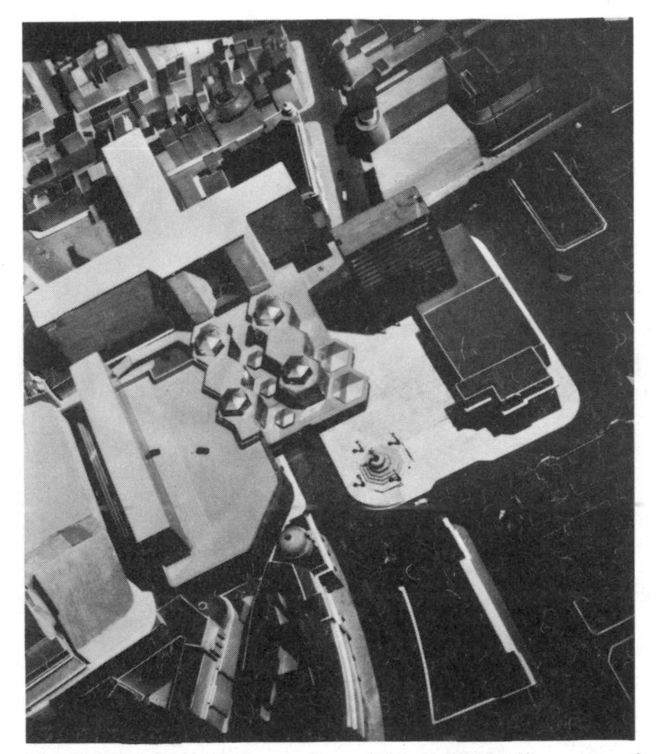

Fig. 7.187. Aerial view of model of Holford's proposals, turned down by the Minister of Transport as inadequate to cope with future traffic flow. The Ministry's standpoint assumes increase in the capacity of all the approach roads. It seems a dubious standpoint.

(i) Buchanan C. D. Report of the Public Enquiry (fully illustrated account). *T.P.R.* Liverpool, January 1961.

Buchanan C. D. Reprieve for Piccadilly. *A.J.* London, 16 May 1960.

Holford W. *Piccadilly Circus*, L.C.C. London 1962.

Piccadilly, Holford's Plan. *A.J.* London, 16 February 1961.

VII. 6.32. *Knightsbridge Green, London*

Design: Morgan & Partners.

A Comprehensive Development Area, a private development with L.C.C. planning approval. Tall offices, 279 ft (85 m), hotel 324 ft (99 m), flats 413 ft (125 m) and a long block of low flats are placed around an open square encircled by a plinth of shops. The hotel straddles a road and extends into an extensive sunken garden, forming a major pedestrian link to underground stations and streets.

"All service vehicles go down to first basement level and service all the buildings. The ground floor piazza is free from traffic but access is provided for fire engines. Vehicular access to buildings is gained at second floor level, as all along the frontage, stopping of vehicles has been prohibited. The solution of the traffic problem on this scheme is therefore a very particular one. It is technically possible to have all taxis and cars serving all the buildings to discharge passengers at basement level and thus free the roof completely of traffic. This solution was discarded because of the rather depressing nature, psychologically speaking, of such an approach."

Fig. 7.188. A view into lower open square, circled by the plinth of shops.

Reference

A.J. London 21 March 1962.

VII. 6.33. *Basildon New Town, Main Centre*

Design: Chief Architect, A. B. Davies.

Population: Regional population of 130,000–150,000, including in-town population of 100,000 approximately. Typical of the new town centres of the Mark I New Town, as designed in Great Britain in the forties and early fifties. Very few pedestrian underpasses enter under ring roads. Pedestrian spaces are surrounded by parking areas, 2400 in this case, based on 39 cars per 100 ft shopping frontage, or 28 for 100 ft office and shop frontage combined. The tall tower block of offices dominates. Detail is carefully considered.

Retail Market, top right, is closely related with shopping.(i) (See also p. 153.)

Reference
A. J. London, 19 February 1962.

(i) See also Fig. 7.4.

Fig. 7.189. Much of the building of the centre is still to be completed.

VII. 6.34. *Town Centre, Elizabeth New Town, Australia*

The layout shows, in this progress photograph taken in 1963 by Roy Glyde, very similar features as Basildon: the vehicle free shopping areas, the vast surrounding parking areas and the ring road; the neighbourhood centres at Elizabeth are likewise similar to their British New Town counterparts.

VIII. Traffic Segregation in Residential Areas

VIII. 1. Scale of Livability Applied to Radburn Principles. **224**
VIII. 1.1. Psychosomatic Criteria: Radburn Possibilities. 224
VIII. 1.2. Psychosocial Criteria. 225
VIII. 1.3. Ecological Criteria. 227
VIII. 1.4. Economic and Administrative Criteria. 227

VIII. 2. Costs. **228**
VIII. 2.1. Comparison of Costs of Streets and Sewer Works. 228
VIII. 2.2. Economic Analysis. 229

VIII. 3. Dwelling Plans. **232**
VIII. 3.1. Analysis of Relationship of Dwelling to Access. 232
VIII. 3.2. Car Storage. 235

VIII. 4. Paths, System of Routes. **236**
VIII. 4.1. Analysis. 236

VIII. 5. Plot Shapes. **238**
VIII. 5.1. Houses. 238
VIII. 5.2. Flats and Deck Housing. 239

VIII. 6. Residential Feeder Roads. **240**
VIII. 6.1. Size and Frequency of Junctions. 240
VIII. 6.2. Feeder Road Design. 240
VIII. 6.2.1. Culs-de-sac. 240
VIII. 6.2.2. Loops. 240
VIII. 6.3. Basic Alternatives. 241
VIII. 6.3.1. Culs-de-sac. 241
VIII. 6.3.2. Loops. 242
VIII. 6.4. Examples. 243
VIII. 6.4.1. Pioneer schemes of the early fifties. 243
VIII. 6.4.2. Projects of the mid-fifties. 244
VIII. 6.4.3. Projects of the late fifties and early sixties. 246
VIII. 6.4.4. Dimensions of service culs-de-sac. 248
VIII. 6.4.5. Connections between feeder road and path system. 249
VIII. 6.5. Mixing Pedestrians and Vehicles in Culs-de-sac. 250

VIII. 7. Superblock Design. **252**
VIII. 7.1. Basic Factors. 252
VIII. 7.2. Application in Redevelopment. 253

VIII. 8. Underpasses and Bridges. **253**
VIII. 8.1. Position and Importance. 253
VIII. 8.2. Design and Cost Alternatives. 255

VIII. 9. Open Space. **256**

VIII. 10. Contours. **256**

VIII. 11. Street Names and Numbers. **257**

VIII. 12. Examples of Traffic-Segregated Residential Areas. **258**
VIII. 12.1. Cumbernauld, nr. Glasgow. 258
VIII. 12.2. Baldwin Hills Village, Los Angeles. 260
VIII. 12.3. Caversham, Reading. 262
VIII. 12.4. Bron-y-Mor, Wales. 263
VIII. 12.5. Huntingdon, L.C.C. 265
VIII. 12.6. Jackson Estate, Letchworth. 266
VIII. 12.7. Ilkeston Road, Beeston, Notts. 266
VIII. 12.8. Albertslund, Copenhagen, Denmark. 267
VIII. 12.9. Eastwick, Philadelphia. 268
VIII. 12.10. Clements Hill, Haverhill. 269
VIII. 12.11. Cité Jardin, Montreal, Canada. 271
VIII. 12.12. Marly-le-Roi, Paris. 273
VIII. 12.13. Baronbackarna, Orebro, Sweden. 274
VIII. 12.14. Lafayette Park Title I, Detroit. 276
VIII. 12.15. Biskopgaden, Göteborg. 277
VIII. 12.16. Willenhall Wood I, Coventry. 278
VIII. 12.17. Kildrum, Cumbernauld, Glasgow. 280
VIII. 12.18. Elm Green I & II, Stevenage. 281
VIII. 12.19. Bedmont, Herts. 282
VIII. 12.20. Prestonpans, Edinburgh. 283
VIII. 12.21. Parkleys, Ham Common. 285
VIII. 12.22. Park Hill, Sheffield. 286
VIII. 12.23. Barbican, London. 288
VIII. 12.24. Flemingdon Park, East York, Canada. 289
VIII. 12.25. South Carbrain, Cumbernauld, Glasgow. 290
VIII. 12.26. Oakdale Manor, North York, Ontario, Canada. 294
VIII. 12.27. Kentucky Road, Toronto, Ontario, Canada. 295
VIII. 12.28. Solna, Sweden. 296
VIII. 12.29. Spon End, Coventry. 297
VIII. 12.30. War Office (A), England. 297
VIII. 12.31. War Office (B), England. 298
VIII. 12.32. Rowlatts Hill, Leicester. 298
VIII. 12.33. Brandon Estate, London. 299
VIII. 12.34. Royal Victoria Yard, Deptford, London. 299
VIII. 12.35. Brunswick Redevelopment Area, Manchester. 300
VIII. 12.36. Primrose Hill, Birmingham. 300
VIII. 12.37. Almhög, Nydala, Hermodsdal, Malmö, Sweden. 300
VIII. 12.38. Mellanheden, Malmö, Sweden. 300
VIII. 12.39. Washington, Durham. 301
VIII. 12.40. Les Buffets, Fontenay-aux-Roses, Paris. 304
VIII. 12.41. Hillfields, Coventry. 304
VIII. 12.42. Highfields, Leicester. 304a

VIII. Traffic Segregation in Residential Areas

VIII. 1. Scale of Livability Applied to Radburn Principles

The function of layout is to make the most of the space given and to fulfil the many varied needs of the foreseeable circumstances of all age groups. It is knowledge and inventiveness which can, without further expense, create a better environment. The segregation of traffic in residential areas is just such an idea, the Radburn Idea, named after the first town which was built in this way. Beyond the points made previously, on the relationship between man and vehicles and the needs of man, it should be remembered that car ownership for every family, now generally assumed, has changed, almost entirely, the problem of housing for the general run of people. The answer to the new clear double problem of providing for man and motor may of course give a better environment than the solutions to the previous muddled demands. As a general criterion any new solution ought to keep and foster old virtues as well as provide well for car and driver.

To gauge the value of any existing residential area I have compiled a scale of livability. This scale can also be used to assess the effectiveness and qualities of an idea, like the Radburn Idea. This exercise will include developments, as well as the original Radburn Idea. Clarence Stein, co-designer of Radburn, still holds that his idea includes three essentials:

1. "*Homes and other buildings face in two directions, one towards peaceful spaciousness the other towards roads and services.*"
2. "*Pedestrians and vehicles are entirely separated.*"
3. "*Superblocks, with internal cores of parks and other open spaces, and surrounded by specialised roads and services result from the essential requirements*"(i)

The development of this idea includes planning the park at the side of a superblock, although still directly accessible by pedestrian-only routes. They include terraces which have pedestrian access only, although this pedestrian access leads to paths on one side as well as the usual road on the other. There may be entire little patio–house–footpath-only colonies within a developed Radburn Idea. The essentials which remain and define the original and the developed ideas are:

1. Homes must have direct access to a footpath system.
2. This footpath system must lead to all the gathering places of the inhabitants.
3. The motor vehicles will be completely separate from the path system, except at points of boarding, and will serve areas through culs-de-sac (or loops with paths under) connected by a collector, or ring road, with the outside world.

(i) Letter to author, 1961.

We are assessing the qualities possible with this idea, as contrasting with housing lay-out, without a comprehensive path system. It must not be taken that each Radburn scheme necessarily has all these qualities, although it is possible to achieve them within any one scheme.

As the Radburn Idea sprang from considerations of living conditions in the motor age it is really not surprising that it lends itself better to the satisfaction of the needs listed, than other forms of lay-out. All this is largely substantiated by objective research and backed by prolonged observations. Why the needs listed are the real criteria of livability is explained in the work of Scott-Williamson, Wilhelm Reich, Henrik Infield, Ian Suttie, Paul Halmos, and others.(i)

VIII. 1.1. *Psychosomatic Criteria: Radburn Possibilities*

(a) *Fresh air:* Separation of noxious petrol and diesel fumes, more forest trees, good air along paths.

(b) *Exercise:* Dogs and children run and play in safety. Adults are tempted to walk more.

(c) *Relaxation:* Less anxiety about accidents, less wearing noise and rushing movement.

(d) *Privacy*: As much as desired or planned for, flexibility possible.

Privacy is not a simple phenomenon. Kuper found that the lady(ii) objected, not to being spied on but, that those opposite thought she was viewing them!

The garden, back-to-back, as in traditional layouts, is mistakenly taken as "private", by shallow thinkers. The overlooking is at a maximum from the first floor.

With the Radburn Idea the car side can be similar to the street side ("front") of the traditional house or, with higher fences more private than either back or front of the traditional. The path side can either be open, giving an urban kind of friendliness to the area, accompanied by the curtains some people like, for the sake of privacy, and rationally, in such circumstances, with the paths passing close by. But, alternatively, the paths can run between two private gardens, both fenced or walled giving greater privacy. Houses orientated in one direction can give complete privacy.

(i) Infield H. *The American Intentional Communities.* New Jersey 1955.
Reich W. *Function of the Orgasm.* New York 1942.
Reich W. *Ether God and Devil.* New York 1951.
Suttie I. *Origins of Love and Hate.* London 1935.
Halmos P. *Privacy and Solitude.* London 1948.
Scott-Williamson. *Physician Heal Thyself.* London 1944.

(ii) Kuper L. *Living in Towns.* London 1953.

VIII. 1.2. *Psychosocial Criteria*

(a) *Functional co-operation:* Objective research by the author(i) into 500 houses, 18 paired replicates, gave clear evidence that motor roads discourage constructive social contact: that families with children tend to move into areas of footpaths: that neighbours are more willing to look after other children in such areas and help in emergencies big and small. Path areas are felt as more friendly by three people to every two who find normal road areas so. There are areas where those who want to make or do something together (i.e. swimming pool) can do so.

(b) *Mothers informal meeting:* More opportunity in paths and places conducive to walking.

(c) *Meeting of babies (established as highly desirable)*(ii): More opportunity in prams and mothers' arms in path environment.

(d) *Toddlers' social play:* Liberation for this socially neglected age group, imprisoned in private gardens.

(e) *Visual harmony:* Pedestrian scale and ground coverage to suit any enclosure can be easily obtained.

(f) *Journeys to school for all ages:* More freedom to walk or cycle, less danger, interest and education in the widest sense.

(g) *Adolescent needs for space adventure, place for love:* Path systems, intimate pedestrian scale giving much variety, incident and enclosure.

(h) *Accessibility to local shops, pub, etc.:* Safe and pleasant routes make distances seem less. Superblocks reduce walks in length—children can be sent safely.

(i) *Accessibility to transport for work and further amenities:* Private cars by dwellings. Public transport reached by pleasant, safe, fast, protected, walks.

(j) *Special need for old people:* Quiet, without isolation from children, shops and pubs, safely reached. Many places to sit and see life.

(k) *Provisions for hobbies—part-time work:* As there is no need to worry about traffic, so in taking and fetching small children to and from school a great freedom, which can also mean economic gain, is achieved. Mothers are not incessantly tied to their children, and vice versa. It is safer for pets (very important in Britain), but in Gothenburg, Sweden, the bottom floor of a 15-storey block of flats showed evidence of the needs of pets. A cat's ladder was provided to allow a route in and out of a window on the ground floor. There are full possibilities for gardening for those who want it, garage by house and service court and path lend themselves to many hobbies.

(i) See p. 27.

(ii) PEARSE and CROCKER. *The Peckham Experiment, a Study of the Living Structure of Society*. London 1943.

(l) *Feeling of belonging and participation of inhabitants:* All the above encourage this; the absence of motor roads brings with it the smaller scale and the feeling of belonging. Statistics show feeling of friendliness found more often in paths so that the stranger in the area becomes more of a "visitor" not just a "stranger", for he has "entered". Unlike the "street" they are not deemed in danger or in the way. Often one finds that toys have been left lying around whilst children are inside their respective houses eating a meal.

(m) *Good impression for visitor: "The pleasure of walking around . . .* [wrote a visitor to Willenhall I Coventry] *is something that must be experienced in reality to be properly appreciated. Free from the curse of traffic (look right, look left, look right again) one can enjoy at a leisurely pace, ones surroundings. Younger children play openly, older ones wander freely for considerable distances, old age pensioners sit outside and talk."*

(n) *Workable dwelling plans:* House plans which make the most of the lay-out have been evolved. Basically they can have the advantage of making full use of two entrances to the house. (See also p. 232.)

(o) *Administration:* Administration in a Radburn lay-out should logically come from within, from the owners, or tenants, because they have something in common to administer. At times, as at Radburn, New York, and Baldwin Hills, Los Angeles, the advantages of Radburn planning are so great that the surrounding large areas, without these amenities, but similar income groups, tend to invade and over-use the facilities and spaces.

Legal protection in such circumstances may be a necessary step, but is a very sad spectacle.

(p) *The greater social possibilities of Radburn lay-out lend themselves to "housing therapy" as described elsewhere,*(i) *to quote:*

> *"Isolation of the tight patriarchal family is one of the symptoms and also one of the origins of the emotional sickness in the social sphere, and we have argued further that there is plenty of unhappy evidence that the solution does not lie in vaguely, if enthusiastically, forming communities, or even opening community centres, which belie their name by their very birth."*
>
> *"The physical lay-out of estates, which form such a very large part of Britain's housing, both municipal and private, in most cases emphasizes family isolation. If the general lack of social contact can be compared with a bodily illness, then moving into a housing estate represents a crisis. In a recent research project the symptoms of emotional strain shown in admission to mental hospitals, visits to the general practitioner with 'nerves', and so on,*

(i) RITTER P. and J. *The Free Family*. Gollancz 1959.

have been found, significantly, very much more widespread in a new housing estate than in other older housing areas; and, in fact, in the estate examined minutely the crises wore off in subsequent years. People vaguely know that loneliness will encompass them on the estate, and this apprehension comes on top of other disadvantages, such as lack of facilities for shopping and entertainment. Estates are attractive only in contrast with the apalling unhygienic and crowded conditions from which their inhabitants often emerge."

"*If, then, moving to the estates is a time of social crisis, and the occupation of an individual new home certainly is, and if we remember the parallel with the physical illness of the body, then to associate both with social therapy at that particular time is right. As soon as we approach the problem of creating a community as a form of therapeutic action, we are in a position to gain some of the benefits of social health, while avoiding many of the pitfalls that await those who idealistically conceive and start communities without regard to the complex emotional patterns involved.*"

"*That the housing estates are as bad as suggested, awful and inadequate can be illustrated by some findings of my own research. For example, eighty-two motor accidents occurred in one estate within its first two years, far more than was to be expected in comparison with other areas not even deliberately planned. It is an invaluable point, giving decisive advantages in arguments, that, unlike the horse-drawn carriage in transport, housing lay-out is substantially the same as fifty years ago and appallingly out of date.*"

"*The idea of housing therapy has emerged from the diagnosis that there is something wrong with society, and that re-housing people is an especially opportune time to do something about it. Components of the idea that there is something wrong socially, the ideas of group therapy and of a housing lay-out which is conducive rather than disruptive of social health, are not new. But it is new to combine the above ideas into something which, because of its integrated plan, is powerful, practicable and likely to be effective.*"

"*Social ill, the loneliness of people, the desire of men and women to keep themselves to themselves in an exaggerated way, all these things are largely due to fear. In the main this is irrational fear, and so much of it can be removed therapeutically, if only by explanation and information. A clear instance of this might be the case of the strangely silent lady down the road, who has men attending her house every evening because she does chiropody, and is therefore thought by the neighbours to be a prostitute. She is felt to be a threat to be feared, and so she is slandered and attacked, unconsciously or consciously, or at best cold-shouldered. Now it would have a therapeutic effect on the social relations between these people, and so on the community as a whole, if the true identity of the chiropodist were established and made known to the neighbours. Suspicions of an infinite variety, all reflecting similar fears of dirt, lewdness, lowness, etc., permeate the social climate of the new housing estate and can poison the atmosphere permanently. In dealing therapeutically by explanation with such instances, it is understood that individual neuroses may well emerge clearly for individual therapy. Just as people who fall in love enter a crisis which makes them more aware of themselves, and which creates enormous energies to deal with difficulties and to improve their lot, so, in re-housing, because the occupation of a new home is a crisis, energies are forthcoming and the therapeutic approach is eminently practicable and timely.*"

"*Contact with future occupants can be made through the housing list, or through some agency which sells or rents the accommodation. Such people can be best approached if you have something to offer that is of advantage to them. You could scare every single person off an estate by saying, in a certain way, that you were going to give them therapy, but there are other ways of conveying the information to people. What is in fact advantageous can sound attractive. Taking a housing list we can at once consider the practical proposition. Those who call to put their names on it could be shown the two alternative approaches: the one traditional, the other, to serve the therapeutic end, experimental. They would be informed that this was no 'guinea pig' experiment but a creative experiment, where all concerned would try with all their energy and by many means to make life in the estate a success, which might even include financial advantages. The very first of the means would be meetings of those to be housed before they even started moving in. Co-operation in details of planning and architecture could then take place, good friends could live near each other, etc.*"

"*And the trump card to attract people to the experiment could be the actual lay-out. It is all to the good that those who would opt for the experiment, the cream skimmed from the list in the way described, would be people not desperately afraid of the new, people still keen to try to improve their life and living, for it is crucial to realize that the lay-out of housing estates can be improved, infinitely, without greater cost, merely by more knowledge. It would then remain only to convince these people of the advantages of the projected lay-out, which might attract even those who would otherwise not very much care to be part of an experiment.*"

"*Briefly, the difference between the traditional and the new would be the complete segregation of footpaths from roads, without denying each and every house a garage and direct road· access. This offers complete safety for children going to school, shops and play-spaces. It takes out of mothers' lives the painful, pram-clutching drags alongside dangerous roads, when their youngsters, eager to run, only at the risk of fearful slaps let loose the handles. The advantage can hardly be exaggerated. My own research, on over five hundred houses, has shown that the danger of the road in front of the house is a major reason for keeping children in. And the result of this keeping in is to emphasize the emotional frictions of the tight family. The research has shown examples that where this type of lay-out exists already, if only on a very small scale, there is more friendship than on normal roads which are in all other ways similar. This is a most decidedly useful finding to townplanners who want to encourage social contact, and the only one sociology has so far had to offer. What is more, I found a greater readiness to look after children for others, temporarily, in the safety of a path.*"

"*To tell mothers that they need worry no more about motor traffic, because like the railways, it is now confined to its safe channels, away from their children (although, as on the railways, they may well be given the chance to watch and view from safety), would be a release from a now permanent and sapping fear and strain. It would be like telling the mothers of a primitive tribe that there were no more tigers in the jungle. The change would be so great that the immediate, full adjustments of behaviour called for to suit the new environment could not be expected to arise by themselves. It would be a very necessary and easily accepted therapeutic emancipation to learn to make full use of the new possibilities, for both children and parents.*"

VIII. 1.3. *Ecological Criteria*

Unity with agriculture: Research(i) has shown that (in Northampton and Grantham) even cattle can be grazed in the immediate vicinity of housing. Footpaths need not disturb the animals. Sheep would lend themselves to give suitable Radburn greens amenity and ecological dimension (in the absence of deer!) with a good educational effect on very young children, adding to a sense of belonging and giving something to do for old age pensioners. Dogs can be trained.(ii) Direct contact with the ground is important for all in this connection. (Note sheep in Cumbernauld town centre. See model.)

Unity with industry: Separation of lorries from pedestrians and shorter walking journeys to work makes it more feasible for non-offensive industry to take its efficient place right by housing. This gives employment (part-time) possibilities and a wider education in everyday life.

(i) RITTER P. Thesis unpublished, Nottingham University, 1957.

(ii) Note also sheep at Hyde Park, London, deer at Richmond, Surrey, and other places.

VIII. 1.4. *Economic and Administrative Criteria*

(a) *Selling price:* This can be right for the category of people for whom it was meant as research tables show on pp. 229, 230, 231.

(b) *Rents:* Can be right, bearing in mind possible subsidies generally applied to housing in any particular area.

(c) *Home maintenance:* Radburn lay-out need not add to this item, but can encourage it as a positive contribution.

(d) *General maintenance of area and running costs:* Given more public open space this may increase in cost. But given narrow private paths leading to open areas accepted as the normal recreation grounds, theoretically there should not be an increase as this also leads to a reduction of damage through the feeling of belonging (author's research). Cheaper dustbin collection has been found in the Radburn lay-out at Wrexham, 75% of the time taken on other estates. Rent collecting has been reported as taking only half the normal time at Stourport. Postmen preferred Radburn Areas at Coventry and Sheffield.

(e) *Capital Expenditure.* House types are as variable as in other lay-outs. Paths provided can cost less than those normally provided by roads (see Tables pp. 229–231). More money must be spent on design time. Detailed design and working drawings for the unroofed pedestrian spaces are essential to ensure creative thought and proper construction. The external spaces should have approximately the same consideration of scale and quality as the inside of dwellings. The parallel extends to the possible contribution, in each case, by the inhabitants.

The installation of services has quite obviously new aspects. It is vital to bear these in mind from the beginning so that statutory authorities co-operate helpfully. In Britain the placing of services so that they have right of access to them is necessary. Where services are to run in paths (and this makes the creation of service ducts possible in some instances), these must be legally adopted by the local authority or other legal provision to allow access must be made.

Whereas electricity, gas and sewers usually run along the path system, water mains may be in the cul-de-sac, so fire-hydrants and washing facilities for cars can be economically placed.

In many local authorities services are normally under roads. In a Radburn layout they can be under paths. In such cases less excavation, omission of concrete protection and heavy duty covers for sewers and manholes are a considerable saving. It has also been found that laying services under paths creates a saving as the subsoil to the roads is not disturbed and so the damage of the road surface through heavy builders' traffic is less likely. Water mains can also be taken up the paths and hydrant for fire fighting placed on branches accessible from the cul-de-sac. Owing to shorter runs, savings on water mains and other services have been recorded (Tranent).

Slender lighting columns are an important economy on culs-de-sac for without paths any extra to the minimum kerb width may have to continue unnecessarily the full length of the road for safety.

(f) *Development:* Any available moneys, landscaping, etc., can be spent to better advantage given the greater possibilities of a Radburn lay-out.

Landscaping continuously and not in dribs and drabs, boldly and without the peculiar sophistication of the parks departments, who fill countless flower beds and roundabouts with ineffective and overwhelming masses of colour, could cover a far larger area for the same cost as now. Hard surfaces should come into their own: the wasteful covering of millions of granite setts with asphalt in all our towns could perhaps yield a material to give hard surface, colour and variety without too great a cost. In Cumbernauld, setts from Glasgow, at the cost of transport only, are being used in a most effective manner.

Beyond the capital cost it is likely that the running costs of landscaping will be less. People who feel for the pedestrian area, will look after it; there will be far less need to replace broken trees. "OURS", the feeling aimed at so often, is easily achieved in pedestrian scaled spaces, closely married to dwellings. But the extent to which such areas could be looked after by residents or their jointly employed gardeners has hardly been exploited. Yet another possibility is the reduction of publicly maintained and landscaped areas to a minimum. Paths can flow in narrow gaps between private gardens. Even where they are hedged the effect is utterly different from the street with its privet fussed-up hedges on both sides. Indeed such narrow paths are an excellent foil, and they articulate the open spaces small and large, formal and informal. Among patio housing clusters the walled, narrow path is another variant without any maintenance cost at all. Communal open space can lead to meaningful use: swimming pools. The lay-out can be easily and functionally adjusted to income groups in speculative housing and the extras obtained are really worthwhile, not useless prestige symbols! (See pp. 41, 42. etc.)

(g) *Investment:* Houses at Radburn in the U.S.A. have risen in value compared with similar neighbouring housing, non-Radburn. (Also Cité Jardin, p. 271.)

Recently, sales-agents in London have agreed that the Radburn lay-out represents an extra value on a house for which it is reasonable to charge a price of over £70 per house (even though it costs nothing extra to the speculator!). On this basis speculative building will very soon have to adopt the idea generally to stay in the market, for good investment.[i]

(h) *A criterion of creative experiment:* If an actual lay-out is assessed by the scale of livability (a–g) the study must reveal why a point is not satisfied: is it the ideas themselves as worked into the lay-out? or is it purely and simply that some detail of administration, some lack of communication, or a misunderstanding lies at the root of a defect? The latter can be put right easily and cheaply, without altering the lay-out, with big results.

It can be seen that the advantages and virtues listed for the Radburn Idea apply to housing schemes where vertical traffic segregation is used. This may be in the form of deck-housing or housing with basement car-storage.

VIII. 2. Costs

VIII. 2.1. *Comparison of Costs of Streets and Sewer Works etc.*[ii]

An excerpt from a Report:

"(*a*) *In order that a fair and proper comparison of the costs may be made of the 'Radburn' layout as against 'traditional' layout, one of the latter has also been prepared on the same area as for the Radburn layout.*

(*b*) *The cost of providing garage sites has been included with the costs for the roads.*

(*c*) *It will be noticed that on the Radburn layout it has been possible to integrate the garage sites within the road framework in a much better form than with the traditional layout.*

(*d*) *The same number of four storey flats have been used in the traditional layout as in the Radburn layout in order that the increased density due to the use of flats, may not give misleading figures, when street and sewer work etc. costs are expressed in terms of cost per unit of accommodation provided.*"

(i) This point persuaded Davis Estates Ltd. to propose the Radburn Idea for their 1600 dwellings scheme at Caversham. (See p.262).

(ii) Billingham U.D.C. *Development at Low Grange. "Radburn" type lay-out.* County Planning Officer, Durham 1958.

TABLE VIII.1. *Billingham cost comparison.*

Summary of Costs of Street and Sewer etc. Works

	"A" Radburn Development £ s. d.	"B" Traditional Layout £ s. d.
"(e) *Roads and Garage Sites*	18,714 2 2	20,180 13 11
(f) *Footpaths*	4,594 7 6	3,633 12 6
(g) *Sewers*	4,355 16 6	4,076 19 11
(h) *Gulleys and connections*	920 1 8	903 9 8
(i) *Service Ducts*	247 10 0	346 10 0
(j) *Lighting, Group B*	2,200 0 0	1,705 0 0
	31,031 17·10	30,846 6 0
(k) *Contingencies, 10%*	3,103 2 2	3,084 14 0
(l) TOTAL	£34,135 0 0	£33,931 0 0
(m) *Total number of houses*	258	216
(n) *Total number of flats (4 storey)*	48	48
(o) *Total number of units of accommodation*	306	264
(p) *Cost of roads and sewers etc. per unit of accommodation*	£111 5	£128 5

(q) *Neither of the two estimates include for site works (bulk excavation and levelling) land drainage or the provision of tress or reservation.*
(r) *Approximate gross density units/acre.*
(s) *It will be seen from these figures that the estimated costs of street and sewer etc. works approximate to each other fairly closely, but that by using the Radburn layout it is possible to increase the density without losing the feeling of spaciousness and good circulation facilities, and so reduce the cost of streetworks per unit of accommodation provided."*

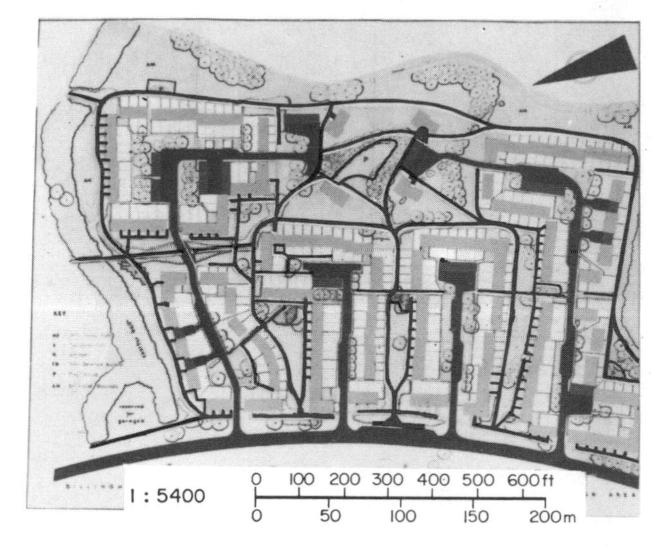

Fig. 8.1. The Billingham Radburn Proposal.

VIII. 2.2. *Economic Analysis*

Table VIII.2 shows the savings in virtually every respect which accrue from using a Radburn lay-out. The great discrepancy between the open areas to be publicly maintained at Beeston is due to the area round a fishpond which was left open. If this rather special amenity were excepted, the overall density would be much greater.

TABLE VIII.2. *Three comparisons between traditional lay-outs and Radburn schemes in terms of lay-out economics**

NAME	*DATA*						*DENSITY*			*AREA ANALYSIS*					
	Area in acres (ha)	*Flats*	*Houses*	*Total dwellings*	*Average hab. rooms per dwelling*	*Other accommodation*	*Dwellings per acre (ha)*	*Hab. rooms per acre (ha)*	*Garage, per cent*	*Sq. ft road per dwelling (m^2)*	*Sq. ft path per dwelling (m^2)*	*Sq. ft public green areas (m^2)*	*Sq. ft road per hab. room (m^2)*	*Sq ft path per hab. room (m^2)*	*Sq. ft public green areas (m^2)*
1. HOUSES 1953															
Average of 48 post-war schemes	24 (9·7)	—	—	338	3·3	Church, etc.	14·2 (36)	47 (116)	—	378 (35·1)	249 (23·1)	—	113 (10·5)	75 (7)	—
Scheme "B" Radburn	24 (9·7)	46	335	381	3·9	,,	15·8 (39·6)	62 (155)	11	257 (23·9)	208 (19·3)	918 (85·3)	66 (6·1)	54 (5)	238 (22·1)
2. BEESTON															
Ilkeston Road	9·8 (4)	74	79	153	3·6	None	15·5 (39)	55 (137·5)	22	486 (45·1)	276 (25·6)	46 (4·3)	135 (12·5)	77 (7·1)	13 (1·2)
Fishpond Radburn	7·2 (2·8)	72	41	113	3·7	Fishpond	15·6 (39)	57 (142·5)	24	306 (28·4)	263 (24·4)	963 (89·5)	82 (7·6)	71 (7·1)	260 (24·2)
3. NORTHAMPTON															
Kings Heath	—	—	—	—	—	—	10·0 (25)	—	10	384 (35·7)	—	603 (56·0)	—	—	—
Dallington Fields	—	—	—	—	—	4 shops	9·0 (22·5)	—	—	304 (28·2)	—	702 (65·2)	—	—	—
Eastfield Radburn	18·1 (7·3)	96	232	328	3·6	None	18·1 (45)	64 (160)	10	228 (21·2)	—	468 43·5	63 (5·9)	—	132 (12·3)

*The author wishes to acknowledge the help received from Derek Lyddon in providing figures for this and for the subsequent tables.

Table VIII.3 shows that there is little difference in terms of first cost between double and single footpath access on the one hand and "Radburn" schemes on the other. It must be pointed out, however, that Radburn schemes have a consistently higher percentage of garage provision and that, whereas single and double footpath systems do not give direct road access to each plot, Radburn schemes achieve this. Direct road access represents a saving to the local authority in dustbin collection, and to the public and to the trade in a number of ways.

TABLE VIII.3. *Four comparisons between single and/or double footpath lay-out and Radburn schemes in terms of lay-out economics.*

NAME	*DATA*						*DENSITY*			*AREA ANALYSIS*					
	Area in acres (ha)	*Flats*	*Houses*	*Total dwellings*	*Average hab. rooms per dwelling*	*Other accommodation*	*Dwellings per acre (ha)*	*Hab. rooms per acre (ha)*	*Garages, per cent*	*Sq. ft road per dwelling (m^2)*	*Sq. ft path per dwelling (m^2)*	*Sq. ft public green area per dwelling (m^2)*	*Sq. ft road per hab. room (m^2)*	*Sq. ft path per hab. room (m^2)*	*Sq. ft public green area per hab. room (m^2)*
1. "HOUSES 1953"															
Double footpath scheme "D"	24 (9·7)	46	337	383	4·0	Church, etc.	15·9 (40)	64 (160)	9	230 (21·4)	205 (19)	1097 (101·2)	57 (5·3)	51 (4·7)	278 (25·8)
Scheme "B" Radburn	24 (9·7)	46	335	381	3·9	,,	15·8 (40)	62 (155)	11	257 (23·9)	208 (19·3)	918 (85·3)	66 (6·1)	54 (5)	238 (22·1)
2. NOTTS. C.C.															
Single footpath	17·3 (6·9)		285	285	3·9	None	16·5 (41·3)	65 (162·5)	9	223 (20·8)	176 (16·4)	122 (11·3)	58 (5·4)	45 (4·2)	31 (2·9)
Radburn Scheme	38 (15·4)			686	3·6	School, etc.	18·0 (45)	65 (162·5)	11	207 (19·2)	210 (19·5)	149 (13·8)	58 (5·4)	58 (5·4)	41 (3·8)
3. STEVENAGE															
Marymead (part)	24 (9·7)			552	4·6	6 shops	13·0 (27·5)	60 (150)	28	279 (26)	188 (17·5)	460 (42·7)	60 (5·6)	40 (3·7)	100 (9·3)
Elmgreen I and II, Radburn	56 (22·7)			707	4·8		12·7 (26·3)	61 (150)	73	400 (37·2)	262 (24·3)	390 (36·2)	83 (7·7)	55 (5·1)	80 (7·4)
4. COVENTRY															
Bell Green (8 or 4)	30 (12·1)	—	—	600	3·4	4 shops	20·0 (50)	70 (175)	12	213 (19·8)	200 (18·6)	404 (37·5)	62 (5·8)	59 (5·5)	119 (11·1)
Willenhall I, Radburn	33 (13·4)	360	248	644	3·4	School, 6 shops	20·0 (50)	70 (175)	50+	256 (23·8)	147 (13·7)	520 (48·3)	75 (7)	43 (4·0)	151 (14·0)

The estates represented in Table VIII.4 are all designed on the Radburn principle. The figures show the high densities which have been achieved. Note the relatively low figures for roads and the high garage provision at Harlow, L.C.C., Huntingdon and Cumbernauld Carbrain. On no other lay out type could so many garages have been provided near the houses.

TABLE VIII.4. *Seven comparisons of Radburn schemes in terms of lay-out economics.*

NAME	*DATA*						*DENSITY*			*AREA ANALYSIS*					
	Area in acres (ha)	*Flats*	*Houses*	*Total dwellings*	*Average hab. rooms per dwelling*	*Other accommodation*	*Dwellings per acre (ha)*	*Hab. rooms per acre (ha)*	*Garage, per cent*	*Sq. ft road per dwelling* (m^2)	*Sq. ft path per dwelling* (m^2)	*Sq. ft public green area per dwelling* (m^2)	*Sq. ft road per hab. room* (m^2)	*Sq. ft path per hab. room* (m^2)	*Sq. ft public green area per hab. room* (m^2)
1. SHEFFIELD Greenhill	14 (5·7)	136	143	279	3·7	School, 10 shops	20·0 (50)	74 (185)	11	265 (24·6)	257 (24·6)	552 (51·3)	71 (6·6)	69 (6·4)	149 (13·8)
2. WREXHAM Area 6	15 (6·1)	144	156	300	3·9	None	21·5 (53·75)	84 (210)	14	203 18·9	193 (18)	473 (44)	51 (4·7)	49 (4·6)	121 (11·2)
3. STOURPORT† Welsh's Farm	16·5 (6·7)	162	174	336	4·2	None	20·3 (51)	85 (112·5)	10	347† (32·2)	48 (4·46)	43 (4·0)	82 (7·6)	11 (1·0)	10 (9·3)
4. HARLOW Area 67	345 (140)	91	424	515	3·4	None	15·0 (37·5)	52 (130)	100	286 (26·6)	90 (8·36)	652 (60·6)	83 (7·7)	26 (2·4)	192 (17·8)
5. L.C.C. Huntingdon (I)	14·8 (6·0)	36	184	220	4·1	None	14·8 (37)	61 (152·5)	70	295 (27·4)	198 (18·4)	593 (55·1)	71 (6·6)	48 (4·5)	143 (13·3)
6. CUMBERNAULD* Carbrain	28·5 (11·5)	149	453	602	4·3*	Shop	21·1 (52·7)	91* (227·5)	100	170 (15·8)	120 (11·1)	—	44 (4·0)	32 (3·0)	—
7. TRANENT Ormiston Road	6·5 (2·6)	—	—	115	3·2	Shops	18·0 (45)	58 (145)	87	270 (25·1)	125 (11·6)	—	84 (7·8)	39 (3·6)	—

*Densities for Cumbernauld have been converted to the higher English equivalent.
†The Stourport figure for roads includes pavements.

VIII. 3. Dwelling Plans

VIII. 3.1. *Analysis of Relationship of Dwelling to Access*

Entrances to houses need to be analysed on a functional basis. If present behaviour patterns are studied and these are projected in a sort of trend planning (as Grenfell Baines did in the fifties, arriving at the need for a "ceremonial and formal" and an "informal" entrance) we are likely to make a mistake. A new lay-out with new immense possibilities can create its own use system, the advantages of which can persuade obviously, through public relations, or subconsciously through sheer clear, simple design.

It is particularly misleading to stick to the traditional nomenclature of "front door" and "back door". The only valid, general and functional distinction with traffic segregation is "street door" and "path door".

There are two distinct aspects to access:
the approach to the house and
the entry into the house.

Thus real needs can be seen to apply to as follows:

A. access for service
(a) waste collection
(b) delivery household goods
(c) letter post
(d) access front and back by window cleaner and others.

B. access for people
(a) going out by car, or to the car
(b) going out on foot
(c) formal visitors
(d) informal visitors, friends
(e) pottering in garden, children in and out

C. ceremonial family occasions
(a) first outing of new baby
(b) wedding
(c) funerals
(d) feasts, parties, processions.

The most important points are probably,
efficiency for service,
pottering in and out without dirtying the house unduly,
bringing visitors into formal room without ostentatiously taking them through the working part of the house. This last does not necessarily enforce a "formal front" to a house.

Discussions between Professor Stephenson, Clarence Stein and Mr. Womersley during the early Radburn experiments in Great Britain suggested that the kitchen should overlook the path side, for the mother will put her children out where she can watch them. This avoids lonely children in gardens. But other subtle possibilities arise.

Arguments have been advanced against gardens on the service side because of the length of carrying service goods. This is not a new problem. Many older houses have long carry distances and some of them obviate the carrying altogether by providing service lockers at the entry point. Far more important is that the garden and the house should be properly insulated from exhaust fumes and noise coming from the service road. High walls with fences or hedges or garages can fulfil this function. To judge raw and new schemes is difficult, as the growth of greenery, even to a small scale, transforms the capacity of an area to absorb noise and smells and to screen, as the American examples show. (See p. 261.)

It is more important to have the garden on the sunny side of the house than it is to avoid, at all cost, visitors or service delivery entering on the private garden side of a house. A hedge along the path can make the remainder of the garden private in any case, as in hundreds upon thousands of existing dwellings from all periods.

It is similarly simple, particularly in broad fronted houses, or those where a living-kitchen takes up the entire width of a narrow terrace house, to allow for those tenants who wish to introduce a curtain. This can cut off from view all but a narrow entrance section of this room, for those who enter on the service side and are to be formally received. This simple, cheap remedy (track optional extra at the building stage), with one blow completely undermines the vehement argument of many architects and planners over the years against the Radburn idea on the grounds that visitors might arrive by car on the service side. This curtain can be drawn at a moment's notice, so there is no question of not being able to tidy up suddenly. It does not prevent the use of the room as a whole. French windows are also effective (if draught proof). Observation shows that the degree of privacy people wish to have varies very greatly. Thus to provide flexible privacy is obviously a good solution.(i) This applies generally.

The dual access allows two major ways of treating ceremonial entrance and exit. The time-honoured spectacle of the bride leaving for her wedding, or the coffin with flowers for the grave, can be given a proper, sensitive expression along a path to the vehicle. If the weather is very bad or privacy is sought then the vehicle can come right to the door on the car side of the house. As this solution applies to those who shun being seen the lack of a public and formality along the feeder road is an advantage. This can be made clear to people on taking up residence.

(i) Willis M. Designing for Privacy. *A.J.* London, 29.5.63, 5.6.63 12.6.63.

The dual access of a Radburn scheme offers many opportunities. It is often forgotten that service through access is not required, a space saving. There are many different types of needs of varying kinds of groups the world over. Each chosen way should be related to real needs and clearly satisfy them within the terms the planner has set himself. The criticism of isolated points of isolated schemes is quite useless and meaningless. The pattern of ideas and the approach must be studied as a whole before proper assessment or criticism or improvements can rationally result.

Not least in the items to be taken into account are the spatial limits set by the government regulations in Britain and other countries, at any particular time.

The Ministry of Housing has published a bulletin in which they try and pinpoint the "performance" of a house. The following questions relate to the relationship of dwelling plan to lay-out.

Questions Implying Performance Requirements.(i)

"Does the kitchen overlook the place where small children are likely to play and where the baby will get an airing in its pram?

Can the garden be used in summer as an extension to the living space?

If you come home in dirty working clothes can you get direct from the main door to a place where they can be kept and to a place where you can wash?

How far does the housewife have to carry the rubbish from the kitchen to the bin store, and can she manage without going through living areas?

Is the fuel store near at hand?

Are there arrangements for storing bicycles and the pram without have to pass through entrance hall and living rooms or kitchen?

Are the refuse bin store and fuel stores arranged so that the collector or the delivery man does not have to enter the house, but so too that the arrangements are convenient to the household?

Can the painter and the window cleaner get ladders to the other side of the house?

Can the meters be read from the outside, and if not, can a reading be taken inside without causing inconvenience?

Is there protection at the front door for casual callers in cold or rainy weather?

Is there sufficient room at the entrance to receive visitors? Is there room to pass?

Can you shut the front door with three or four people in the hall?

Fig. 8.1a. View from kitchen into courtyard with play space, Clement's Area, Haverhill.

Is there adequate spaces for hats and coats?

Can you get from the garden into the house without passing through the living room?

Can large garden tools be shifted from where they are kept to where they are used without going through the house?

Can you easily get from the kitchen to the washing line?

Is there somewhere safe for children to play?

Does the siting so far as possible avoid—

kitchen doors opposite each other with no barrier in between?

overlooking of the private garden by neighbours or passers-by?

people passing to close to windows?

nuisance and hazards from traffic?

Is the car space or garage near to the house?

Can refuse be collected and fuel delivered easily?"

(i) *Space in the Home*. Ministry of Housing and Local Government. Design Bulletin 6. H.M.S.O. 1963.

One early solution adopted at Sheffield uses a through living room with a french window on the road side. The kitchen is placed on the path side with a hall dividing it from living room. Delivery from the road to the kitchen is through a large store-room.

Another solution used at Coventry has a through living room and a through dining kitchen. These two through rooms solve the orientation problem. The division into one work–play–eating space and one sitting–entertaining space is rational, given the restricted areas in public housing. A pilot survey carried out at Coventry showed that a significant majority of householders like this plan and minor variations have been tried. One of these includes a porch for the pram on the path side.

The house plans which we have been considering so far imply a relatively narrow frontage. With traditional housing lay-outs, where the house faces on to the street, the narrow frontage has been considered inevitable if building costs were to be kept low and densities high. In Radburn planning this logic does not apply and at Cumbernauld, Carbrain, a house is used with a frontage of 30 ft (9·14 m) and a depth of 15 ft (4·52 m). A general advantage of a wide front is that it gives more privacy in the garden. Since then broad fronted houses have been planned in a number of offices including those of the L.C.C. at Andover, Clement's Hill and Haverhill. (See also, and for dwelling plans, pp. 248, 259, 261, 271, 282, 284, 287, 301.)

The connection between the two entrances required in Radburn planning is well managed with a through space in the centre, only 15 ft (4·57 m) long. The placing of all windows (except for one high-level window in the bathroom) on one front, apart from giving privacy, enables the terraces to be close, only 35 ft (10·60 m) apart, on a southerly slope at Cumbernauld; and a further advantage of this plan is that it lends itself to corner siting and thus avoids the need to put corner flats in this position. Garages, with this plan, are placed either against the house on the north side or along the southern end of the garden. The lay-out at Cumbernauld gives the surprising density of 23 dwellings to the acre (57·5 per ha), three-quarters of which are houses. Though the plan chosen owes its form, to some extent, to the steeply sloping site, the principles it brings to light are of general application,[i] particularly for terraces which stretch in a north–south direction, which can be placed "back-to-back" on to a service lane (car port).

(i) See Carbrain in examples given.

Investigation in Sweden has shown that in flats, kitchens should likewise face the children's play areas (Bruno Alm). If tall blocks do not allow direct aural contact, sight of the child may still be valuable. (The isolation imposed by simple point blocks and slabs will surely soon limit the building of them. They are like rabbit hutches not because of squared elevations but because, beyond the flat itself, there are no habitable spaces which create a sense of belonging—a street, a path, a deck seem essential.)

Entrances to flats should be from the pedestrian side of the lay-out, or at least from both sides. Children will gather, as surveys in Sweden have shown, immediately outside the exit, where cars and the roads threaten. By bringing father from work, and mother from the shops and the child at play to the pedestrian side, the path gains in attraction and the children will play where it is safe and healthy and where it has been planned for them to play.

Fig. 8.2. Main exit from flats leading to cars is wrong.[i]

Fig. 8.3. Main exit leading to path system is right.[i]

(i) Wohlin H. *Studier or Färskolbarns Lekvanor I Modern Bostads belyggelse.* Stockholm 1960.

VIII. 3.2. *Car Storage*

The following are alternatives each with advantages that can justify its use if appropriate.

(a) Individual garages at end of garden (or car port). (Willenhall Wood.)

(b) Individual parking place within plot (with or without heating point(i)).

(c) Individual garages or car ports, in groups at distance from homes. (Parkley's Common.)

(d) Individual parking place as one of a group at distance from home (with or without lockable heating points for winter). (Marly-le-Roi.)

(e) Individual garage or car port within the confines of the house at ground level. (Bermondsey.)

(f) Individual garage below ground level with direct access to home. (Flemingdon Park.)

(g) Communal garages in one large building, single or multi-storey above ground level. (Baronbackarna.)

(h) Communal garages below ground level. (Solna.)

(i) Communal parking at sunken level. (Lafayette.)

There is an evident advantage for the environment if the cars are kept at a distance from the residential areas, especially while they produce petrol fumes, poison and noise.

The advantage in placing each garage next to the house it belongs to lies in the ease with which the garage can be used as extra storage or as work and play space (beyond the obvious convenience). When each garage is so placed, the garage number should correspond to the house number, as confusion is caused when the garage number is different.

The bad design of English garage is remarkable even where house design is good.

Relieved of paths the floor of the garages court should be punctuated only by bollards and rails protecting the corners of garage blocks and by indication of parking and traffic areas.

Visitors parking (50%), trade delivery and turning for commercial vehicles can be provided separately or combined with residents car storage.

Hydrants should be provided. Trees, if used, should not be associated with grass so that there should be no confusion with the path side and at night the difference might well be enforced by the use of different coloured lighting.

(i) Heating points: Particularly in Sweden (Stockholm demands 1·8 parking space for each dwelling) it is now recognized that the car which withstands all weathers moving should do so standing. The most vulnerable points in the engine can easily be kept warm by connection to small elements built instead of garages with a locking device against theft of current. They are foot high concrete sockets.

If car manufacturers could only be influenced to produce weather resistant finishes, rather than chrome, the problem whether to protect cars or not would become much clearer.

VIII. 3.3. *Cost and Density*

With an increase in density the provision of car space for every family becomes more difficult. Above 100 ppa (250 ppha) the problem becomes tricky. Although multi-storey and underground provision is expensive the intense use of the ground and the fine environment gained counterbalance the cost. We are already at the stage where underground parking for flats, and even houses, is being built by private enterprise and local authorities, e.g. flats by John Madin on the Calthorp estate Birmingham, and by Womersley, at Norfolk Park, for the City of Sheffield; and houses by Irving Grossman, and Klein and Sears in Toronto. The car in the basement of flats with lift connection is very convenient.(i)

Fig. 8.4. Willenhall Wood II, Coventry; privacy, and cleanliness. Garages in no-fines concrete and pebble dash to match houses.

Fig. 8.5. Park, Cumbernauld, Scotland.

Fig. 8.6. Carbrain, Cumbernauld, garages parallel to service road.

(i) WOMERSLEY J. L. Planning for the Motor Car. *J.T.P.I.* London Sept.-October 1961.

VIII. 4. Paths, System of Routes

VIII. 4.1. *Analysis*

The requirements of paths in general have been described in the chapter on the needs of man. The design, conscious design, of path environment is so new that factors particularly relevant to housing should be stressed.

Traffic engineering for pedestrians is more difficult than engineering for vehicular traffic.

The shortest possible route to any goal must be the route along the footpath system. "Short-cuts" which mean crossing roads should be made a detour so that they do not tempt people. Where there is one main goal, at the end of a superblock, the paths ought to point towards this. If the goals are divided then the right angled junction between branch path and main path will be the optimum.

Desire lines, or as the crow would fly towards a goal, should be approximated as nearly as possible. The pedestrian should be drawn to the path by the life and interest that has been planned into it.

The sculptural, colour and social value of post boxes, telephone boxes, electric transformers, waterboard signs, fire and police alarm boxes, advertisement boards or kiosks and signs, slot machines, wall-bracket lights, and lamp-posts should be considered and brought into use. (See Fig. 8.10.)

How people select routes to take on foot is an important and little understood phenomenon. Derek Lyddon[(i)] has listed the mental image of a route, that which is encountered *en route*, habit and danger, as determinants over and above the mere directness of any given route.

In other words, particularly for the generation that is still used to walking along roads, the footpath system must be simple to envisage, the route must attract by its promise. Each time, habit of use must be established at the outset of any plan. Danger, in little used residential roads, is not recognized by those who have never known anything else. However, surveys of Willenhall Wood show that people who have lived side by side with the genuine safety of traffic segregation do appreciate it and ask for it where it is not provided.

No one has done more original work on the subject than Derek Lyddon. Designer of Willenhall Wood I, Coventry, and at Carbrain, Cumbernauld, he has given much thought to this subject, and writes,

> "*It must be recognised that in an urban area teeming with life some convenient journeys (one hopes the individual ones of neighbouring) will inevitably involve crossing a road or even moving along a road. This will be true unless vertical separation can be afforded.*"
>
> "*To argue that because these minor journeys are undertaken involving the danger of conflict a separated system does not work is obviously expecting perfection in a sphere of human activity where balance of conflicting criteria can be the only aim, and perfection is an unbalanced ideal.*"
>
> "*On the other hand it is equally false to ignore the minor cross currents in a pedestrian network, and fondly hope that people will take the long correct way round. The balanced solution will result from,*
>
> *correct prediction of conflict points and cross current routes,*
> *limitation of vehicle speeds,*
> *pedestrian sight lines,*
> *alternative safe route which can be taken if danger of shortest route becomes apparent.*"

This opens the question to what extent pedestrians should be channelled. Given dual access naturally people will make contacts and wander in all directions to some extent. But we need not take this as inevitable. If roads and culs-de-sac are made extremely inconvenient and people are thereby determined by the plan to make their contacts and their desire lines in the preferred direction of the footpath system, then we still impose less limitations than are imposed on millions of houses whose gardens adjoin: these people are also hiddenly persuaded to make their friends and contacts on the access side, with the exception of the immediately adjoining neighbours. Nobody has thought this evil. The positive channelling of people in a meaningful way towards *their* safe and convenient path system is far better than arbitrarily making them use the road outside their house and that alongside.

Where trespassing, illegal or merely stupid, on to roads occurs, the danger should be made very obvious indeed. That way maximum safety is regained.

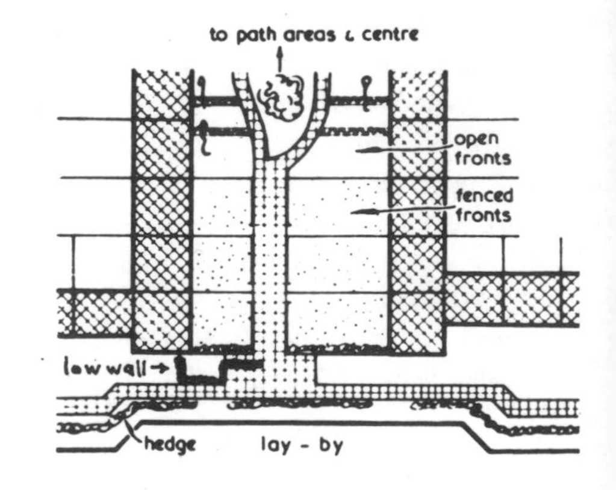

Fig. 8.7. Where path meets road: marked as a transition to danger, and should provide a place for "car watching".

(i) Letter to author, 28 March 1963.

Crossings of minor roads or culs-de-sac can be marked comparatively easily and are the most dangerous points. Walking along culs-de-sac is less dangerous and less likely if the routes are indirect as Derek Lyddon's diagram shows.

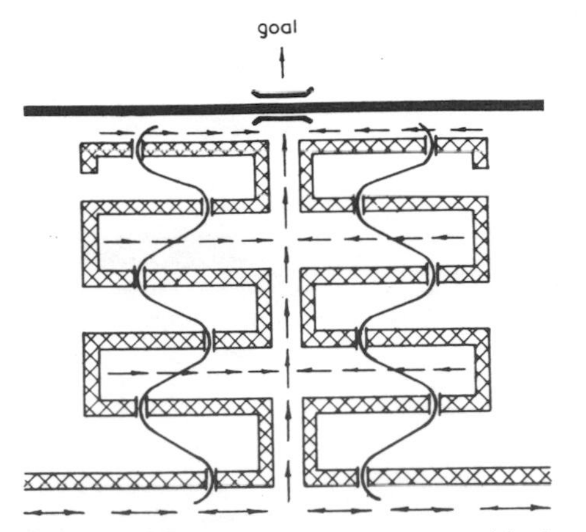

Fig. 8.8. Each path, where it crosses any vehicular traffic, should be marked, red and clearly. In that way a clear picture of the different realms, pedestrian and vehicular, is made vivid.

Fig. 8.9. Federal Housing, Minneapolis.

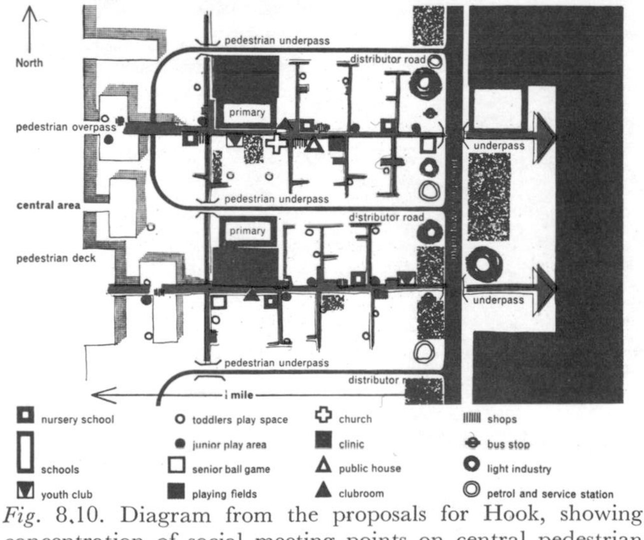

Fig. 8.10. Diagram from the proposals for Hook, showing concentration of social meeting points on central pedestrian way.

The greatest difficulties arise when a main cul-de-sac with minor culs-de-sac is combined with frontage on to the major cul-de-sac. A considerable road network makes this the most direct route to many points. This is in itself a good reason for avoiding this arrangement if possible. To aim for simple and well-defined arrangements eases this and other problems. In such cases, and in fact when other advantages outweigh the disadvantages (i.e. limited access to ring road) and the cross traffic is expected, it should be channelled with dangers made obvious by every means available. Where intimate paths cross culs-de-sac there is a special danger point: often there are no sightlines and even slow vehicles would be lethal to a child running along an alley obviously made for running along. Guard rails can to some extent lessen dangers at such points, and make the traffic obvious.

The design of the point of contact between ring road and path areas is important. At such points the following needs must be fulfilled in the given manner or some similar way:

A. Paths in front of any houses which face on to the ring road must lead into the pedestrian areas. This can be done by placing barriers at the cul-de-sac crossings and all along the road front (hedge or bank or low fence).

B. The pedestrian area should have a clearly defined end where it meets the road so that small children can recognize it easily and the visual enclosure of the area is more complete. An increase of private front gardens at such points of entry might be used effectively. Gates are desirable but very difficult. The name should be clearly shown. A look-out where children can watch cars safely should be provided. (See Fig. 8.7).

C. At such points adequate lay-byes and space are required for visitors coming by car, hearses, wedding cars, and sometimes to serve as a bus stop.

D. Where the point coincides with an underpass, as at Carbrain, Cumbernauld, as each main footpath serves a large population, as does each road junction, the planning of bus stop, steps to the road, steps continuing the path and ramps for prams, result in a major design problem worthy of much thought.

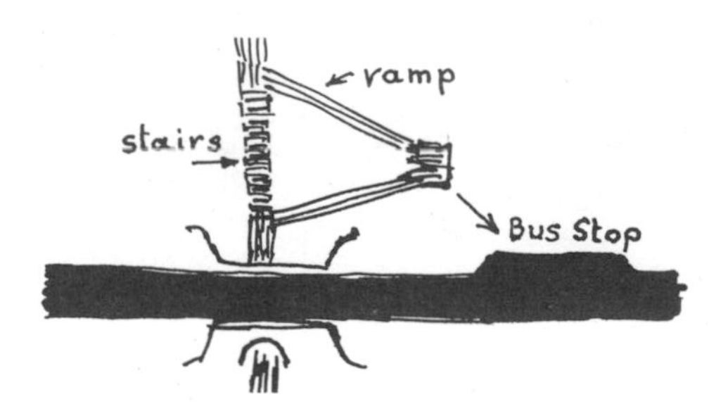

Fig. 8.11. Underpass, and Bus-stop point at Cumbernauld.

VIII. 5. Plot Shapes

VIII. 5.1. *Houses*

Fig. 8.12 On steep slopes or small sites, collective car space, leaving an area free from road, arranged informally for housing, can be successful.

Fig. 8.14. Pedestrian-only access housing in terraces, within a Radburn lay-out, is distinguished decisively: the paths lead to a main path and to goals on one side as well as to the service road on the other. Car storage is necessarily collective away from dwellings. As the alternatives show the path may lead:

(a) between the open "front" of one house and the enclosed garden of the next row,

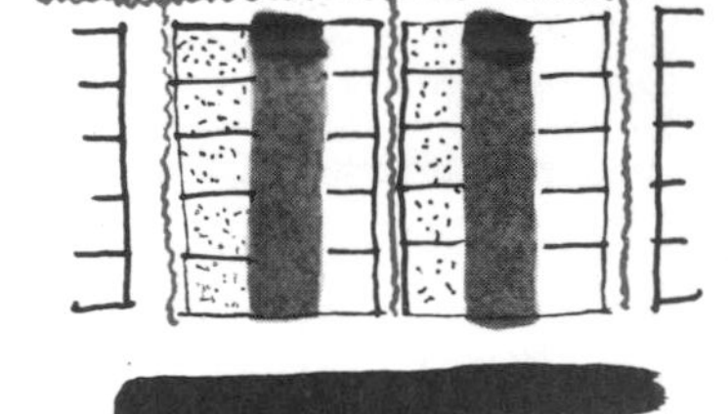

(b) between gardens, fenced, walled or hedged on one side only,

(c) between private gardens on both sides of the house, with different (d) or similar garden use.

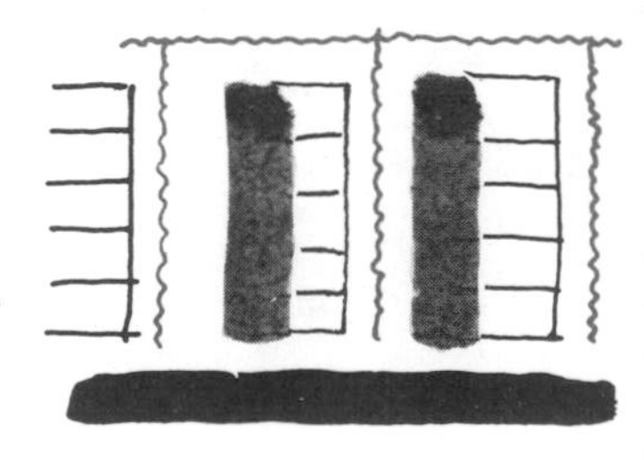

(e) between open "fronts" facing onto path, and other combinations.

Fig, 8.13. Seafar II, Cumbernauld, informal arrangement of housing on northerly slope as in Fig. 8.12. Heather used as overall ground cover. Design team leader, Roy Hunter.

Fig. 8.15. In Radburn culs-de-sac proper, each plot is directly accessible by car. Car storage may be provided with each plot or collectively. In the latter case it is of obvious advantage to make the collective storage at the entrance to reduce traffic and road need for further than the minimum distance of the cul-de-sac.

(a) Similar garden use, and size, can be given to all dwellings round a cul-de-sac. Path and street sides can be fenced giving only narrow public strips each side.

(b) Similar garden use may be planned for the car side but the path side may be open on both sides; giving a wide public pedestrian environment.

(c) The street side may be openly related to the houses. Private gardens face to the pedestrian side making it narrow. Grouping of garages must be carefully arranged so that it will play a most important part in filling the area between houses. The space necessarily achieves functions beyond those of a car port, more so than in the other alternatives. Reversing of cars in such seemingly "safe" places is a special danger. The segregation of children and cars becomes a complicated but still essential consideration.

(b) and (c) can be used with alternate, and not symmetrical and adjoining similar garden use as shown, i.e. large and small gardens may adjoin the road on alternate sides and similarly on the footpath side. The alternate arrangements are dictated by orientation and are virtually necessary in north–south terraces of flats.

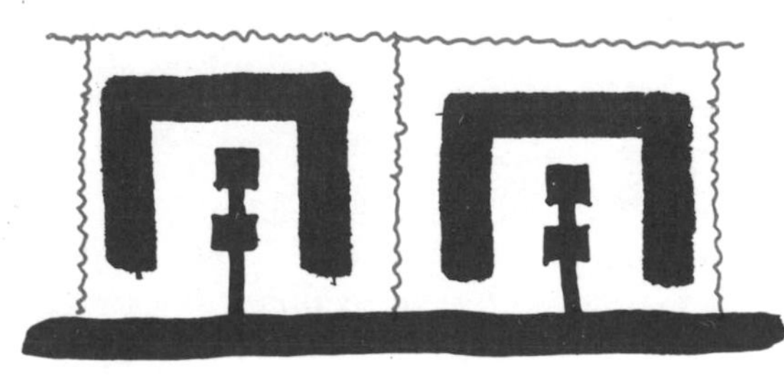

(d) Open to car port and to pedestrian side.

VIII. 5.2. *Flats and Deck Housing*

In lay-out with flats there is the possibility of continuous development in touch with the ground, i.e. four-storey maximum, or multi-storey development with separate deck path systems. Point blocks and disconnected slab blocks are not wholly satisfactory for any age-group from the point of view of the primary need for some path system or corporate life, directly accessible from the dwelling.

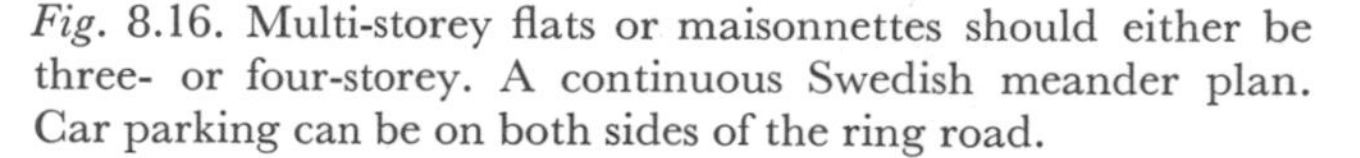

Fig. 8.16. Multi-storey flats or maisonnettes should either be three- or four-storey. A continuous Swedish meander plan. Car parking can be on both sides of the ring road.

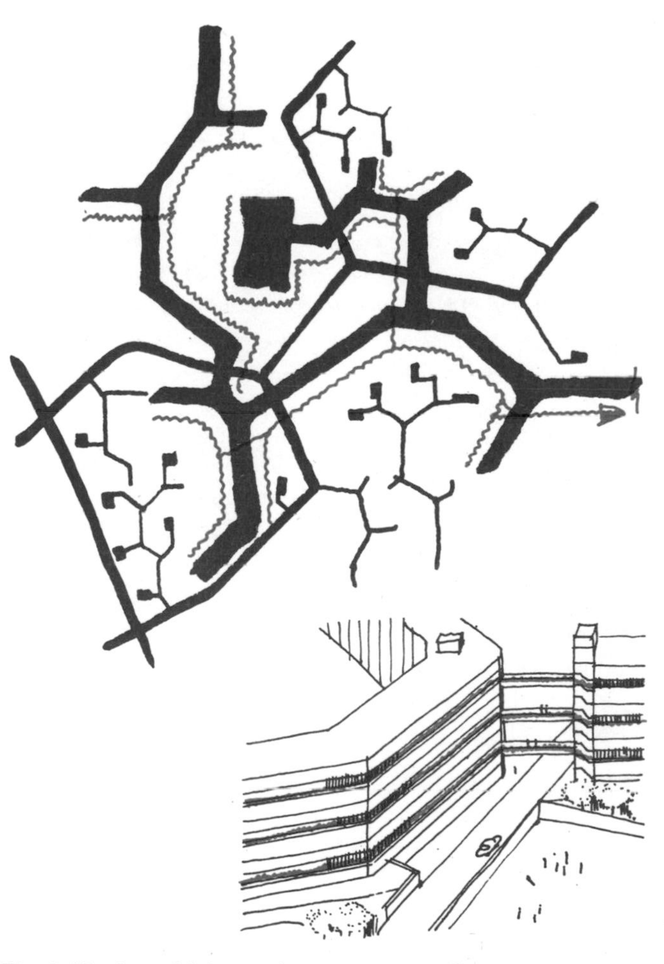

Fig. 8.17. A multiple vertically continuous deck system for pedestrians creates its own plot shapes.

VIII. 6. Residential Feeder Roads

VIII. 6.1. *Size and Frequency of Junction*

Cumbernauld Development Corporation has evolved certain conclusions on the size and duties which the residential feeder roads or culs-de-sac should perform, on the basis of their master plan main road system.

A report considers cul-de-sac design procedure:[(i)]

"1. *Frequency of Access to Town collection-distributor road.*
The C/D is a three lane road and at access points the centre lane is reserved for turning vehicles and a minimum of 150 *ft is desirable for both the acceleration and deceleration movements in order that merging and diverging with the main streams may be made at minimum speed differentials. Thus a minimum of* 300 *ft is desirable between accesses.*

"2. *The widths of housing feeder roads and culs-de-sac are dependent on the number of families served. It may vary according to income grouping and car ownership but in general may be taken as follows:—*
18 *ft width with occasional parked vehicle:*
Instantaneous peak capacity = 300 *v.p.h.*
Average peak hour capacity = 200 *v.p.h. say*
No. of families on 68% *car usage* = 294 *families*
24 *ft width with occasional parked vehicles:*
Instantaneous peak capacity = 500 *v.p.h.*
Average peak hour capacity = 350 *v.p.h.*
No. of families on 68% *car usage* = 515 *families*

In either case it is recommended that not more than 300 *families be served off one road and that if more families are planned for an area then a number of culs-de-sac be fed from an internal, frontage-free, collector, which in turn connects to the town collector-distributor road. The width of the internal collector being determined from the total number of families served at any point.*"

A further report[(ii)] considers that the misuse by pedestrians crossing, the occasional parked vehicle, and the capacity of junctions, are the real determinants. This report finds that 300 v.p.h. peak hour is the maximum to ensure safe crossing times for the pedestrian, even if he is where he should not be. This is interesting to compare with the work of Harlakoff.[(iii)] He arrives at the same figure of 300 v.p.h. allowing safe crossing times. But he uses this as an argument against traffic segregation in his extensive thesis on the most economic ways of serving housing with roads.

The figure which allows for the parked vehicles and the waiting likely at the busiest junctions on the Cumbernauld ring and collector road is also in the region of 300 v.p.h.

Thus it is clearly established, for the circumstances of that new town, that 300 v.p.h. is the optimum number to plan for.

A final interesting point relates to a frequency of junctions. The minimum distance apart was set by the first report. But the second report adds that secondary collector roads within a housing group may be preferable from the motor traffic point of view and the economic point of view. Junctions to major roads are so much more dangerous than minor road junctions and so much more expensive that the extra road area that would be required by an internal collector may well be compensated for.

(i) Cumbernauld Development Corporation, Traffic Engineering Section, 8 March 1963.

(ii) Cumbernauld Development Corporation Traffic Engineering Report, 1963.

(iii) Harlakoff N. *Wirtschaftliche Erschliessung von Wohngebieten.* Berlin, June 1961.

VIII. 6.2. *Feeder Road Design, General*

VIII. 6.2.1. *Culs-de-sac.* Under the influence of Radburn ideas and, more particularly, as a result of the steady increase in the percentage garage provision, the traditional cul-de-sac design (street + pavement + turning circle) changes into a "garage court".

Some local authorities still insist on a pavement, but this conflicts with the logic of the Radburn Idea. It is not necessary, it is expensive, and it is visually less satisfactory. Where there are proportionately few garages, the gardens cannot be screened except by high walls or fences.

With a garage provision above 30%, the garages themselves will take up most of the frontage and the whole cul-de-sac can be treated as a garage court. It will normally be much wider (garages facing one another should be 25 ft (7·62 m) apart) and as the garage aprons will be nearly continuous there is no question of a footpath or side walk. The floor can be uninterrupted, while the back gardens will be screened by the garages. Courts need not be regular in shape. When the area is large, it is wise (as at both the Willenhall sites and at Carbrain) to use a complex plan so that the long lines of garages are broken up. Wide frontage allows garages to lie parallel to each plot, reducing width required. Materials and colour of garage doors offer opportunity for careful design. (See p. 235.)

VIII. 6.2.2. *Loops.* Where tight turning spaces discourage vehicles, dust-carts and the like, it has been found that they will reverse out, an exceedingly dangerous procedure. Large efficient clear turning spaces are an obvious but expensive answer. The problem relates directly to the alternatives of loops rather than culs-de-sac.

The disadvantage of loops is that they tend to create many more vehicle–people conflict points than culs-de-sac. In many instances and by several techniques this can be guarded against (see Fig. 8.20). But the important distinction is between two-way loops connected adequately for emergency and occasional service only (see Section VIII.10.3, Caversham) and the one-way loop which brings with it advantages to set against the increased pedestrian and vehicular difficulties. In many instances, and with the help of contours, the loops may increasingly appeal as more satisfactory. One-way systems and such loops cut out the difficulties and danger and space needs of turning circles, they cut out the increased danger of two-way traffic in residential areas and of accidents at junctions. They simplify delivery, if only in certain respects, and make possible deeper development without additional roads where this is desirable. The planning of parking and garaging, where in groups, is simpler and safer, with one-way traffic parking for each individual house more easily achieved. Road frontage will be increased with extra cost, of course, compared with culs-de-sac (see Kitimat, Section VI).

A one-way street will be narrow. As such and with the psychological effect of a one-way street seeming significantly and obviously for vehicles only, it would probably discourage misuse by pedestrians.

As a certain amount of residential jay-walking and risk-taking is bound to take place, very careful balance of various factors should precede decision in each case. Whichever solution is adopted, cul-de-sac or loop, there will be efficient and inefficient ways of implementing each one.[i]

(i) The danger of reversing vehicles was borne out by the only fatality recorded by Clarence Stein in his work, a reversing dust cart; also by a newspaper report of a milk float killing two women, slowly reversing out of a cul-de-sac. *Nottingham Evening News*, 29.8.1963. Similar reports appear with tragic regularity.

VIII. 6.3. *Basic Alternatives*

VIII. 6.3.1. *Culs-de-sac*

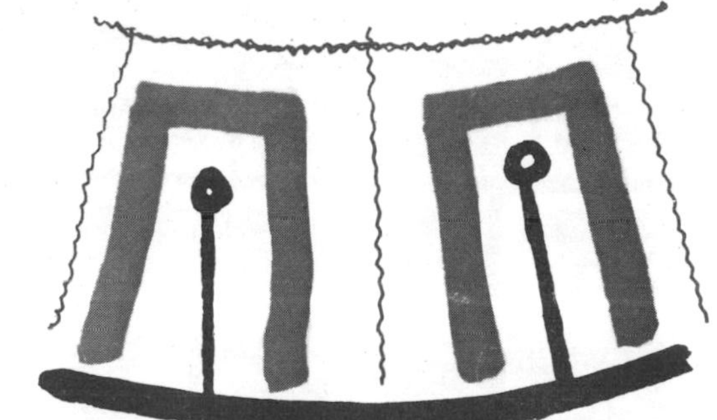

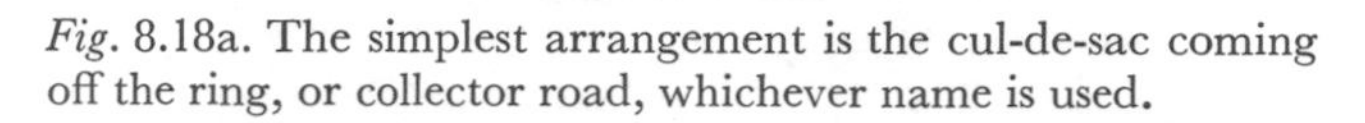

Fig. 8.18a. The simplest arrangement is the cul-de-sac coming off the ring, or collector road, whichever name is used.

Fig. 8.18b. Such a cul-de-sac can serve footpath-access terraces, serving a larger area.

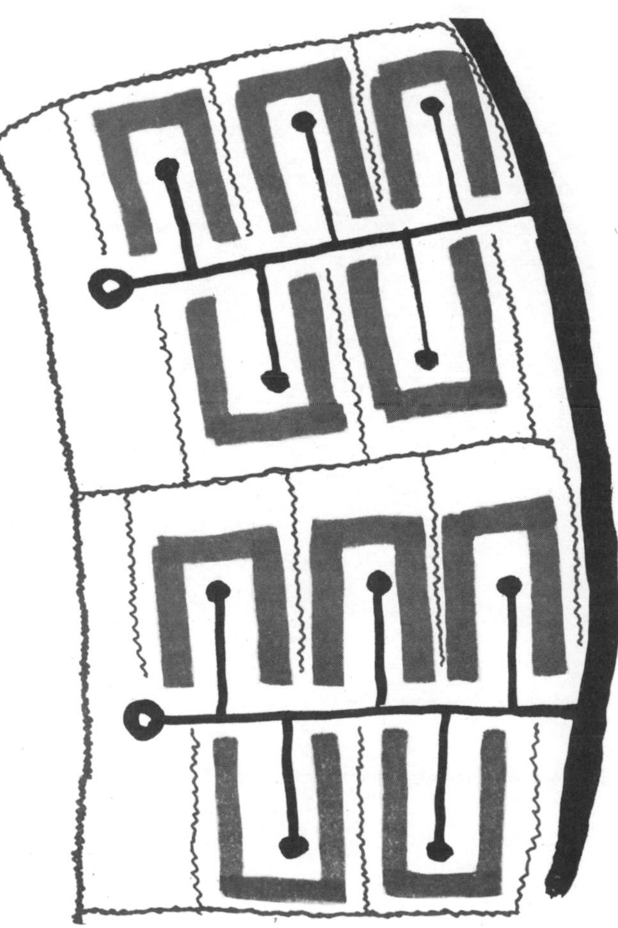

Fig. 8.18c. The cul-de-sac itself may become a collector road to branches which are the car port culs-de-sac. An even larger area is served that way, with car access to every plot.

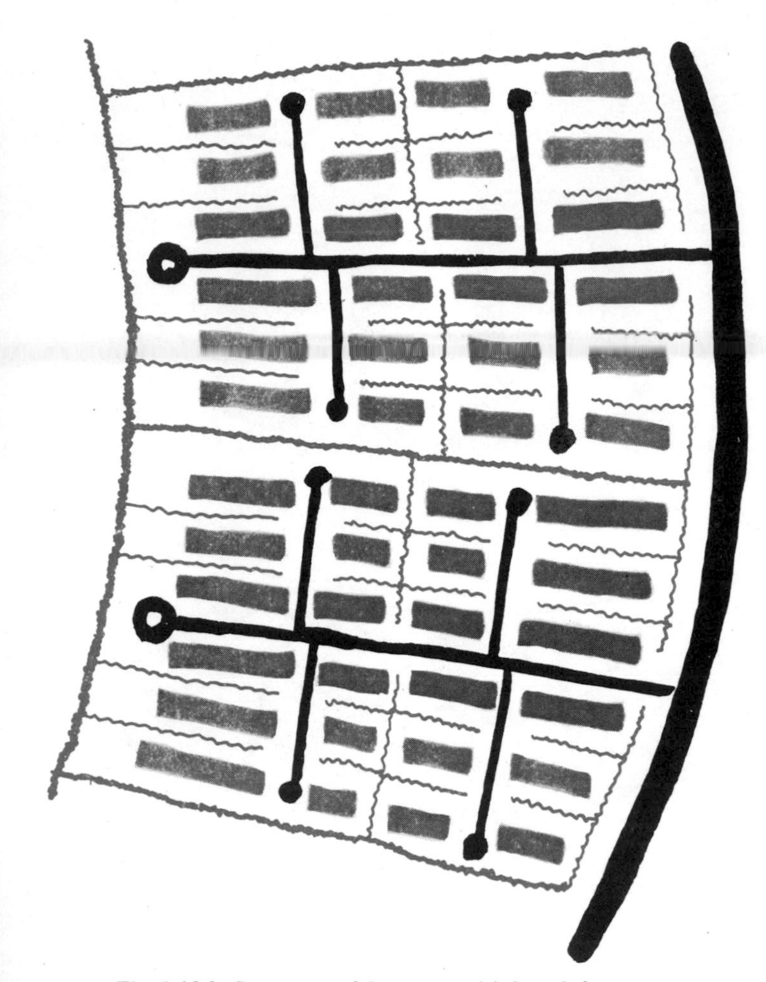

Fig. 8.18d. Areas served by one multiple cul-de-sac, combined with path access. In this system the danger of paths forming continuous short-cuts across culs-de-sac must be guarded against. Capacity of junction of ring-road and cul-de-sac limits the number of dwellings served by one multiple cul-de-sac.

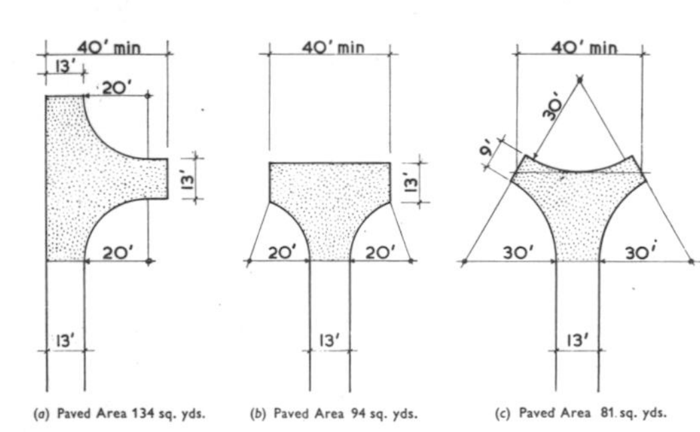

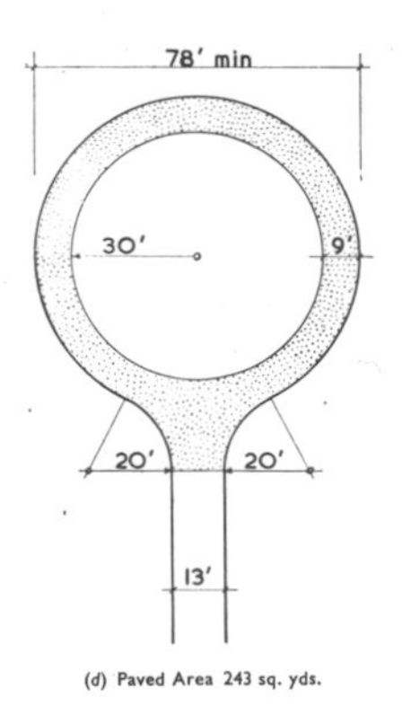

Fig. 8.1.9 Turning spaces in culs-de-sac with 13 ft carriageways. The areas of paved surfaces are exclusive of parking space. With 9 ft carriageways comparable results are achieved.

VIII. 6.3.2. *Loops.* Two-way traffic loops, meant for cars in emergencies only, can be slowed down to walking speed on either side of the strip which has in it the path leading to the main path. Turning space must be provided as if it was a matter of two culs-de-sac. (See p. 262.)

Fig. 8.20 Detail of effective slowing device.

Fig. 8.21. Loop—slow or emergency connection only—turning space as if cul-de-sac.

Fig. 8.22. One-way traffic is important and right in loops intended for through use, particularly if they are to serve many dwellings. Path crossings need underpasses or at least slowing of motor traffic. One-way traffic is safer and does away with the noise and fumes generated particularly through reversing. Road width can be reduced.

VIII. 6.4. *Examples* (Sections 6.4.1–6.4.4.)

The following plans represent a documentation of cul-de-sac and garage court planning over the last 15 years in Britain. They are divided into four groups: pioneer projects of the early fifties; the mid fifties; New Town schemes of the late fifties and sixties. In all drawings the houses are shown in dark outline and the garages are shaded. Arrows represent ways through to the path side. The Table VIII. 5 gives statistical and dimensional data of all plans.

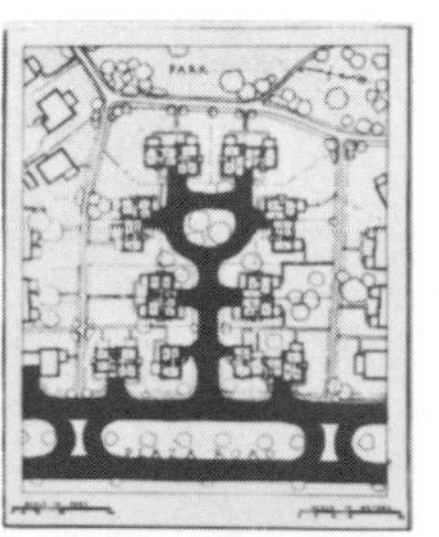

Fig. 8.23. An original "Radburn" cul-de-sac end (1928).

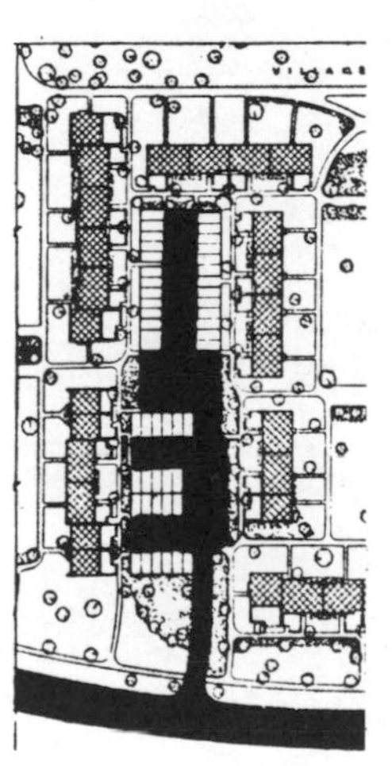

Fig. 8.24. A Baldwin Hills car port, with 3 car-spaces per dwelling (1941) for comparison.

VIII. 6.4.1. *Pioneer schemes of the early fifties*

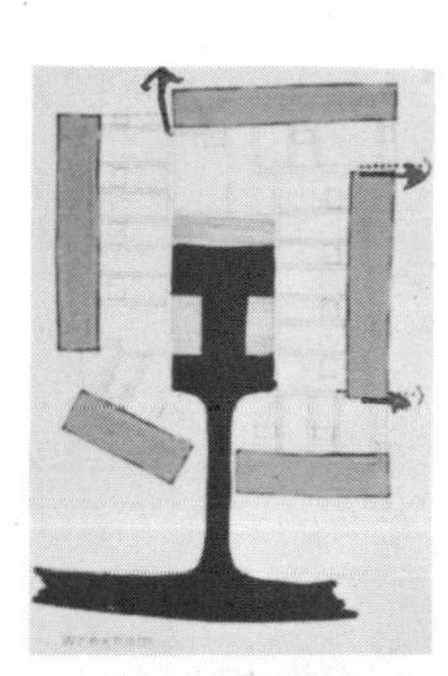

Fig. 8.25. Wrexham. This shows a very small garage court. For reasons found in research[(i)] this scheme is a social failure largely determined by post war restrictions. (See also pp. 325, 326.)

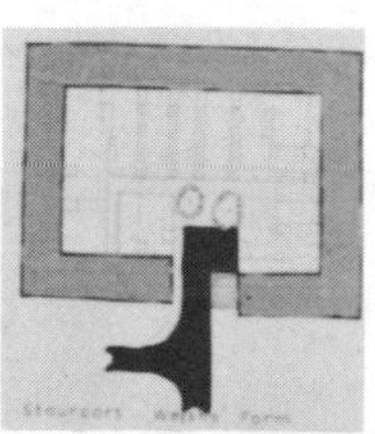

Fig. 8.26. Stourport, Welsh's Farm. A minimum cul-de-sac. The small area enclosed by the houses (90 ft by 150 ft) makes for noise. A footpath enters the cul-de-sac to give access to garden gates. Fruit trees planted at outset may reduce noise and give more privacy. No way through from cul-de-sac to pathside. (G. Stephenson.)

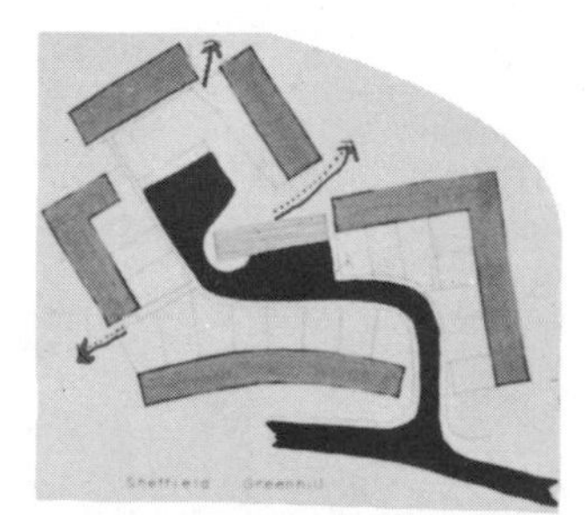

Fig. 8.27. Sheffield, Greenhill. A very free plan, but paths on both sides of roadway make this example more akin to the cul-de-sac type than to the garage court. Post and wire fences give no privacy. (See also p. 326.) (J. L. Womersley.)

Fig. 8.28. Aerial view from Towerblock onto cul-de-sac at Greenhill. (See also p. 326)

(i) Ritter P. Thesis, unpublished, Nottingham University 1957.

VIII. 6.4.2. *The mid-fifties*

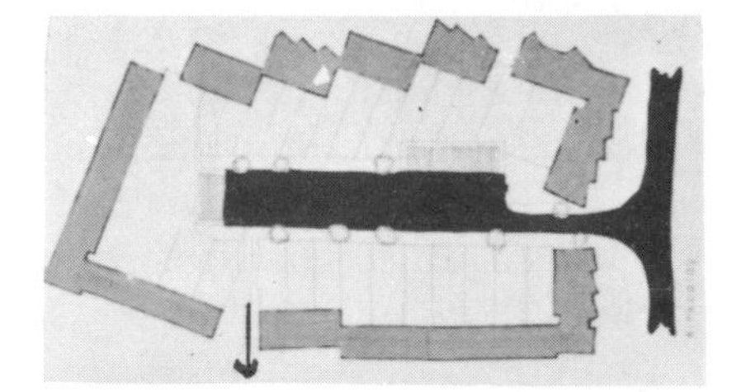

Fig. 8.29. Kirkcaldy. This is a very spacious lay-out (average dimensions between houses 170 ft by 300 ft), but garage court character is diluted by the use of the wide flanking paths.

Fig. 8.30. Aerial view of Willenhall Wood I.

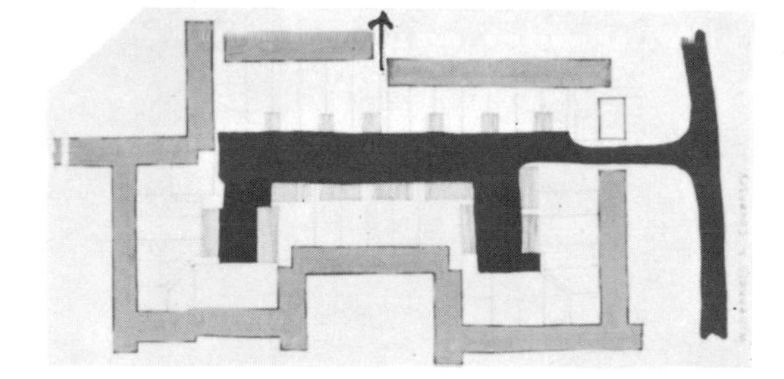

Fig. 8.31 Coventry, Willenhall Wood I. With a higher garage provision, the garages are mostly linked with the houses and screen backs and gardens. Uniform treatment of floor. Long stretches of garage doors give dull effect. Some fences kept low for cheapness. (See also pp. 235 and 278.)

On the road side, and in the garage courts, one surface to cover the whole breadth, robust bollards, anti-fog lighting etc. must be seen as a background to large shiny car-bodies of many different kinds and colours. Where garage courts are enclosed by garage doors or high garden fences to give privacy the same boldness of colouring might extend to them.

VIII. 6.4.3. *Projects of the late fifties and early sixties*

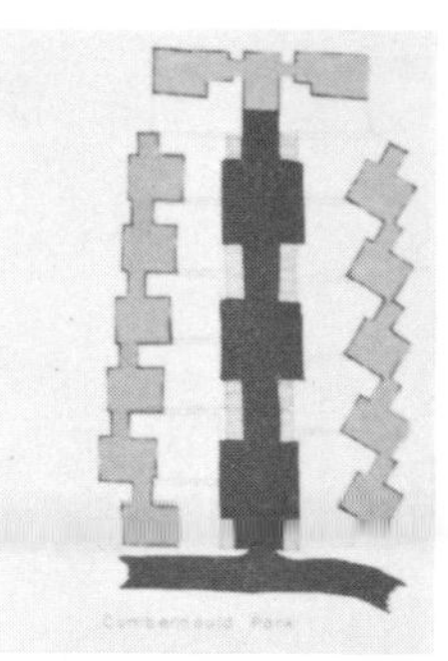

Fig. 8.32. Cumbernauld Park. Quite the best garage courts in the country, with complete screening. Note lay-bys for service deliveries between garage groups. (Hugh Wilson.)

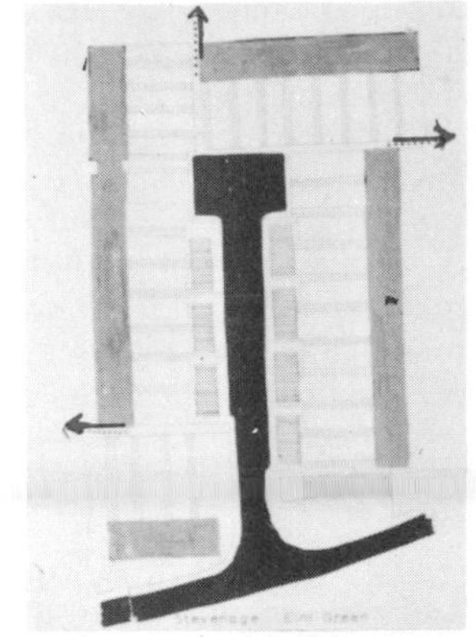

Fig. 8.35. Stevenage, Elm Green. A spacious plan with good screening by garages. Generous turning space. (See also p. 281.) (L. Vincent.)

Fig. 8.33. Basildon, Lee Chapel North. A tight plan with some screening of "backs" by garages. Uniform floor. (A. Davies.)

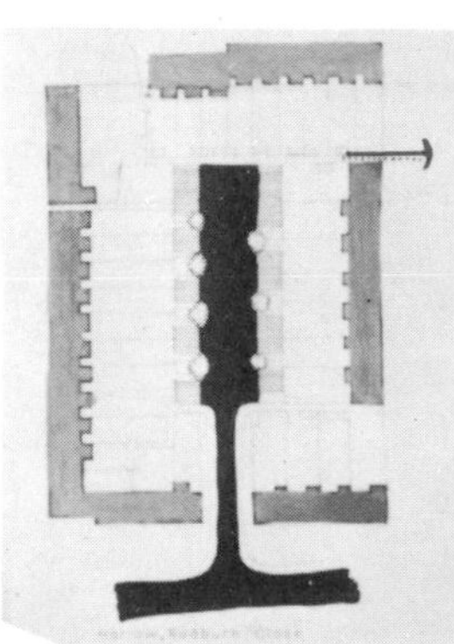

Fig. 8.36. Harlow, Radburn Close. Almost complete screening of backs by garages with trees between to soften harshness of lines. (F. Gibberd.)

Fig. 8.34. Aerial view of Radburn Close, Harlow. August 1962.

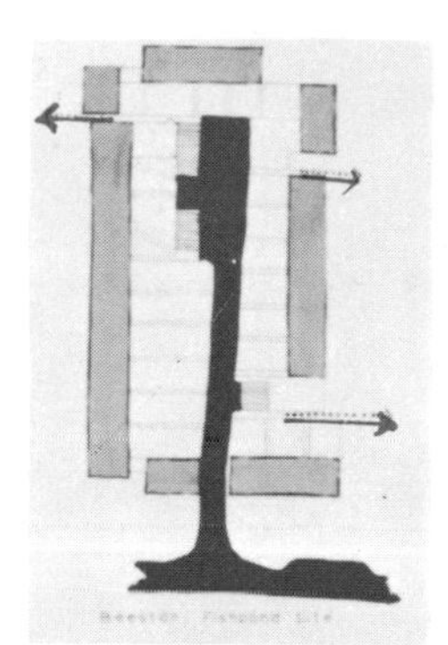

Fig. 8.37. Beeston Fishpond. Minimum space and low garage provision. (H. Chadwick.)

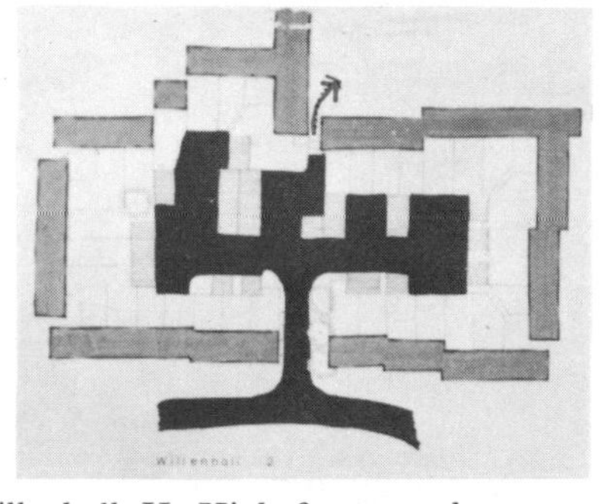

Fig. 8.38. Coventry, Willenhall II. High fences and garages between them give complete privacy. Garages grouped to break up area and avoid monotony. Gable facing on to road entrance improves the sky line and the architectural detailing is generally of a high standard. (See p. 235.) (A. Ling.)

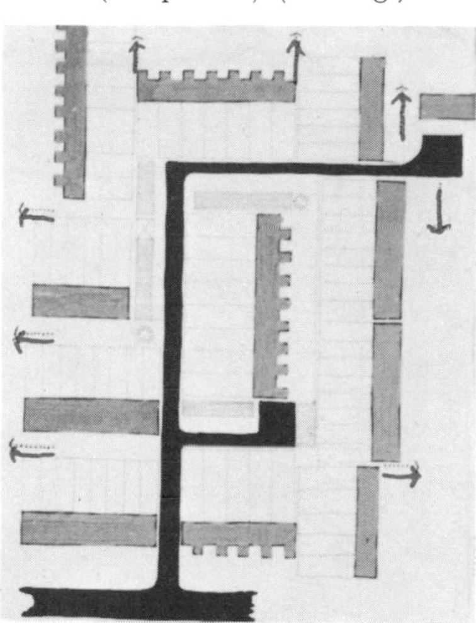

Fig. 8.39. L.C.C. Huntingdon (see p. 265). The complex culs-de-sac lead to batteries of garages not associated with the houses, an arrangement which is common where the garage provision is low, but not where it is as high as this (70%). Some of the houses (e.g. those in the centre, top) do not have direct access to the cul-de-sac, but are reached along footpaths.

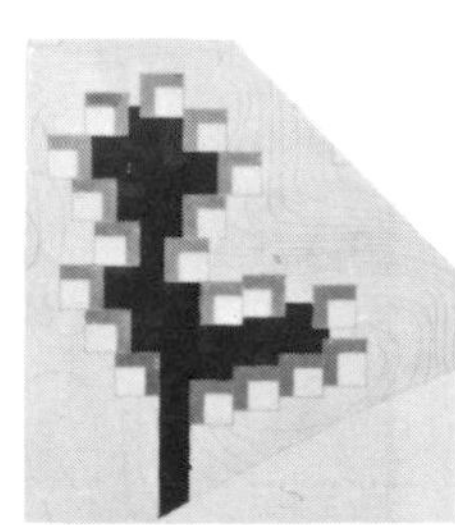

Fig. 8.40. Elsinore, Denmark. Each patio house has a garage romantic cluster. (J. Utzon.)

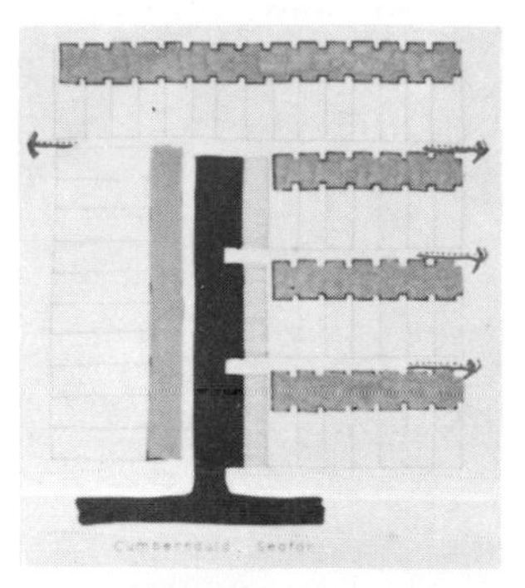

Fig. 8.41. Cumbernauld, Seafar. Maximum density without corner dwellings, grouped garages. (Hugh Wilson.)

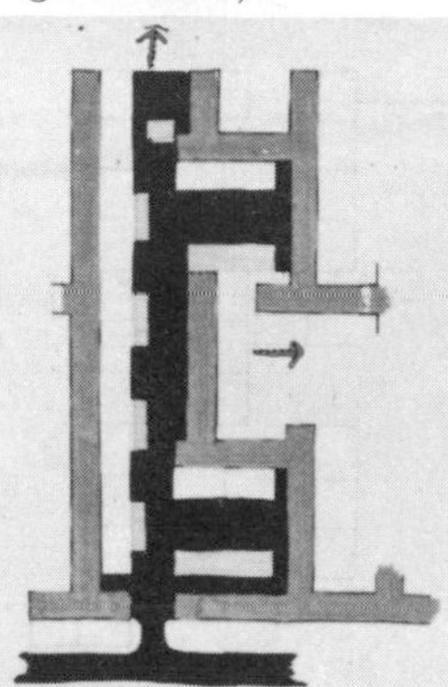

Fig. 8.42. Cumbernauld, Carbrain. A complex garage court used in conjunction with very small gardens. Distance of less than 50 ft between houses made possible by fact that all windows face one way (see discussion under house plans, page 232). Wide frontages with individual garages sited parallel with the garden fronts in some areas. (Hugh Wilson.)

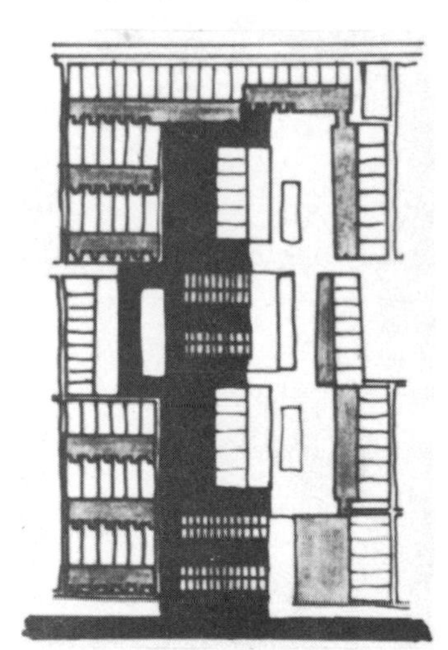

Fig. 8.43. Hook, 100% car provision plus space for parking (from L.C.C. book on Hook).

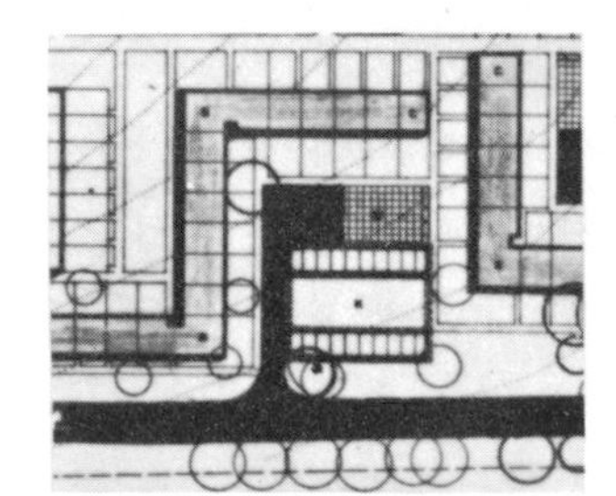

Fig. 8.44. Leicester. A garage court within the cul-de-sac, and play space behind. (K. Smigielski.)

Fig. 8.45. Aerial view of Carbrain, Cumbernauld under construction. (See also p. 116.)

VIII. 6.4.4. *Dimensions of service culs-de-sac*

TABLE VIII.5. *Dimensions of service cul-de-sac*

	Density: dwellings per acre of whole scheme	Percentage garages	Area at back (minimum width in brackets) (ft)	Average depth of garden (back) (ft)	Width of road in cul-de-sac (ft)	Maximum	Depth of cul-de-sac (ft), example shown	Width of entry (ft)	Width of ring road (ft)
Wrexham	21·5 (54)	10	200×140 (70) (61×46 (21))	40 (12)	13 (4)	600 (183)	200 (61)	13 (4)	16 (4·9)
Stourport	20·3 (51)	14	90 × 150 (80) (27 × 46 (24))	40 (12)	Hardly enters	350 (107)	120 (37)	13 (4)	18 (5·5)
Northampton	18·1 (45)	10	110 × 180 (34 × 55)	35 (10·7)	13 (4)	320 (98)	225 (225)	13 (4)	22 (6·7)
Sheffield Greenhill	20·0 (50)	11	100 × 200 (80) (30 × 61 (24))	35 (10·7)	13 (4)	400 (122)	300 (91)	13 (4)	16 (4·9)
Kirkcaldy	Low	10	170 × 300 (155) (52 × 91 (47))	55 (16·8)	40 (12)	300 (91)	300 (91)	13 (4)	18 (5·5)
Coventry, Willenhall	20·0 (50)	50–80	150 × 400 (125) (46 × 122 (38))	35–50 (10·7–15)	35 (10·7)	600 (183)	480 (146)	13 (4)	22 (6·7)
Coventry, Willenhall II	High	66	—	35 (10·7)	35 (10·7)	500 (152)	300 (91)	13 (4)	22 (6·7)
Cumbernauld Park	Low	100	100 × 250 (30 × 76)	30 (9)	20–50 (6–15)	250 (76)	250 (76)	13 (4)	18 (5·5)
Basildon, Lee Chapel North	17·0 (42)	35	150 × 200 (80) (46 × 70 (24))	40 (12)	40 (12)	300 (91)	300 (91)	13 (4)	18 (5·5)
Stevenage, Elm Green	12·7 (32)	73	170 × 300 (170) (52 × 91 (52))	50 (15)	40 (12)	350 (107)	300 (91)	13 (4)	18 (5·5)
Harlow, Radburn Close	15·0 (37)	50	175 × 250 (150) (53 × 76 (46))	45 (13·7)	40 (12)	300 (91)	250 (76)	13 (4)	22 (6·7)
Cumbernauld, Carbrain	23·0 (57)	100	50 × 350 (15 × 107)	20 (6·0)	18–35 (5·5–10·7)	1050 (302)	350 (107)	18 (5·5)	18 (5·5)
Beeston, Fishpond	15·6 (39)	24	100 × 250 (30 × 76)	35 (10·7)	16–30 (4·9–9)	300 (91)	300 (91)	16 (4·9)	18 (5·5)
Huntingdon L.C.C.	14·8 (37)	70	Footpath access	50 (15)	18–25 (5·5–7·6)	850 (259)	640 (195)	18 (5·5)	25 (7·6)

VIII. 6.4.5. *Connections between feeder road and path system.* No research has been done to date on culs-de-sac without connection to the path system except through the house or garden itself. This is a rigorous way of enforcing traffic segregation and is, theoretically, reasonable. We need careful observation of such examples, (Stourpost p. 243, Basildon, p. 246; Highworth by Eric Cole and Partners) to note the factors important for success.

The Ministry's plan in Bulletin No. 6. (p. 250), suggests gates to keep small children out of the culs-de-sac. Once that area has all the life in it, as that plan suggests it is likely to need more than gates to keep children out.

Links between the culs-de-sac and the paths must not be opposite. If they are, they will be used for dangerous short-cuts. The only possible exception to this rule is where a passage leads along the far end of a garage court or cul-de-sac. There is no objection to this provided that the path is railed off from cars and access gates provided.

Fig. 8.48. At Carbrain with its secondary culs-de-sac system, a path parallel to the primary footpath crosses the former demonstrating the dilemma of the system. Such crossings should be very clearly marked to avoid danger. (See also p. 258).

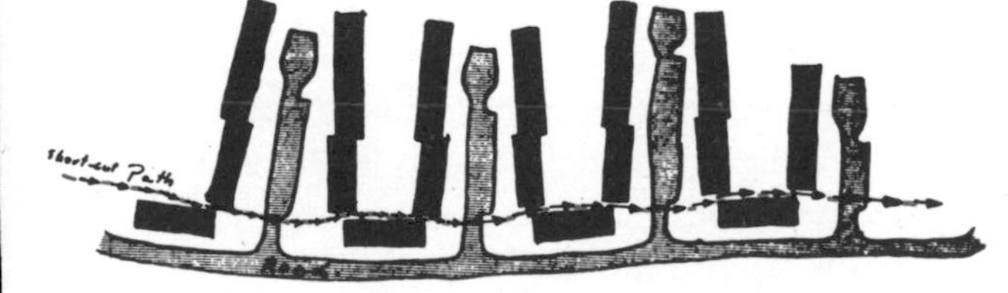

Fig. 8.46. Plan of Racksta showing dangerous short-cut.

Fig. 8.47. View through dangerous short-cut, cars beyond, Racksta.

Fig. 8.49. The narrow way through to the service cul-de-sac, not leading to anything on the other side, purely a local convenience as it should be. Hässelby Strand.

VIII. 6.5. *Mixing Pedestrians and Vehicles in Culs-de-sac*

There is another school of thought on culs-de-sac planning in Radburn type lay-out.

Its origins can be traced to dissatisfaction with the untidiness of early Radburn prototype schemes in Britain. The aim of this approach is to create a better visual impact. Examples to date, seen on paper only, are the Harlow Competition, by Culpin, the *Design Bulletin No. 6* of the Ministry of Housing, and some municipal schemes.

"*Vehicular access (V) is to one side of the houses, pedestrian access to the other. As well as the cars, most tradesmen and visitors, and the family itself, would usually approach the house from the road side: the pedestrian approach, linking to the private gardens, would be used mainly by children or adults when not using cars. The pedestrian approach gives safe access to children's play-spaces (CP) and, if connected up in a larger layout, to schools and shops as well. There is the advantage in this layout of a direct link between the private gardens and the children's play spaces which are sited somewhat apart from the dwellings to minimise nuisance from noise. The pedestrian way could be either open to the public as a whole or, in certain conditions of layout, it could be private to each group of houses with gates (G) provided to increase children's safety.*"

"*Storage for cars, bicycles, and motor-bikes is provided on the vehicular access side, and for garden equipment, toys and tricycles on the other side.*"

"*The ground surface treatment on the road side would need careful and imaginative detailing. On both sides of the houses the layout shows provision for the planting of forest trees, and for the screening of private gardens.*"(i)

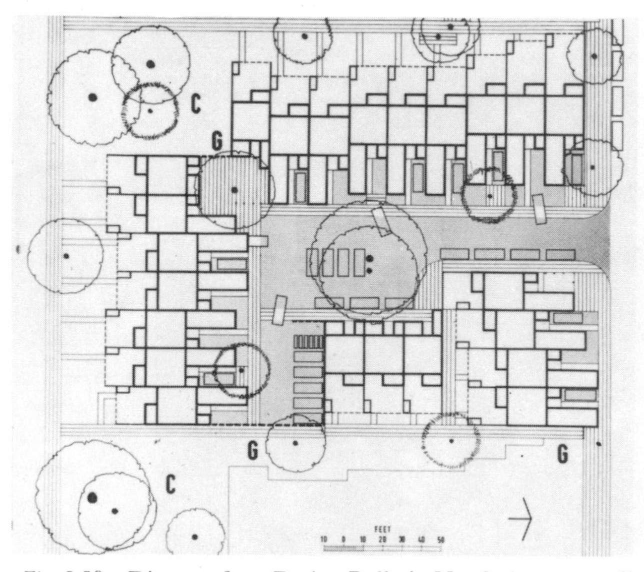

Fig. 8.50a. Diagram from Design Bulletin No. 6. An extraordinary mixture of cars and pedestrians.

(i) Min. of Housing & Local Govt., Space in the Home, *Design Bulletin No. 6*, H.M.S.O., London, 1963.

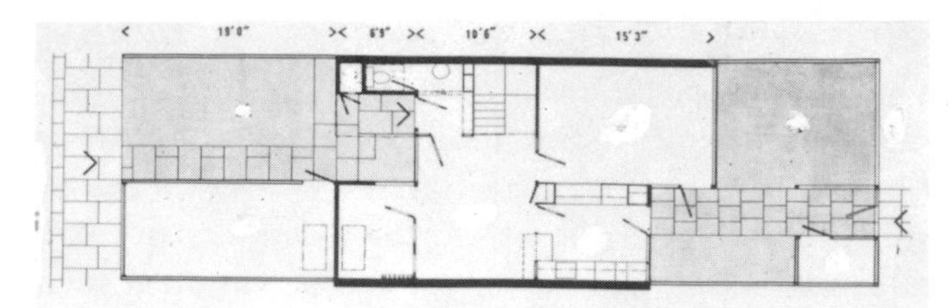

Fig. 8.50b. House plan.

The criteria of such designs are:

(a) the main access ("front") door, is on the culs-de-sac or car side.

(b) the cul-de-sac therefore has specially to accommodate pedestrians and, like the old fashioned road, has pavements each side. At Baldwin Hills, it is interesting to note the paths on the car port side run behind the banks of garages, thus providing virtually a second pedestrian way leading to the cars.

(c) the attractiveness of the cul-de-sac is stressed, hard-surface play areas are provided in very close contact with the road surface.
The Ministry Bulletin says "*the ground surface treatment on the road side would need careful and imaginative detailing*". (Why more than the path side?)

(d) the private garden is often on the path side and individual garages open direct onto pavements.

The Ministry Recommendation, as in Fig. 8.50a seem contrary to established principles:

(a) To mix traffic deliberately is dangerous.

(b) To provide, as shown on the cul-de-sac side, attractive hard surface is likely to bring with it:

(A) children, damaging cars and endangering their lives;

(B) the completion of this side of the lay-out at the expense of the other; in practice, and with very limited funds, this is a very real danger;

(C) a consequent lack of use of the path side, in spite of the position of toy and cycle storage (easily transferred to garage) and of private gardens;

(D) this being re-inforced by the drawing off of much of the vitality and everyday life from the path system on to cul-de-sac pavements;

(E) the use of funds on the car side to the exclusion of the paths. (Note Ministry stress quoted above.)

(c) Swedish research in Vällingby and Gothenburg has clearly shown that main entrances to dwellings should face towards the path side, or children will not play there. This applies particularly where the car side is as fascinating and with as many hiding places and interesting things as shown in the Ministry Bulletin.

(d) The very necessary orientation of a path-system as a whole, to provide shortest routes to shops, school or bus is, to a very large degree, jettisoned.

(e) The very essence of traffic segregated design is to let car requirements, scale and function determine the look of one side and the pedestrian scale and human landscape the other.

(f) Specifically the Ministry diagram shows blatant dangers in garage doors in line with pavements, parking along pavements.

All this arises out of a lack of differentiation of the functions of man and motor. It is certainly desirable that experiments should be made. As this idea seems to contradict general principles of traffic segregation in residential areas as established by research it is vital to make its adaption deliberately and carefully and study it critically. Any defects will increase as car ownership rises.[i]

Fig. 8.51. Perspective of Culpin's prizewinning scheme for Harlow. A car port is given pride of place. This, together with the dangerous blind corners, similar to the Ministry proposals, on the extreme left, illustrates the confusion in this kind of planning for "traffic segregation". It is dangerous.

Fig. 8.52. A play area next to a cul-de-sac and blind corner separated only by bollards: a dangerous design for a provincial city, to be changed on second thoughts by the planners.

(i) The author was consulted on a whole report to be prepared by the Ministry on the Radburn Idea. This promises to take more careful note of past research.

Fig. 8.53. Photographed on the path system, Cumbernauld, February 1963.

VIII. 7. Superblock Design

VIII. 7.1. *Basic Factors*

Design of a superblock is determined by the interaction of a number of factors:

position and number of centres or attractions, goals for the path network;
size of site; slope of site;
character—urban or rural core;
existing road capacity;
convenient walking distances to future public transport;
contours;
need for underpasses or bridge in connection with, or avoiding rails, main roads, rivers, etc.;
density;
dwelling types.

The following key to the four diagrams explains the basic possibilities in superblock design.

A is the shopping centre and public transport.
B schools
C open country
D corner shop, or minor social attraction.
E employment, industry, etc.
x the distance apart of culs-de-sac.
y length of culs-de-sac.
x times *y* the area served by a cul-de-sac.
z the total width of the superblock.
alpha α the angle between ring and feeder cul-de-sac.
beta β the angle between branch path and main path.
a, *b*, *c*, *d*, the respective sides of the cul-de-sac.

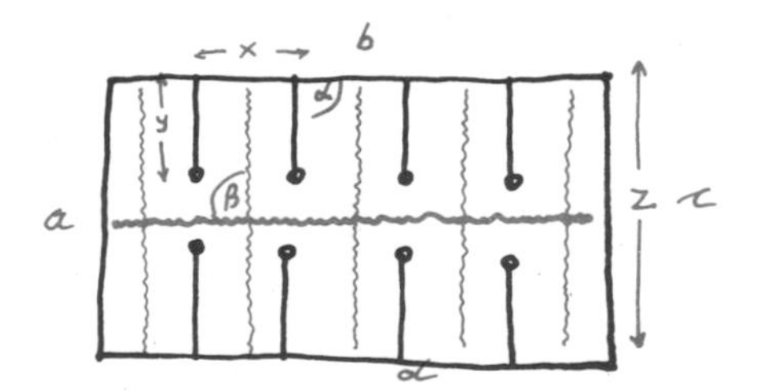

Fig. 8.54. Key diagram.

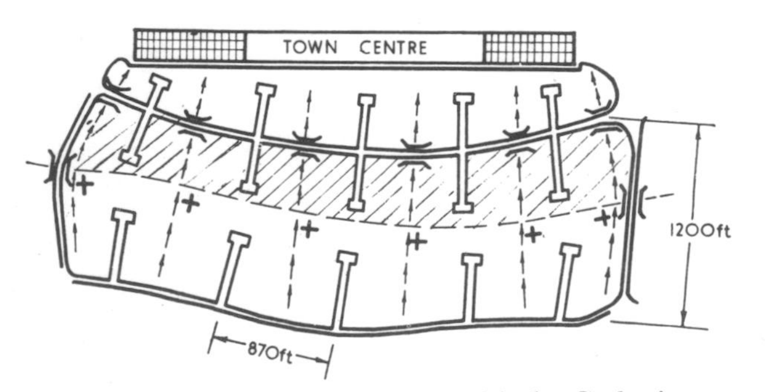

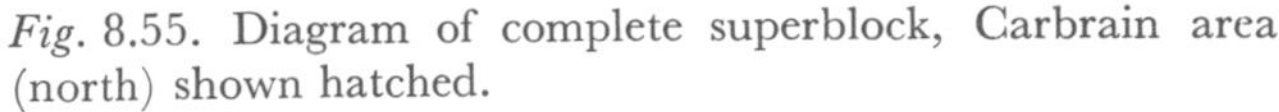

Fig. 8.55. Diagram of complete superblock, Carbrain area (north) shown hatched.

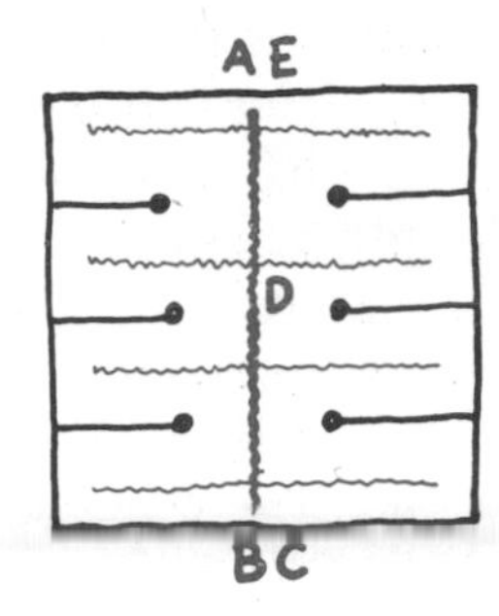

Fig. 8.56a. In this case shopping and employment are one goal at one end, and schools and open space another goal at the other end. Paths are therefore logical at right angles to optimize the convenience for all pedestrian journeys. Minor attractions give weight to the axis, where it may be weak.

Fig. 8.56b. In this case public transport, shops, school and employment are in one direction and only open space in the other. Here the urgent journeys are all in one direction and only leisure walking in the other. Thus directing footpaths towards the desire line is justified. The detour of roads is of no consequence to the driver. However, where the roads join the collector or ring road, there ought to be at least an angle pointing in the preferred direction of drivers.

In this case *c* may not be a road at all, so that there are two major culs-de-sac with minor ones and the superblock connects directly with the open space.

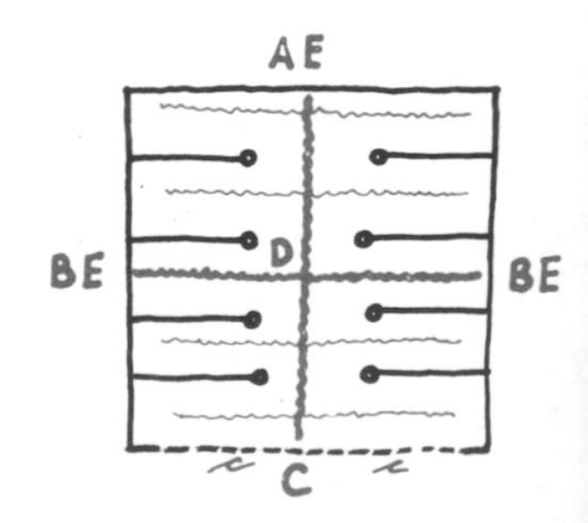

Fig. 8.56c. In this case there are three major and one minor goal. Shopping, transport and employment are one, schools and employment two others and open space the fourth. Such is the case in Cumbernauld where schools are used in a mixed fashion from all over the town and reached via a main ring route. This is nearly as important as the paths to shopping and employment. Where the two routes meet is an obvious place for minor gathering, corner shop, etc.

Whether *c* is a road or not depends on site, and in Cumbernauld it is not, and many such units form the 150 acre (60·7 ha) superblock at Carbrain.

Footpaths are at an optimum of convenience at right angles on the side of the shopping. Further out they might be directional with regard to desire lines.

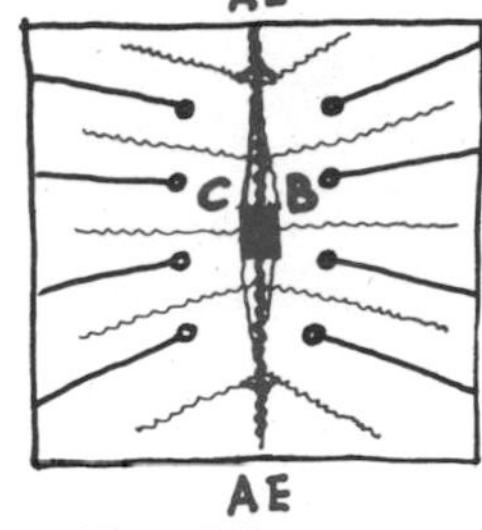

Fig. 8.56d. In this example two goals, with public transport employment and shopping, occur and this would be the case in a long superblock, perhaps between two radials, leading into a town having two centres, each with public transport, which would be essential. In this case the heart of the scheme might well hold the schools, the open space and other minor attractions might lie between these and the centres, at the junction of minor and major paths. The latter might weave in and out of urban stretches into openness. The paths may be directional to the centre or to both ends, or neutral depending on contours etc. Roads would surround such a superblock.

If in such a long superblock the "centre" were in the middle, the attraction of a bigger "centre" via public transport at the extremities, would starve the central shops of customers. Where children walk to school there is not (here in the "centre") the same draw on mothers to fetch from the school and so shop in that direction.

VIII. 7.2. *Application of Superblock in Comprehensive Redevelopment*

Given the narrow wedge, normally found between the radials in old cities near the centre, for comprehensive development, the core should be developed in a dense urban manner with the main walkways leading directly to centre. Cushioning this from the radials might be green strips. Existing parks or grounds can be incorporated usefully.

Where the wedge becomes too wide to allow easy walking distances, the path spine might turn its direction towards sub-centres, giving public transport to the centre, yet connecting its path system with the other superblocks, nearer the centre for pedestrian connection.

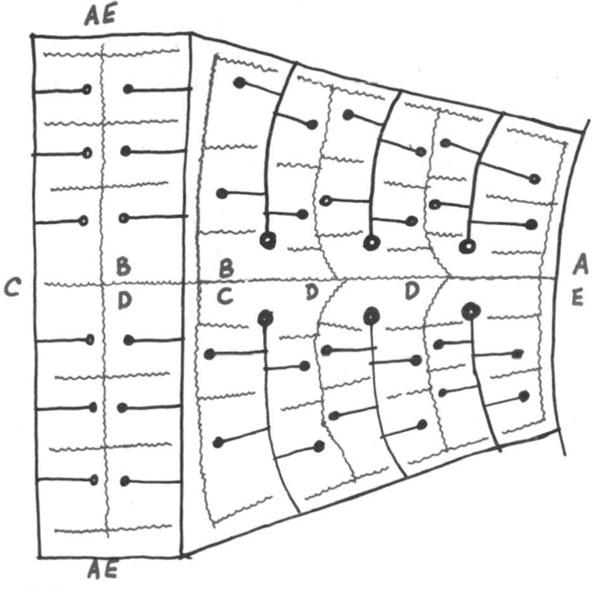

Fig. 8.57.

VIII. 8. Underpasses and Bridges

VIII. 8.1 *Position and Importance*

In one Scandinavian city the administrative strategy, a new one but essential to the world of planning, has been perfected. The planners indicate the path system first. The road engineers subsequently put in the roads. There is an embargo on crossing paths with roads so the road engineers have to plan bridges or underpasses. This is a comparatively easy matter because it is taken for granted that road construction requires large funds, whereas to demand an underpass for a footpath may be met with scepticism and doubt. A similar problem will exist in most countries for years.

If we look at a plan and follow railway lines we expect to see bridges and tunnels where they cross pedestrian traffic. The same attitude must be cultivated for car traffic and pedestrian traffic.

However, the position of major elements of an existing environment, including traffic arteries, will in turn affect the planning of the path system and so the form of a superblock.

Underpasses should be placed with a view to making use of existing contours and directing the pedestrian traffic. The gradient should be as gentle as possible so that pram pushers are not put off by it. Powered vehicles should do the climbing where possible.

The use of underpasses as covered play spaces should be borne in mind. Ramps and steps leading to them should be designed for this. Precast main drain components may be suitable and adequate in diameter to form many interesting underpass shapes, halving construction cost.(i)

Much research remains to be done on this problem; very little has been tried. A recently visited underpass at Cumbernauld,(ii) still under construction, used pipes, but threatens with too steep a gradient on one side to make pram pushing impracticable. This could be a serious point in a town planned for walking. In another new town I found an engineer's dislike of narrow underpasses because "they are dark and might lead to attacks". This must be exposed as irrational: fluorescent tubes and advertisements could, for example, make them points of special brightness and interest, if required or desired.

(i) Since which time it has been reported from Cumbernauld that *in situ* underpasses have been constructed as cheaply as precast ones. The latter can be moved, however, if this is necessary. In Stockholm the pipe underpass has been abandoned because it is dark and is used as a w.c., and because its narrow aperture is dangerous to skiers. A prefabricated beam system is now in use, and it has a number of armoured plate-glass light fittings.

(ii) May 1961.

Fig. 8.58. Pipe underpass, Stockholm.

Fig. 8.59. Precast beam underpass, Farsta.

Fig. 8.60. Carbrain, Cumbernauld, main radial path leads through terrace to underpass beyond.

Fig. 8.61. Underpass entrance, Carbrain, Cumbernauld.

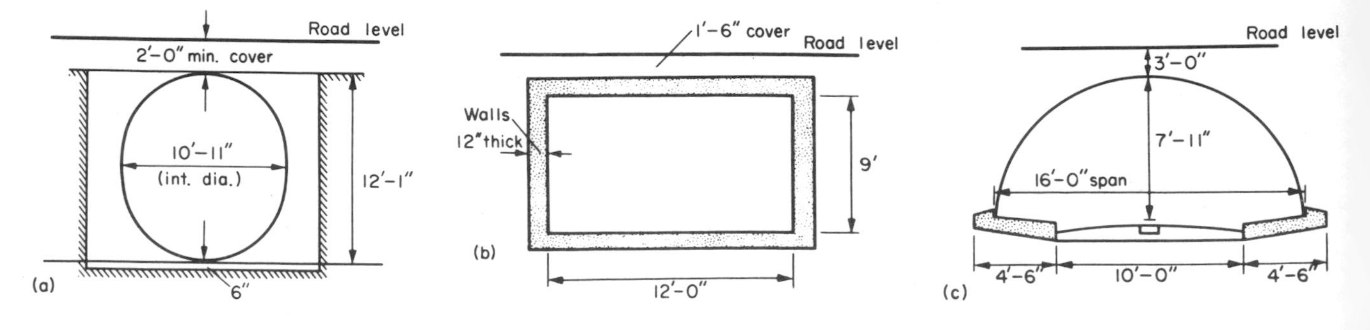

Fig. 8.62. Six underpass types as used at Cumbernauld.

VIII. 8.2. *Design and Cost Alternatives (to be read with Fig. 8.62)*

TABLE VIII.6. *Cumbernauld Development Corporation; pedestrian underpasses.* 1962 *rates based on local findings of* 10% *increase in rates/annum cost per ft run based on cost of underpass divided by roof length*

Type	Minimum cover	Length	Cost of underpass alone
a) Armco multiplate pipe	2 ft 0 in.	Roof length 56 ft 6 in. Base length 91 ft 0 in.	£2680 (1960) £3215 (1962) £57 per ft
b) Reinforced concrete box	1 ft 6 in.	Roof length 59 ft 3 in. Base length 86 ft 6 in.	£2315 (1961) £2546 (1962) £43 per ft
c) Armco multiplate arches	3 ft 0 in.	Roof length 59 ft 0 in. Base length 83 ft 0 in.	£3185 (1962) £54 per ft
d) Bridges		Overall length 38 ft 0 in.	£6000 (1958) £8500 (piled construction) £8400 (1962) £11,900 (1962, piled construction)
e) Armco liner plate	9 ft 6 in.	Roof length 60 ft 0 in.	Tunnel £8590 (1962) Ancillary £2324 £143 per ft
f) R.C. Portal frame	1 ft 0 in.	Length 115 ft 0 in.	£12,833 (1962) £114 per ft*

TABLE VIII.6. cont.

Type	Total cost inclusive of underpass	Comments
a) Armco multiplate pipe	£6000 (June 1962)	Construction time for main Armco pipe. One labour gang of 5 men. Excavating 5 days Assembling 5 days Backfilling 2 days 12 days under normal working conditions
b) Reinforced concrete box	£8220 (June 1962)	Construction time for main structure. One labour gang of 5 men and 2 steel fixers. Excavating 4 days Base slab 6 days Walls 6 days Roof 7 days 23 days under normal working conditions
c) Armco multiplate arches	£3300 (1962)	No steps to bus lay-by constructed at this time.
d) Bridges		
e) Armco liner plate	£11,000 (January 1962)	Carried out under existing trunk road with traffic maintained at all times.
f) R.C. Portal frame		*This price includes railings and head walls.

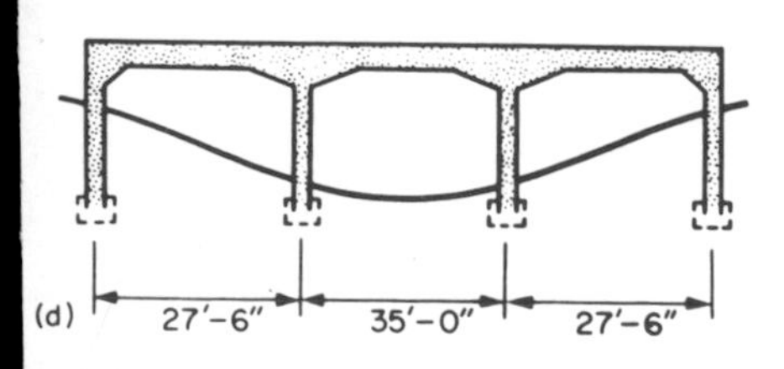

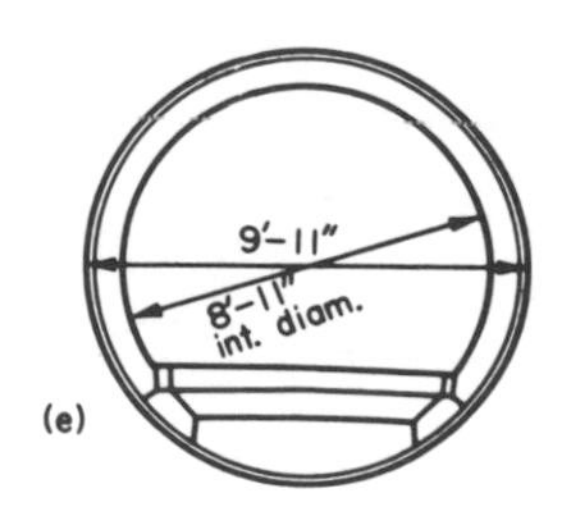

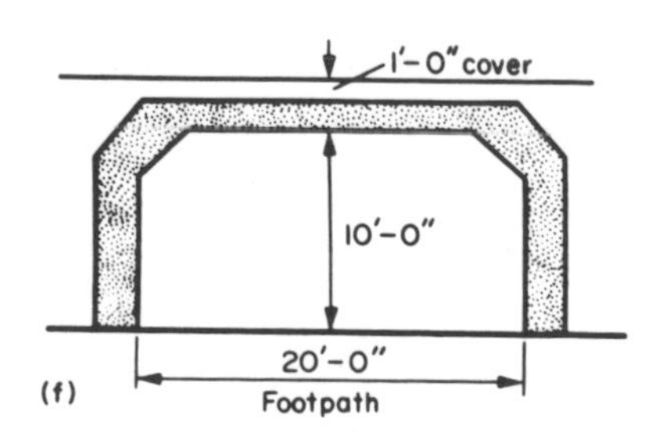

K

VIII. 9. Open Space

Open space may be coupled with school fields or not. It may be around the main path, as in the original Radburn conception, or to the side of the residential development, something that may be particularly appropriate to make the most of the situation in an existing town or to make the most of particularly fine natural features on a virgin site.

Close to town centres in urban renewals it is particularly appropriate to bring density and path into the urban environment and to use existing recreation areas and the areas along the main radial motor routes as green spaces and buffers, properly designed to fulfil their function, not just planted with a few trees. Not barriers to social contact but connecting pedestrians by bridges or underpasses. The scale of the buildings affects the width of any channel for pedestrian movement. High flats cannot enclose the narrow alleys that may be appropriate with two-storey housing. On the other hand, high flats may provide their own intimate deck system looking out on to the broader areas of separation in rural green style.

Fig. 8.64. View inside superblock, or Vällingby centre.(i)

Fig. 8.63. Open space around main path centre of superblock at Racksta, Vällingby, Stockholm.

Fig. 8.65. Eric Lyon's kind of open space.

VIII. 10. Contours

From many points of view, and particularly in terrace housing, to align with the contours has great advantages. Roads along contours are cheaper and paths along contours easier to negotiate.

This will work out in terms of Radburn development into a simple or a complex branching of the culs-de-sac, so that any direction of terraces is possible in relation to ring roads.

A fall in the ground should be used to separate roads, visually and functionally, from gardens or path even if the height differences are moderate, say only 3 ft 6 in (1 m) at each terrace.

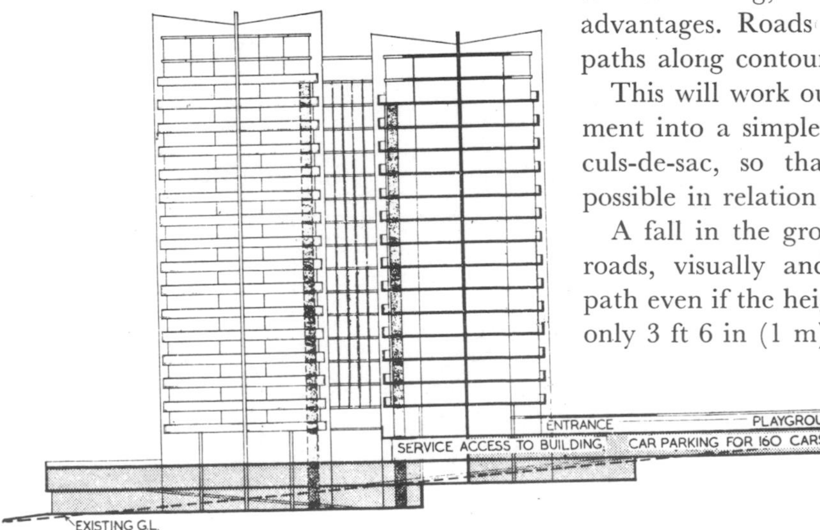

Fig. 8.66. Norfolk Park, Sheffield, use of contours.

(i) See under Examples.

VIII. 11. Street Names and Numbers

One minor incidental problem which arises from Radburn planning is that of choosing a method for naming the streets and paths and numbering the houses. This is an important detail because naming and numbering, if well conceived, can help reinforcc thc Radburn idea and, if badly conceived, can both make nonsense of it and lead to much practical inconvenience. Careful, car-scale and pedestrian-scale signposting is in any case very necessary, as the lay-out is unusual.

The diagrams which follow illustrate four possibilities. Each pattern represents a unit bearing a common name.

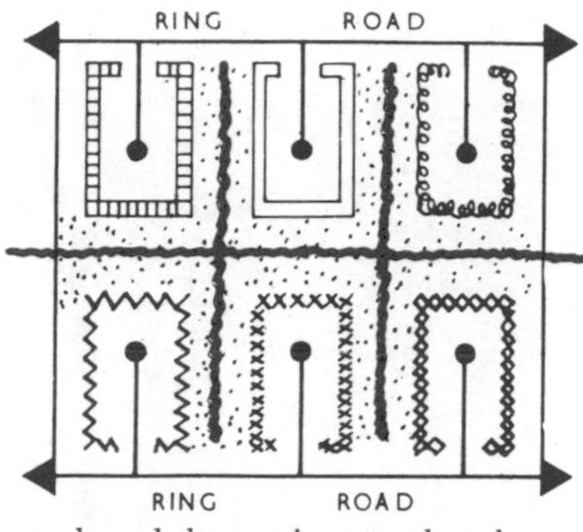

Fig. 8.67. In the first, each cul-de-sac is named and numbered individually. This system is used in Coventry, Willenhall, and at Wrexham. It leads to a lack of identity: path areas carry different names on opposite sides, and the pedestrian side of a square will show four different names.

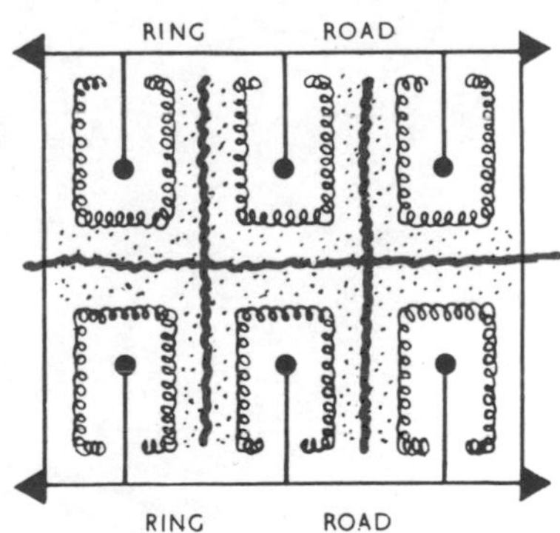

Fig. 8.68. In the second example, the whole of the path area of a superblock is given a name and the numbering is continuous. This gives a lack of identification on both road and path side.

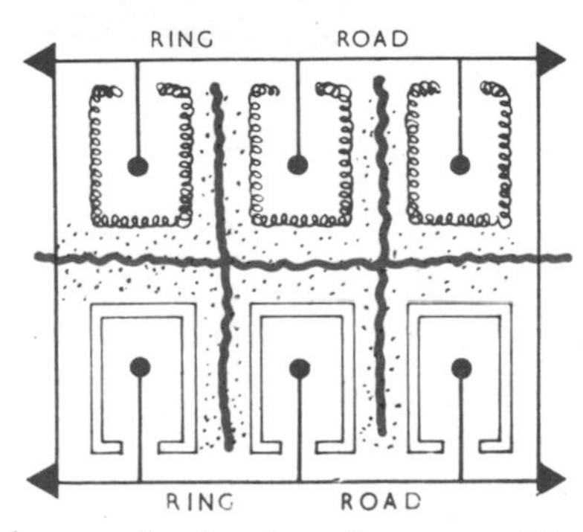

Fig. 8.69. In the third example (used at Stevenage, Elm Green) the ring roads carry the name and the numbering is continuous. This results in the opposite sides of path areas having different names.

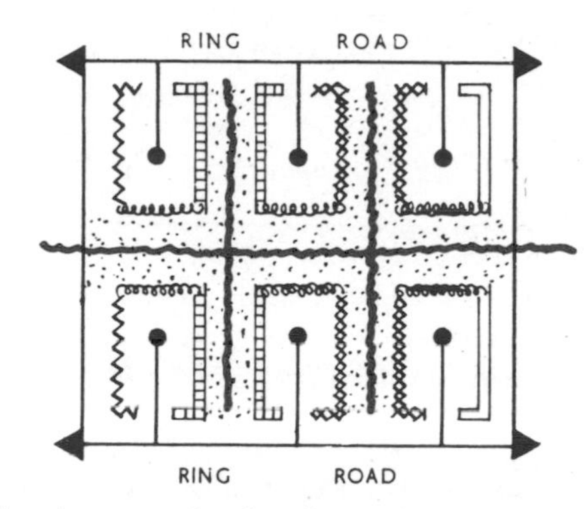

Fig. 8.70. In the fourth example the footpath areas are named individually and numbering follows as in the traditional grid-iron road pattern. This makes deliveries difficult in culs-de-sac where names are different and numbers non-continuous.

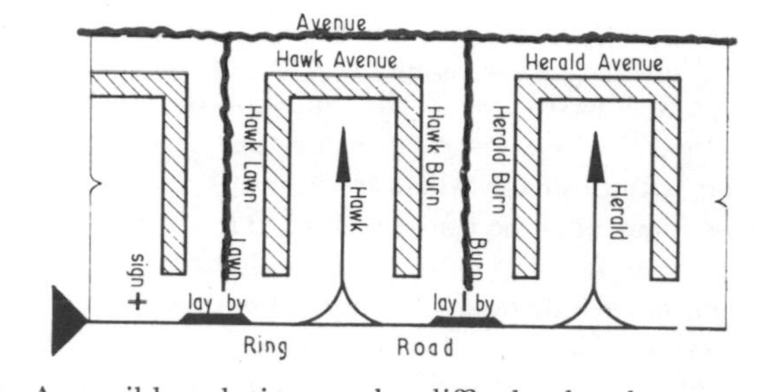

Fig. 8.71. A possible solution to the difficulty has been put forward by Mark Robertson. This involves double-barrelled names. One will apply to the footpath areas and will comprise names associated with walking (in the example "avenue," "burn," "lawn") and the other will apply to the cul-de-sac. House names will be doubled and will show on which walk and on which cul-de-sac they are sited, i.e. No. 6 Hawk Burn, No. 10. Herald Burn.

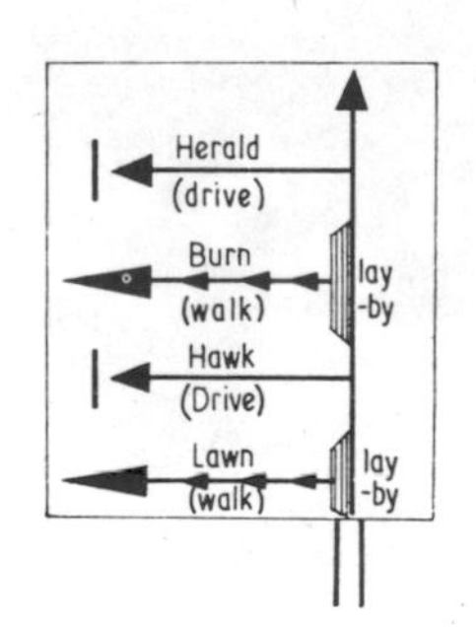

Fig. 8.72. Suggested signpost for ring road.

Fig. 8.73. Road sign used for turning somersaults, at Willenhall Wood I. Signs must not only be very sturdy but also imaginative.

VIII. 12. Examples of Traffic-Segregated Residential Areas

VIII. 12.1. *Cumbernauld, nr. Glasgow*

Carbrain 1, 2, 3, 4.

Chief Architect, Hugh Wilson. Group Architects, D. Lyddon and D. Garside.

Begun in 1961, the scheme is to provide 1038 dwellings, over 60% houses, the rest in maisonettes and flats with 100% garage provision. Density, 100 ppa (250 ppha). Much of the population comes from slums in Glasgow. Part of an immense superblock.

This is a new and important concept of a broad-fronted single aspect, dual access house. These can replace corner flats, and make a shallow plot feasible. The houses all face the sun, answering the sunshine hours requirement of the Scottish Department of Health. There is no overlooking. The two-storey development is the highest density of any post-war development of the kind in Great Britain. The climate and contours have specifically influenced overall and detail design. The very large superblock fits clearly into the segregation pattern of the whole town and is an original, important development of the Radburn Idea. The "man-sized" spaces are fully used for play by very young children and the larger ones seem to lend themselves to modest games of football. The uniform exterior of the dwellings have lost some of the character shown in the design drawings, but the individuality of the play spaces orientates effectively. Further thought is to be given to some unresolved difficulties linked with the service access roads. There is a great deal of reversing. (See also pp. 41, 43, 248, 250, 254.)

Fig. 8.75. One of the many small and imaginatively enlivened toddler play spaces. Even in winter hard surfaces allow concentrated use.

Fig. 8.76. Vertical division between houses shown in the design sketch is omitted in the executed terraces. Service access, maisonettes beyond marking radial main path leading to the Town Centre.

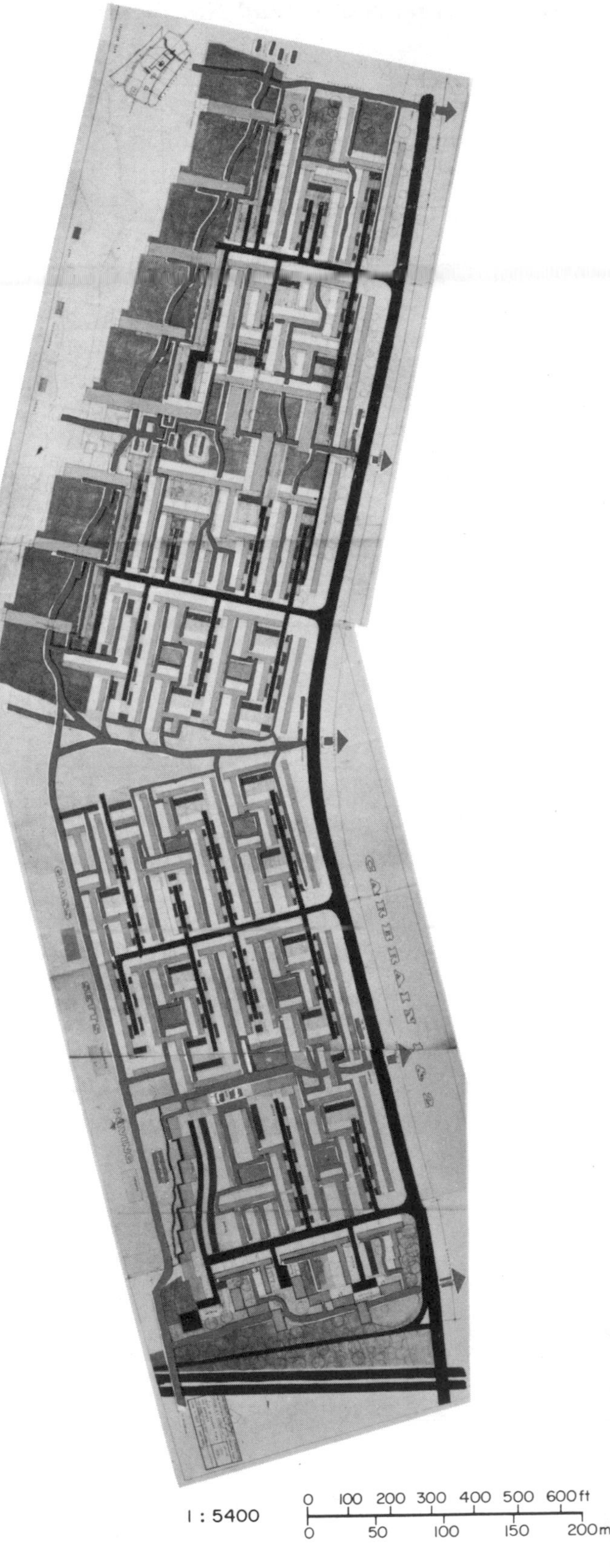

Fig. 8.77. Plan of Carbrain, 1, 2, 3, 4.

KEY
Red —Pedestrian Areas.
Black —Motor Roads.
Dark Grey —Car Storage.
Grey —Dwellings.

Unless otherwise stated this key applies to each plan.

Plans have not been redrawn to retain character and variety. The red colour, denoting pedestrian areas, is therefore shown in a number of ways; either solid red, or edged with red, or main red paths leading through squares or greens, or public buildings and shops about pedestrian areas are solid red or edged in red.

Fig. 8.78. View from top floor of maisonnettes.

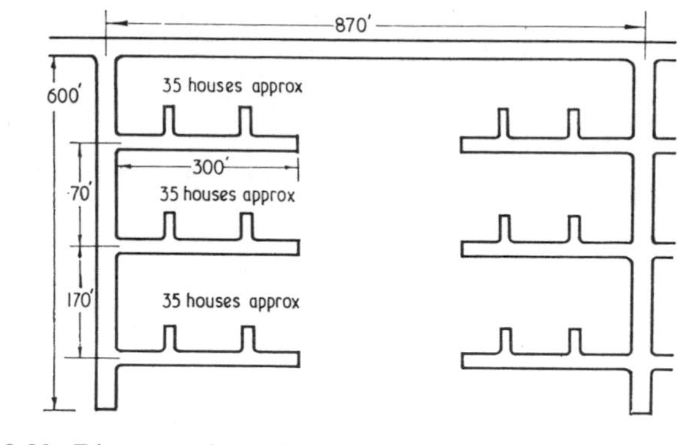

Fig. 8.82. Diagram of cul-de-sac area.

Fig. 8.79. Concrete blocks used imaginatively to give identity to similar spaces

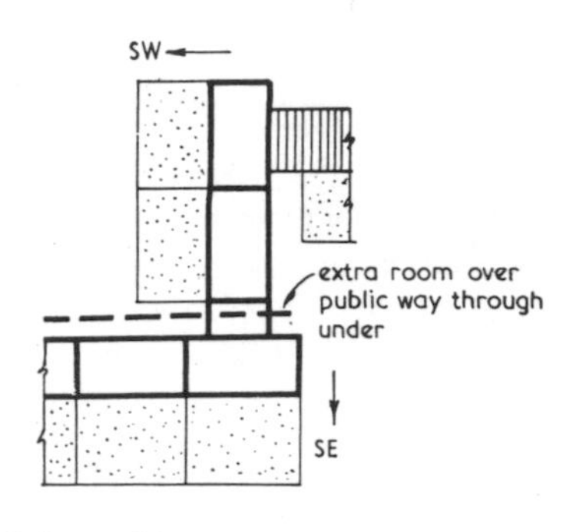

Fig. 8.83. Corner with broad-fronted house.

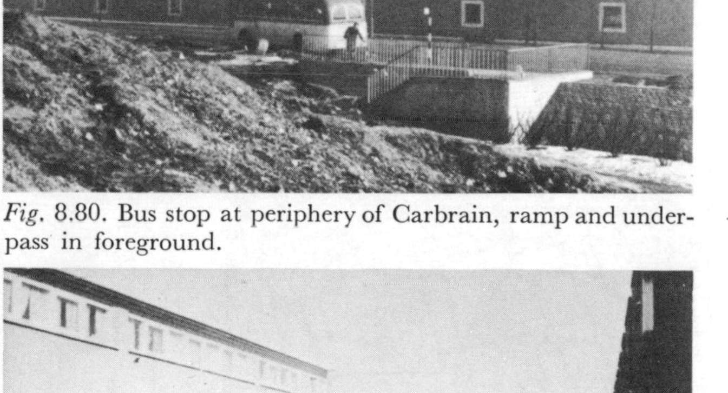

Fig. 8.80. Bus stop at periphery of Carbrain, ramp and underpass in foreground.

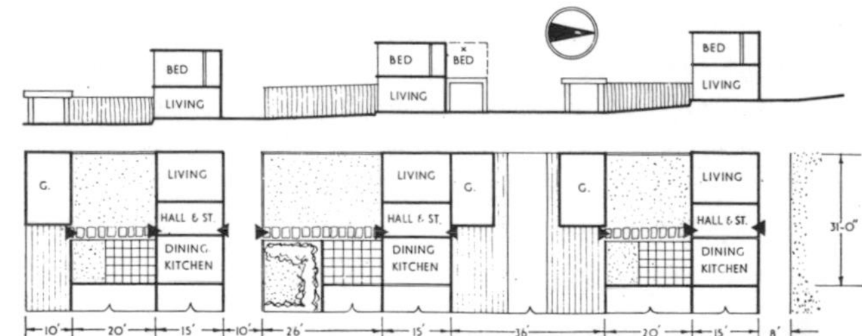

Fig. 8.84. Broad-fronted house arrangements.

Fig. 8.81. Pedestrian way. 35 ft between single aspect houses. Note sunlight (March) on house front. Only working kitchen faces north. Snow on the path (1963).

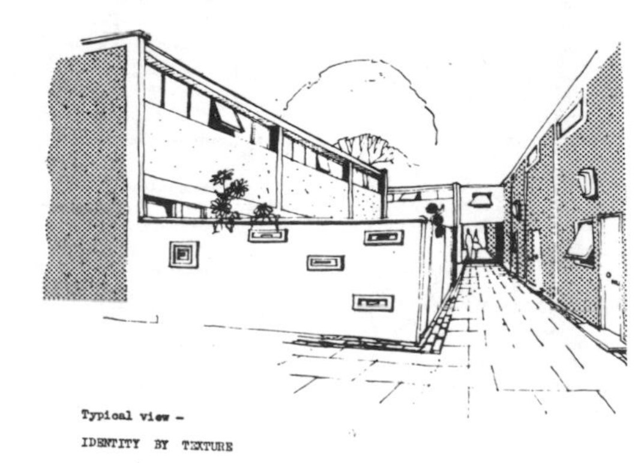

Fig. 8.85. Another aspect.

Reference

Cumbernauld Development Corporation Reports 1–5; Carbrain, Cumbernauld, 1959. See also many other illustrations in the book.

VIII. 12.2. *Baldwin Hills Village, Los Angeles, California*

Architects, R. D. Johnson, Wilson, Merril, Alexander. Consultant, Clarence Stein.

A private enterprise venture and a success in spite of massive opposition from the Federal Housing Agency, Official planners and engineers against the superblock of 80 acres (32 hectares) and 20 acres (8 hectares) park. It provides 627 apartments, mostly in terraces, car storage in carports, three cars per family, currently to be increased (!) by the New England Life Insurance Company, who bought the "village" as an investment in 1949, covering all previous investments with back interest. Completed in 1941. 100% let all the time.

The architects lived there themselves. Exceptional house accommodation, amenities and communal services are provided cheaply. Density is not as low as is thought. Firstly, Clarence Stein told me that the calculations in his book are faulty, decreasing the figures. Secondly, it includes all the village greens which might normally be taken out of net density calculations. Thirdly, if the corners of the meander were filled in with the type of corner block developed since the war in Britain, and other such possibilities were taken into account, then, together with the 300% car park provision (many of which are in garages), the density is in fact high, very high for California, about 20 dwellings per acre 50 (per hectare) net. This is stressed because this particular scheme, architecturally more pleasing to the British architect than any of the other Radburn schemes, was rejected again and again as a prototype because it was asserted it had too low a density for Britain. Our New Towns were meanwhile built with lower densities and less sense and sensibility.

The full impact of Baldwin Hills is felt on approaching in hot weather over the oil-derrick-ridden, parched and lifeless "Hills" themselves. Thoroughly depressed by the harsh and hard surfaces and the speed and inhumanity of the super-colossal Los Angeles road system, the Village is truly an oasis in which mind and body are refreshed. In spite of recent rumours to the contrary reprinted in books, children do play on the greens. Activities are perhaps limited by the immaculate state of the landscaping, in itself an object lesson. The ivy ground cover in the forecourts and garage courts take the "new" ideas for Hook garage court design much further than any European example to date. The scheme is so full of excellent detailed planning that a careful study is of the utmost advantage to planners for man and motor. The tragedy of the disconnected shopping centre, across Sycamore Avenue, which planning law did not allow to be turned into culs-de-sac is written in the many accidents at this "death trap", so named by inhabitants.

Amenities of Baldwin Hills contrasting with surrounding housing leads to protective legislation and rules as seen on not very "welcome" sounding notice boards. However, the clear signposting is an important lesson to be followed in any lay-out on new principles.

Fig. 8.87. Aerial view contrasting Baldwin Hills Village with surrounding traditional housing and bare hillsides.

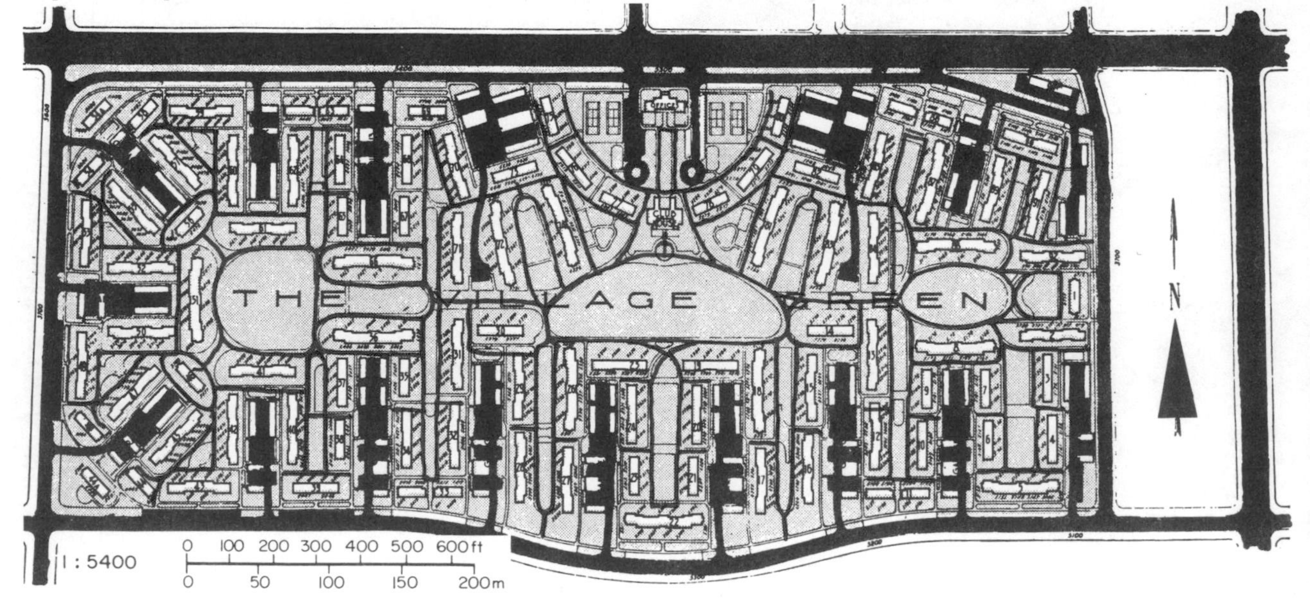

Fig. 8.86. Plan of eighty acre site showing the road which cuts Baldwin Village off from the shopping centre in the vacant area.

Fig. 8.88. Fine landscaping and fine architecture make the most of a flat site.

Fig. 8.91. Garages do not detract from the total composition.

Fig. 8.89. The archway, accentuating the spaciousness of what is beyond, a form found in traditional British housing and much repeated in progressive terrace housing in Britain.

Fig. 8.92. Garage court showing cars and garages on the left, and patio fencing on the right.

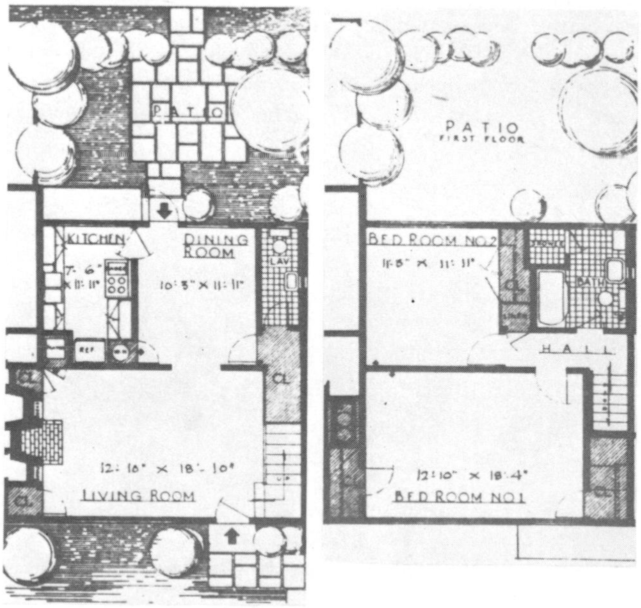

Fig. 8.90. Enclosure of patios facing the garage courts. Ivy covered walls originally built. Serpentine walls added by new owners of Baldwin Hills for upstairs tenants. It gives privacy and protection against vehicular noises and smells which, as seen from the plan, are close by.

PATIO
KITCHEN
DINING ROOM
7'-6" × 11'-11"
10'-5" × 11'-11"
12'-10" × 18'-10"
LIVING ROOM
PATIO
FIRST FLOOR
BED ROOM NO.2
11'-5" × 11'-11"
BATH
HALL
12'-10" × 18'-4"
BED ROOM NO.1

Fig. 8.93. One of the house plans showing access from both sides. See also cul-de-sac detail.

VIII. 12.3. *Caversham, Reading, Berkshire*

Design: Diamond Redfern & Partners. F. Briggs, A. Redfern. Partners in charge. D. Ticker, Job Architect. K. R. Gibson, Group Architect. Planning Consultant, Paul Ritter.

Origin: A private speculative development by Davis Estates Ltd. The site is 140 acres (57 hectares) of superbly wooded parkland, in a valley and on its banks on the outskirts of Reading. The Western Boundary faces largely on to parkland and the eastern one on to the green belt. The development will comprise 1600 dwellings, each with garage, 16 dwellings per acre (40 per hectare) a primary school, twenty shops, pubs, a church and a community centre. Standardization of house construction components and production, and really private and usable outdoor spaces are aimed at. The site is regarded by the planning authority as suitable for dense housing and areas of urban character.

The scheme represents the first very large, suburban, Radburn, private enterprise scheme in Britain. The site lends itself to one large superblock although some of the culs-de-sac are long and emergency road connections have to be provided at their extremities. Pathways will have clear priority at such points and cars will be slowed down by ridges each side (see Fig. 8.20). The north road is fast and is to have limited access. The valley represents a fine path spine and the rather long superblock is crossed, via a bridge, by one road. This, also, leads to the school and is indirect. Shops with parking are placed at the extremities of the valley. This ensures their early construction, their use by those going to public transport stops at the radials leading into Reading, at each end. It also serves those living in areas adjacent. The centre is enlivened by the school, church and community centre and large recreational areas. The sales agents felt that both the Radburn lay-out and the new house types would help sale of the dwellings, a remarkably enlightened point of view in a notoriously conservative profession. Pedestrian access to the roads will be restricted. Corner shops will be provided within the area. It is anticipated that this breakthrough of the Radburn Idea in speculative housing will lead to a general acceptance of the idea in this field. The very real advantages of safety the efficiency for the vehicle and the pleasant environment for man, at the same price, can not be forfeited in competitive markets.

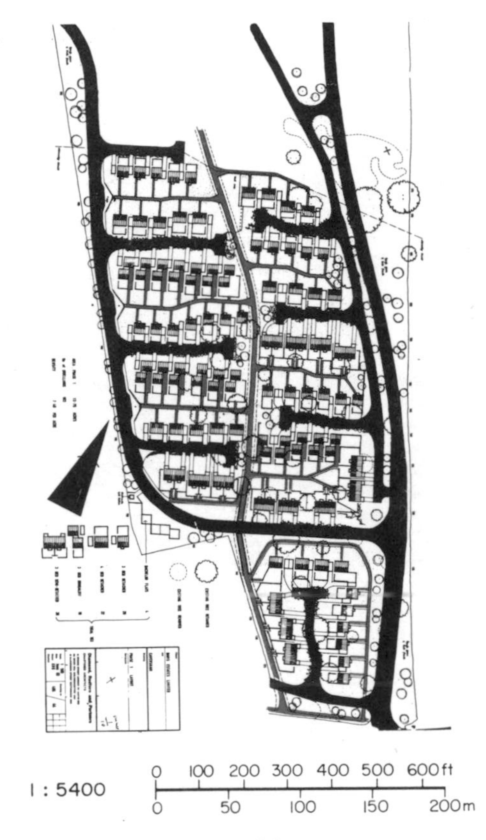

Fig. 8.95. Housing lay-out of first stage.

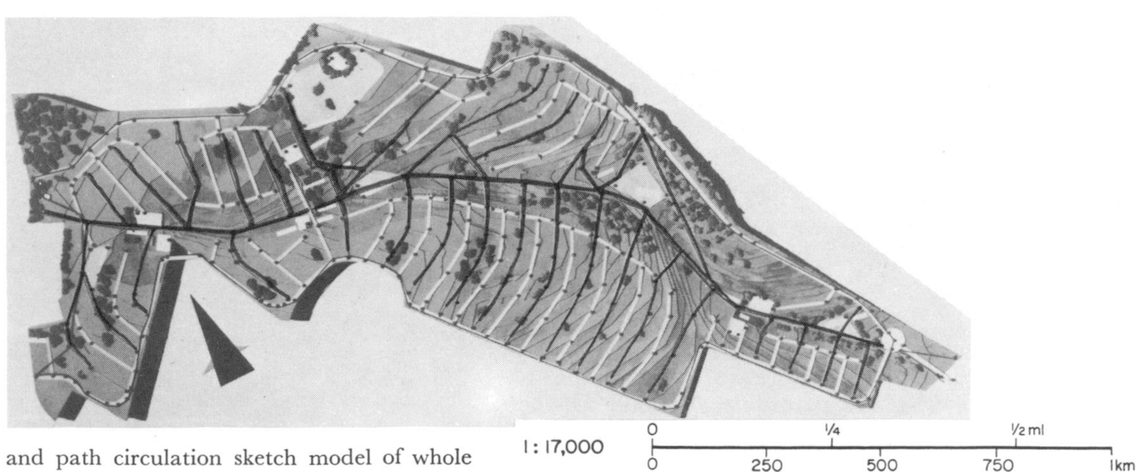

Fig. 8.94. Road and path circulation sketch model of whole area.

Fig. 8.96. High density path area.

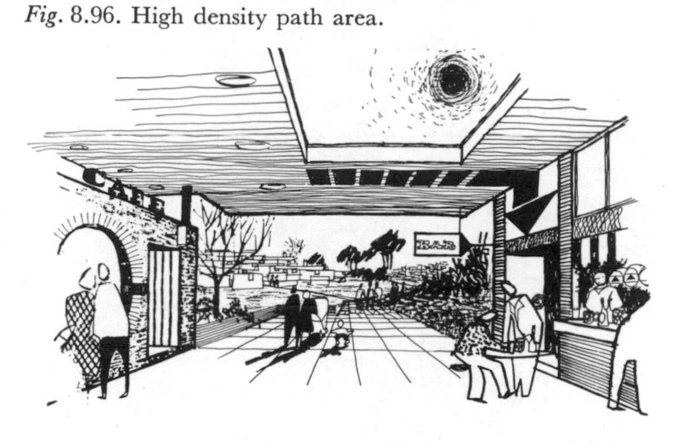

Fig. 8.97. View on to spine path from lower shops.

Fig. 8.98. Underpass, showing main spine path under cross road.

Fig. 8.99. Suggestion for service road treatment.

VIII. 12.4. *Bron-y-Mor, Mid-Wales*

Architect: John Madin.

The building of Bron-y-Mor began in 1962. It is to cover 40 acres (16 hectares) providing 769 dwellings for permanent and seasonal resort use for 2768 persons. There are to be 16 dwellings and 58 gross ppa (40 dwellings and 145 ppha). 30 shops, a motel and an entertainment centre are also to be provided by a miniature pier.

This is a finely planned and thoroughly studied design. 2000 parking places are being provided, 400 below the centre and 400 in a multi-storey park. The character of the pedestrian spaces has been carefully conceived. The motel links up with the footpath system and the roads to the centre are bridged by pedestrians. Bron-y-Mor is an important experiment and the very first traffic-segregated, large lay-out privately financed in Britain.

Fig. 8.100. The pedestrian environment.

Fig. 8.101. Play pool.

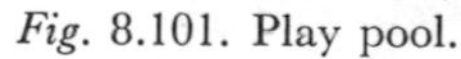

Fig. 8.102. Sea-side environment—boats in view.

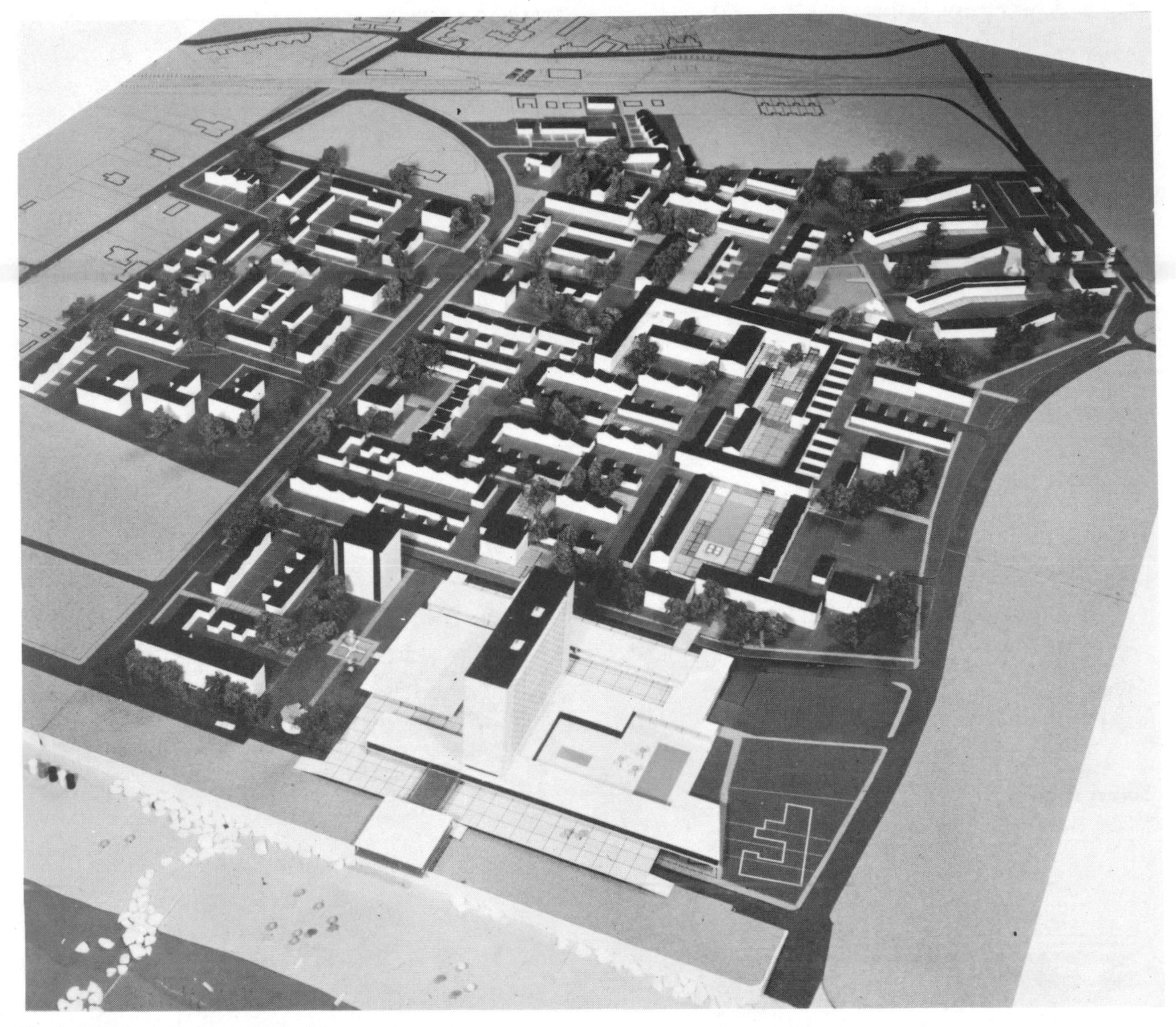

Fig. 8.103. Aerial view of model.

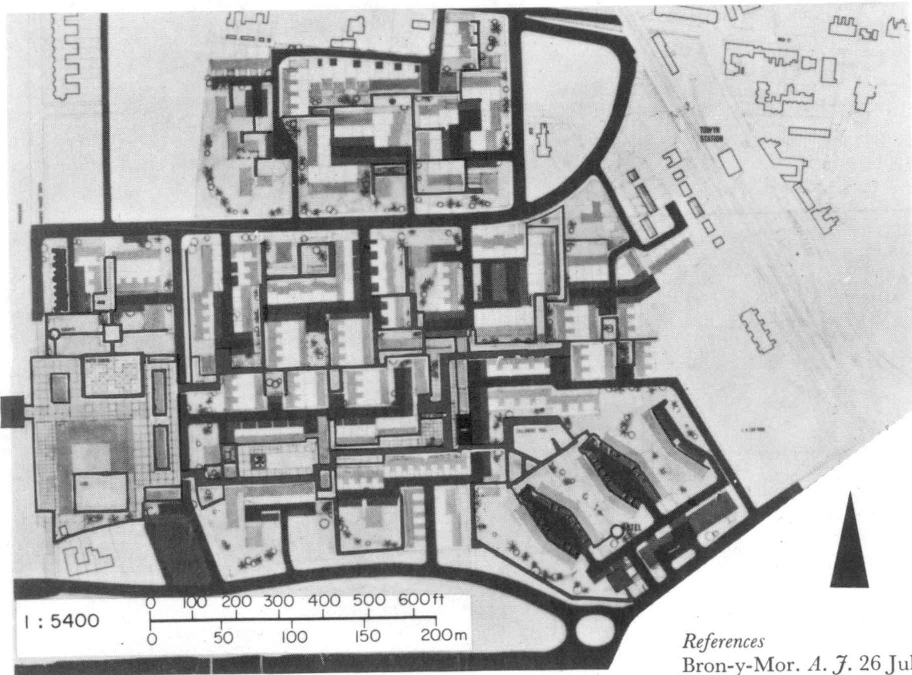

Fig. 8.104. Plan.

References
Bron-y-Mor. *A. J.* 26 July 1961; 30 August 1961.
MADIN J. Bron-y-Mor. Report by the Architect. 1961.

VIII. 12.5. *Huntingdon, Housing for the L.C.C.*

Architect to the Council, Hubert Bennett.

Huntingdon is the first application of the Radburn Idea by the L.C.C. The first section of 220 dwellings was begun in 1959. There is 70% garage provision and a net density of 61 rooms per acre (152·5 rooms per hectare) over 1500 dwellings with shopping and entertainment centre are planned. The main paths run along ancient well established lines.

This is a large scale combination of footpath access to terraces from Radburn type culs-de-sac. The very long culs-de-sac have secondary branches, total length up to 850 ft (359 m). It is a very large scheme with complete segregation intentions. But, from the plan, it looks as if roads will be crossed by short cuts. Underpasses are needed in strategic places.

Fig. 8.106. Cul-de-sac model.

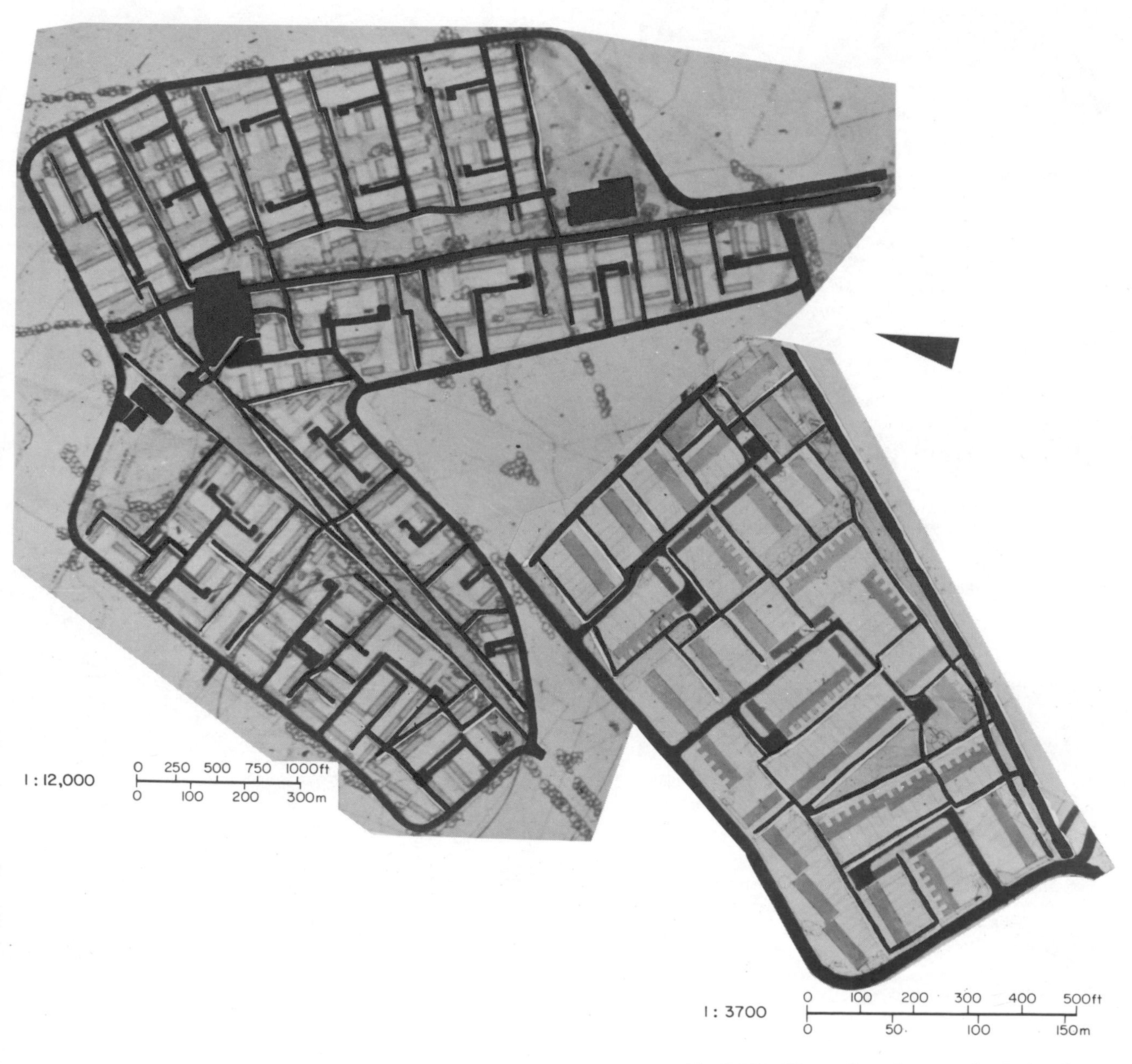

Fig. 8.105. Plans: first phase, and whole scheme.

VIII. 12.6. *Jackson Estate, Letchworth, Hertfordshire*

Design: William Barnes, Leonard Brown, Martin Priestman, architects. Assistant in charge; Michael Williams. Planning Consultant, Paul Ritter.

1641 dwellings, 61 acres (25 hectares). The first phase has 14 dwellings per acre (35 per hectare). Construction began 1962.

The subsequent larger area with the loop is to be revised from the preliminary sketch plan and model shown, to give even greater and more efficient traffic segregation, better positioning of the bus route and stops, and to increase density and incident along pedestrian routes.

The scheme is an L.C.C. and Letchworth Council extension. Car parking provision for each dwelling is required.

The scheme will include centres, church, open areas, shops, schools and old people's dwellings.

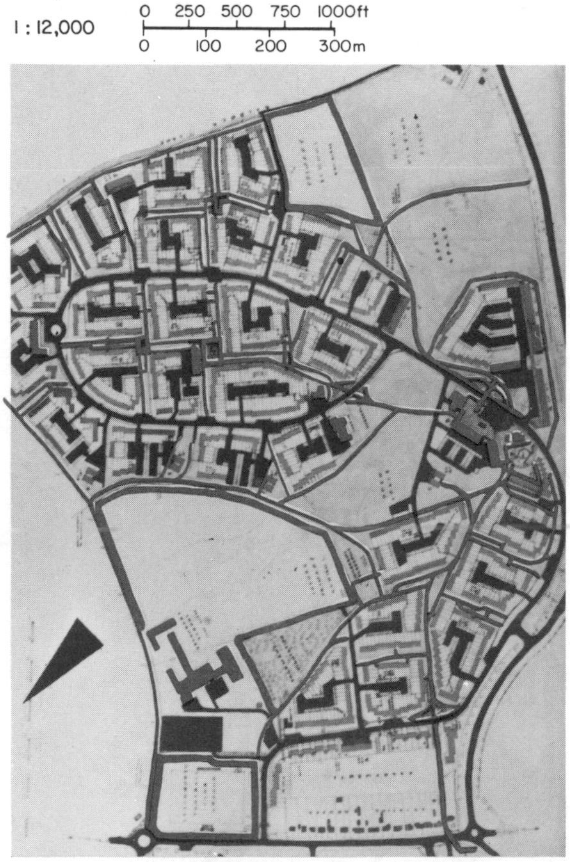

1 : 12,000 0 250 500 750 1000ft / 0 100 200 300m

Fig. 8.107. Plan showing early loop scheme with extremely economical road pattern.[1]

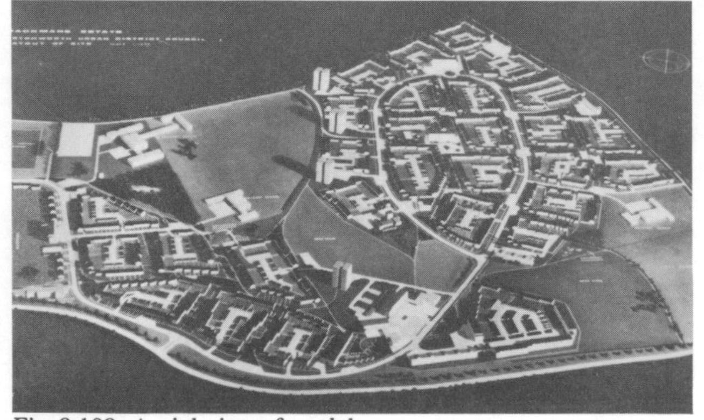

Fig. 8.108. Aerial view of model.

(i) HARLAKOFF N. *Wirtschaftliche Erschliessung von Wohngebieten.* Berlin 1961.

VIII. 12.7. *Ilkeston Road, Beeston, Notts*

Design: Housing Architect, F. Hayes: Job Architect, H. Chadwick.

A scheme, due to begin in 1961, to provide 476 dwellings. Garage provision 22%, future provision 70%, net density 18 dwellings and 64 rooms per acre (45 dwellings and 160 rooms per hectare).

This was the first local authority in the world to pass a resolution that all its future housing shall be based on the Radburn Idea.

The underpass is strategically positioned using slope of site and connecting shopping area with recreation area. The Ministry of Housing doubted the need of the underpass and the authority is to be congratulated on its insistence.

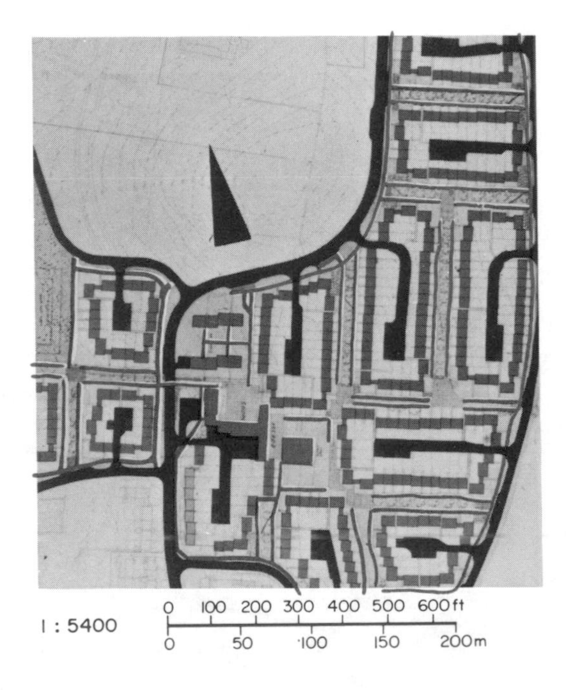

Fig. 8.109. Plan.

Fig. 8.110. Aerial view of model.

References

Beeston and Stapleford U.D.C. Ilkeston Road. Future Housing Estates. 30 October 1959.

Housing Architect. *Report on Radburn Principles.* Beeston 1959.

VIII. 12.8. *Albertslund, Dormitory Town for Copenhagen, Denmark*

Design: Architect, Knud Svensen; Consultant, Peter Bredsdorff. also Nørgård, J. Elback, M. H. Rasmussen, K. Rasmussen.

Begun in 1961, the scheme is to provide 2100 dwellings, 800 single storey, 700 $1\frac{1}{2}$-storey and 600 2-storey, with parking for each dwelling. Two schools, churches and shops, are to be built.

This is a private development. It is part of the "Finger Plan" for Copenhagen, providing high density housing within a greenbelt of its own. It is the first large-scale traffic-separated scheme in Denmark to be built. There is very careful and rational provision for pedestrians and cars, the paths leading directly to shops, station and playgrounds, along pleasant routes. The threatened mechanical repetition seen on the plan would not be experienced in walking along the development, given size and detailed arrangement.

Fig. 8.113. Path in patio housing.

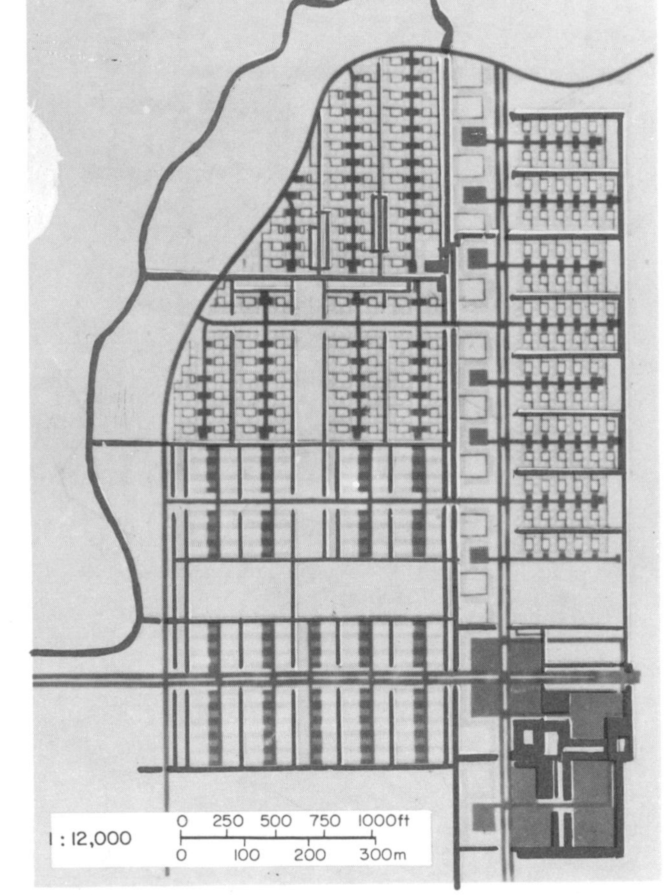

Fig. 8.111. Master plan.

Fig. 8.114. Main path to shopping centre along canal.

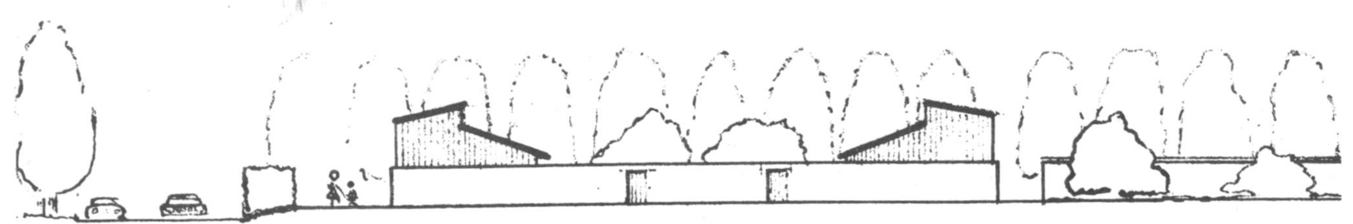

Fig. 8.112. Section through one-and-a-half storey housing.

VIII. 12.9. *Eastwick, Philadelphia, Penn. U.S.A. Residential Redevelopment with Industrial Sector*

Design: Doxiadis Associates. 360 acres (145·7 hectares) close to the City, 7500 inhabitants, 100% garage provision, 23 dwellings per acre (57·5 per hectare). Planned 1958. Building begun 1962.

Origin: Development by a private company.

The scheme is carefully worked out from house type to large community facilities and pedestrian and vehicular traffic. The area was a typical urban wedge between radials, filled with a grid-iron street network. The proposals show a continuous pedestrian path system leading to the local and main centre.

The acceptance of the scheme depended on the City Planning Commission waiving its ban on culs-de-sac over 500 ft in length.

This seems to have been all right until the police objected that they could not hunt criminals on cars in the intimate spaces created. The latest news from the office seems to be that much of the original concept has been lost in the modification of the scheme to counter these objections.

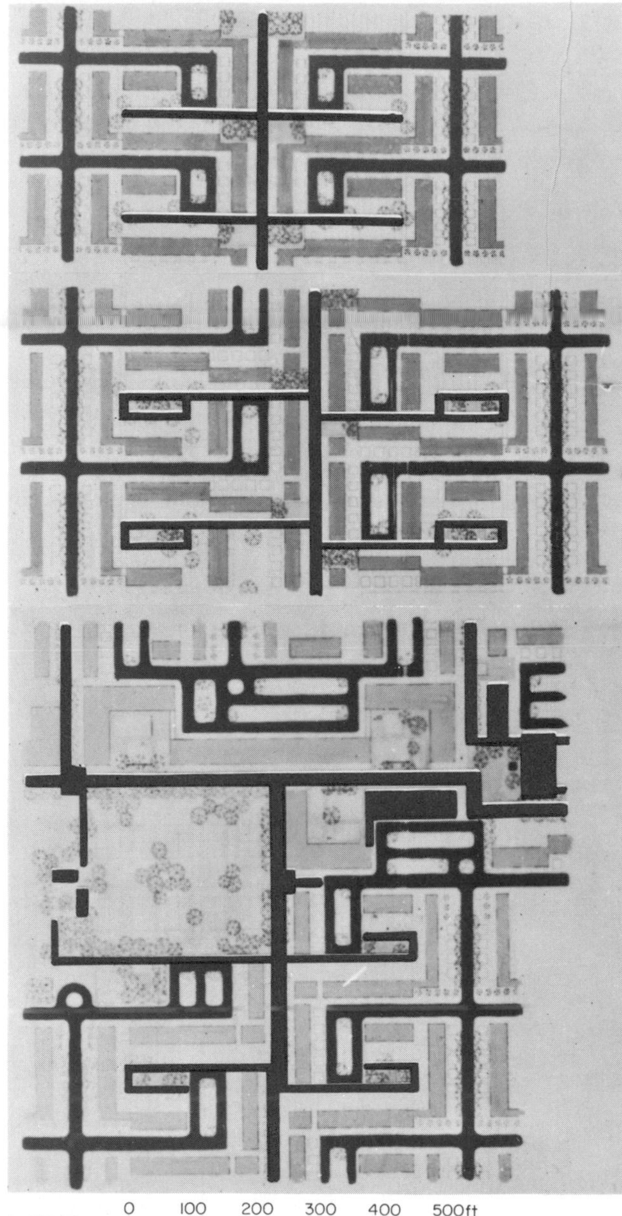

Fig. 8.116. Detail of lay-out.

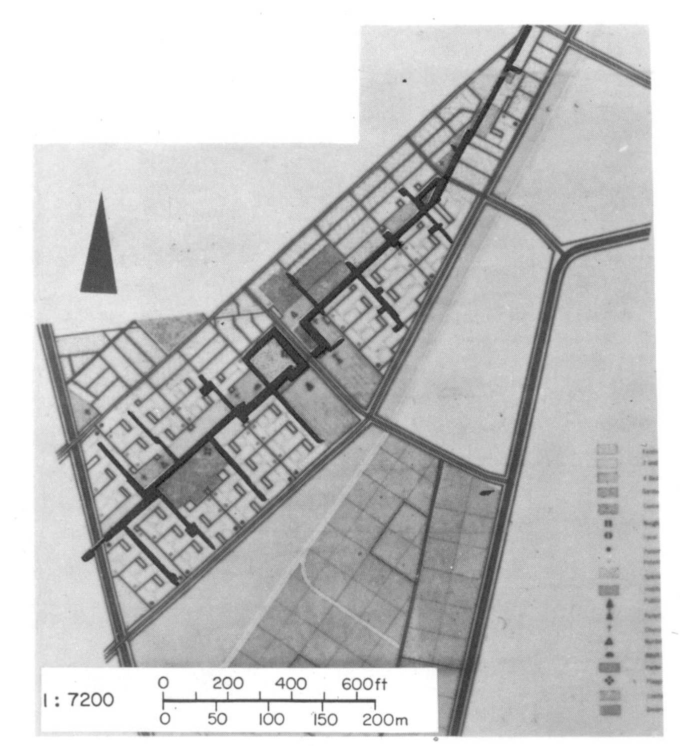

Fig. 8.115. Plan Eastwick, Philadelphia.

Fig. 8.117. View of lay-out.

References

Doxiadis Associates. *Eastwick Development Project.* Philadelphia 1959.

Bacon E. Philadelphia. *A. Rec.* May 1961.

Doxiadis Associates. *Eastwick Philadelphia* 1960.

VIII. 12.10. *Clements Area, Haverhill, Suffolk*(i)

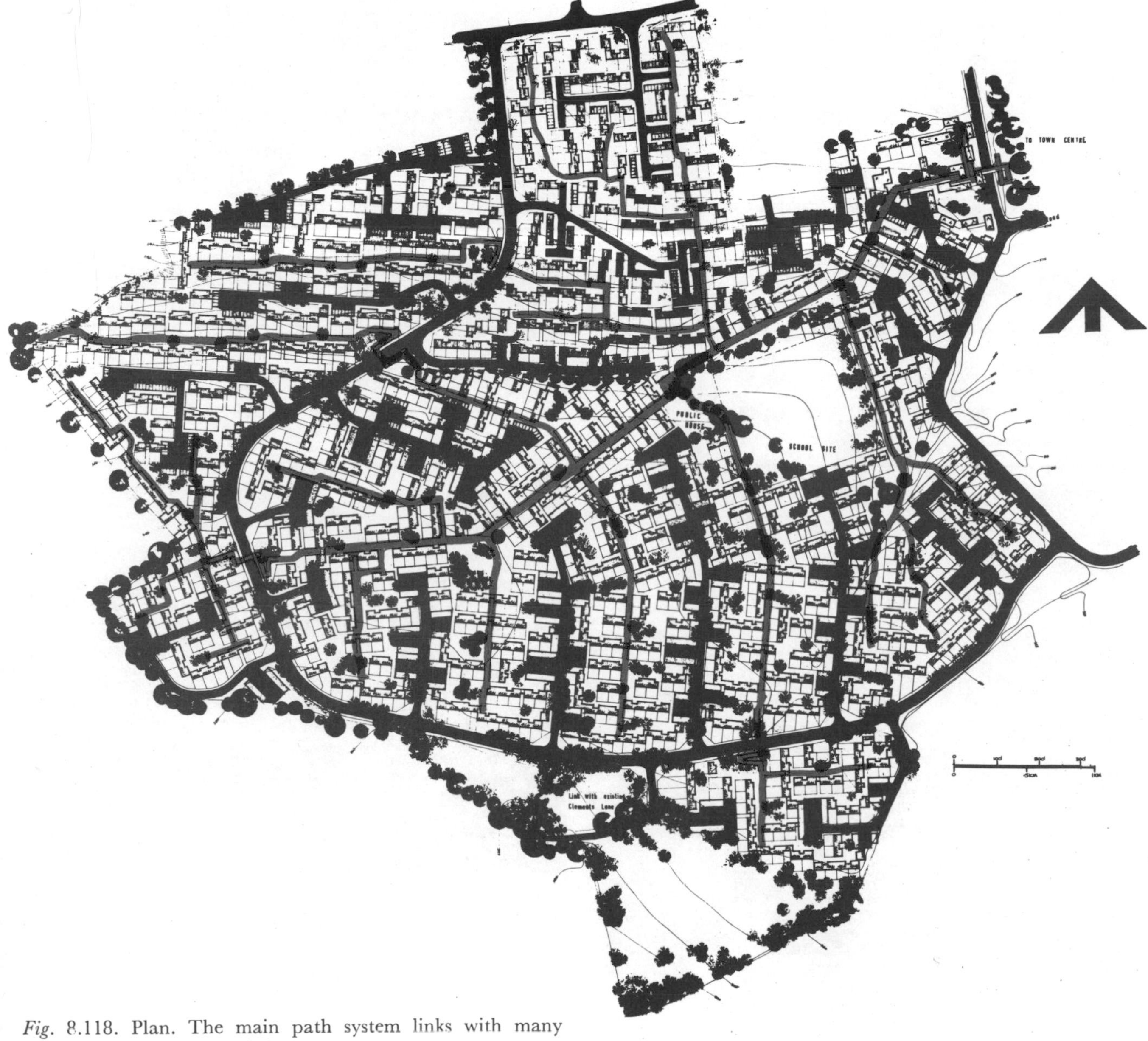

Fig. 8.118. Plan. The main path system links with many minor ones leading to each house.

Design: Architect to the Council, Hubert Bennett. F. G. West; D. C. H. Jenkin; P. E. Jones; J. Stedman; C. H. Davidson; M. Pickering; E. Ribet.

Origin: Town expansion housing by L.C.C. for Haverhill. Building begins 1963. About 1300 dwellings, 100% garage provision 20% visitor parking. Mostly two-storey houses and some 50 old people's dwellings mixed with development. 75 acres (30·3 hectares), density 17·5 houses per acre (43·8 per hectare). Cost kept to limits set by economic but conventional previous standards.

Nature of Scheme: "*All trees and other natural features will be retained as far as practicable. There will be complete segregation between road and footpath systems. The footpath system will provide the most direct route from all houses to infant school and town centre. Houses will be in clusters associated with a court or place. These courts will link up and be part of a pedestrian footpath system joining a main pedestrian way containing corner shops, pub and other public buildings*".

(i) Housing at Haverhill, Suffolk. *A.J.* 7 November 1962.

"*All houses have their main living space windows facing south or west overlooking small enclosed gardens. The layout of the public spaces for pedestrians is treated as a series of small semi-enclosed courts.*"

"*The courts and footpath system will be generally urban in character and provided with seats, children's play equipment etc., in the hope that they will be used as a social meeting place as many streets once were before the advent of the motor car.*"

"*All gardens will be private, will not be overlooked or used as access to the house. All living-rooms will be linked to the garden and have south or west orientation and some winter sunshine. All front and store doors of houses will lead on to the footpath system. No front door will be farther than one hundred feet from a road but no pedestrian court or place will be directly connected to the road system. As far as possible garages will be within one hundred feet of the front doors.*"

This seems a most promising scheme which should be regarded as a valuable experiment from which results ought to be collected to help substantiate and modify the hypothesis.

Fig. 8.119. One of the many semi-enclosed courts.

Fig. 8.120. The character of the path system from the model. It promises an intimate pedestrian scale.

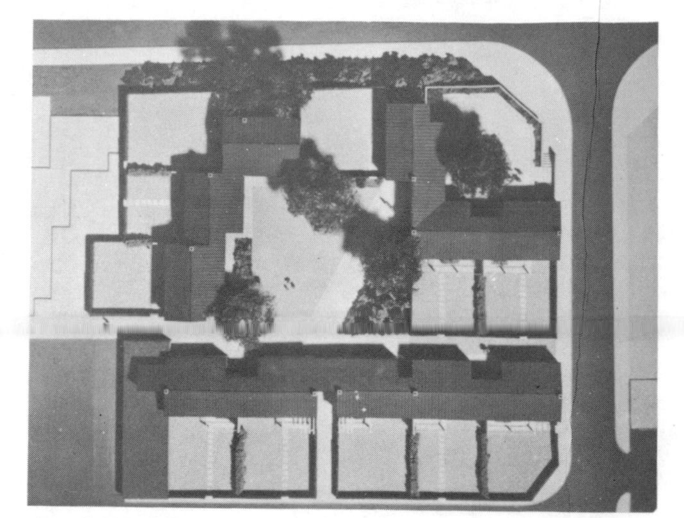

Fig. 8.122. Plan photo of one court.

Fig. 8.123. A close up of old people's dwelling amongst the houses.

Fig. 8.121. Section through housing showing underpass.

Fig. 8.124. Aerial view of one cluster about a private court.

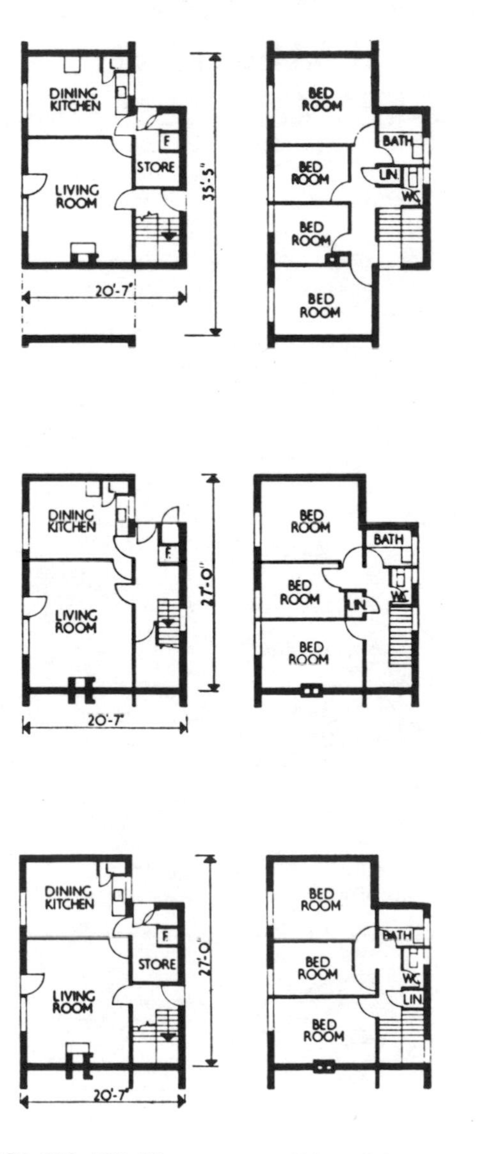

Figs. 8.125, 126, 127. House types, all broad-fronted.

VIII. 12.11. *Cité Jardin, Montreal, Canada*

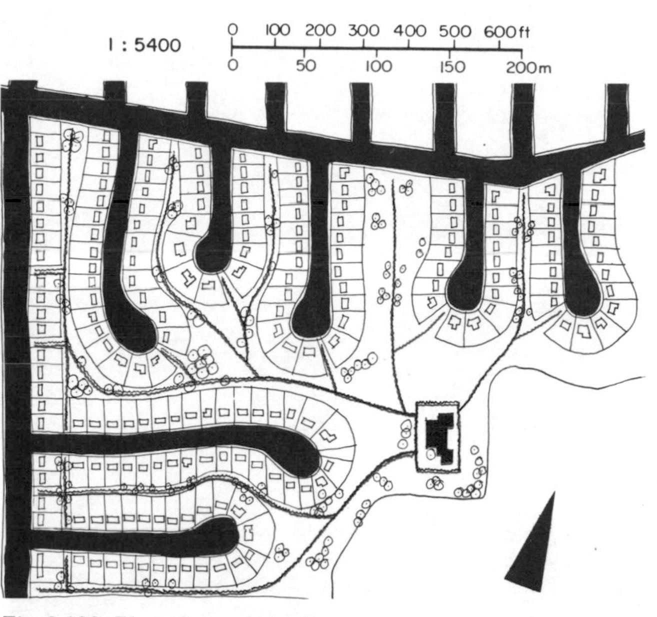

Fig. 8.128. Plan of area as built.

Design: J. Auguste Gosselin and R. P. Jean d'Autenil Richard.

Origin: L'Union Economique d'habitation was founded by a priest with the help of a lawyer. Without any professional designers they envisaged what they felt to be healthy surroundings for the working classes of Montreal in 1940. The social, economic and moral basis was co-operation. By subscription, members joined and houses were built as quickly as the war conditions allowed it. 425 active members moved in from 1942 onwards. Sadly, tragically, the scheme met so many difficulties that it failed to progress as planned. This would have given cities all over Canada such communities. A central building providing a clinic, school, bank, grocery store, cafeteria and directors office, was working well for many years and is now being rehabilitated after some years of disuse.

Remarkably these lay men, starting from first principles, hit on the Radburn Idea. I paid a prolonged visit to this scheme, completely unknown to planners, beyond one architect in Montreal, to whom I am deeply indebted for showing me the area. It works as well, or better, than any other traffic-segregated scheme I have seen in any country. The path areas are an informal, well-used space in which members have built swings, planted rockeries and in which in the autumn they burn leaves after children have romped in the piles. Swimming pools would be built in the same spirit only Montreal by-laws make it necessary to fence these off because of danger to small children. Although unlit, the path system is in no way ever regarded as dangerous in the dark and in twenty years there has been no attack on anyone. The culs-de-sac get their name and individuality from the species of tree planted along them, different with each one . . . larch, planes, chestnuts, cedars, spruce, oak.

These culs-de-sac have no laid footpaths along them in the main. They are worn. The area, as so many Radburn areas, has gained in value beyond the gain shown by other housing over the years. The turnover of residents is much smaller than in Montreal in general and a few strolls around the area makes one realize why. I met Paulette Thivierge, twenty, who had lived all her life in the area. She showed a great deal of understanding and enthusiasm for the advantages of the lay-out. There was a great deal of friendly and helpful contact.

Fig. 8.129. One of the culs-de-sac with its own species of tree.

Fig. 8.130. During winter in Canada the pedestrian areas are of great recreational value. The mountain of snow replaces the heap of leaves.

La CITÉ NOUVELLE

Vol. 1 Nos 7 - 8 LA CITE-NOUVELLE Juillet - Août 1945

LE MILIEU PROPICE
À L'ÉPANOUISSEMENT DE LA FAMILLE

Fig. 8.132. Cover of journal run by the builders of Cité Jardin.

Fig. 8.131. Gardens spill informally into common green. Children have bonfires. Maintenance is at minimum.

VIII. 12.12. *Marly-Le-Roi: Estate near Paris*

Design: Architects; Lods, Honneger brothers, Arsene Henry.

A scheme to provide 1500 dwellings covering 78 acres, (31·6 hectares). There are 27 blocks of flats in groups of three housing about 600 persons in each group. Density is 19 dwellings per acre, 67 persons (47·5 dwellings per hectare, 167·5 persons). There is a shopping centre for an area of 6000 people within superblock.

The above densities are gross and therefore high. There is parking provision for about 900 cars in the residential area, 200 parking places in the centre. There are 49 shops, cafés, services, good play provision for children and seating for adults. The schools are situated within the superblock. Of note are the curving paths, contrasting with dwelling and inviting walking.

References

LODS, HONNEGER, Shopping Centre, Marly-le-Roi. *C. du CSTB*, Paris, October 1960.

LODS, HONNEGER. Grand Terres, Marly-le-Roi. *C. du CSTB*, Paris 1959.

Some impressions of French Housing. *H.R.* December 1960.

Fig. 8.135. Children's play sculpture enlivens the scene.

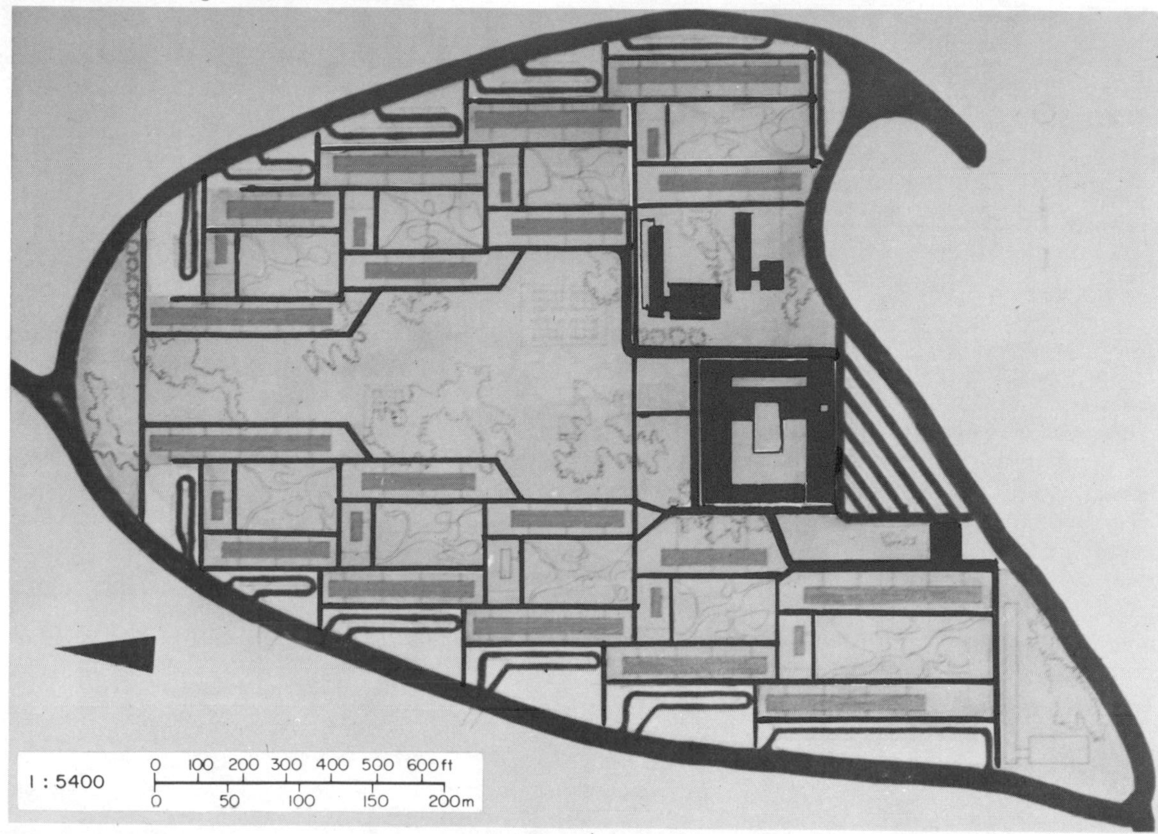

Fig. 8.133. Plan.

Fig. 8.134. A main path leading to shopping centre.

Fig. 8.136. View of shopping centre, sculpture and landscape in the making.

VIII. 12.13. *Baronbackarna, Orebro, Sweden*

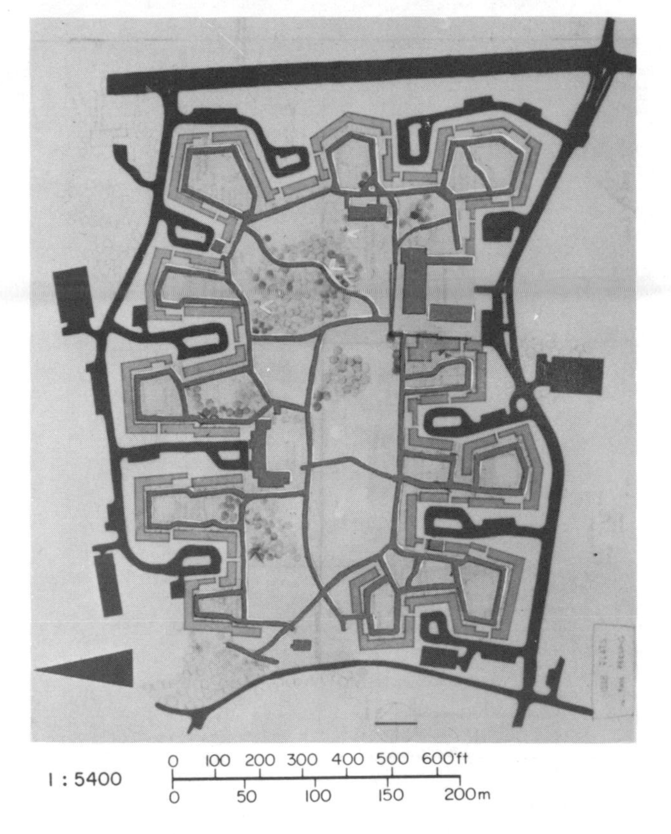

Fig. 8.137. Plan

Fig. 8.139. Balconies give mothers direct contact with toddlers.

Design: Architects, A. A. Ekholm and S. White.

This is a scheme to provide 1200 flats, at a density of 51 rooms per acre (127·5 rooms per hectare), schools, shops and other facilities within the superblock. Completed 1957.

The courtyards are used as open air rooms. Father comes home, mother goes to grocer, children to school, grown-ups meet and kitchens overlook these courtyards. Privacy is found within the flats. The central free area has running through it cycle tracks to the factories and schools. There is no temptation to take short cuts across roads. Visually the open space is terminated by the wooded hill in the east.

The architects believe in limiting family dwellings to four storeys so that mothers can talk to their children below. The tall blocks which marks the shopping centre is for single people and couples. Invalid flats are charmingly incorporated at ground floor level. Nursery and primary schools are within the site, the architecture of the second similar to the flats. One of the most successful housing schemes in the world.

References

ALM B. Markhocken, Plan 3.62. Stockholm 1962.
RICHARDSON *Orebro*. Orebro 1960.
EKHOLM and WHITE. Baronbackarna, University of Manchester. *A.S.J.* No. 8.
FALK K. Plan 6.60. Stockholm 1960.

Fig. 8.138. Aerial view of the superblock showing the large park in the middle made possible at this density by the Meander Flats around the edge.

Fig. 8.140. Enclosure for nursery school within superblock, secondary school at rear.

Fig. 8.141. Summer use. Green space with opportunity for many sports.

Fig. 8.142. View from within.

Fig. 8.143. Bicycles are used sensibly without trouble.

Fig. 8.144. Winter use of centre of superblock.

Fig. 8.145. Single person flats by shopping centre.

VIII. 12.14. *Lafayette Park Title I, Detroit*

Design: Architect, Mies van der Rohe.

This is a mixed development of houses and flats covering 72 acres (29·1 hectares) involving 1800 dwellings. 200% car port provision is provided. The density is 25 dwellings per acre, (62·5 per hectare). The first stage was completed in 1961, the second is under construction.

The previous grid-iron plan was in contrast with the plan of superblock. The car level is sunk 3 ft 6 in. (1·07 m) so that these areas do not dominate. The scheme is a redevelopment of slums in Detroit. It is an attempt to bring the middle classes into the centre. Within the superblock community feeling is reported strong and the inhabitants have formed a school, and grow tomatoes in flower beds in the public space. The rectilinear geometry of paths here contrast with Marly-le-Roi.

Fig. 8.147a. View from house window showing cars at 3 ft 6 in. (1·07 m) lower level and so unobtrusive.

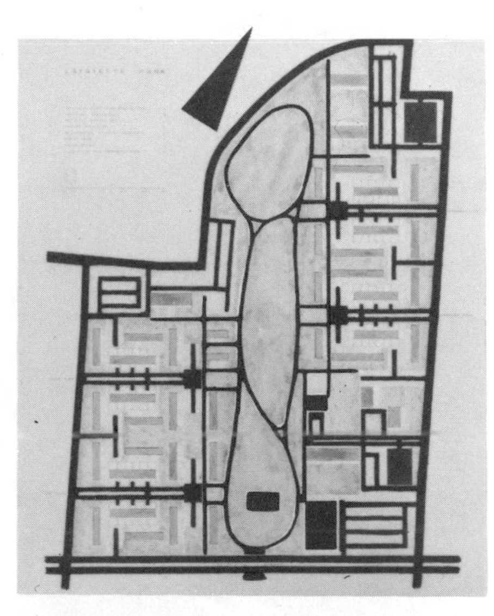

Fig. 8.146. Plan.

Fig. 8.148. Slab block of first phase.

Fig. 8.147. View along footpath. The grass by the walls make this less useful as play-equipment, to throw a ball against.

Fig. 8.149. Model view of whole scheme, second stage under construction.

Reference

MIES VAN DER ROHE. Towers and row houses Detroit. *A.F.* May 1960.

VIII. 12.15. *Biskopsgaden, Göteborg*

Design: Architects, P. A. Ekholm, S. White, Alm and Falk.

A scheme to provide 583 flats for 2277 inhabitants, at a density of 119 rooms per acre (297·5 rooms per hectare).

In this scheme there is an underpass connection to shopping centre, school and station. A very fine booklet was produced to explain the use of the lay-out to its users. It is one of a number of schemes which connect to the facilities mentioned above by paths. It is the first of several high density schemes of this nature using a prefabricated system, designed by the architects.

Fig. 8.151. In spite of high density, scope for children's play.

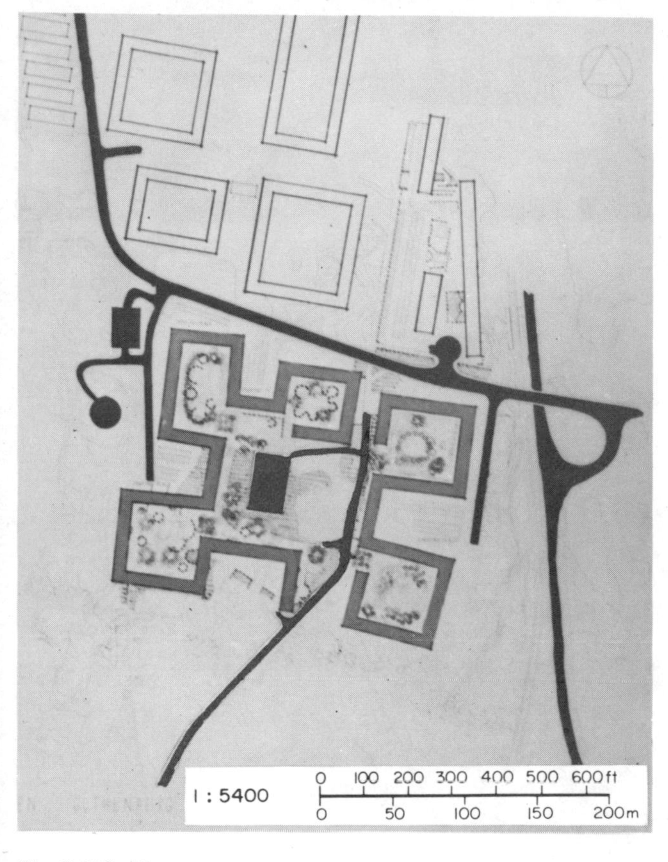

Fig. 8.150. Plan.

Fig. 8.152. Superbly landscaped courts.

Fig. 8.153. Car park at a sunken level at the exterior of block at each indentation.

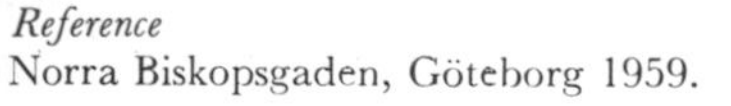

Reference
Norra Biskopsgaden, Göteborg 1959.

VIII. 12.16. *Willenhall Wood I, Coventry*

Designer: City Architect, Arthur Ling. Job Architect, D. Lyddon.

Willenhall Wood I, completed 1960, provides 644 dwellings, net density 70 rooms per acre (175 per hectare), 50% garage provision, six shops and a nursery school.

It was the second Radburn superblock planned in England and the first to have its own school. It is the first Radburn scheme in England to treat the culs-de-sac as garage courts, and to use high fencing to give privacy. The landscaping is successful and there is proper contrast of pedestrian side and street side lighting. Tenants were informed of the Radburn Idea by a leaflet on moving in and this has contributed to the rational and proper use of the lay-out, very much appreciated by the inhabitants. The scheme cost no more than previous economic municipal housing at Coventry.(i)

An underpass to link the smaller half of the scheme with shops and school was not allowed although an existing pit made it practicable. Mothers have already noted the danger. This shows how standards of safety rise as people appreciate a separate path system.

Willenhall Wood II, the nearby housing built subsequently, incorporates a porch to be able to leave the pram and baby outside, and uses more devices such as gable ends and aluminium garage doors with brightly coloured back garden gates to give the culs-de-sac a finished and refined look. The garage courts are better looking, constructed with the same pebble-dash finish as the no-fines houses. (See also pp. 43, 233, 244, 245.)

Fig. 8.155. View towards shops and canopy leading to single storey junior school.
For aerial view see Fig. 8.13d.

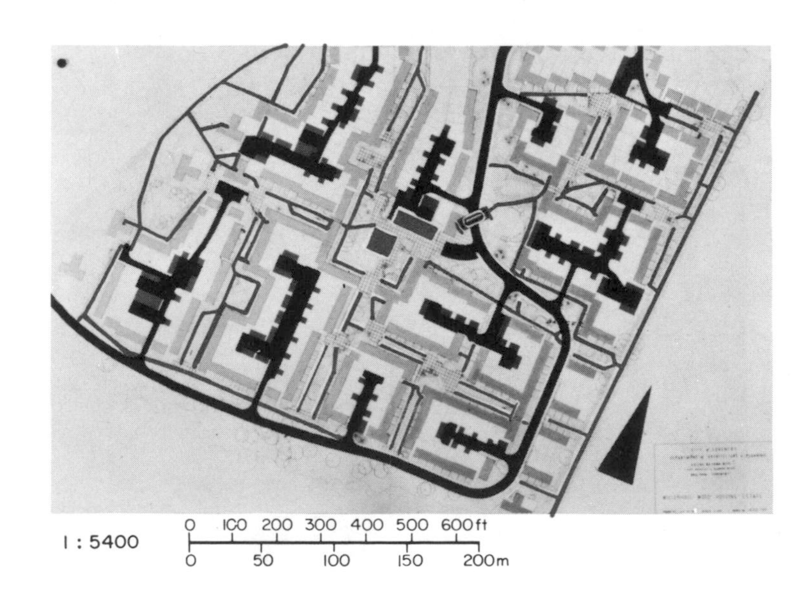

Fig. 8.154. Plan, showing point where an underpass is needed.

(i) See Tables VIII.2 and VIII.4.

References
TWEDELL N. The New Town Village. *A.R.* September 1960.
City Architect, *Report on Willenhall I.* Coventry 1957.

Fig. 8.156. Grandmother can look after child in safe area, doctor makes routine visit and chats in the open. Lamp-post the right scale.

Fig. 8.159. Pleasant narrow lanes at edge of site.

Fig. 8.157. Small hard areas with benches serve for play. Mature trees have room between houses.

Fig. 8.160. Service cul-de-sac, Willenhall No. 2.

Fig. 8.158. Tree-trunks attract spontaneous expression in congenial surroundings.

Fig. 8.161. Nursery school in square with shops and maisonnettes.

VIII. 12.17. *Kildrum, Cumbernauld*

Design: Chief Architect, Hugh Wilson. Group Architects, J. R. Hunter, with L. W. Buckthorpe, W. C. Thomson.

The scheme provides 226 dwellings, with 100% garage provision or hardstanding. Density is 90 ppa (225 ppha) 1958.

This was one of the first areas to be built at the New Town. A short survey showed the following: that the density is not objected to, that paved network of paths is used intensely by children and shoppers, that hard surfaces throughout allow use in wet weather and that ramps, steps and walls are used in children's play. The slatted fencing gives a feeling of space to walking: but privacy to those sitting in the gardens. For passing one sees into the gardens through the gaps, but inside, stationary, the eye does not register what passes. Kildrum has the first circular garages in Great Britain. The area is adjacent to the main pedestrian ring route and two schools are built, as shown, along this route in this area. (See also p. 43.)

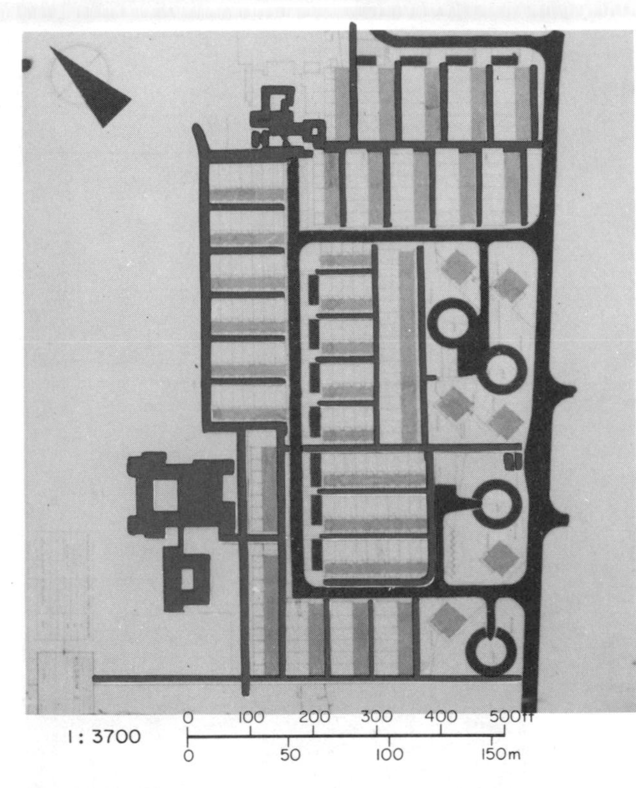

Fig. 8.162. Plan.

Fig. 8.164. First circular garages in Britain.

Fig. 8.163. Terrace access as collector path. The sweeper, it was noted, attracts children's interest. Note use of wall as seat by child behind sweeper. Corner shop closes vista.

Fig. 8.165. Terrace access, apron in front of house door gives variety and scale right for children's play.

VIII. 12.18. *Elm Green I and II, Stevenage, Herts*

Design: Chief Architect, Leonard Vincent.

707 dwellings are provided, 12·7 dwellings per acre (31·25 per hectare), 61 habitable rooms. There is 73% garage provision and shops. 1959.

In this scheme are two unrelated superblocks, in one the path system leads to the shops. In the other the paths do not seem to lead anywhere. This and, in a positive way, the regularity and tidyness of culs-de-sac is typical of the early use of the Radburn Idea in British New Towns, Mark I, as also at Harlow and Basildon.

Postal address numbers follow the same name of the ring road, in and out of each cul-de-sac. The people who live in the area like this scheme which underlines the tragedy of the housing lay-out of the Mark I New Towns in their lack of provision for walking.

Stevenage has begun another very large scheme, Pin Green, on the above principle. Here no houses will front on to the ring road, a good innovation. First houses 1963.

Fig. 8.167. Looking up a subsidiary path.

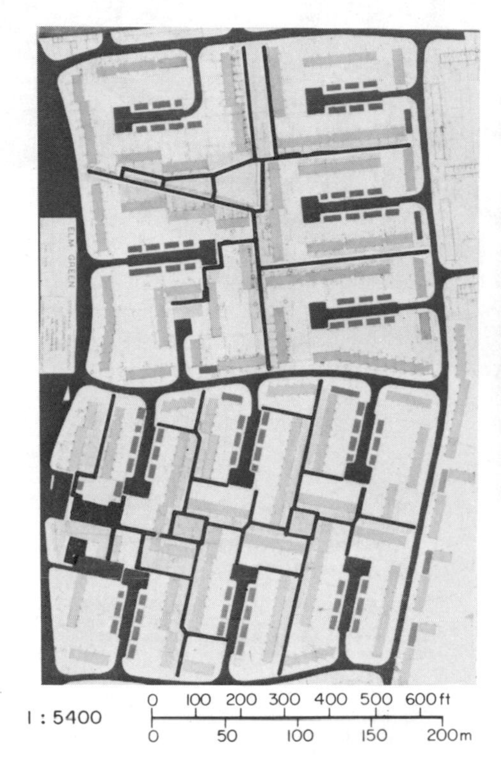

Fig. 8.166. Plan.

Fig. 8.168. Looking through archway to local shopping arcade.

Fig. 8.168a. Garages screen virtually all gardens.

Reference
Stevenage Development Corporation. Revised Proposal, June 1962.

VIII. 12.19. *Housing at Bedmont, Hertfordshire*

1:1800

0 50 100 150 200 ft
0 20 40 60m

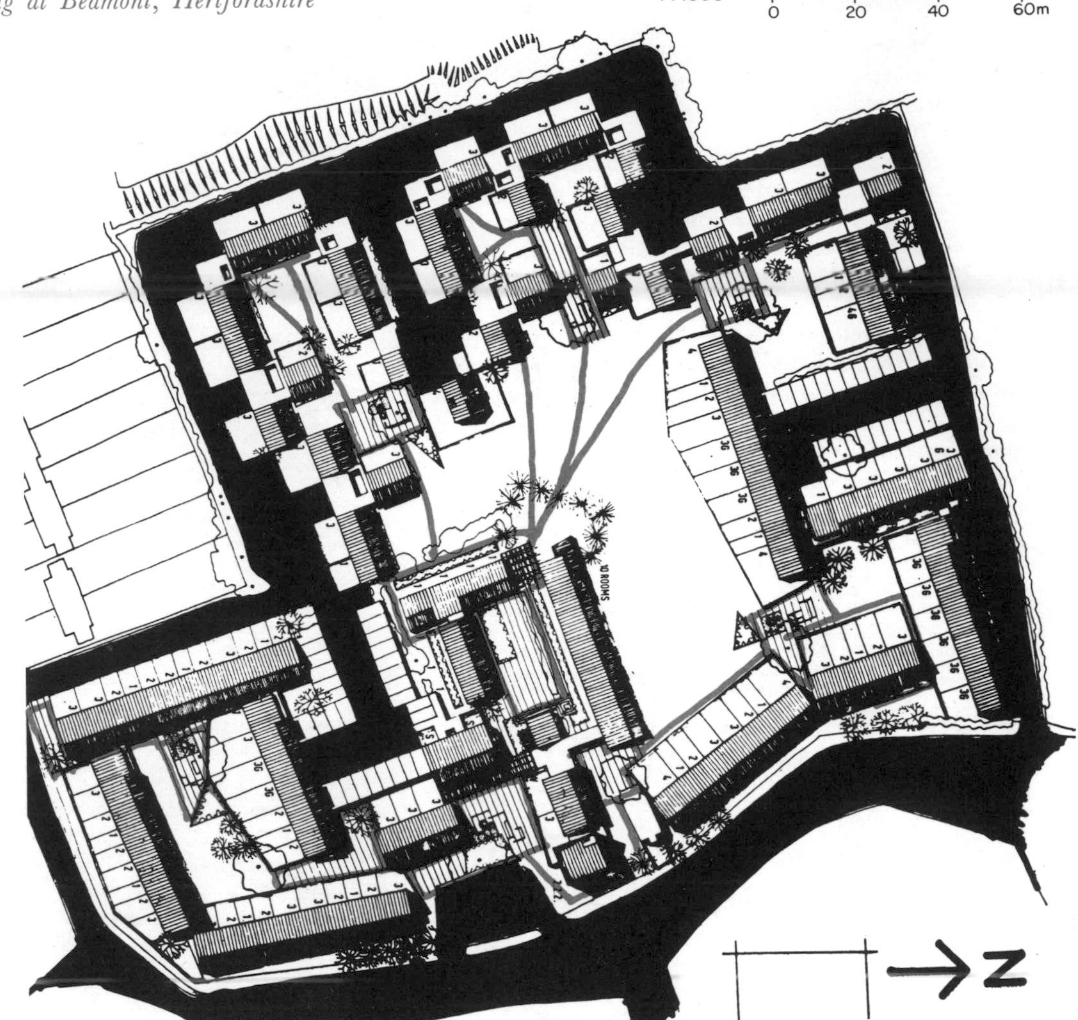

Fig. 8.169. Plan.

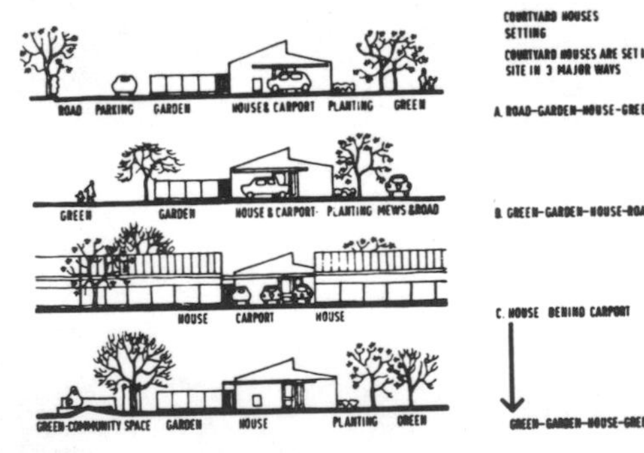

Fig. 8.170a. Courtyard houses, setting.

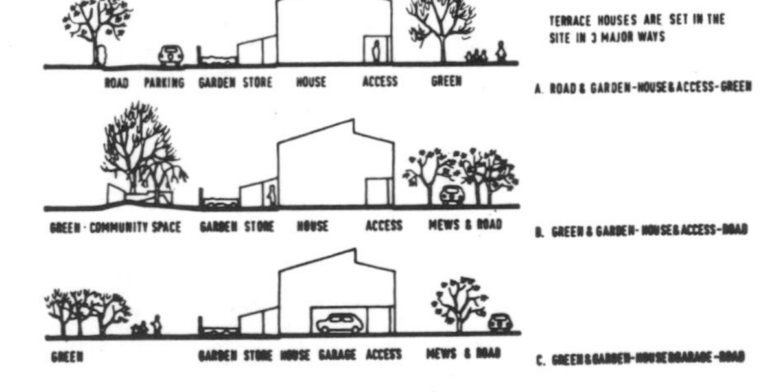

Fig. 8.170b. Terrace houses, setting.

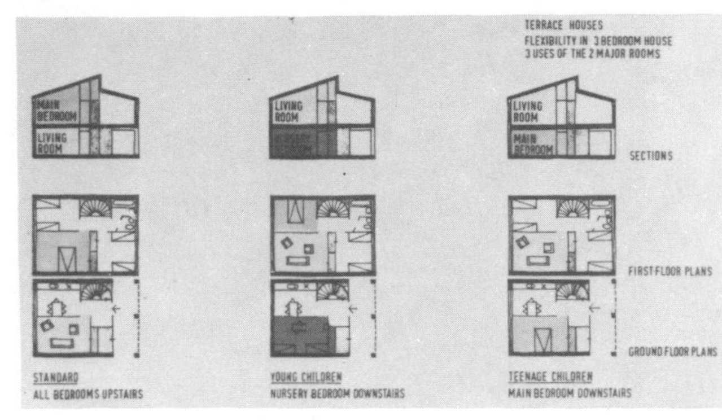

Fig. 8.170c. Terrace houses.

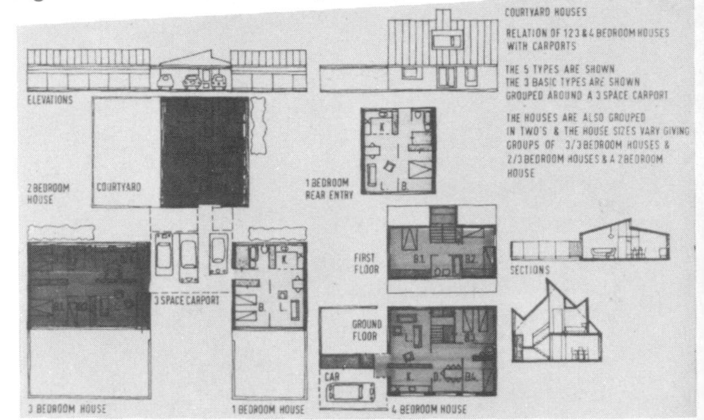

Fig. 8.170d. Courtyard houses.

Design: Michael Calthrop.

Origin: First prize in competition organized by Watford Rural District Council with the aim of improving the general standard of their housing programme.

Nature of Scheme: A Radburn superblock with 131 dwellings, including ten old people's rooms, two shops, garaging for 84 cars and parking for 140.

The approach described by Michael Calthrop in his report is quoted below as generally as well as specifically relevant.

"*General Approach.*

The Competitor's approach to the problem has been to redevelop at village scale, close to the ground. This is an attempt to provide the intimacy and identity of village life, to build into the layout those factors which attract private individuals to restore old properties in existing communities.

It has been the competitors endeavour to relate these qualities to contemporary life to their mutual advantage in an existing community. Factors considered in pursuing the general aim were as follows:—"

"*Layout*

1. *Continuity of village street with formal terraces along main road of the site.*
2. *Informality in courtyard clusters along inner area of the site.*
3. *Community buildings set in central position beside main road, opening to both site and village, with one identifiable tall building to be seen from both site and village—the scheme a part of the village.*
4. *A central common green extending into smaller local areas through a ring of miniature squares, with paving, a bench, a big tree, a sandpit and something to look at or climb on.*
5. *A one way traffic system running along the periphery, giving segregation within the site. The car travels the longer distance, the pedestrian the shorter.*
6. *Private outdoor space for every house, with as many gardens as possible.*
7. *Cross-section of family sizes throughout the site so that while one may predominate, no group is exclusively of one type.*
8. *Apart from the one four-storey tower, no building to be over two storeys.*
9. *Covered access and close spacing for a tight, close-built relation of buildings.*"(i)

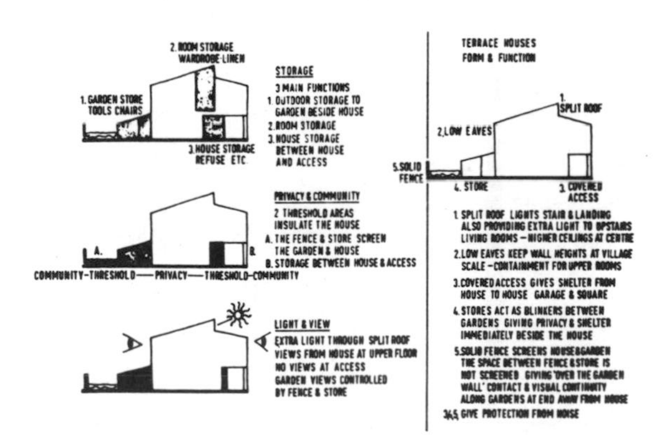

Fig. 8.171. Explanation of terrace house type and functions.

(i) CALTHROP M. Housing at Bedmont, Herts. Housing at Bedmont. *A. J.* 3.1.1962.

VIII. 12.20. *Inchview Development, Prestonpans, East Lothian, Scotland*

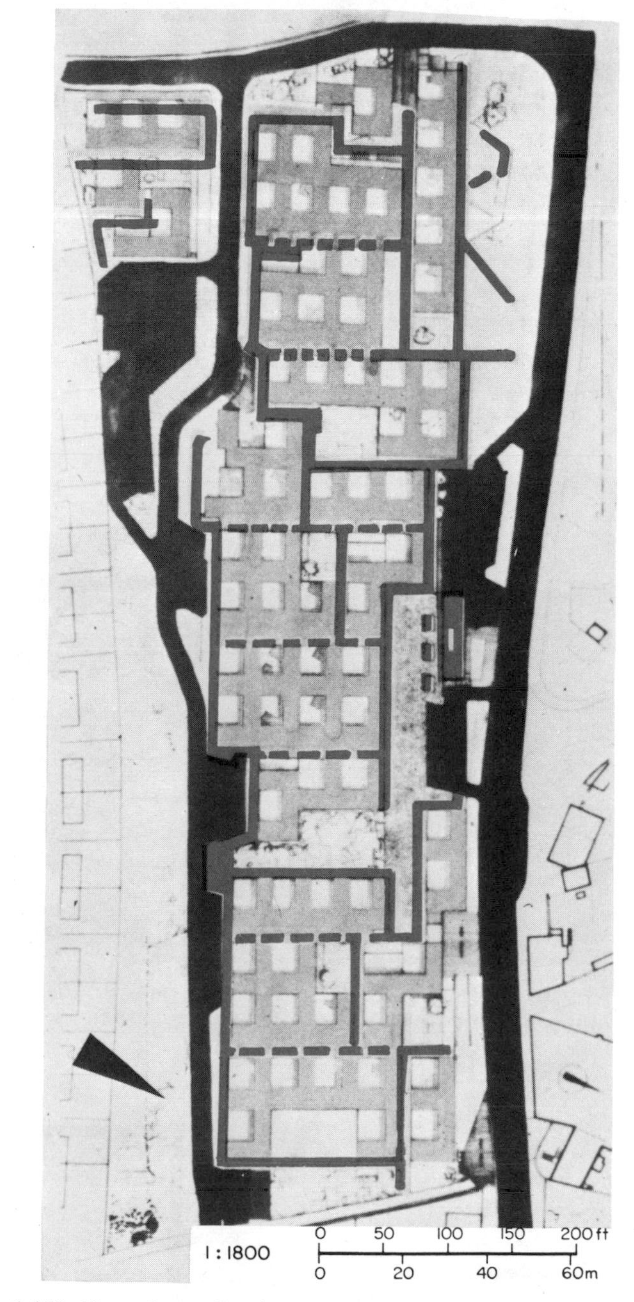

Fig. 8.172. Plan of completed project.

Design: Professor Sir Robert Matthew and Housing Research Unit, University of Edinburgh.

Origin: Commissioned by East Lothian County Council as an experiment.

First stage completed 1963, 45 houses. Final stage 90 houses. Ultimately 12·4 dwellings per acre (31 per hectare), 53 persons and 5% garage space and 116% parking. 2–3 shops, launderette.

This scheme is a cluster of patio houses of varying size similar in character to housing by Coderch and Vallis on the Costa Brava, keeping motor traffic right outside. Site, in grim surroundings, overlooks sea front but all windows from houses turned into private courts except for some bedrooms in 3-bedroom types which are placed at the periphery. There are high level windows from living room and kitchen on to the neighbouring patio. The experimental purpose and hypothesis is described as follows:

"1. *To develop a form of courtyard housing which would permit the examination of the suitability of a courtyard house to*
(a) a rural community,
and (b) the scottish climate.
2. *To apply a method of Cost Planning and Comment on its suitability.*
3. *To record the design process to enable checking of any design data or assumptions at a later date.*"

"*Hypotheses.*

That when a courtyard development is envisaged and the houses are (a) sited adjacent to each other, and (b) physically linked by covered ways, then a community feeling will be developed amongst the tenants, and that shared services and facilities will reduce to a minimum costs for the type of development."(i)

The tenants for this development are selected from the housing list.

Fig. 8.175. Covered way.

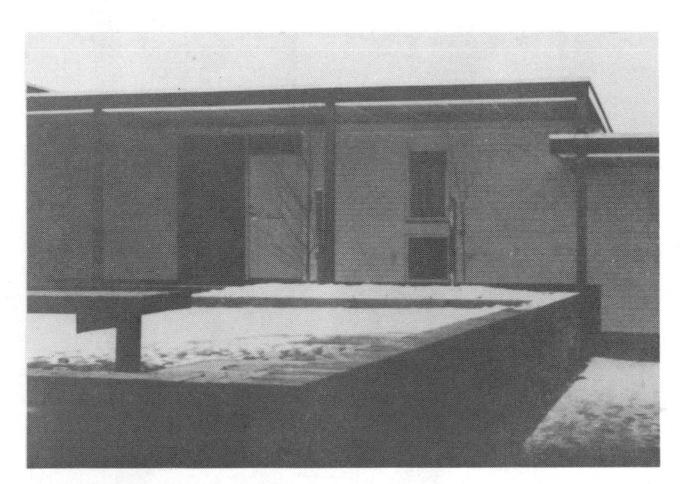

Fig. 8.173. Hard play area raised to stop too boisterous ball games and next to covered way.

Fig. 8.176. Model of first stage.

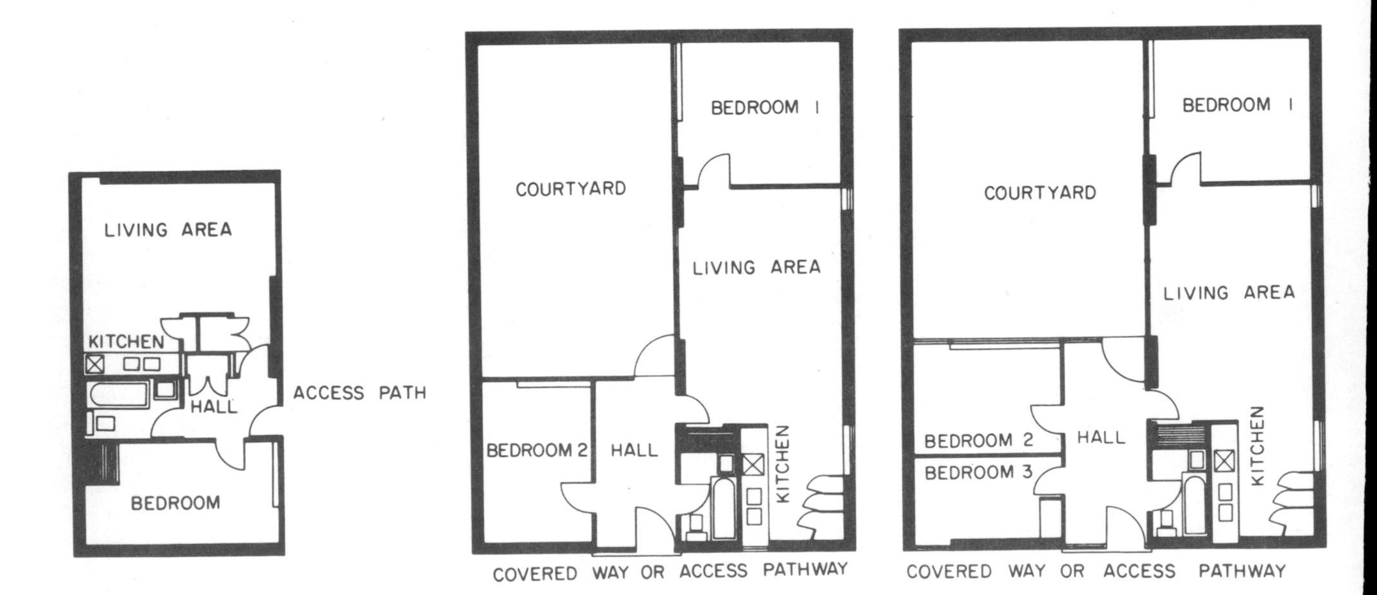

Fig. 8.174. Dwelling plans scale $\frac{1}{16}$ in. to 1 ft.

(i) Letter from Prof. Sir Robert Matthew, Dept. of Architecture Research Unit, University of Edinburgh. February 1963.

VIII. 12.21. *Parkleys, Ham Common*

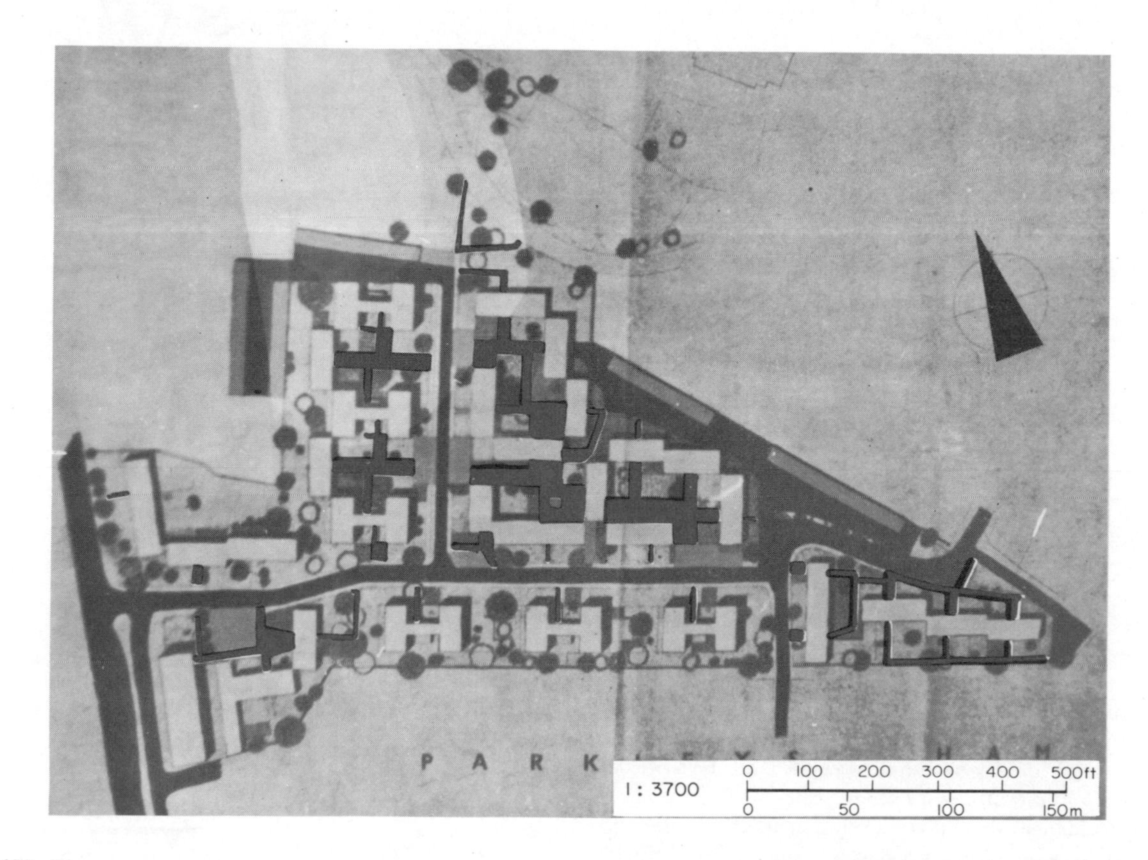

Fig. 8.177. Plan.

Speculative, private enterprise housing for sale

Design: Span Development, Architect, Eric Lyons.

The aim is to provide better class housing with a garage to each dwelling and a density of 50 ppa (125 ppha).

Eric Lyons pioneered private housing for sale in a good environment. This is achieved by creating a holding company to deal with maintenance and a 999 years lease. There is prejudice in Britain against leasehold particularly in the provinces. Traffic separation is limited to court sanctuaries but there is no continuity in the paths to make a system. (See also p. 256)

Fig. 8.178. Pedestrian court, gravel and paved surface, pool and flower beds (photo. Sam Lambert).

Fig. 8.179. Pedestrian approach; bollards demarcate ends of vehicular area.

VIII. 12.22. *Park Hill, Sheffield*

Design: City Architect, J. L. Womersley. W. L. Clunie, B. F. Warren, J. Lynn, I. S. Smith, A. V. Smith, W. J. Winder, J. Snow, T. Mark, G. I. Richmond.

This is a scheme to provide, initially, 944 dwellings and space for 244 cars.

It includes the district shopping centre, and local shops and pubs at the bottom of the lifts. The slope of the site can be seen from the increase of from 3 to 14 floors, with a horizontal roof line. There is access to accommodation at every third floor along a 10 ft (3 m) wide street or deck, running the full length of the development completely avoiding the enforced isolation of tower and slab block flats. The first stage was occupied in the Autumn of 1959.

A further similar development Hyde Park, on even higher ground and up to 18 storeys in height is partially completed. (See plan 1963.)

This is an extremely important experiment, the first of its kind in the world, occupied by working-class people.

Though the decks provide well for many things, as a preliminary number of visits have shown, two findings are relevant to any future use of the idea. People seem to prefer to walk at ground level in Park Hill. The decks are left for the ground as soon as possible. The reasons have not yet been established. Secondly there ought to be focal points, points of arrival on the deck. Niches, recessed, a seat, anything to mark certain points, give individuality and break monotony. A very interesting phenomenon is the placing at each front door of a small piece of patterned lino by each and every tenant. This breaks the monotony and establishes a human scale.

There is evidence that this development shows greater identification as a neighbourhood than any other Sheffield housing.

A research project on Park Hill, by the author, will compare this with point blocks on the basis of scale of livability (see also pp. 44, 45).

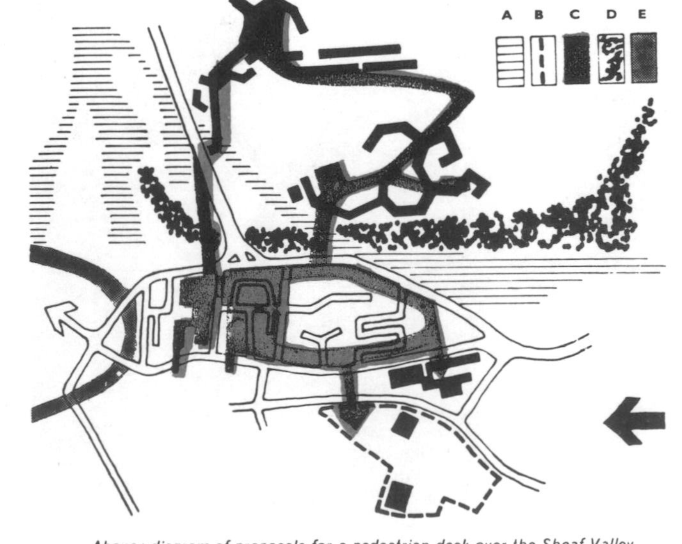

Above: diagram of proposals for a pedestrian deck over the Sheaf Valley
*Key: **A** rail **B** proposed civic centre area **C** pedestrian deck **D** new woodlands **E** River Don*
Right: pedestrian shopping centre, Park Hill 1

Fig. 8.180. Section showing pedestrian deck connecting Park Hill with town centre.

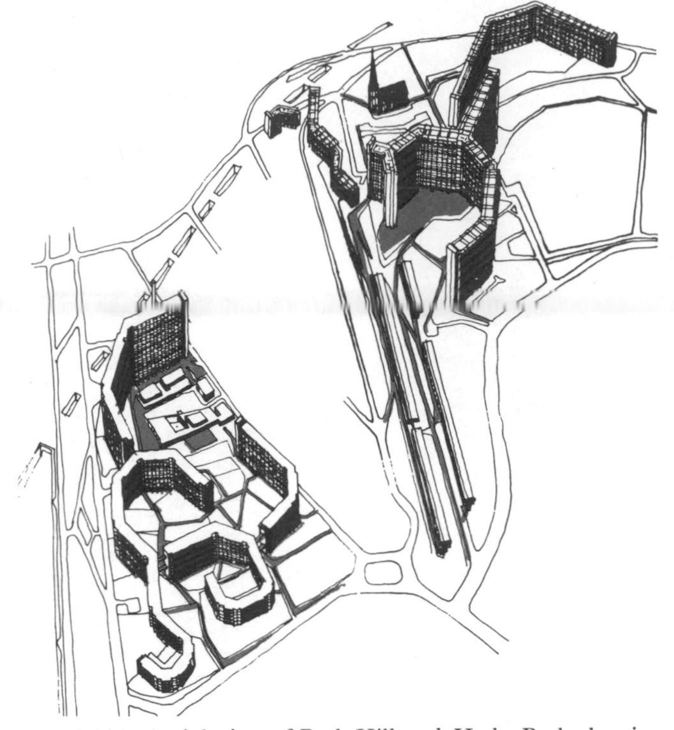

Fig. 8.181. Aerial view of Park Hill and Hyde Park showing "corner shops" and pubs at strategic points at bottom of vertical circulation, shopping areas, schools, path network and deck.

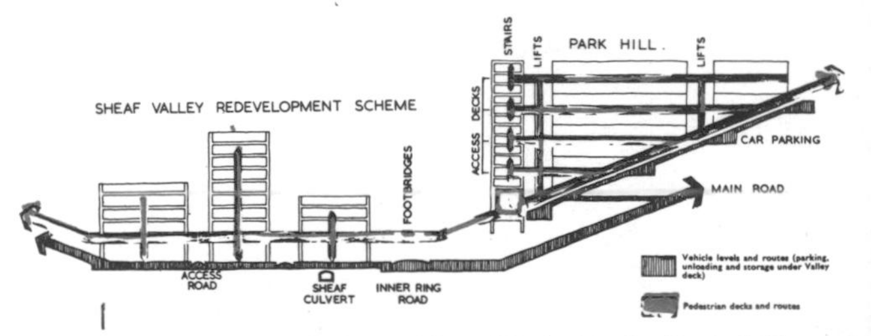

Fig. 8.182. Plan of deck linking Hyde Park, Park Hill and town centre.

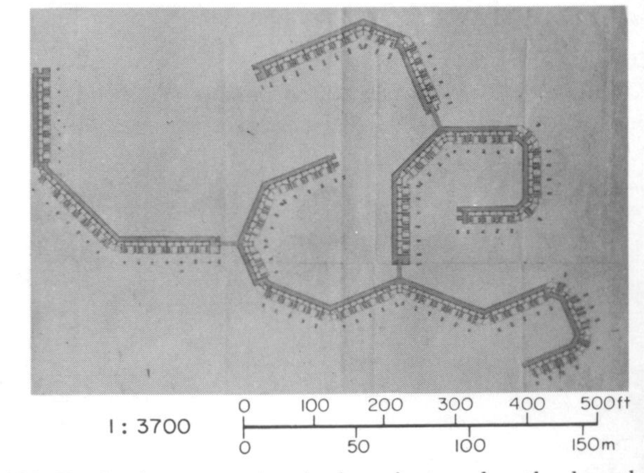

Fig 8.183. Deck plan at top level where it stretches the length of the whole scheme.

References

Park Hill Redevelopment. *R.I.B.A.J.* December 1962.
Park Hill Redevelopment. *A.J.* 23 August 1961.
Park Hill Redevelopment Part I. Report on Amenities and Management. City Architect's Department. Housing Department.
Park Hill Survey. City of Sheffield Housing Department. September 1962.
Park Hill, an Urban Community. City of Sheffield Housing Development Committee 60–61.
Crooke P. Sheffield, special number. *A.D.* September 1961.

Fig. 8.184. The fenced area for football, most popular and essential amenity, if misuse of other areas is to be stopped. (Photo, Roger Mayne.)

Fig. 8.185. Even when its wet children stroll along the decks.

Fig. 8.186. Shop built into ground floor. Note low level lighting and concrete rails.

Fig. 8.188. From below one watches life on the decks. Carpets often hang over the rail, adding colour.

Fig. 8.189. View of main shopping centre of Park Hill seen from the entrance. Its intimate sheltered character using levels well make it attractive.

Fig. 8.190. Father coming home from work, women chat, children play on deck . . . monotony of deck can be seen, although the continuous view relieves this somewhat. Note lino strip outside each front door, to give sense of identity. In several cases neighbours have combined to purchase one piece of lino to go across adjoining doors. (Photo. Roger Mayne.)

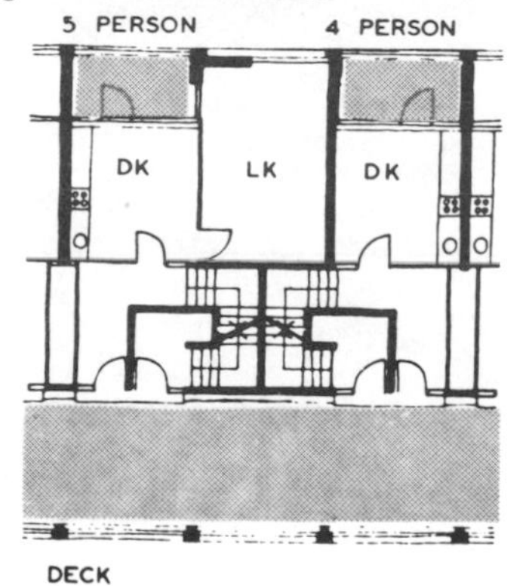

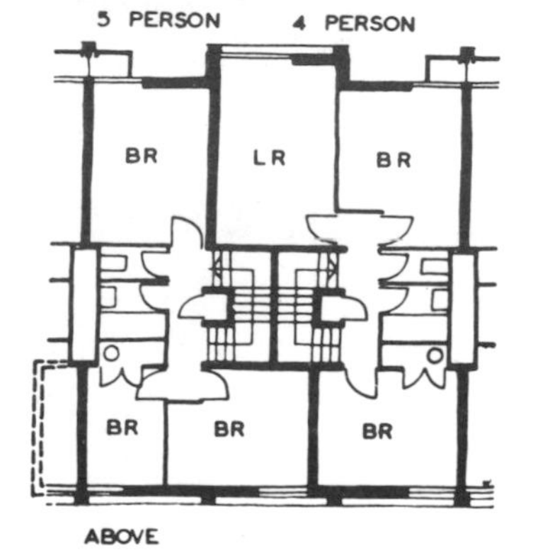

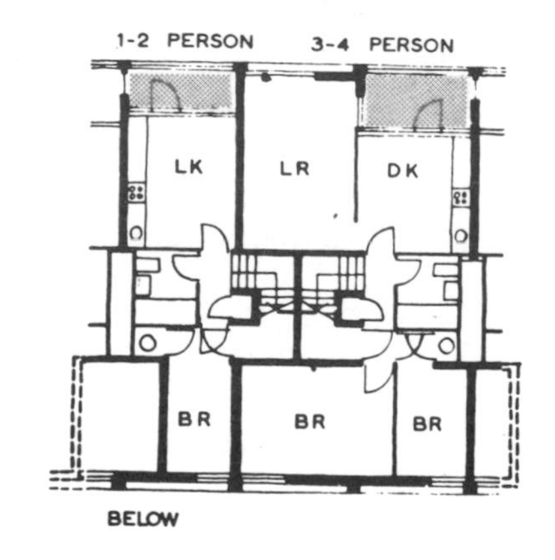

Fig. 8.187. Dwelling plans.

VIII. 12.23. *Barbican, London*(i)

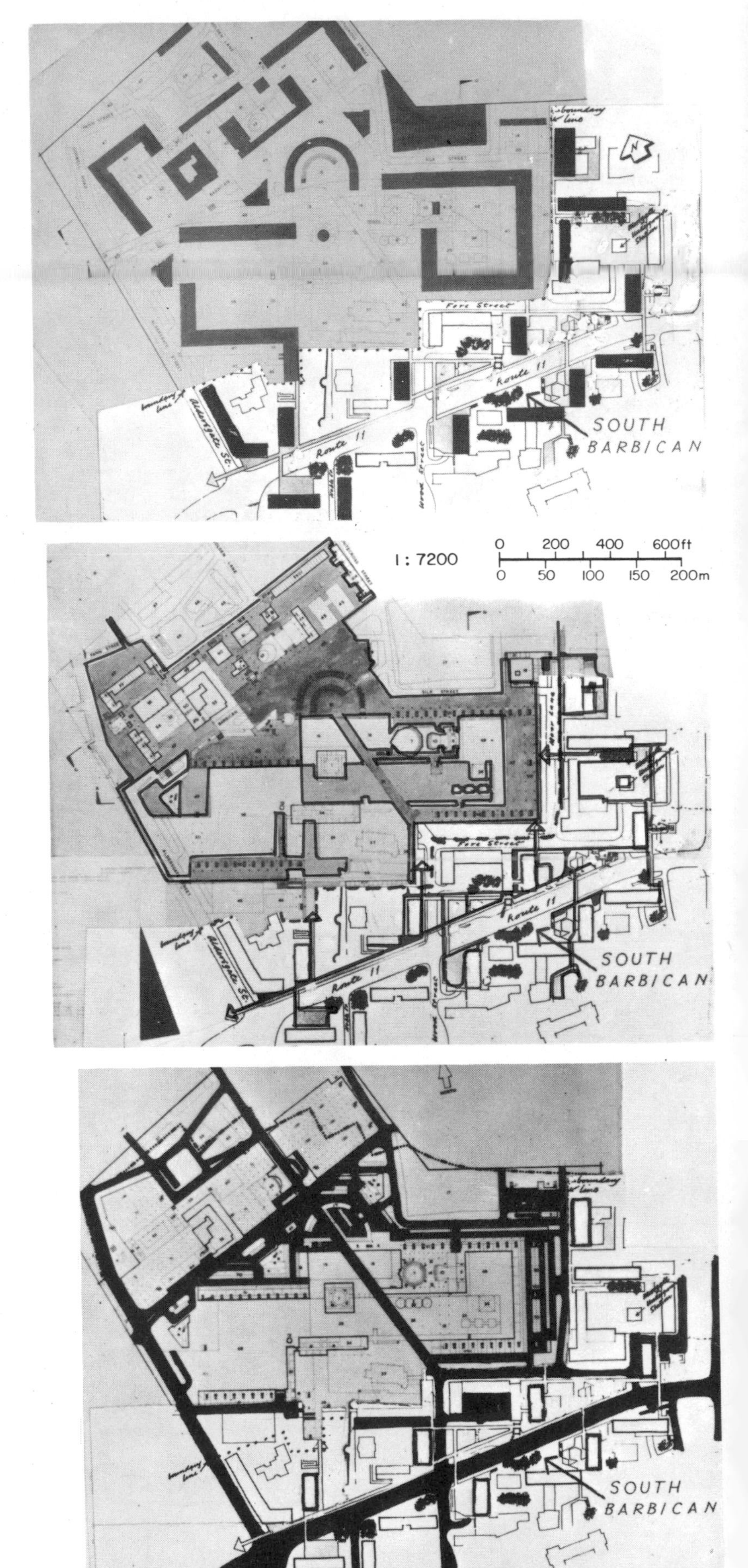

Fig. 8.191. Plan at main vehicular level.

Fig. 8.192. Plan at pedestrian level.

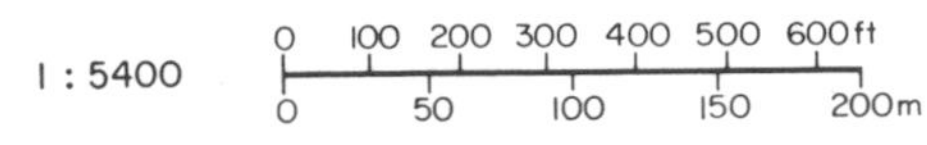

Fig. 8.193. Plan at upper level.

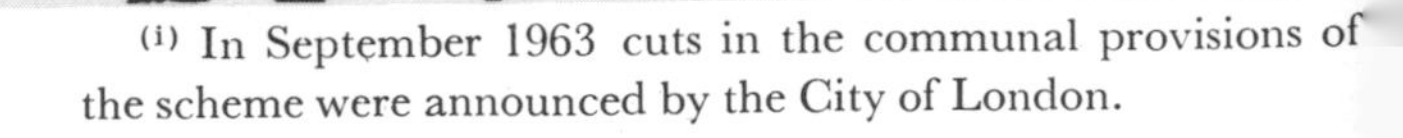
(i) In September 1963 cuts in the communal provisions of the scheme were announced by the City of London.

Design: L.C.C.; City of London; Chamberlin, Powell, Bon.

A scheme to provide 2150 flats, in six-storey blocks and three 37-storey towers with provision for 2600 cars. A group of shops, a school and a restaurant are also included.

This scheme is the result of the policy of the Corporation of London to create a "genuine residential neighbourhood" even if this means "forgoing" a more remunerative return on the land. Pedestrian areas and roadways are segregated on to separate levels which will connect with neighbouring commerce developments of great size. There is order without monotony, high density without congestion. Attractive living accommodation in the centre of the City has been achieved. The historical Coal Exchange was retained in the plan but since, in spite of public protests, has been destroyed. A magnificent illustrated report was published by the Corporation of the City of London.

Fig. 8.194. Scheme by City of London for the south section with office blocks.

South Barbican was the first part of the development undertaken by the staff of the L.C.C. and the City of London. The pedestrian decks 18 ft (5·5 m) above ground level, crossing route 11 connecting the office slab blocks, set the circulation pattern adopted in the Central and Northern areas. The scheme prepared by the City of London Architect was rejected as it had less residential accommodation than the adopted scheme. The Northern area for mixed commercial and residential use is being designed jointly by Chamberlin, Powell and Bon, and the City of London. The central section is purely residential.

Fig. 8.195. Residential scheme for centre and north sections by Chamberlin, Powell and Bon.

References

Browne K. South Barbican. *A.R.* London, May 1960.

Chamberlin, Powell, Bon. *Report on Barbican.* London 1959.

The Barbican Committee. *Residential Development within the Barbican Area.* April 1959.

VIII. 12.24. *East York, Canada, Flemingdon Park*

Design: Architect, Irving Grossman. Client, Webin Communities, a private company.

This is part of a large high density project. Garaging and car-parking for every dwelling below ground, probably the first such example ever built.

Air-locks ("*the idea came from submarines*"(i)) and the high density make possible underground garaging for terrace housing. Asphalt wastelands are eliminated, for the first time in private development on a comparatively flat site with terrace housing. The distance between terraces at times is only 35 ft (10·7 m). Detailed planning of the house ensures privacy, sunshine, etc. These houses, with direct access from cars and safety for children, were easy to let. "*The situation was unique . . . one owner and management . . . a team of experienced personnel . . .*".(i)

A very fine intimate character is obtained through good use of levels and materials.

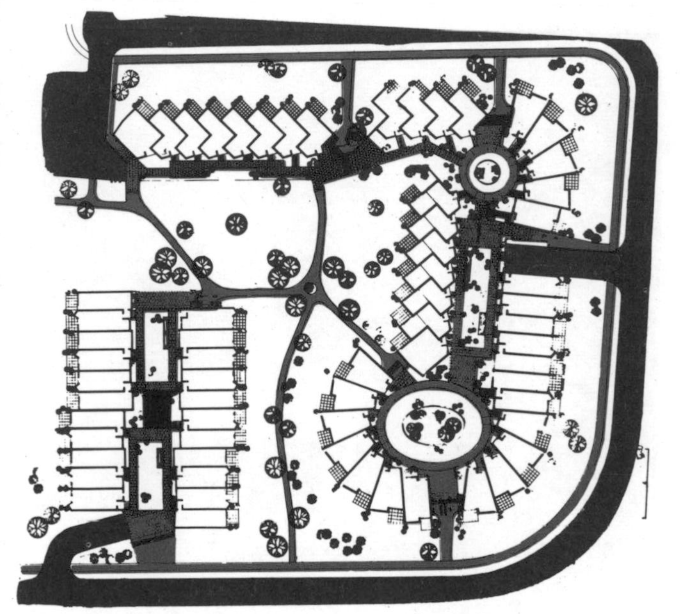

Fig. 8.196. Plan at ground level.

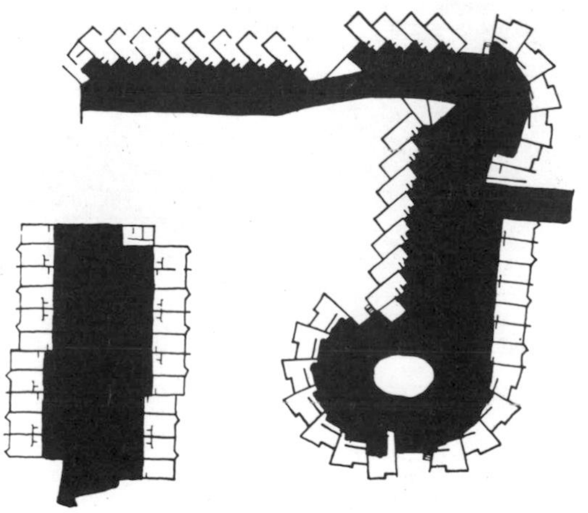

Fig. 8.197. Plan of garage basement.

(i) Grossman I. *The Canadian Architect.* May 1961.

Fig. 8.198. Entrance for cars below, people above.

Fig. 8.199. One of the pedestrian courts.

Fig. 8.200. Aerial view showing entrance to car storage on the right, and in the foreground centre, with pedestrian courts.

Fig. 8.201. Section.

VIII. 12.25. *South Carbrain, Cumbernauld*

Design: Chief Architect, Hugh Wilson. Group Architect, R. A. Barlow. Design Team, J. Robertson, J. Latimer, L. Sheach.

Origin: Scheme for the New Town by the Architectural Department of the Development Corporation.

Nature of Scheme: Rather than create very dense pockets of 120 ppa (300 ppha), this scheme deals with the area comprehensively at 100 ppa (250 ppha). It was thought that the steep slope to the south would make possible underground garaging, i.e. multiple use of land, and better sun penetration than elsewhere. The higher dwellings at a low level would not cast shadows of consequence.

There are chiefly two types of dwellings, two- and three-storey housing with direct access to the ground and open space for families with children and "stub" blocks for the remainder of the population, totalling 6300 for the Carbrain South Area. The path routes are set by the ones which lead through Carbrain, 1–6 (see VIII. 12.1) to the town centre and which are built or under construction. The ring roads are also previously determined. As the house types and the use of levels is original and interesting with general relevance, plans and section and descriptions are given from the report of the Development Corporation.

Proposed Lay out

Low-rise mixed development

This proposal which is recommended provides for the whole of the families with children to have houses with some private open spaces and the remainder of the population to be accommodated in maisonettes or flats in walk-up stub blocks.

This arrangement which seems to ensure suitable amenity and cost standards can be achieved by the adoption of the following principles:

1. The multiple use of land; for example the use of garage roofs as play spaces, two level garages, and houses over garages.
2. The adoption of a narrow frontage three-storey house in terrace form.
3. The use of the hillside to enable houses to be entered at first floor level and stub blocks of flats and maisonettes at second floor level.
4. In addition, the pedestrian–vehicle separation is maintained both on the main downhill and lateral footpaths and on the more intimate meshing of roads and footpaths close to the dwellings. Where the footpaths cross the culs-de-sac, underpasses or bridges are proposed.

Fig. 8.202. Lay-out—working model photograph. Taken from Development Corporation report.

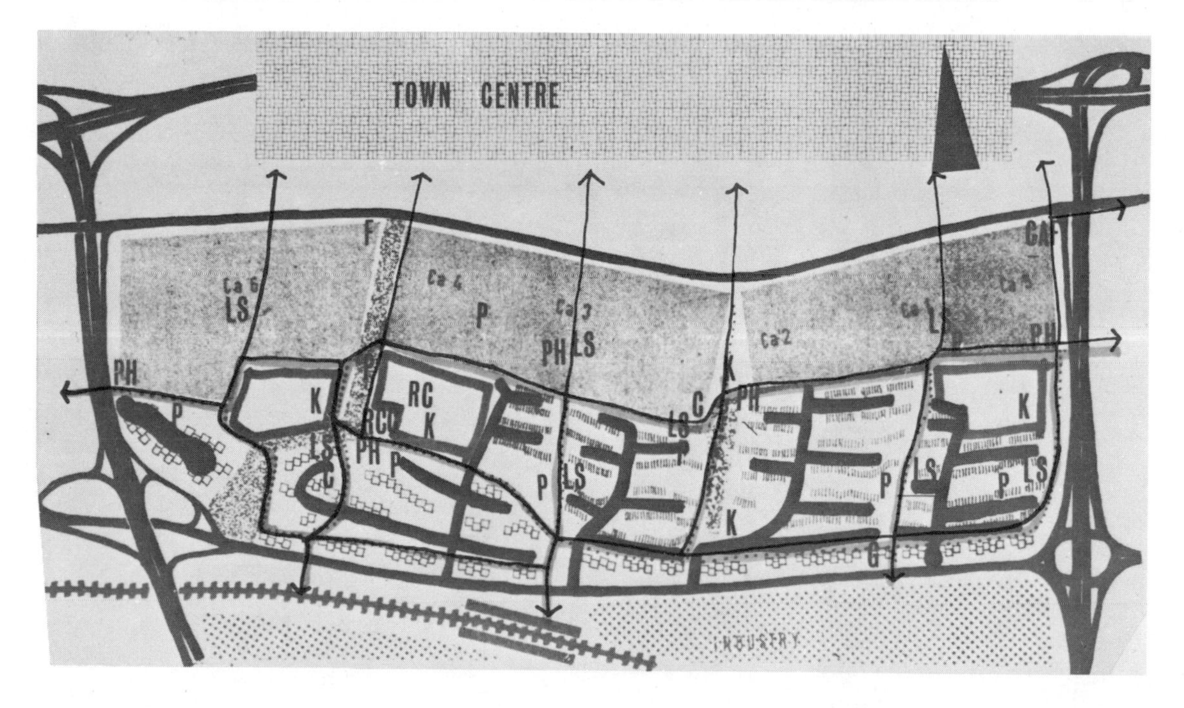

Fig. 8.203. Plan of whole area related to Carbrain North and the Town centre.

LAND USE

PRIMARY SCHOOL	PS
PRIMARY SCHOOL R C	RC
CHURCH R C	RCC
LOCAL SHOPS	LS
PUBLIC HOUSES	PH
CHILDRENS PLAY AREAS	P
KICKABOUT PITCHES	K
COMMUNITY ROOMS	
COMMUNITY CENTRES	C
PETROL FILLING STATION	G
FLATTED FACTORY	F
CAFE	CAF

Building form

On the regular slopes of the hillside three-storey terrace houses are used in staggered rows that closely follow the contours.

The houses are double aspect, privacy being safeguarded by spacings between terraces of a minimum of 70 ft (21 m); this permits the use of screen planting between terraces.

All living and dining rooms look towards the sun and view, and are only faced by bedrooms at a distance of 70 ft (21 m).

Non-residential uses

There are three primary schools (one Roman Catholic) in the area as indicated on the sketch.

Local shops are sited on the pedestrian ways and it is proposed that consideration be given to the expansion of these facilities in two cases to provide a focal point with a community hall with smaller rooms in suite. Two sites for public houses have been provided, one in each of the focal points previously referred to.

It has not been considered necessary to provide flatted factories in the area in view of the proximity of the industrial area across the Ring Road.

Car provision

(a) *Terrace housing*. The first stage garaging is provided in the podium under the two-storey houses and forms 66% of the total provision for terrace houses.

The second stage provision in the form of covered or open parking is positioned on the opposite side of the road to the first stage.

Visitors' parking and space for service vehicles is provided at the end of the cul-de-sac.

(b) *Stub blocks*. In some cases first-stage garaging is carried under the block with an elevated footpath above. In other cases a large communal garage is provided elsewhere with the roof used as a play area. Total first-stage provision is approximately 60%.

Second-stage provision is in the form of two level garaging adjacent to the stub blocks with the roof being used as open space or play space.

(c) 6–7-*Storey stub block*. These blocks are entered at second- or third-floor level making use of the slope of the site.

There are several permutations of dwelling within the block including three appartment maisonettes and flats and four apartment flats. Children's play spaces are also proposed at intervals under the blocks.

The dwellings in the block have a refuse chute with central bin chamber accessible at road level.

The block has a three-floor walk-up from pedestrian level to the top maisonette which has an open terrace or patio. It is appreciated that previous Corporation policy has been to confine walk-ups to no more than two floors, but it is considered that the maisonette is sufficiently attractive to compensate for this.

If a reduction of height of walk-up to two floors were applied to the site a loss of density of 12 rooms per acre, (30 rooms per ha) (accommodation for 650 persons) would result, or about 10%.

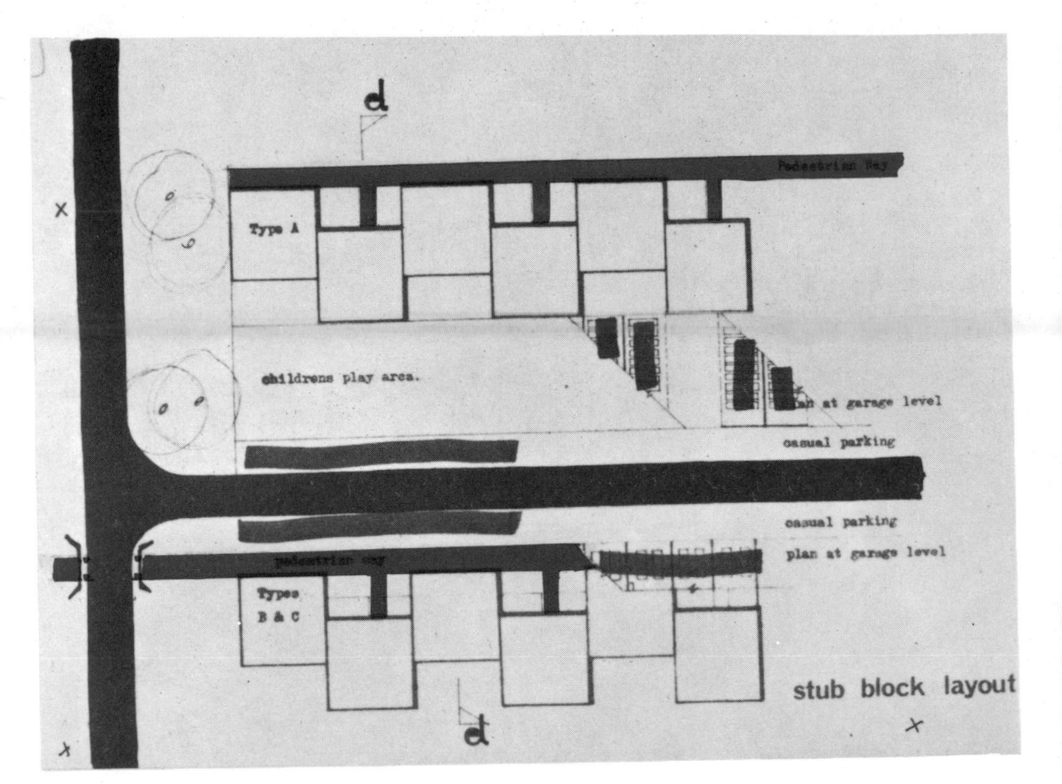

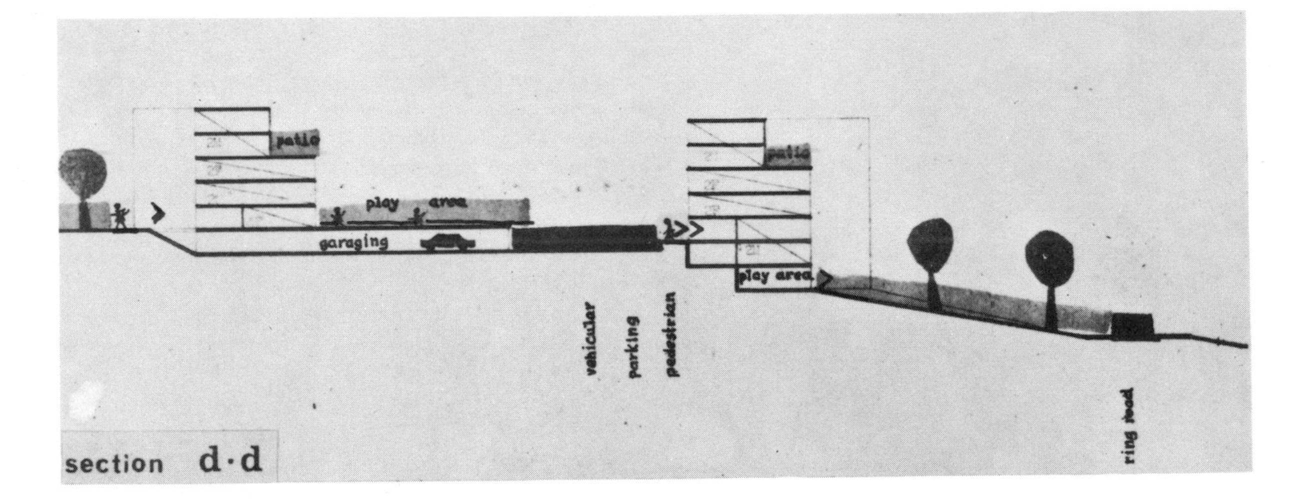

Fig. 8.204 and *Fig.* 8.205. Diagrammatic stub block lay-out. Plan and section.

Where the access road occurs at the entrance side 50% garage provision is possible within the block itself. The roof of the garages is used as a footpath and it is from this level that the three-floor walk-up occurs. Occupants arriving by car have the added amenity of access direct to the main staircase at the lower level from the garages.

Six- or seven-storey stub blocks of flats and maisonettes are used where the steeped slopes occur and the ground is irregular.

It is also proposed to use these blocks to define certain areas of the site and emphasize the ground form.

Garages have been confined to the mouths of culs-de-sac and are positioned in communal podiums under two-storey terrace housing. The roofs of the garages form patios and paved areas for the houses over them.

This form of garaging, together with two-storey garaging and garages adjacent to stub blocks, is first-stage provision and forms approximately 65% of the 100% total.

An attempt has been made throughout the whole layout to make the garaging as unobtrusive as possible.

Open space

Adequate provision for toddlers and children's playgrounds has been made within the layout.

Toddlers' play spaces are placed close to the family terrace houses and will be provided at the end of each double terrace adjacent to the pedestrian way and well away from traffic. Each play space will serve approximately 60 houses.

Equipped play spaces are located to minimize noise disturbance, adjacent to the main pedestrian ways. Six of these have been provided.

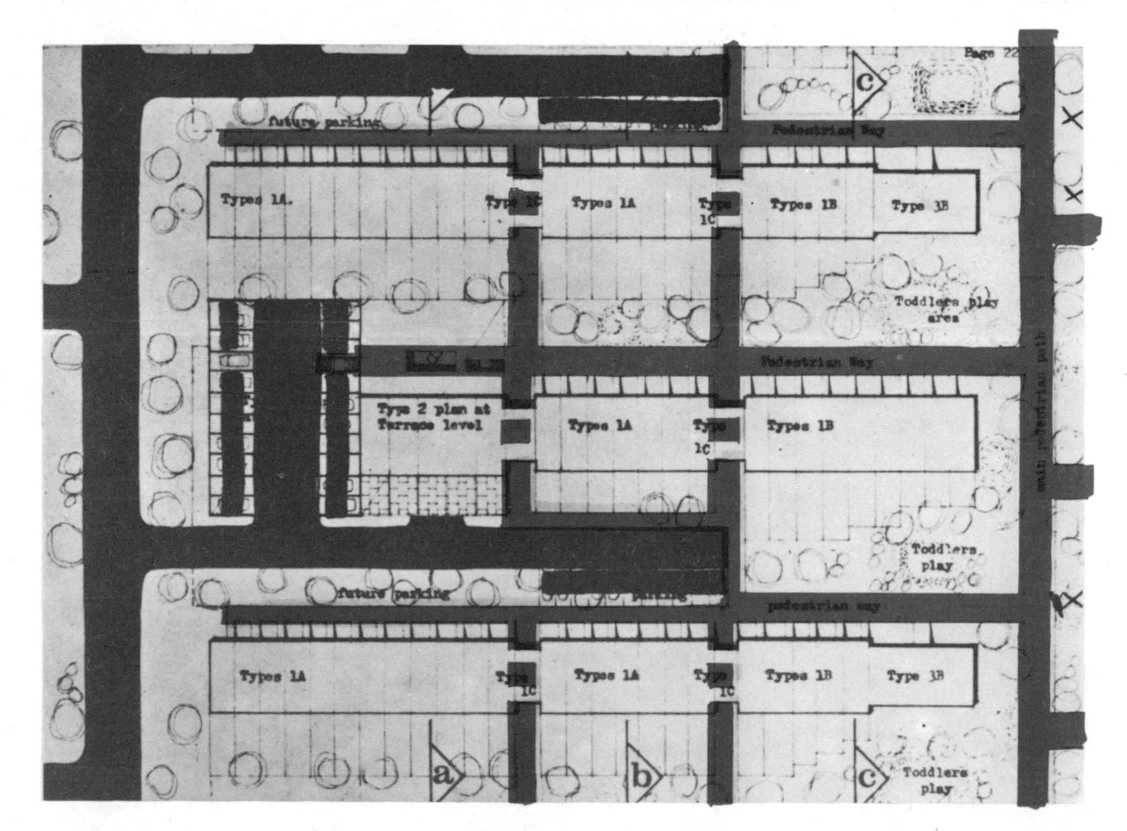

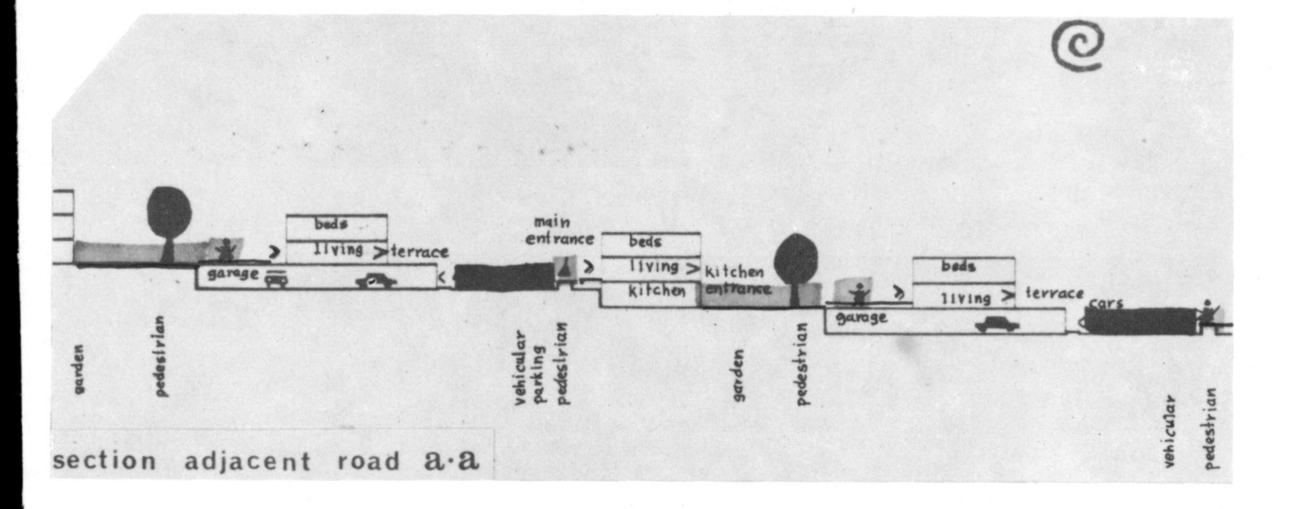

Fig. 8.206. and *Fig.* 8.207. Diagrammatic. Terrace lay-out. Plan and section.

Kick-about pitches have been allowed for in the layout but these facilities would be greatly enhanced if the play spaces of the three schools proposed could be made available outside school hours.

The provision of playgrounds and kick-about pitches totals 3·5 acres, (1·4 hectares).

Incidental public open space is provided among the terrace housing between the access footpaths and the private gardens. These spaces could be treated in a variety of ways but generally include tree and shrub planting to provide screening.

VIII. 12.26. *Oakdale Manor, North York, Ontario, Canada*

Design: Jack Klein and Henry Sears.

Origin: A scheme of speculative houses, to let. Completed early 1962 and fully rented at an early stage. 41 row houses on 2½ acre (1 hectare) site. Underground parking for 48 cars. Two staircases connect dwellings with parking basement.

"*Because there were no significant features surrounding the steeply sloping site the units were grouped to provide visually interesting internal spaces and to fit naturally into the complex contours of the site. Three basic units were developed and arranged with connecting links over the two pedestrian entries into the project. The spiralling form creates public and private spaces and masonry screens give privacy to gardens of units facing the main street. All units are of standard construction—load bearing masonry walls, wood joists.*"(i)

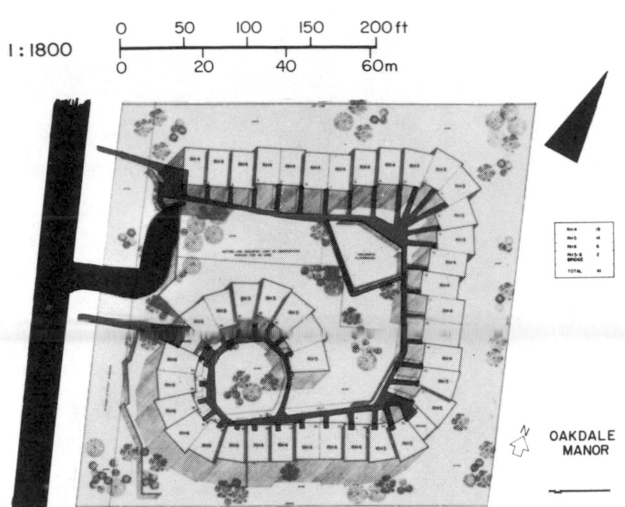

Fig. 8.210. Plan with garaging below dotted.

Fig. 8.208. Courtyard.

Fig. 8.211. Entrance to garaging.

Fig. 8.209. Pedestrian entry.

(i) Oakdale Manor, North York, Ont. *The Canadian Architect.* September 1962.

VIII. 12.27. *Kentucky Road, Toronto, Ontario, Canada*

Design: Jack Klein and Henry Sears.

Origin: Proposals by Revenue Properties Co. Ltd.

247 dwellings, 238 off-street parking of which 156 are underground. A density of 27 dwellings per acre (67·5 per hectare). The majority are family houses.

It is one of a number of schemes proposed or under way which the architects have designed providing an economic mixture of surface and underground parking.

Fig. 8.212. Plan.

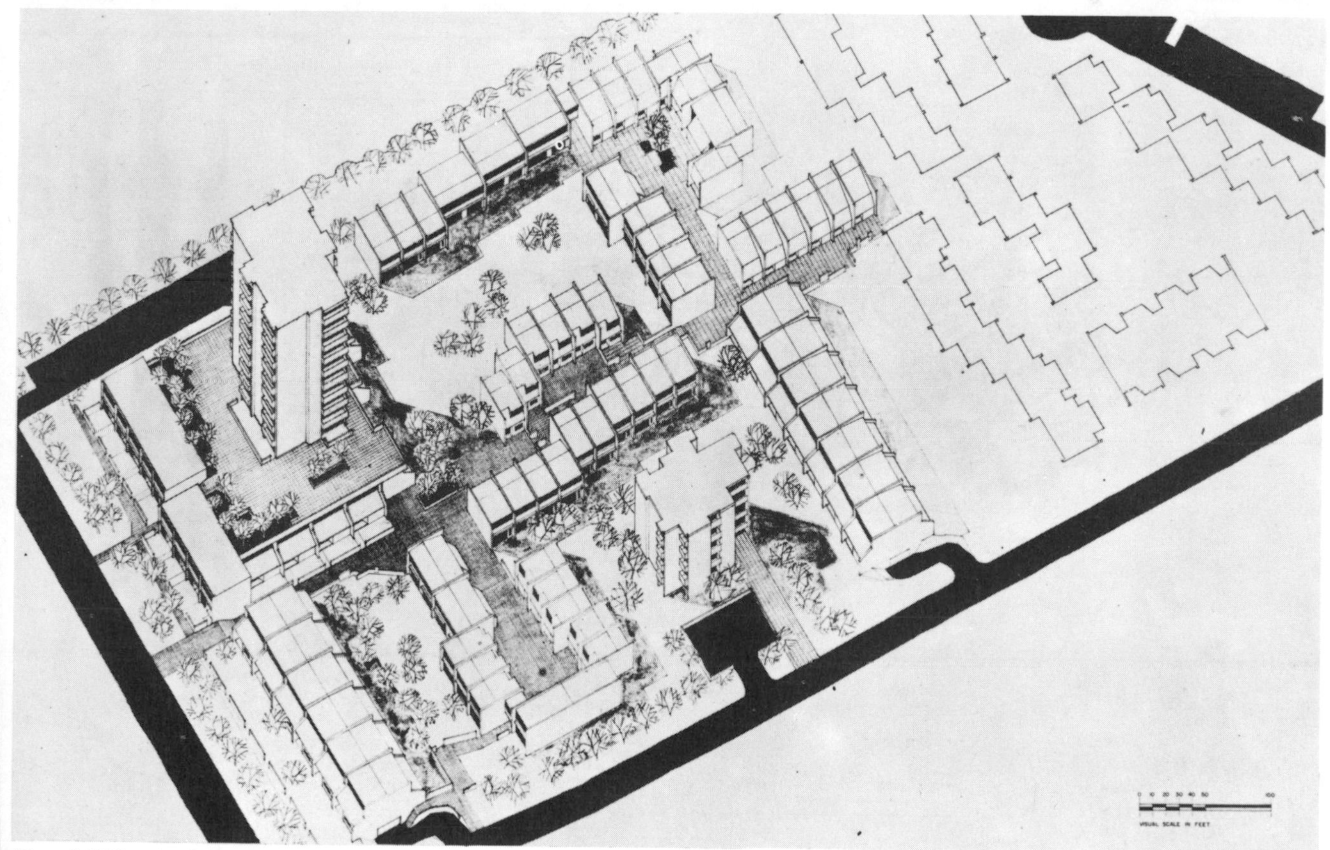

Fig. 8.213. Aerial view showing paved urban side of dwelling contrasting with green side.

VIII. 12.28. *Solna, Stockholm, Sweden*(i)

Design: Forsman, Snellman, Brolin, Arkhammar.

Origin: Competition, first prize.

This scheme introduces, in 1962, on a large scale the concept of underground garaging in residential areas in Stockholm, where a parking provision of 1·8 per dwelling is the highest enforced in Europe.

Just as Vällingby, Baronbackarna in Orebro, began an epoch with clear superblocks, so this may be the beginning of a new approach to Swedish dwellings.

The roofs of the garages, partly buried in the fall of the ground, are used as walkways and play spaces. There is a continuous footpath system leading to all the main goals. Unlike many Swedish schemes, traffic segregation is a functional principle, not just a fashion to follow. It should be added that it is only in Sweden that the principle has, as yet, become a fashion. But other countries should guard against this danger.

Fig. 8.214. Decks above garages.

Fig. 8.216. Aerial view of scheme from the lake.

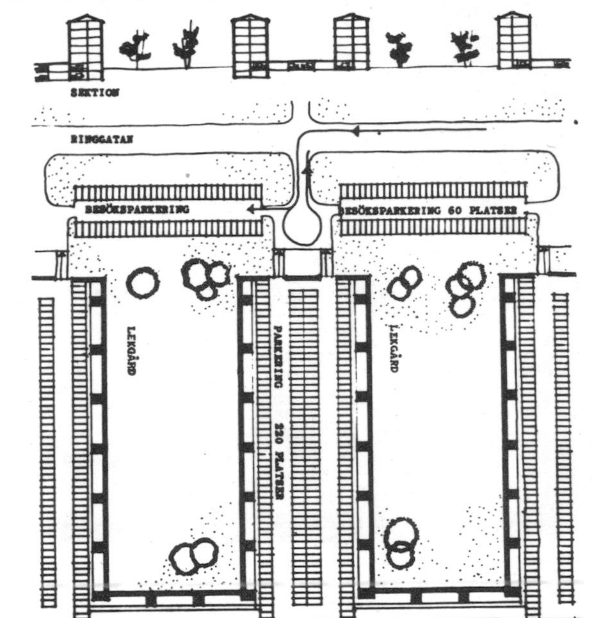

Fig. 8.217. Plan of garages below deck.

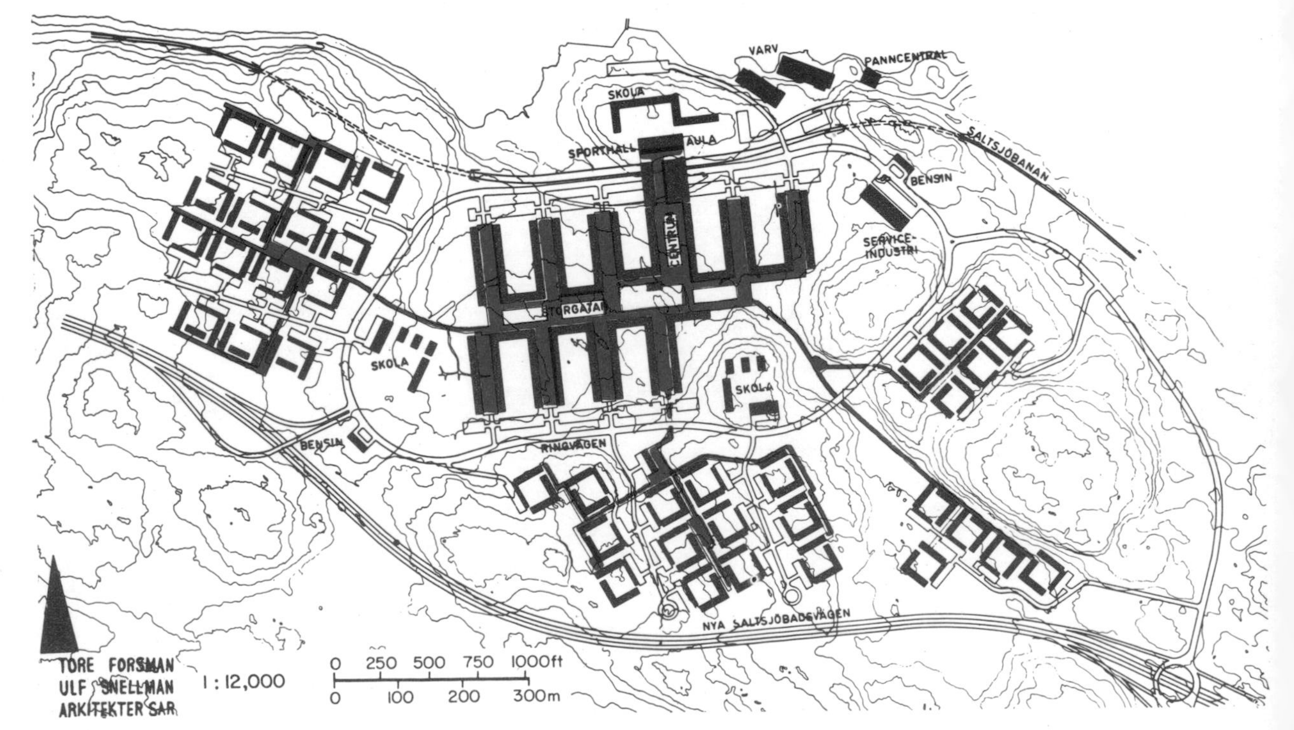

Fig. 8.215. Plan.

(i) No statistics of this very large scheme have been sent by the architects.

VIII. 12.29. *Spon End, Coventry*

Design: City Architect and Planning Officer, Arthur Ling.

A scheme to provide dwellings for 2100 inhabitants approximately, 69% of population in flats and 31% in houses, with one car place per dwelling. A density of 100 ppa (250 ppha) has been planned, building started in 1962.

This is one of two comprehensive redevelopment areas approved as such in 1957. None of the existing industry is to be retained and many of the eighty existing shops will be extinguished. Clubs, churches, schools, etc., are to be provided. The present population is 1300, the population in 1950 was approximately 2000, as aimed at in the new development. The path system is to connect, via underpass, directly with Coventry's main pedestrian centre.

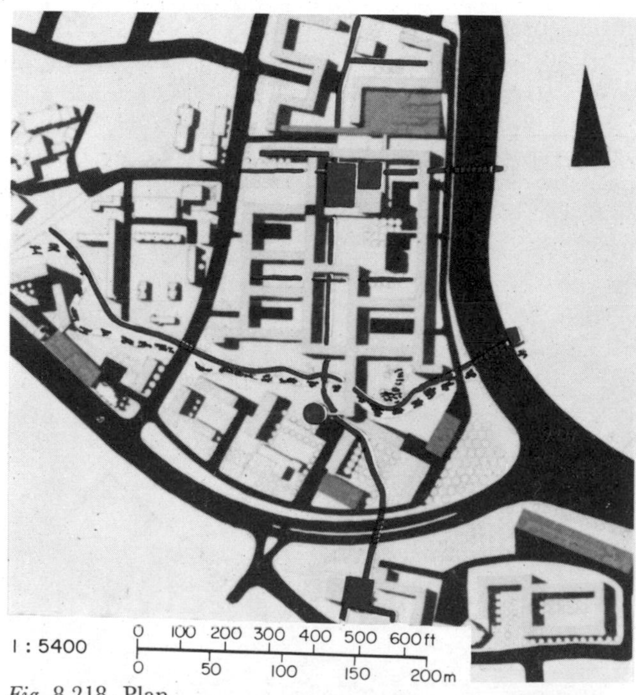

Fig. 8.218. Plan.

Fig. 8.219. View of model.

Reference

City Architect. Spon End. C.D.A. Coventry 1961.

VIII. 12.30. *Married Quarters, British War Office (A)*

Design: Morrison and Partners. Partner in charge, Minter. Job Architect, Cook; Planning Consultant, Ritter.

Gross density 10·8 dwellings per acre (27 per hectare). 317 houses, 76 garages and a shop. The main pedestrian route runs through a continuous green belt towards the shop. It approximates the desire line and people would therefore gravitate towards it naturally. (Design 1962.)

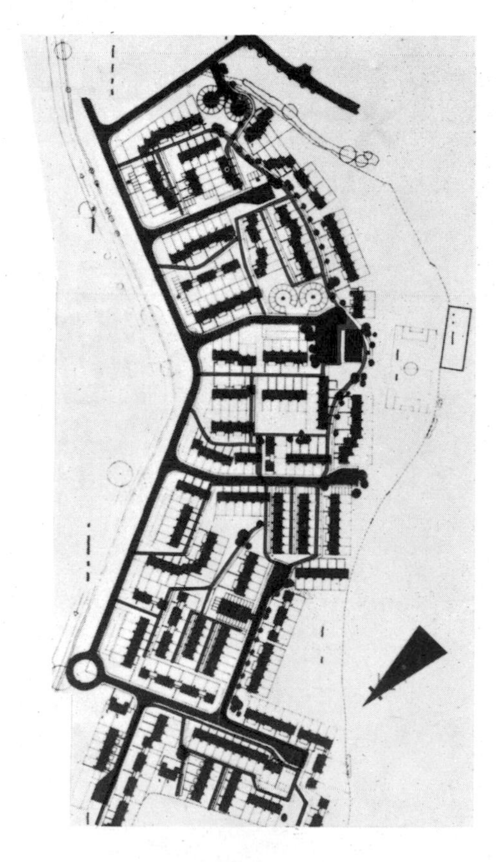

Fig. 8.220. Plan.

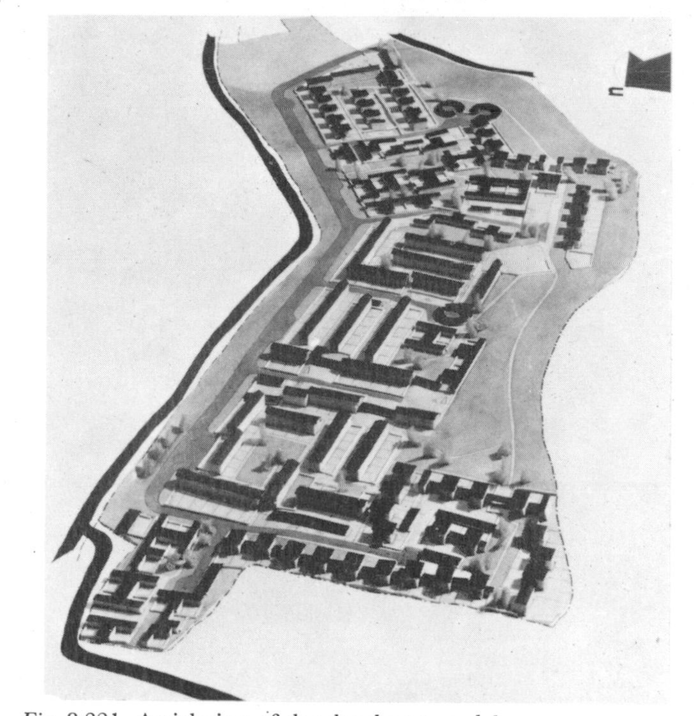

Fig. 8.221. Aerial view of sketch scheme model.

VIII. 12.31. *Married Quarters, British War Office (B)*

Design: Morrison and Partners. Partners in Charge, Minter and Cook. Job Architect, Spring. Planning Consultant, Ritter.

Gross density 14 dwellings per acre (35 per hectare), 159 houses, 40 garages. Close to an existing camp, conflicting route needs. Threading through the dwellings the path system has a focus in the children's playground in the middle of the development. Vehicular access to existing electricity substation and borehole complicated the requirements. (Design 1962.)

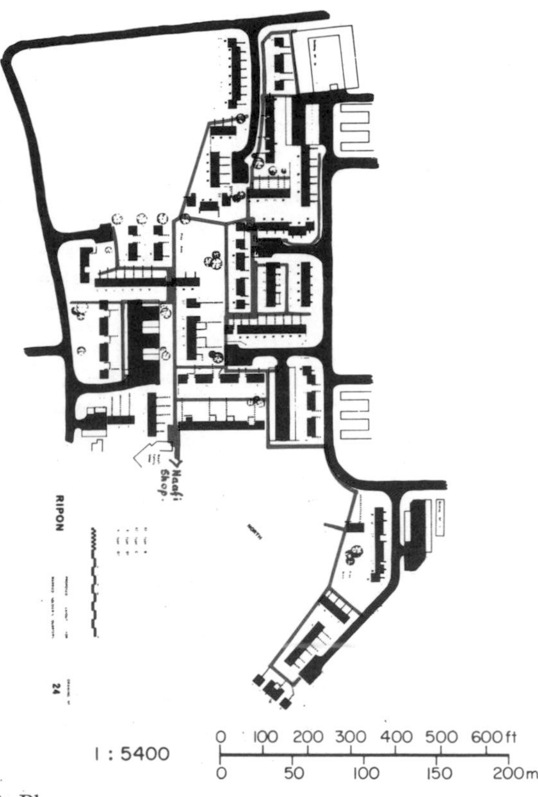

Fig. 8.222. Plan.

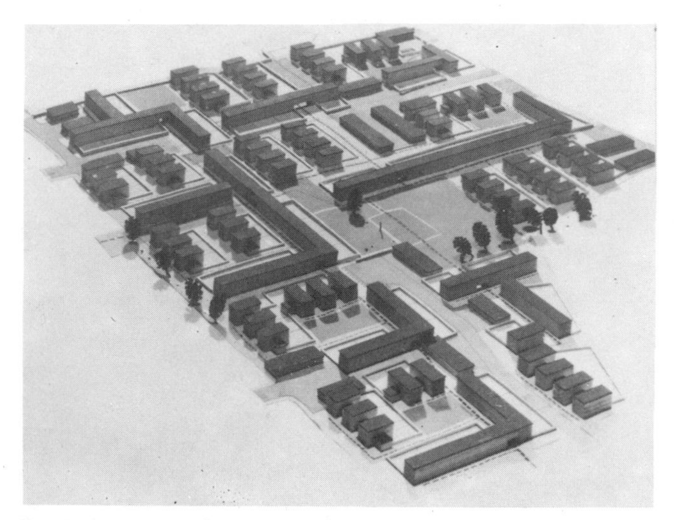

Fig. 8.223. View of sketch model.

VIII. 12.32. *Rowlatts Hill, Leicester*

Design: City Planning Officer, W. K. Smigielski. Design Group, Blachnicki, Baldwin, Cropper, Fountain, Hayes, Neal, Winter.

Construction to begin 1963. 408 dwellings with over 130 old people's dwellings demanded by the brief. Density 59 ppa (147·5 ppha), garages for all dwellings except old people's. Sloping site, 150 underground or under deck garages for 144 flats in the tower blocks. This avoids tarmac on car roof deserts. Existing mature forest trees are dominant. Orchard trees remain. The scheme replaced a traditional plan in a very few weeks, designed by the newly staffed City Planning Office. The original treatment of the culs-de-sac seems to contradict the Radburn Idea of having children on the path side, away from the vehicles. Here father, so the theory runs, can get help from children for cleaning (!) the car. It will be of vital interest to study the use of this project. A more definite stop between culs-de-sac and play space is anticipated. A cost comparison showed that the new Radburn lay-out gave a cost of £192 per dwelling, whereas the original traditional scheme was £242 for roads and services, and had fewer amenities. (See also p. 249.)

Fig. 8.224. Aerial view of model. The brief for this scheme asked for 50% old peoples dwellings. Stepping up the density has decreased this percentage. Without garages, these form the core of the superblock.[1]

(i) Ritter P. Rowlatts Hill. *J.T.P.I.* May 1963.
Ritter P. Radburn Superblock. *A.J.* 10.4.63.

VIII. 12.33. *Brandon Estate, London, England*

Design: Architect to L.C.C., Hubert Bennett in succession to Sir Leslie Martin. Job Architect, E. E. Hollamby, D. Stamp, D. Gregory-Jones.

1059 New dwellings, 3742 persons at a density of 41 dwellings per acre (102·5 per hectare), and 145 ppa (363 ppha), with 12 shops, only 88 garages (150 spaces in all) and other facilities associated with large rehabilitated area.

Most of the new development with the 18-storey blocks connect directly with the pedestrian centre by footpath. There are nine toddlers' play spaces, 4 fitted playgrounds which attract children from too wide a radius (!)

Brandon Wood Park which is also connected directly by footpath is to be enlarged considerably, converting existing housing to park.

Reference (with full list of credits).
Housing at Brandon Estate, Southwark. *A.J.* London, 1 November 1961.

Fig. 8.225. Shopping centre. High flats in point blocks, accessible under slab, seen in the background.

VIII. 12.34. *Royal Victoria Yard, Deptford, London*

Design: Architect to L.C.C., Bennett. Housing Architect, Campbell. Assistant Housing Architect, Whittle. Assistant Senior Architect, Hollamby. Layout and Design, Knight, Westwood, Gilmour and Streatfield. Scissors blocks, D. G. Jones, Grove, A. C. C. Jones, Hampson.

L.C.C. preliminary model of high density residential scheme on the banks of the Thames. 45 acres (18 hectares), 1300 dwellings and an old people's home. 140 ppa (350 ppha), 50% garages mainly under the eight-storey blocks. Deck housing, eight-storey "scissor plan" decks running internally connecting the lower blocks. Covered walkways lead to shopping and other communal facilities. Riverside walk making use of eighteenth century warehouse. (Planned 1961.)

Fig. 8.226. Aerial view of model.

References
Redevelopment of Royal Victoria Yard Site, Deptford. L.C.C. April 1961.
Thames Side Development at Deptford. *A.J.* 13 April 1961.

VIII. 12.35. *Brunswick Redevelopment Area, Manchester*

Design: Austen Bent, Director of Housing.
Planned 1962.

165 acres (67 hectares) close to the city centre. 90 habitable rooms to the acre (225 per hectare), 60% of the dwellings with ground level access. 39 shops, schools and other facilities. Traffic separated path system gives direct access to all goals. Amenities are concentrated along paths. 100% garage provision. Plan for major roads, secondary roads and access roads to serve area.

Fig. 8.227. View of model.

VIII. 12.36. *Primrose Hill, Birmingham, England*

Design: City Architect, Sheppard Fidler. Deputy, Ash. Housing, Griffith. Architect in charge, Fenter.

3048 persons in 740 dwellings, half two-storey houses. Net density 76 ppa (190 ppha). Total site 76 acres (31 hectares), open space 34 acres (14 hectares).

Pure Radburn Superblock, planned as neighbourhood, with neighbourhood shopping, schools, old people's homes and 17-storey flat block as focus of path system. Garage per dwelling. Sloping site (up to one in ten) used to store cars below houses in split level plans, and in podium under tall block with landscaped surfaces. Many dwelling types. Building starts 1963.

Fig. 8.228. Aerial view.

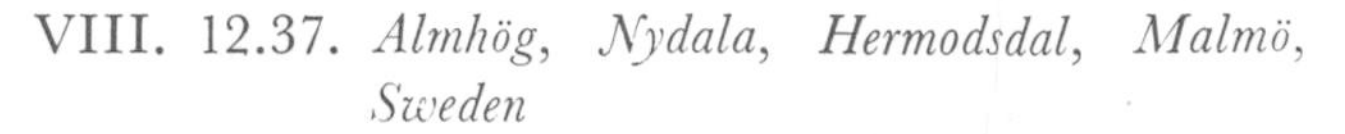

VIII. 12.37. *Almhög, Nydala, Hermodsdal, Malmö, Sweden*

5700 flats mainly in four-storey, eight-storey and some higher blocks of flats. Pedestrian systems, in each of the three clearly defined superblocks, lead to schools parks, and shopping centres.

The architects of Almhög (foreground) are Roos, Thornberg, Ekstrand. Work began in 1959.

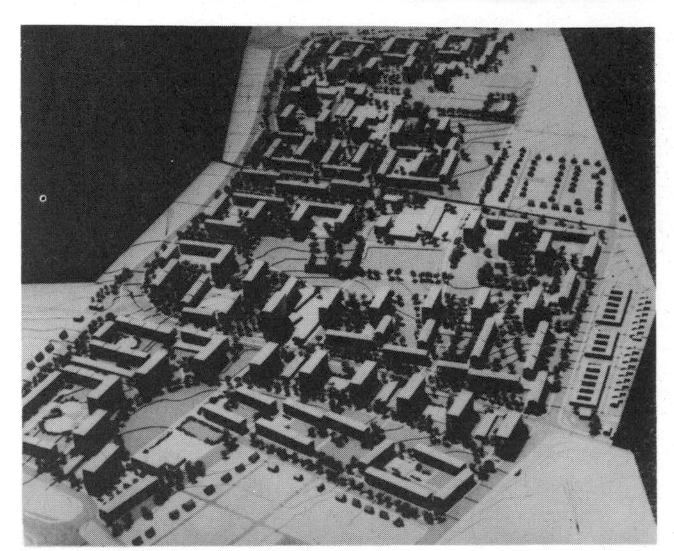

Fig. 8.229. Aerial view of model.

VIII. 12.38. *Mellanheden, Malmö, Sweden*

This estate of 2150 flats, in superblocks including schools, sportsfield and an underground garage, designed by Thorsten Roos, is one of several schemes in Malmö built within the town plan of Gunnar Lindman, pioneering the principle of traffic separation in Scandinavia and Europe, from 1950 onwards. There are several different kinds of schemes by such architects as Jaenecke, Samuelson, Thornberg and others. It is surprising that it took twelve years for the ideas to get through to Copenhagen only 17 miles (27 km) away, albeit across the water and the border. The first large municipal scheme of this nature is now under way in Copenhagen.

Reference
Malmö 1862–1962. Malmö, September 1962.

Fig. 8.230. The first hexagonal housing in Sweden.

VIII. 12.39. *Usworth Development, Co. Durham.*

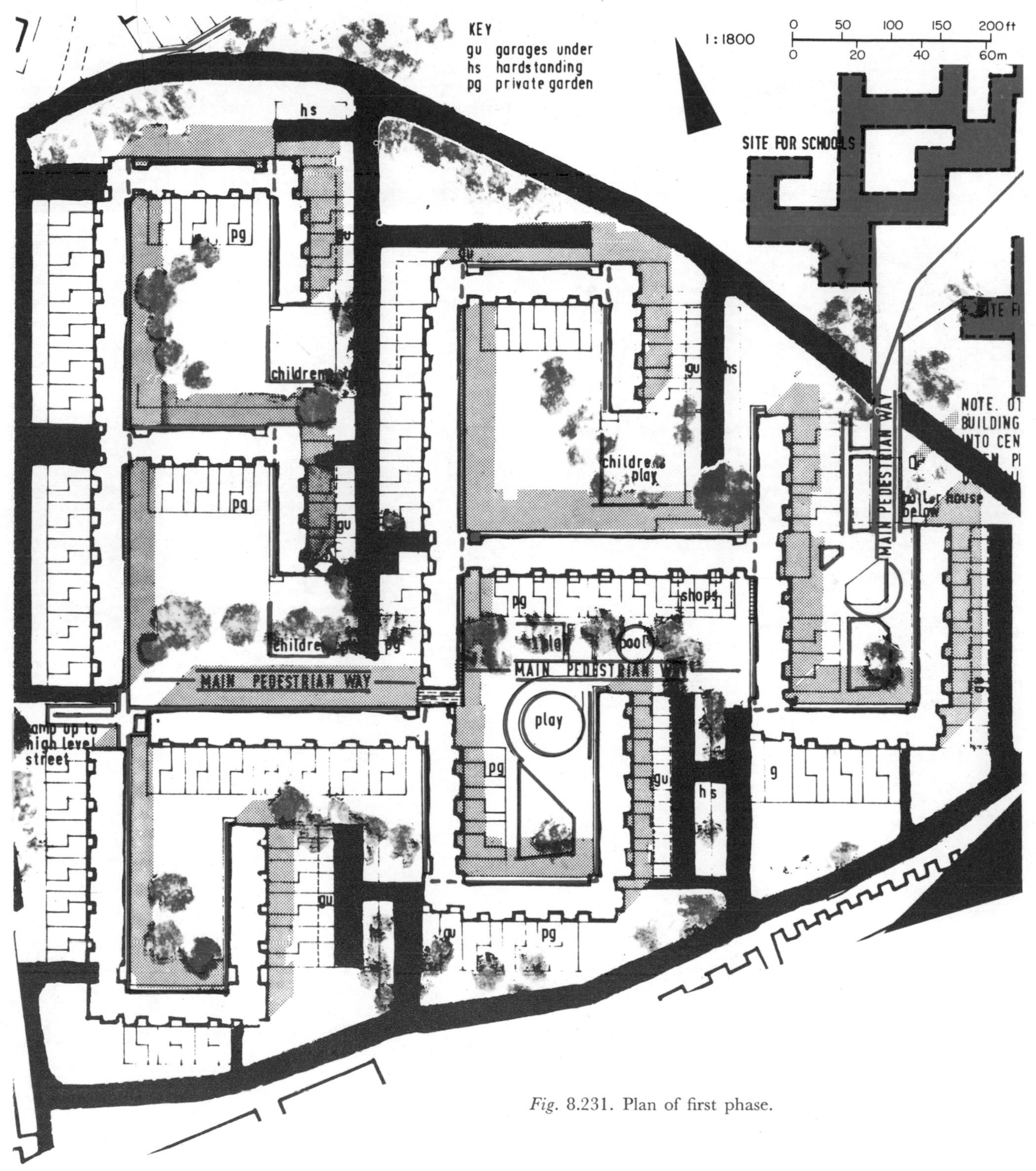

Fig. 8.231. Plan of first phase.

Design: J. H. Napper & Partners; Project Team, A. Collerton W. Barnett, E. Scoffham and R. A. Allott.

This development for Washington U.D.C., will include district heating using piped deliveries of local coal with high temperature hot water mains distribution at roof level, and is planned for 100% car provision.

The first phase of the scheme, illustrated here, is designed for a density of 135·75 ppa (340 ppha) with 650 dwellings accomodating 2172 people. This phase consists of terrace housing: later phases are to provide schools, a church and communal buildings, and more housing. The terraces are laid out to provide a series of enclosed play areas for children. At ground level terraces consist of family maisonnettes with their own gardens, while upper floors and corner units consist of small flats with direct access and delivery route by means of a high-level pedestrian street. Advantage has been taken of the sloping site to vary the height of the terraces from three to five storeys, the high-level street remaining level throughout its length.

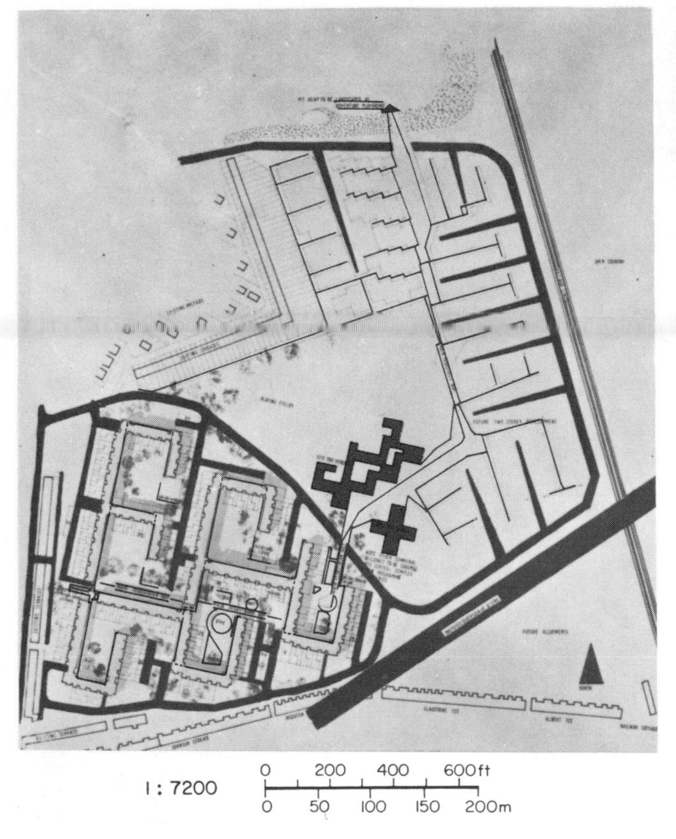

Fig. 8.232. Plan of lay-out for whole scheme, showing connection to slag heap to be turned into adventure play ground.

Fig. 8.233. View along deck.

Fig. 8.234. Pedestrian area at ground level.

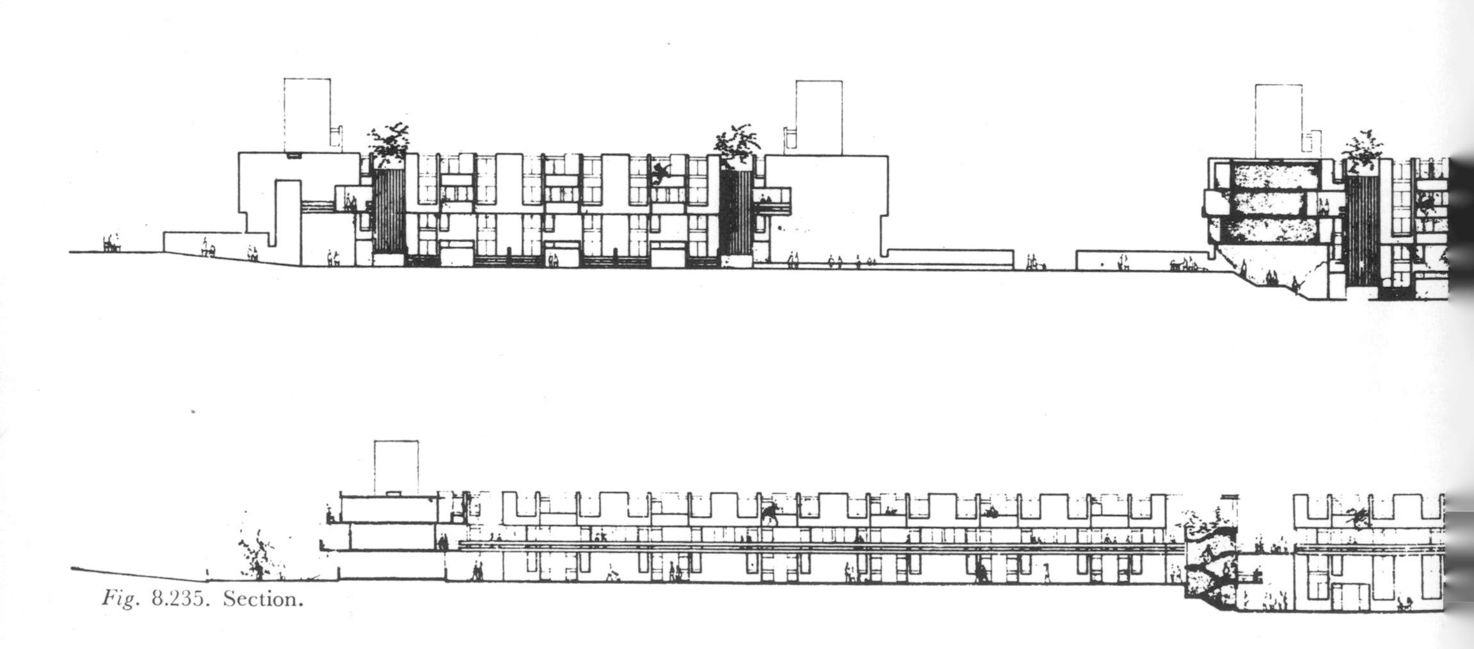

Fig. 8.235. Section.

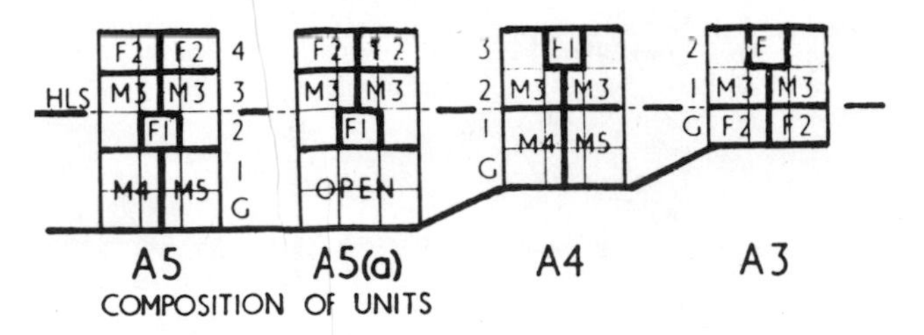

Fig. 8.236. Composition of Units and Plans.

Key to floor plans: bic, built-in curtain track; c, cooker; cp, cupboard; d, duct; dc, drying cabinet; dn, down; his, high level storage; ka, kitchen appliances; s, sink; st, store; ts, tenants' store; w, wardrobe; HLS, high level street.

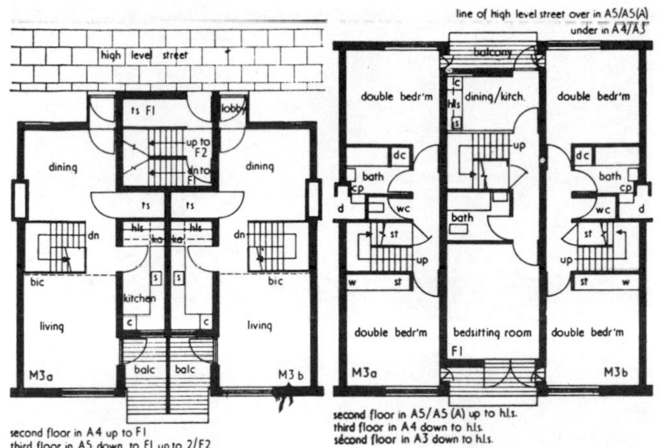

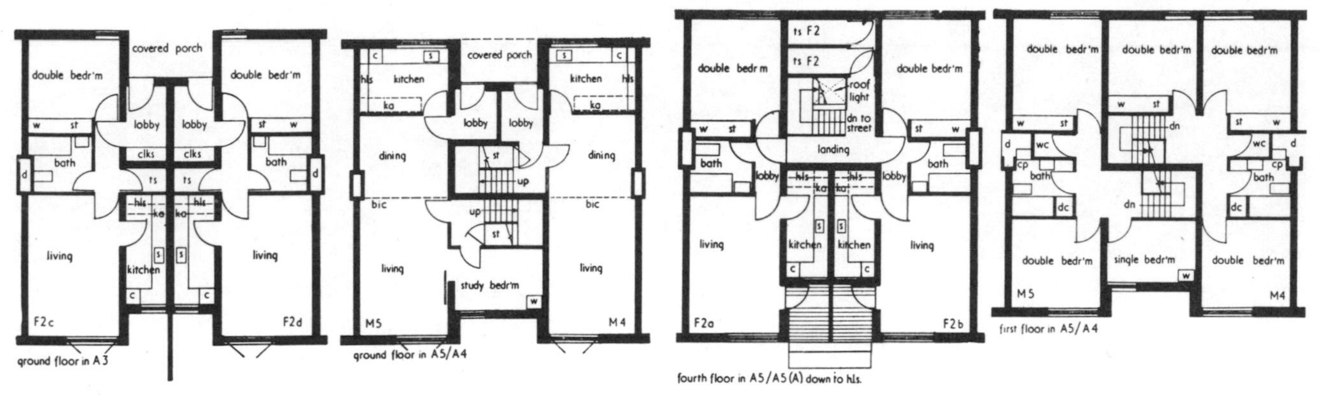

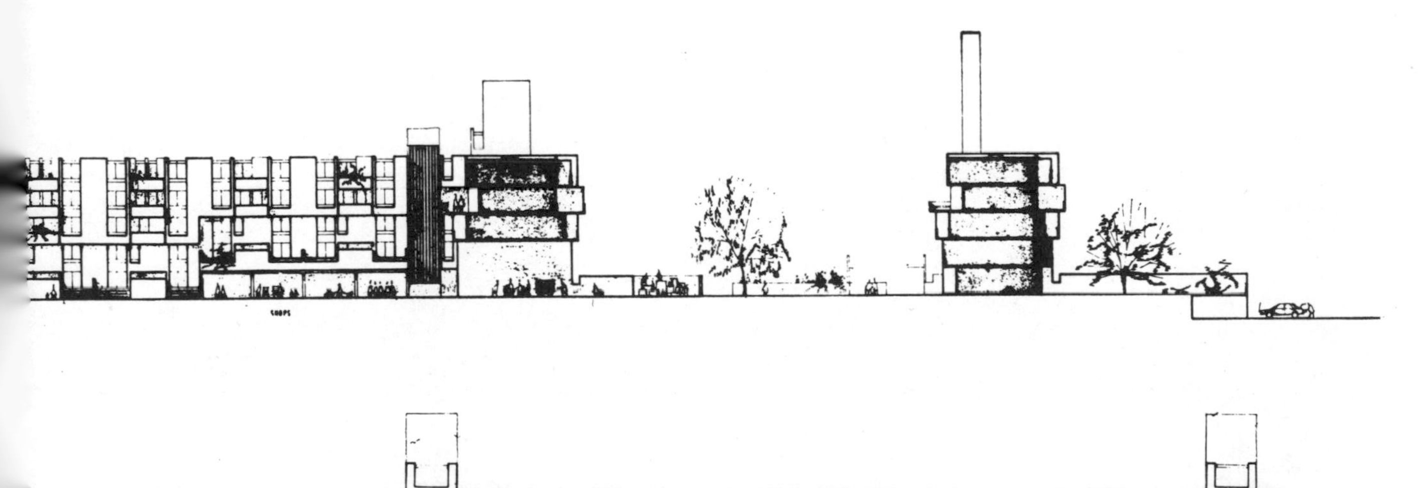

Fig. 8.237. View from the edge into the development.

VIII. 12.40. *Les Buffets, Fontenay-aux-Roses, Paris*

Design: Lagneau, Weill, Dimitrijevic, Perrotet.

900 inhabitants on 7·5 acres (3 hectares) (density 120 ppa, 300 ppha), 200 flats in 11 identical blocks placed around a pedestrian area with play spaces. Access with parking space each side of each cul-de-sac, keeping the vehicles to the edge of the superblock. This prizewinning scheme has homes adjustable to individual needs (5 on each floor) and common television aerial. The landscaping shows a fine balance of organized play area and informality.

VIII. 12.41. *Hillfields, Coventry*

Design: City Architect, Arthur Ling.

This is a comprehensive development of an area near the centre of the city began 1961. Housing is at a density of 117 ppa (292·5 ppha), 40% houses, 60% flats. Industry and shopping is to remain within the area. 100% storage provision for cars is provided and 100% traffic segregation. The development is for a total population of 5900.

Maisonettes, flats, patio houses and terraces are all to be included, with special provision for housing old people. The path system is to connect with Coventry's main pedestrian centre. (See also p. 45.)

Reference City Architect, Hillfields C.D.A. Coventry 1960.

Fig. 8.238. First stage showing covered walkways linking blocks.

INCLUDED AT PRESS STAGE

VIII. 12.42. *Highfields, Leicester*

Design: W. K. Smigielski, City Planning Officer.
Design Group: H. Blachnicki, K. Baldwin, R. Cropper, E. W. Fountain, D. Hayes, J. Winter.
Realization Group: G. H. Fanstone, J. W. M. Manners.

Origin: 29 acres (11·7 hectares), slum clearance area on the fringe of the central area, partly flat, partly undulating.

Nature of Scheme: 881 dwellings. Total pedestrian vehicle segregation. A series of inter-connected residential squares, human in scale, linked by a footpath system at ground level and an upper level walkway. The corners of the squares are open to avoid a feeling of claustrophopia. The squares consist of four-storey maisonette blocks approached directly from the footpaths or the upper level walkway. Separate identity of individual squares is achieved by different landscaping and different colour schemes. In the centre of the site a junior school (2 acres, ·8 hectare) linked with a district shopping centre, marked by a tower block of flats. Buildings served from outside by a service road with attached service culs-de-sac. Garage provision—approximately three garages for four dwellings provided in the basement of blocks. Open car-parking brings the provision to 100% and there is additional generous provision for visitors' parking and in the shopping centre. Net density 113 ppa (282 pha). A phased 3½ years' building programme starts in 1964.

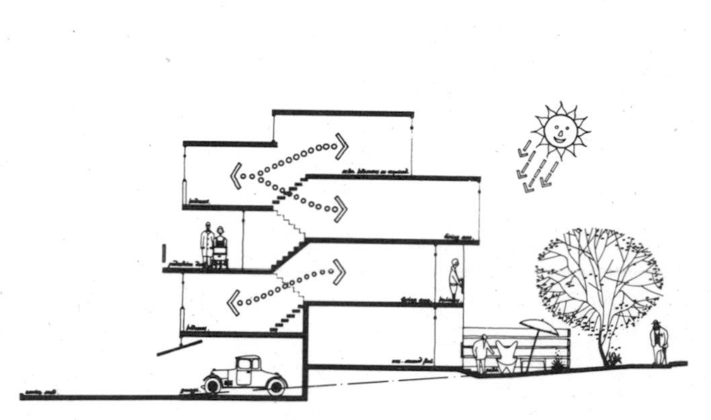

Fig. 8.239. Cross-section through a typical residential block, showing the split-level maisonettes and the upper level walkway.

Fig. 8.240. View of the Model. In the centre, junior school linked with a shopping centre and a tower block of flats. Residential squares with upper level walkways.

Fig. 8.241. Perspective view of a residential square with upper level walkways providing access to individual dwellings.

IX. Residential Renewal

IX. 1. General Principles. **306**
XI. 1.1. Research into Grid Iron Plan Accidents. 306

IX. 2. Examples. **308**
IX. 2.1. Street Improvement Scheme, Pershing Field Area, Fulton Neighbourhood, Minneapolis, U.S.A. 308
IX. 2.2. Mill Creek Redevelopment Area, Philadelphia City Planning Commission, 1954. 308
IX. 2.3. British Bye-Law Housing, Nottingham. 309
IX. 2.4. Clearing Space at the Back of Flats. 310
IX. 2.5. Interwar Housing, Nottingham. 310
IX. 2.6. Transformation of a Metropolis, Chicago. 312

IX. Residential Renewal

IX. 1. General Principles

Under the above heading come two kinds of development:

(1) The gradual rebuilding and restructuring of grid-iron layout, mainly of nineteenth-century origin, where much housing is structurally unsound, both in Europe and America.

(2) The improvement of later residential areas, where rebuilding can have no place for many years (e.g. "Garden City" housing estates), and these may include modern grid-iron areas in the U.S.A. which has attracted intolerable through-traffic.

The areas of application of residential renewal, from a traffic point of view alone, are vast. It is quite true to say that it applies to perhaps 90 % of all housing of all eras! And in all countries. The evil is twofold. The danger to the driver from other drivers and stationary objects is distinct from the danger of mixing pedestrians and vehicles. To the danger of the latter must be added the uncivilized way of life described in detail as suffered by the overwhelming majority.

One of the oldest principles of traffic segregation relates to the nuisance of traffic in residential areas(i) and stipulates "precincts" from which through traffic is kept away by the indirectness of service roads and by limiting access to surrounding main traffic arteries. It is this old principle which awaits application in an ever-increasing number of areas as the main roads become blocked and residential roads, more or less direct, are used as quicker detours. Reichow worked out an early scheme for Düsseldorf on this basis in 1948.(ii) In the U.S.A. this is one of the most prevalent causes of blight in residential areas. The grid-iron plan, the wide streets and the number of cars all aggravate the problem.

However, wherever such areas are regenerated, both evils, driver to driver danger and the lack of a civilized pathway system, ought to be remedied. The two can be suitably combined as will be shown.

To summarize, the ills to be cured are:

(a) a dangerous road system for vehicles;
(b) lack of pedestrian routes that give safety or pleasure for walking;
(c) lack of play facilities for all children;
(d) lack of hygiene, particularly in the first;
(e) and lack of garages.

(i) Tripp A. *Road Traffic and its Control*. London 1938.

(ii) Reichow H. B. *Autogerechte Stadt* 1959.

In the older areas the density is up to 100 houses per acre (250 per ha), 560 ppa (1400 ppha). In the other areas the density is very low.

IX. 1.1. *Research into Grid Iron Plan Accidents*

Research in Los Angeles established clearly that grid-iron planning is more dangerous. It was done for the sake of those who complained that anything else was not as convenient. The conclusions were devastating:

"*The study included eighty-six residential subdivision tracts with a total developed area of* 4370 *acres, representing a population of* 53,000 *persons. It embraced* 108 *miles* (173·8 *km*) *of residential streets including* 660 *intersections.*"

"*The study period included a five-year accident history for each tract. Since right angle collisions represent approximately* 84 *per cent of all vehicular intersection accidents within subdivisions, this initial study has been limited to right angle collisions at intersections of local streets.*"

"50% *of all intersections in grid-iron subdivisions had at least one accident during the five year period ... only* 8.8% *of intersections with limited access experienced accidents during this period ... this is particularly significant since there were* 65 % *more intersections in the limited access areas ...* "(i) Eight times as many accidents occurred each year in grid-iron lay-out.

The research supplied some clear evidence for Reichow's point that three-legged are safer than four-legged intersections! 56% of four-legged intersections in grid-iron lay-outs have at least one accident in five years. In limited access areas only 27%. What monstrously unnecessary danger we have built into our cities is shown by the figure of 3% for three-legged intersections.

The diagram of the author's solution to the regeneration and making safe of grid-iron areas is very similar to that now built in Minneapolis and also planned or built in many other places.

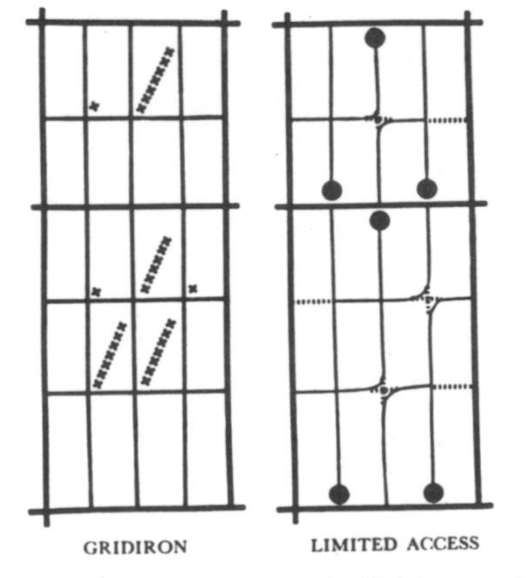

Fig. 9.1. Grid-iron sub-division, redesigned by Marks.

(i) Marks H. Sub-dividing for Traffic. *Traffic Quarterly*. Saugatuck July 1957.

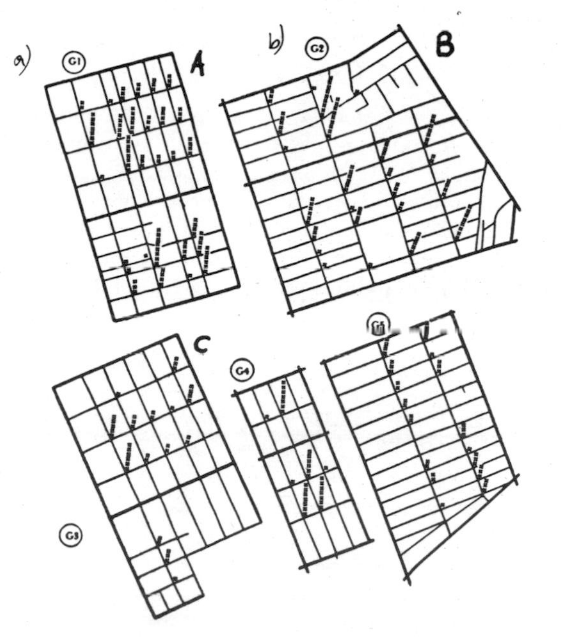

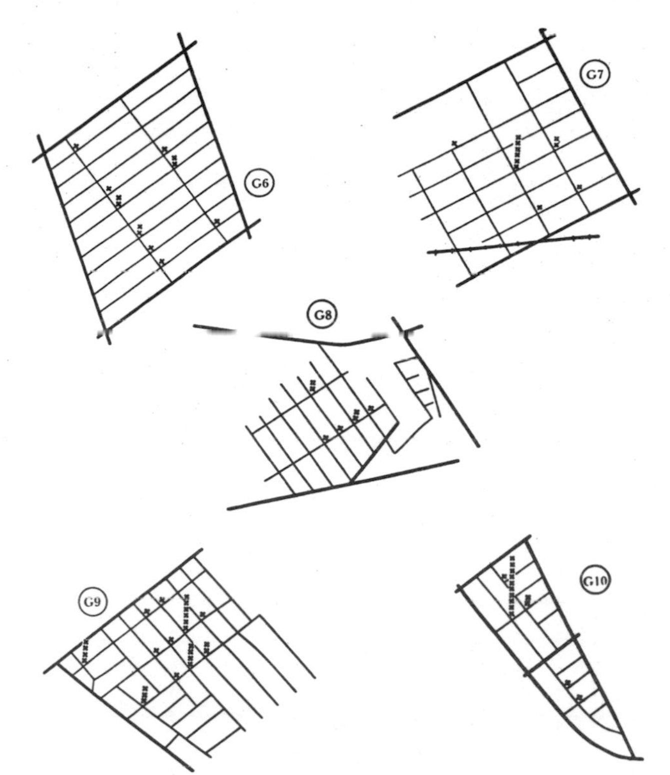

Figure 2. Subdivision Design: Gridiron

Fig. 9.2. Grid-iron.

"Figures 1 and 2 illustrate the accident pattern for a group of grid-iron subdivisions. Each x *represents one vehicular accident during the five-year study period. The heavier lines are major streets bordering the subdivisions which were not included in this phase of the accident analysis."*

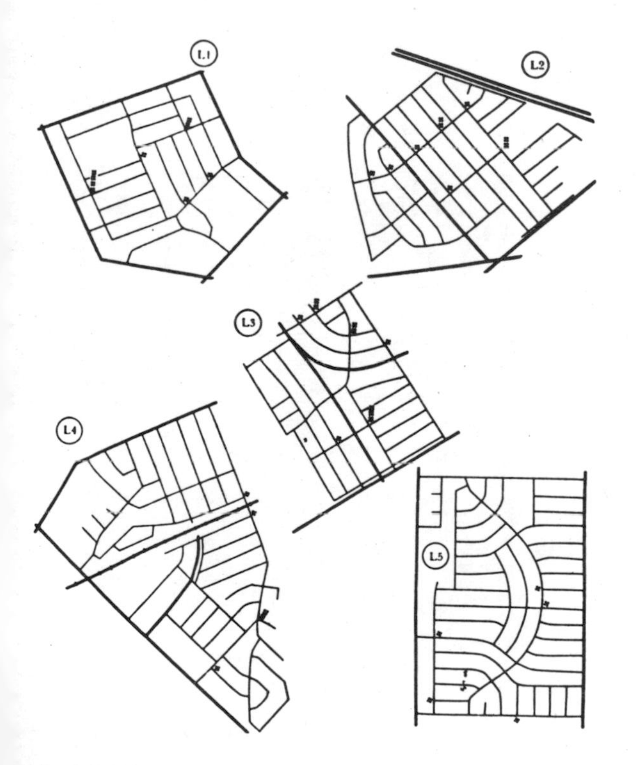

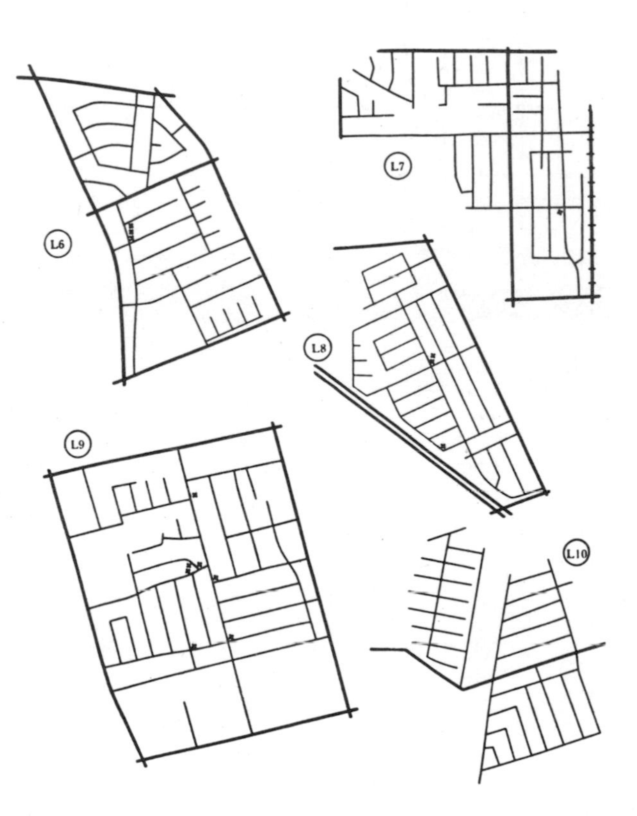

Fig. 9.3. Limited access.

"Figures 3 and 4 illustrate the accident history for a series of limited access tracts for the same five-year period. Particularly interesting is the relative number of accidents in these tracts as compared to the grid-iron tracts, and also the distribution of these accidents. There no longer appears to be a uniform pattern of distribution but rather concentrations at a few intersections. Here again, an accident pattern appears to develop wherever two continuous streets meet at a four-legged intersection. Especially significant is the large number of T intersections with practically no accident record.[(1)]"

(i) MARKS H. Sub-dividing for Traffic. *Traffic Quarterly*. Saugatuck July 1957.

IX. 2. Examples

IX. 2.1. *Street Improvement Scheme, Pershing Field Area, Fulton Neighbourhood, Minneapolis, U.S.A.*

This scheme is to provide for increasing traffic (Minneapolis forecasts a 100% increase from 1960 to 1980), without allowing it to destroy the "safety, value and peaceful quietness" of the roads. Through traffic has been using residential roads, particularly in the grid-iron plans of such cities as Minneapolis, and it has been found that this deteriorates the areas. Thus in limiting and chanelling such through traffic a solution is sought. Many dangerous cross-roads remain after this mild and very cheap system of amelioration.

The first of the junctions was completed by the end of 1962.

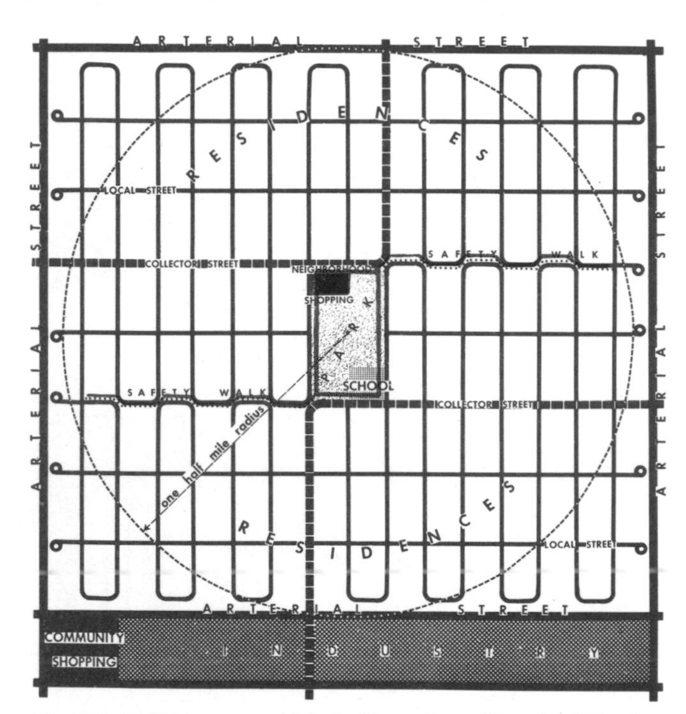

Fig. 9.4. Taking areas within half a mile radius the following neighbourhood scheme is worked out, with "safety walks leading children to school, shops and park". Safety is sought by making the sidewalk zig-zag without crossing a road. The reorganization of the grid-iron system keeps through-traffic to the perimeter of the area.

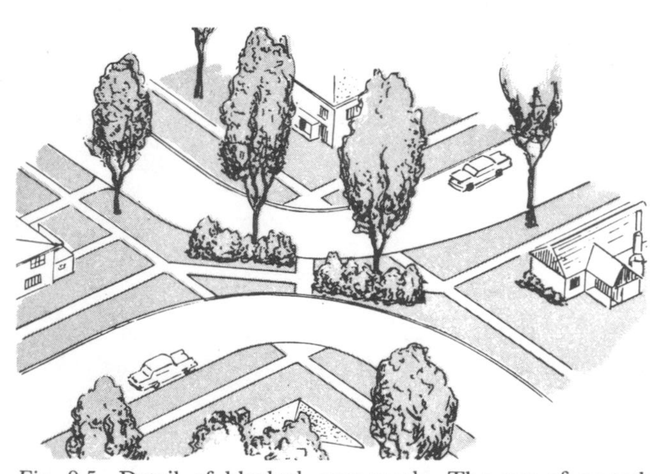

Fig. 9.5. Detail of blocked cross-roads. The use of smooth curves and planting of trees and shrubbery gives the illusion of two curved streets. The sidewalks on the diverter between the two curves serve a dual purpose, both as walkways for everyday use and for emergency passage of firetrucks.

IX. 2.2. *Mill Creek Redevelopment Area, Philadelphia City Planning Commission*, 1954

In this example the formation of loops and abolition of crossroads is accompanied by planting a strip taken from the previous frontages of blocks on one narrow side. The walkway leads to shopping and connects with other parts of a walkway system. The application of this principle is likely to be effective in many grid-iron areas. However, the "corner shop" is not transformed to Greenways but stays on its street corner.

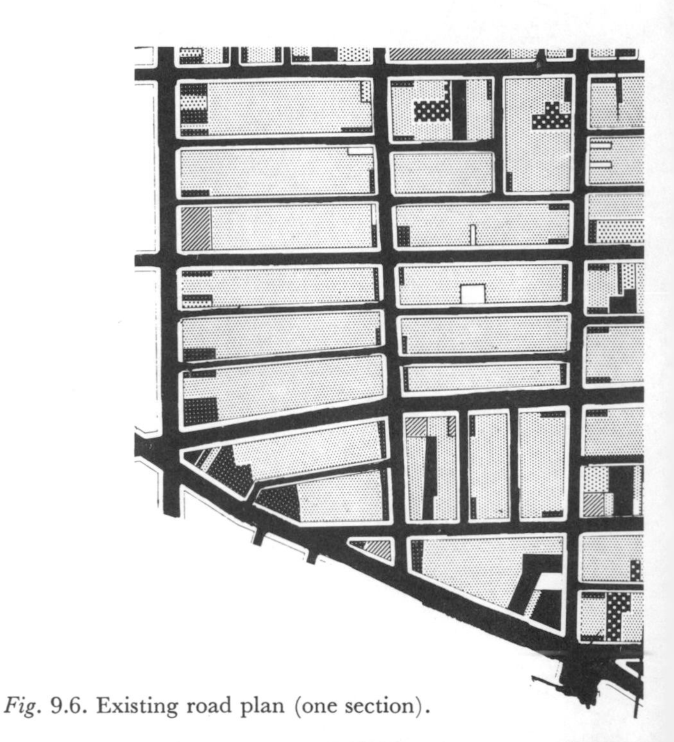

Fig. 9.6. Existing road plan (one section).

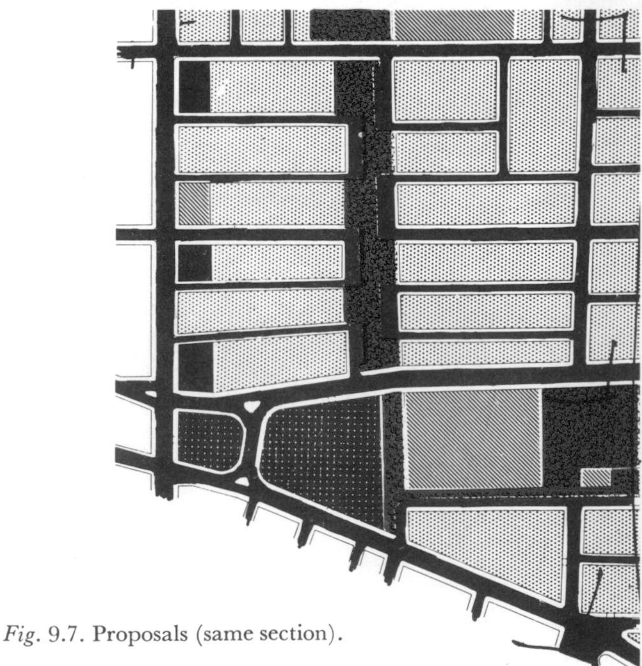

Fig. 9.7. Proposals (same section).

KEY
Red —Pedestrian Areas.
Black —Motor Roads.
Dark Grey —Car Storage.
Light Grey —Dwellings.

Unless otherwise stated this key applies to each plan.

Plans have not been redrawn to retain character and variety. The red colour, denoting pedestrian areas, is therefore shown in a number of ways; either solid red, or edged with red, or main red paths leading through squares or greens, or public buildings and shops about a pedestrian area are solid red or edged in red.

IX. 2.3. *British bye-law Housing*(i)

Proposals by Paul Ritter for Nottingham.

As a measure parallel to the American ideas but suitable for the far narrower block formed by bye-law housing, the shutting of every other street for traffic has been suggested. The pedestrian street becomes the play-lots and sitting areas for old people. Trees can be planted in the middle of the space, far enough away from houses not to darken them unduly, and garage space is made by demolishing some of the houses, or closing-off streets.

Where an extensive area of grid-iron planning is to be tackled, minor roads can connect to a major spine leading towards centres, bus stops or any of these. The junctions should be arranged alternately for safety of driving. Parking places should be provided at the end of the pedestrian access terraces. Access to the houses along the pedestrian spine could be along the back alleys which exist in most cases.

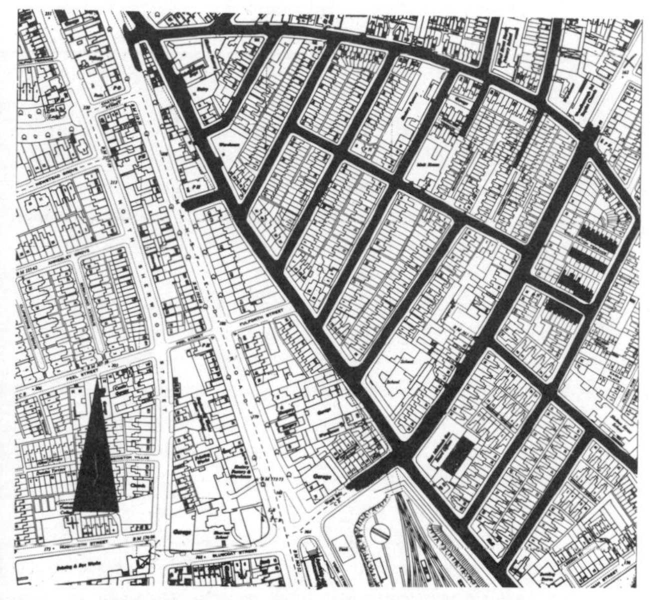

Fig. 9.8. Existing street pattern of bye-law housing, Nottingham.

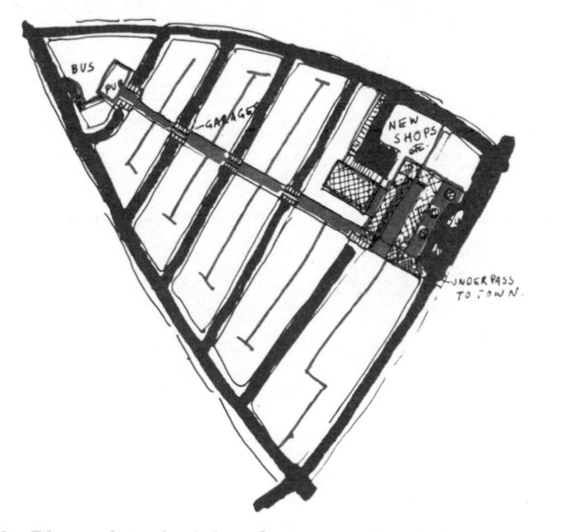

Fig. 9.9. Plan of typical bye-law area treated as suggested: there the through roads are cut by garage blocks to form culs-de-sac. The road in the centre of the spine becomes pedestrian, articulated into play spaces by the garage blocks. The path system created links bus stops, new shopping precinct and leads via underpass through next section to centre within walking distance. When comprehensive redevelopment occurs, the path can remain in its functional position.

(i) *A.J.* London, 10 February 1955.

Fig. 9.10. Play streets, are a useful first aid, found in places as far apart as London, and tiny Cornish villages.

Fig. 9.11. Much bye-law housing could be made to look like Sunnyside, built by Stein and Wright in New York in 1927.

IX. 2.4. *Clearing Space at the Back of Flats, City of Copenhagen*

This example from Copenhagen, which ends up in a kind of "Sunnyside" as built by Clarence Stein in 1924, has much to commend it for many old European towns. The parklike interior contrasts with the city streets on the outside. The density is high but human. Connections between blocks in important directions might be contrived.

Fig. 9.12. The old inhuman town. (560 ppa, 1400 ppha.)

Fig. 9.13. Humanity takes command.

IX. 2.5. *Inter-war Housing, Nottingham*

Proposals by Paul Ritter.

Enormous areas of inter-war housing in Great Britain and elsewhere, with a long life in front of them, have to be rehabilitated to cope with private cars and create a better environment. The proposal for a typical estate in Nottingham, described below, can be applied to much local authority housing and also some private estates.

The separate footpath system follows in the main the existing but deliberately disconnected back-access alleys. It represents a shorter route to shops, school, bus stop, cinema for most people.

Through roads are cut into culs-de-sac. This in itself, according to research quoted (Marks, H), make cross-roads much safer. The closure is affected by garage blocks (providing 12% dwellings), badly needed, facing into the widened end of the cul-de-sac, using up overlarge corner gardens. This takes through traffic off some streets and safety is further increased by taking parked vehicles off the street, around which children play at present to the detriment of the vehicles and the great danger of the children.

The connected path system makes possible other use of overlarge gardens created to satisfy sight lines or fill sites not accessible previously. Thus, four extra dwellings per acre, (10 per ha), can easily be included. It is suggested that these might be homes for old people and single persons. This would have the sociological and economic advantage of providing some dwellings the right size for such groups in the area, which they now have to leave if they are to find accommodation the right size. At crucial points a corner shop can be added to the existing corner house and provide a point of attraction. This may well combine with the play areas as suggested, by the garages. Their walls and roofs can be used as play-equipment, and could include viewing points on to the roads.

The advantages of these proposals are economic, sociological, aesthetic. The crying need for garages is leading to the use of the few open spaces in such areas for garages while the all-over density remains depressingly low and gardens are so large that people are regularly evicted for not keeping them up to local authority standard.

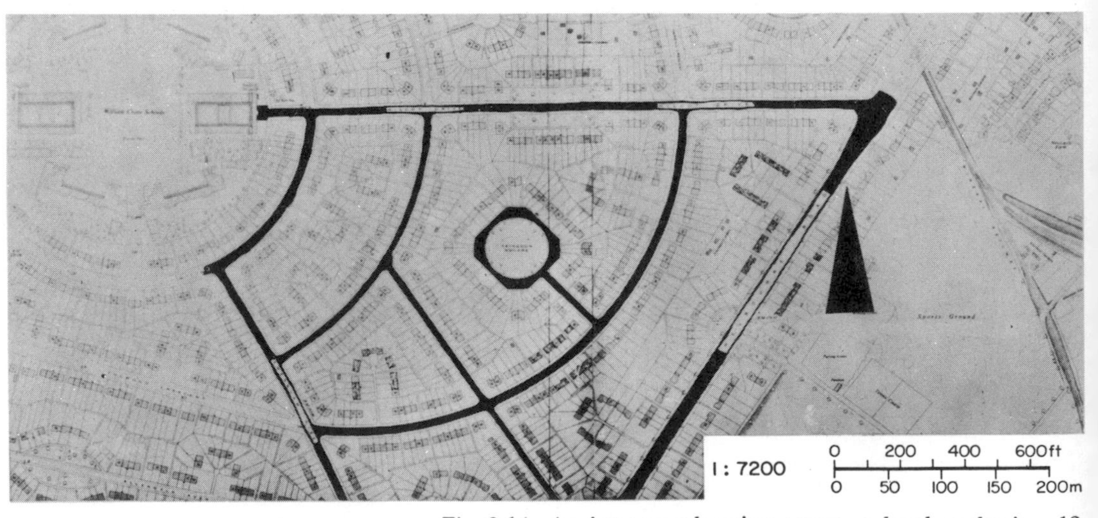

Fig. 9.14. An inter-war housing estate, a local authority, 12 dwelling to the acre (30 per ha) no garage provision, no play areas. The large area in centre (300 sq ft, 28 m²) is entirely fenced and not to be used for play. It may be built up for garaging. Dangerous roads.

Fig. 9.15. Existing boring wide dangerous roads.

Fig. 9.16. Pedestrian cul-de-sac as existing, showing how path system would look.

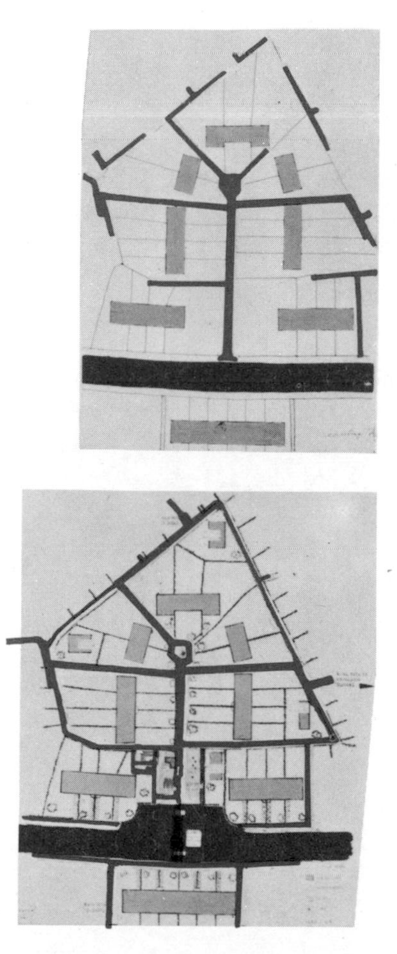

Figs. 9.18 and 9.19. Existing pedestrian cul-de-sac, with existing back alleys. Proposals for improvement and development of area in detail about one of the existing cul-de-sac paths.

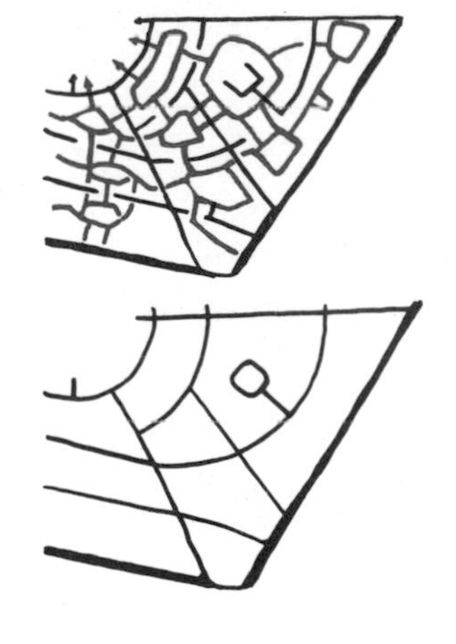

Fig. 9.17. Road system as existing in the area, part of a large area. Road and path system as proposed, applicable to the whole area.

Fig. 9.20. View of play space as proposed by new corner shop and against garage side walls, using garage roofs. Path system leads through between garages. (Drawing by Eric Lee.)

IX. 2.6. *Transformation of a Metropolis. Chicago.*

Proposals by Ludwig Hilberseimer. (1944)

This was a remarkably early and comprehensive proposal, the principles of which are still valid in many instances as the basis for long term planning.

Taking the grid iron city Hilberseimer divides it into individual units of something like 50,000–70,000 inhabitants, which are largely self-sufficient as far as work, schools, shopping and recreation are concerned. The units are separated by park strips and the whole is linked to the lakeside. Little-used railway lines are removed and the road system is worked out in a manner that would be comparatively easy to realise and would also work relatively well, certainly very much better than a grid iron road system. It is intended that walks to work, shops, schools and recreation would cut down car traffic.

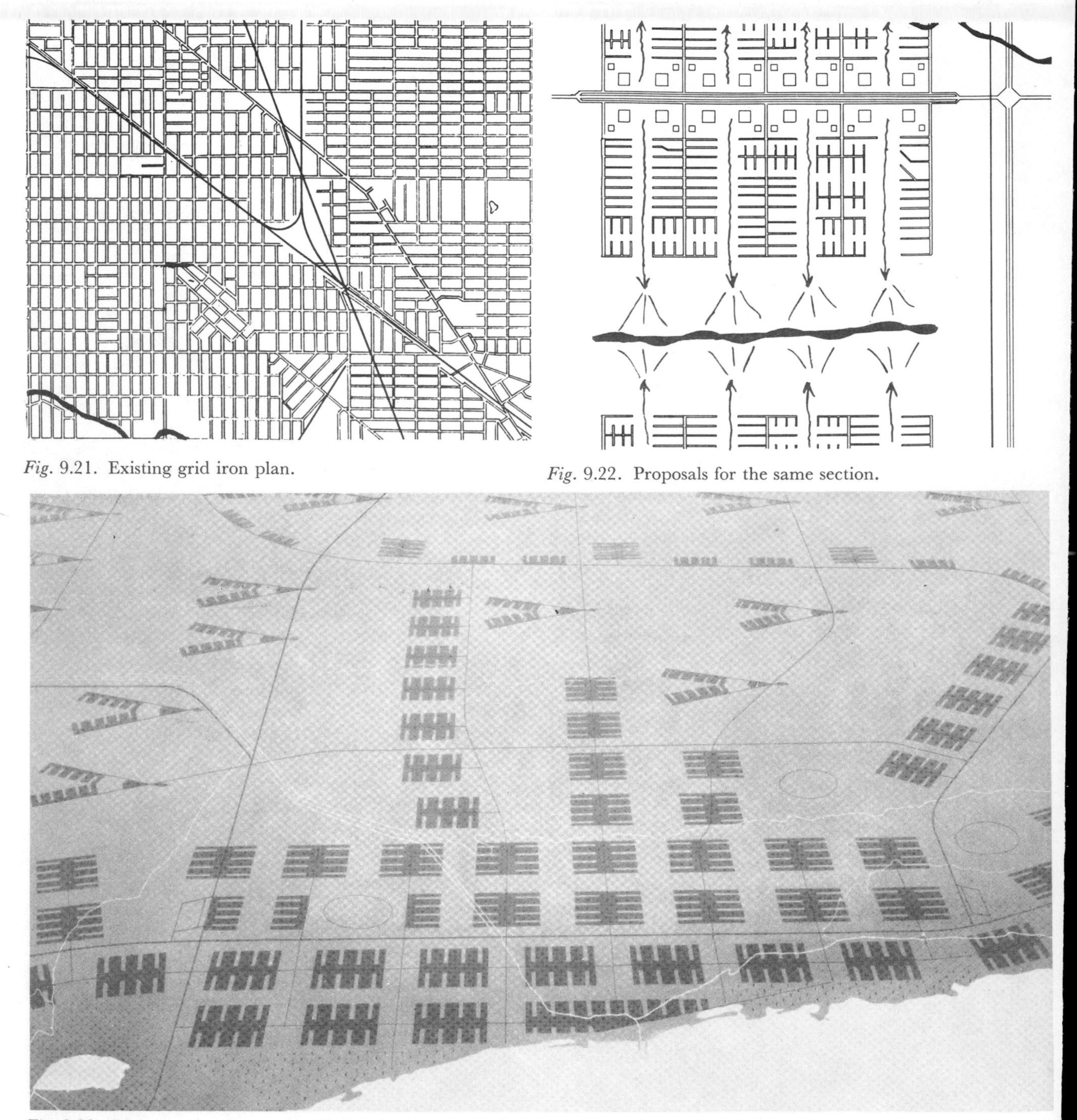

Fig. 9.21. Existing grid iron plan.

Fig. 9.22. Proposals for the same section.

Fig. 9.23. Diagram showing proposal for the transformation of the whole city.

References

VOGLER and KÜHN, Medizin und Städtebau, Berlin, 1957.
HILBERSEIMER LUDWIG. The New City, Chicago, 1944.
HILBERSEIMER LUDWIG. Raumdurchsonnung und Siedlungsdichtigkeit, moderne Bauformen, Stuttgart, 1936.

X. History of Traffic Segregation

X. 1. Summary. **314**

X. 2. History of Traffic Segregation, Illustrated. **314**

X. 2.1. Examples Pre-dating Petrol Engines. 314

X. 2.2. Planning for "Life in Spite of the Motor car". 319

X. 2.3. Planning for Traffic Segregation in Europe. 322

X. 3. Chart Summary of Dates in the History of Planning for Traffic Segregation. **328**

X. History of Traffic Segregation

X. 1. Summary

From Pompeii onward through history and long before the advent of any powered vehicle there are examples of planning for traffic segregation, horizontal and vertical.

> "*The traffic problem was not invented this century. Charles II, the merry monarch, had some sour things to say in 1660*
>
> *'Charles R.*
>
> *Whereas the excessive number of Hackney Coaches and Coach horses, in and about the Cities of London and Westminster, and the suburbs thereof, are found to be a common nuisance to the Publique Damage of Our People, by reason of their rude and disorderly standing, and passing to and fro, in and about our said Cities and Suburbs, the Streets and Highways being thereby pestered and made unpassable, the Pavements broken up, and the Common Passages obstructed and become dangerous, Our Peace violated, and sundry other mischiefs and evils occasioned.'*
>
> *The King's remedy was no external parking. However, lacking parking-meter attendants, enforcement was left to the Lord Mayor, the Aldermen and 'all other Our Officers and Ministers of Justice'.*"[i]

The invasion by railways was very much controlled. Saturation of needs was easily obtained. The petrol engine brought a new severity to the problem. Was "Cité Industrielle" by Garnier, designed 1901, with its path systems, a first sane effort with motors in mind? From two chief sources, the Regional Planning Association in the U.S.A. (with the work by Wright and Stein and the writings of Mumford and others) and Le Corbusier in France, there emerged a realization of the far-reaching nature of the problem and a multitude of solutions, most of which are now applied in various forms all over the globe.

Their immediate success was very limited, but they inspired a number of secondary pioneers in the U.S.A. and particularly in Europe. Hans Bernhard Reichow in Germany, Stein Eller Rasmussen in Denmark, Sven Markelius in Sweden, Van den Broek and Bakema in Holland, Gordon Stephenson, P. Johnson-Marshall and Sir Donald Gibson in Britain, gave the idea a new impetus with their works, often against enormous opposition. Traffic segregation became finally accepted in many influential quarters in the early sixties. Public and private offices design accordingly and speculative housing invests in it.

Opposition to the idea of traffic segregation has been continual and largely irrational, as is the case with all original ideas. This has meant the loss of countless lives and limbs and robbed humanity of much pleasure. The tragedy of bad planning is its continuing effect.

(i) *Sunday Telegraph*. London, 19 May 1963.

X. 2. History of Traffic Segregation, Illustrated

X. 2.1. *Examples Pre-dating Petrol Engines*

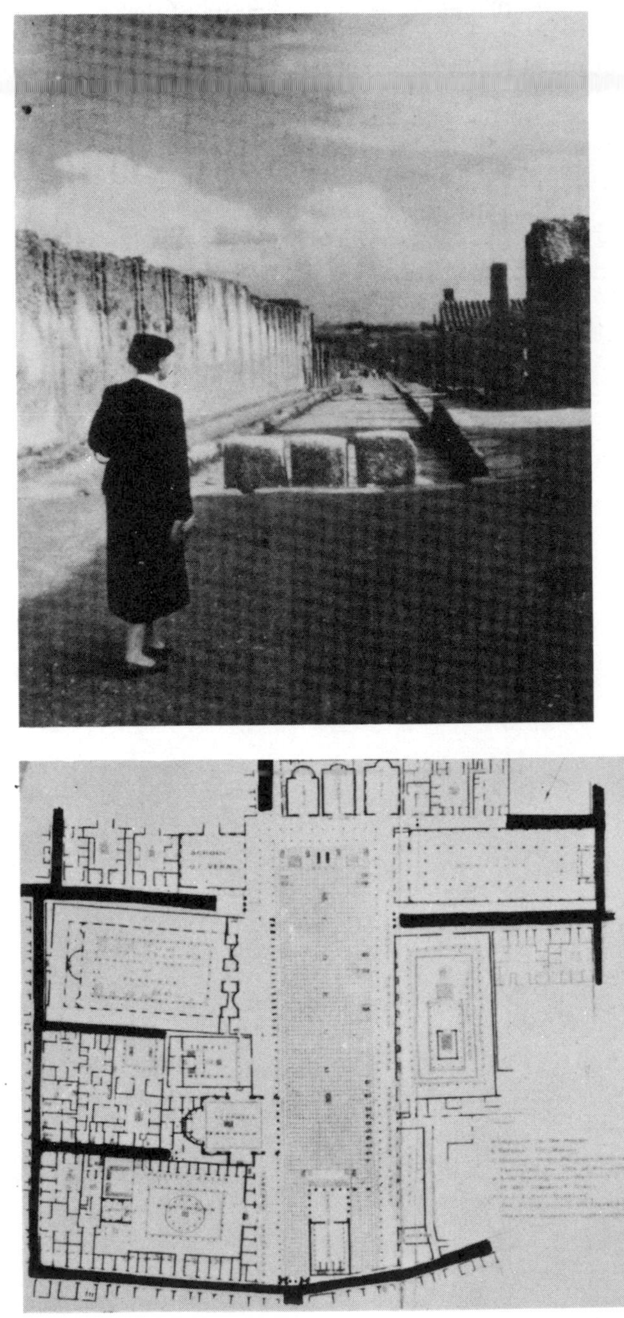

Fig. 10.1. and 10.2. Pompeii rebuilt 1900 years ago shows clearly that even non-motorized vehicles were consciously kept out of the forum. It was designed as a superblock with 7 culs-de-sac closed by bollard-like, upright slabs. The size, speed and scale of chariot and horse were already incompatible with certain social gatherings in the open. In Rome Caesar decreed segregation by time in his *Lex Julia municipalis*: heavy wagons were forbidden "within the limits of continuous habitation" from dawn to dusk.[i]

(i) Illustrations from Reichow H. *Autogerechte Stadt*. Ravensburg 1959.

Fig. 10.3. A model showing Leonardo da Vinci's town planning scheme, with vertical traffic segregation, as developed in his sketch books. (A.D. 1500 approximately.)[(i)]

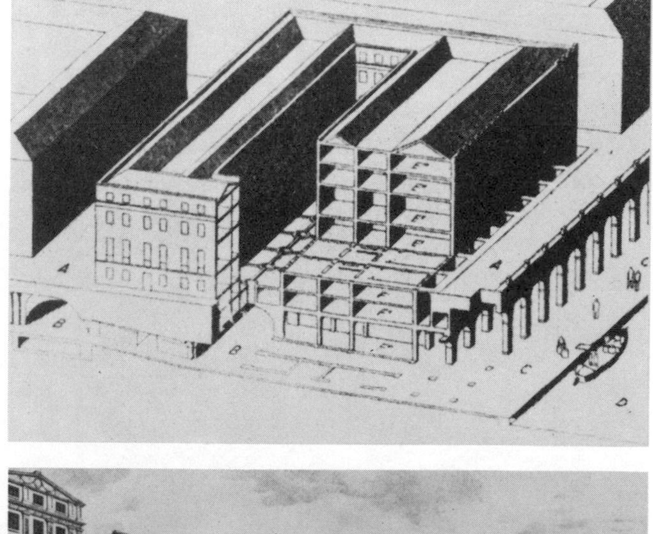

Fig. 10.4 and 10.5. The Adam brothers' design for the Adelphi along the Thames in London (A.D. 1760). The slope to the river is used for a multilevel traffic scheme.

Fig. 10.6. Henry Alken's engraving, inspired in 1831 by the first steam vehicle, foreshadows fairly exactly the mixture of bubble car, scooter, bus, tractor, lorry, car, motor-bike with side-car and endless similar contraptions: he only had the fuel wrong.

(i) See also Medieval Chester, pp. 25 and 208.

Fig. 10.7. In 1865, Thomas Dunn & Co. of Manchester introduced this idea in a sales catalogue. The need to find aids in crossing roads, in this case Ludgate Circus, was relevant in Victorian times. Substituting lifts, escalators, or paternosters, for the spiral stair, would give a space-saving and practical idea for many present day crossings, where long ramps or stairs still discourage use and are a great inconvenience to pedestrian.

Fig. 10.8. A scheme to create tunnels for goods-traffic, when it was still horse-drawn, to take congestion out of city streets. The first underground railway was opened between Paddington and Farringdon, in London, in 1863, but, alas it was the people who were thus pushed underground, not the goods or private vehicle. This precedent continues to influence planning.

Fig. 10.9. The accommodation of vehicles in the existing congested street presented problems and brought forth ideas similar in the nineteenth and twentieth century, e.g. Oxford Street proposals, London.

Fig. 10.10. Bringing in trains. A multi-level solution. It is very fortunate that this idea of running lines along main roads did not catch on. There is a similarity to mono-rail proposals of today as used at Seattle.

Fig. 10.13. Early Victorian housing facing wide green areas, with service roads at the back, quite frequent in England (Nottingham).

Fig. 10.11. Victorian Parks were quite frequently crossed by road bridges leaving the park paths safe, continuous and quiet. Olmstead at Central Park, New York, in 1856, produced a system which separated, by bridging, through-traffic, local traffic, equestrian and pedestrian movement.

Fig. 10.14. Pre Bye-law housing, Nottingham. 25 ft apart, density 100 *houses* per acre. Such densities did not allow for road access and tiny gardens.

Fig. 10.12. The Arboretum Nottingham. An elegant and much used Victorian park way leads efficiently under the road. Approach and stone arch size are similar to the famous Radburn underpass.

Fig. 10.15. Factory at back gives interest and work near dwellings. (Smoke too, in the coal age.)

Fig. 10.16. Countless such little terraces are tucked away double banked behind terraces facing roads.

Fig. 10.17. Site development was haphazard. The path on the lower level ends in a wall, and above. ...

Fig. 10.18. Lush greenery with quite well kept better class terraces.

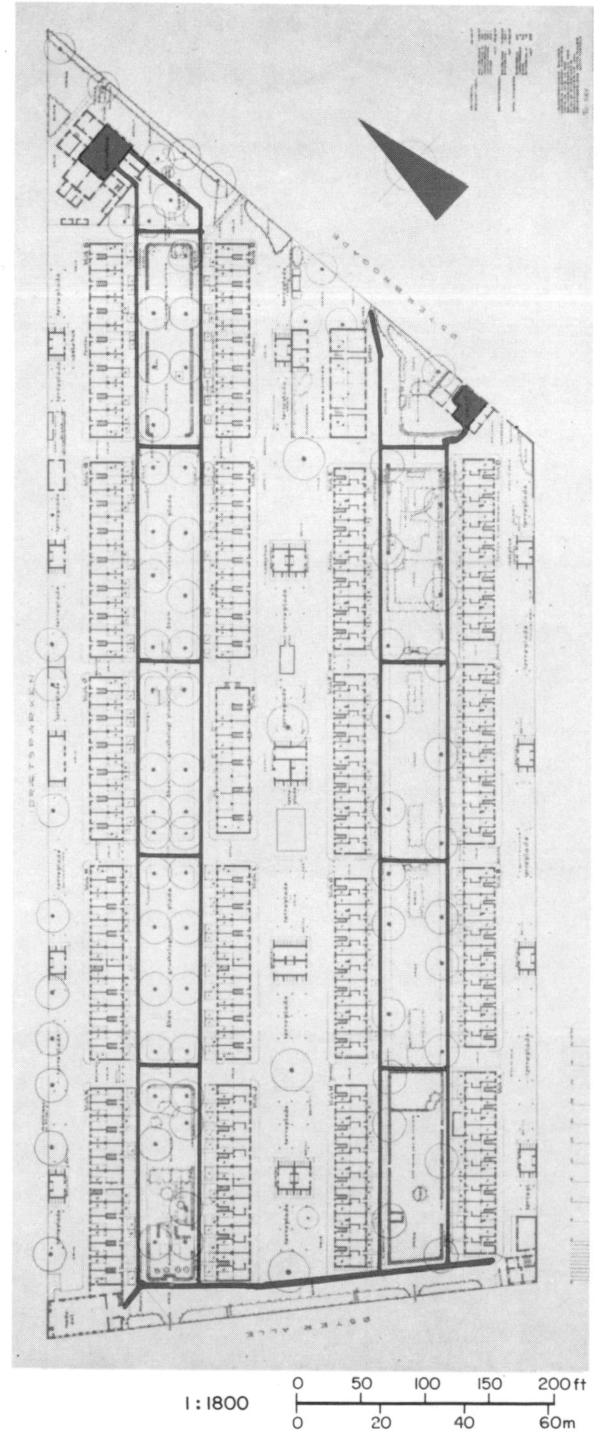

Fig. 10.19. Plan.

In 1853 the Danish Medical Association(i) built a community of houses in Copenhagen as an example of hygiene following a plague epidemic. In separating service traffic from the pedestrian and garden side, private and communal, this remarkable development has a charm and practicability today which is astonishing. The "latest" developments of the Radburn Idea are very similar.(ii)

(i) FISKER and MILLECH. *Danske arkitektur strømninger 1850–1950.* Copenhagen 1951.

At present there is a national controversy about the future of these unique dwellings. The present Medical Association wishes to sell the land for redevelopment and many societies and newspapers are opposing this vigorously.

(ii) See Hook Report. *The Planning of a New Town.* London County Council 1961.

Fig. 10.20. Leaves clothe all (above), Winter exposes the many *ad lib.* additions.

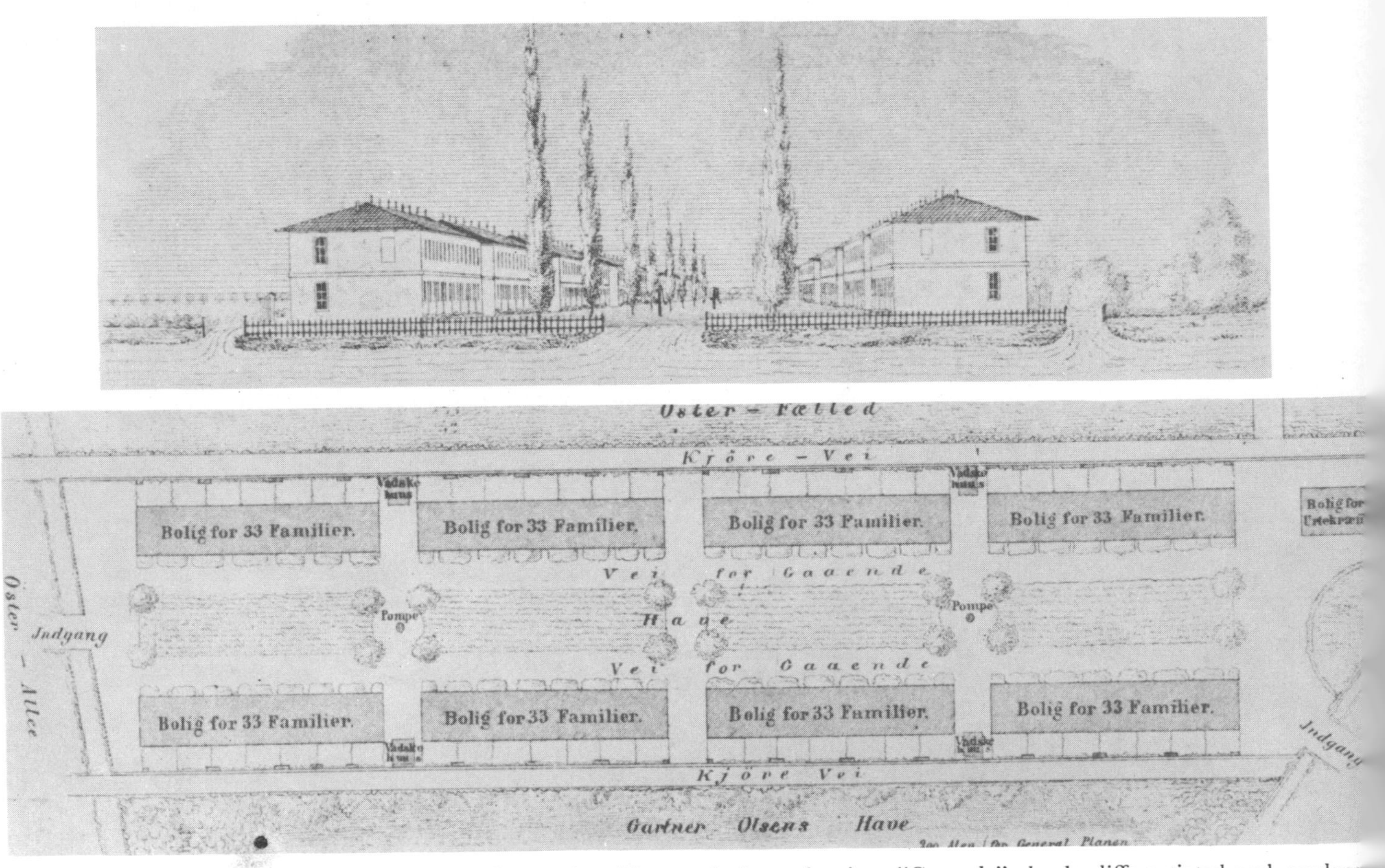

Fig. 10.21. Original drawings of the first section. Note path for pedestrians "Gaaende" clearly differentiated and gardens and pumps as social centres.

Fig. 10.22. The garden and path side.

Fig. 10.23. The service side.

Fig. 10.24. Summer use of the path side.

X. 2.2. *Planning for Life in Spite of the Motor Car in the U.S.A.*

Figs. 10.25–10.33. The work of the Regional Planning Association, founded 1923(i) is described fully in Clarence Stein's book *Towards New Towns in America*, published at Liverpool in 1950 when Professor Stephenson was there. Clarence Stein and Henry Wright planned for "life in spite of the motor car" before anyone else, in isolated garden court blocks, in New York and whole New Towns. High numbers of vehicles now fairly common in Europe were developed much earlier in the U.S.A., in 1928 21,308,159 vehicles were registered. The Association was not aware of the work of le Corbusier until long after the first projects in 1929.(ii)

Stein, Wright, Ingham and Boyd, designed a residential area of three superblocks, Chatham Village, which has attracted an almost entirely professional population.(iii)

Fig. 10.25. Sunnyside Gardens, by Stein, Wright and Ackerman. The City Housing Corporation under Mr Bing allowed these members of the Regional Planning Association, formed in 1923, to start in 1924 to build some blocks of flats and terrace houses with green pedestrian interiors in New York. The photo taken after about twenty years shows the trees planted at the beginning matured and immeasurably enhancing the environment.

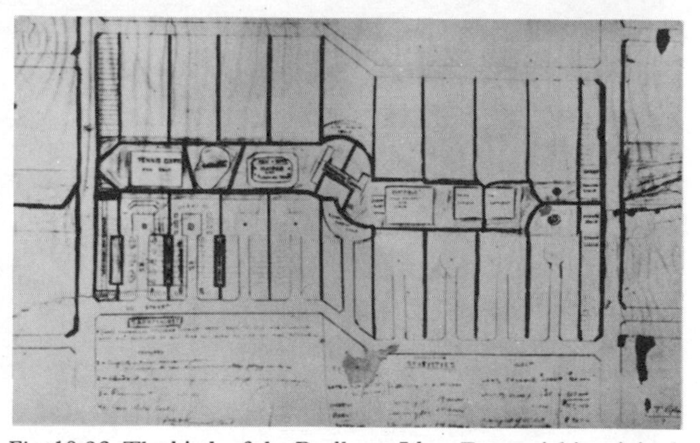

Fig. 10.26. The birth of the Radburn Idea. Emmerich's original concept, arrived at through consideration of basic needs, on the back of an envelope (1927, December). It is interesting that Emmerich, as also the originator of Cité Jardin in Montreal, was a layman.

(i) Composed of F. L. Ackerman, Frederick Bigger, A. M. Bing, John Bright, Stuart Chase, R. K. Kohn, Benton MacKaye, Lewis Mumford, G. Stein, C. H. Wright, Herbert Emmerich.

(ii) Letter to author from Clarence Stein, 4 October 1961.

(iii) According to Jacobs J. *Death and Life of Great American Cities.* London 1962.

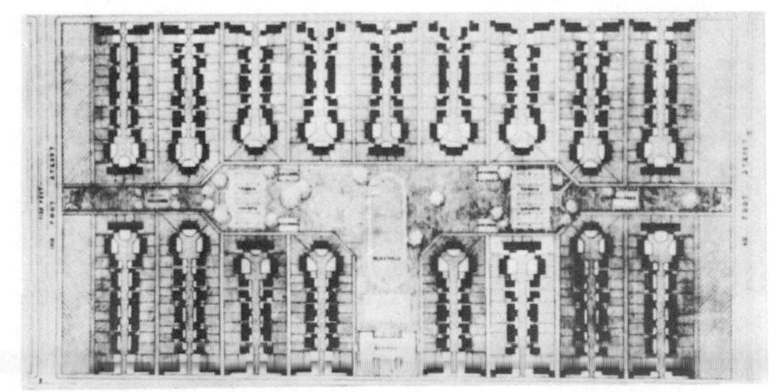

Fig. 10.27. Clarence Stein's interpretation of the diagram (January 1928).

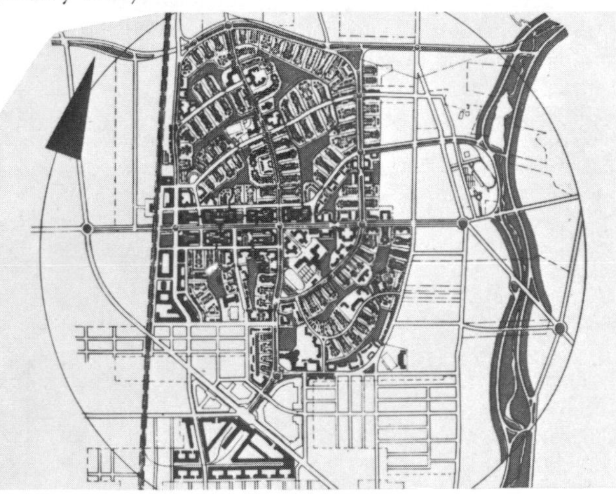

Fig. 10.28. Entire plan—not realized. Circle is 1 mile radius.

Fig. 10.29. The sales brochure issued for Radburn is still a model. Much of today's sales literature ignores the importance of making vivid the advantages of traffic segregation to the general public in non-sophisticated terms which they can readily grasp.

Fig. 10.30. These little paths have individuality and charm.

Fig. 10.33. One of the greens, children at play, Sunday morning in October, 1962.

Fig. 10.31. Criticism of the cul-de-sac is, ironically that they are too attractive, literally, to those who theoretically ought to be on the pedestrian side.

Fig. 10.34. This underpass is well placed and really used, unlike underpasses in many places. Whereas, Radburn as a community is almost ultra-respectable, the graffiti in the underpass offer a delightful contrast and a sign of life. The swimming pool nearby, and the other amenities create a considerable social problem: children who share schools, cannot share the swimming pool because, although each child is allowed a certain ration for friends, the demands of all the surrounding amenityless, deprived families are far too great to satisfy.(i)

Fig. 10.32 Taken in 1955 this view shows clearly the fine trees which have matured in Radburn making the greens truly park-like. Although not meant for organized ball games, for which provision is made elsewhere, many children playing informally can be seen on the greens during non-school hours and particularly during the week end. The greens are truly successful in this sense in allowing adults a fine place to walk in. The value of the house has gone up more steeply than that of comparable traditional housing near by. The only explanation why this has not made the speculator use this idea is the political implication; the Greenbelt Towns were part of the Roosevelt New Deal. The organization, for looking after the common greens, is too much for most people to organize and arrange for legally.

Fig. 10.35. Even in 1940 the land around Radburn was not built on. By 1955 the community had become an oasis in a sea of "the usual". The folly of man in the face of good example.

(i) See also Baldwin Hills, in Section VIII.

X. 2.3. *Planning for Traffic Segregation in Europe*

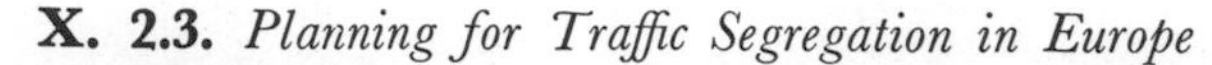

Fig. 10.36 and *Fig.* 10.37. Cité Industrielle by Tony Garnier at the very beginning of the century provides pleasant connected path environment. But is is difficult to ascertain whether he had the motor-car in mind.

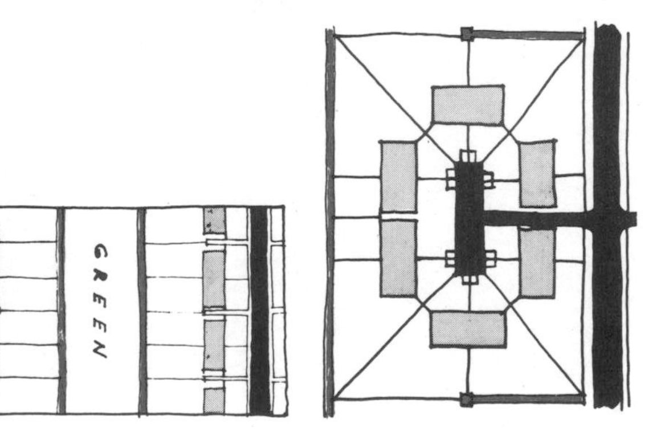

Fig. 10.38. Three variations of separate footpath and service access for residential areas from Longstreth Thompson's book, *Site Planning In Practice*, London 1923.

Le Corbusier, from November 1922 (Une Ville Contemporaine), designed a variety of projects which incorporated many ideas for traffic segregation. It is very interesting how the twin influence of The Regional Planning Association and Le Corbusier have between them covered before 1930 almost all that can now be said or written on the subject.(i)

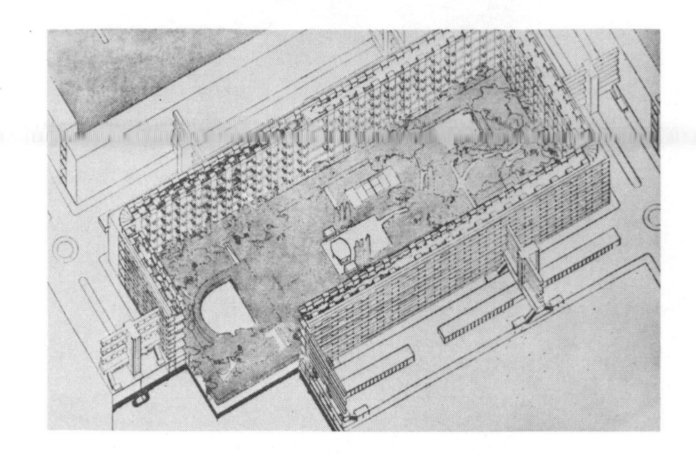

Fig. 10.39. Immeubles-Villas 1925 has superblocks with green interiors. In a grid-iron pattern. The long side of blocks have garaging in mid-street. In all four directions, and on every second floor, there are bridge connections to the other superblocks. Complete traffic segregation.

Fig. 10.40. Plan Voisin for Paris 1925. Here is a realistic appreciation of the scale of roads needed to accommodate the traffic of private cars in a large city. Here also is the appreciation that junctions have to be grade-separated and the whole lifted onto its own separate plane. In spite of the lyrical writing, and in spite of the far-sighted planning of an underground railway to link the vertical cores of the vast blocks, 3200 ppha, (8000 ppa) might take up, with their cars, even on his own model, most of the "verdure", the whole point of the exercise.

(i) Le Corbusier und Pierre Jeanneret 1910–1929. Zürich 1937. Le Corbusier and Pierre Jeanneret 1929–1934. Zürich 1935.

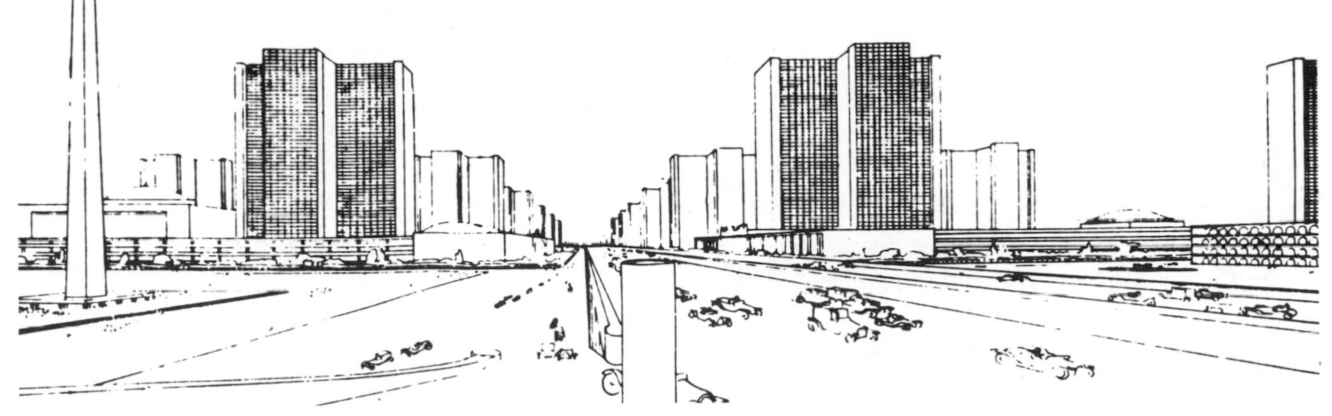

Fig. 10.41. Le Corbusier's Ville Contemporaine, 1922, the "autostrade de grande traverse". The scale of the road is a frighteningly correct vision. The pedestrian dots on the left pathetic.

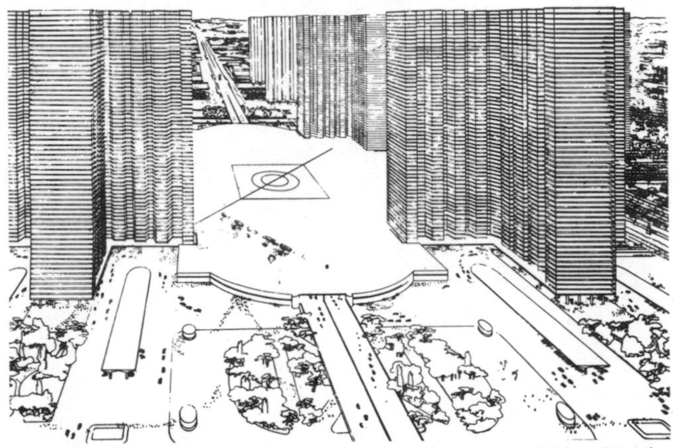

Fig. 10.42. The area around the Central Station of Plan Voisin, Paris 1925, underestimates the influence of the car. One can imagine the poor "dot" of a pedestrian today, in such a vast tarmac area, dashing hither and thither among the cars, as he crosses where the drawing clearly shows this intention.

Fig. 10.43

Fig. 10.43 and 10.43a. Roads on the roof tops of flats running continuously for very long distances have been imitated in a suggestion, since this scheme for Algiers in 1930.(i) The complex of housing in the background with its deck system visible is comparable with that at Parkhill, in Sheffield, built 1960.

(i) JELLICOE G. A. *Motopia*. London 1961.

Fig. 10.44. Part of Copenhagen, by Bennet Windinge, 1930. Professor Steen Eiler Rasmussen received one of the original Radburn Sales Brochures with the effective cover(i) in 1929 from the U.S.A. from a lay person who thought he might be interested. As Professor at the Royal Academy of Fine Arts at Copenhagen, this influenced his teaching decisively, as the student's work in Figs. 44 and 45 show.

Fig. 10.45 This superblock for Køge by Thygge Baggesen is particularly interesting for it shows, in 1933, the variety of layout possible within a superblock, similar to those suggested by the L.C.C. for Hook recently.

(i) See page 320.

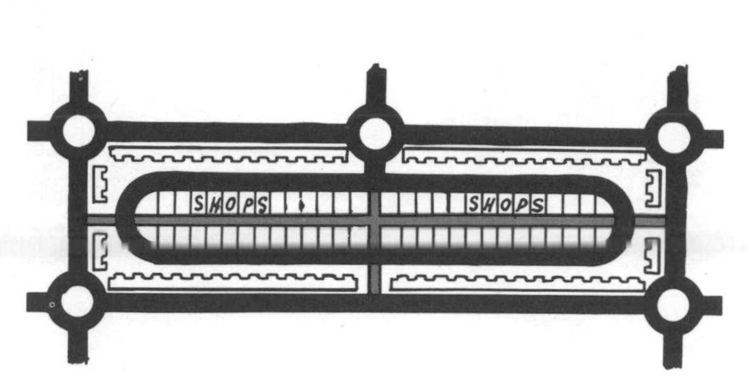

Fig. 10.46. Alker Tripp's proposal for "shopping way"(i); naive indeed, but a beginning.

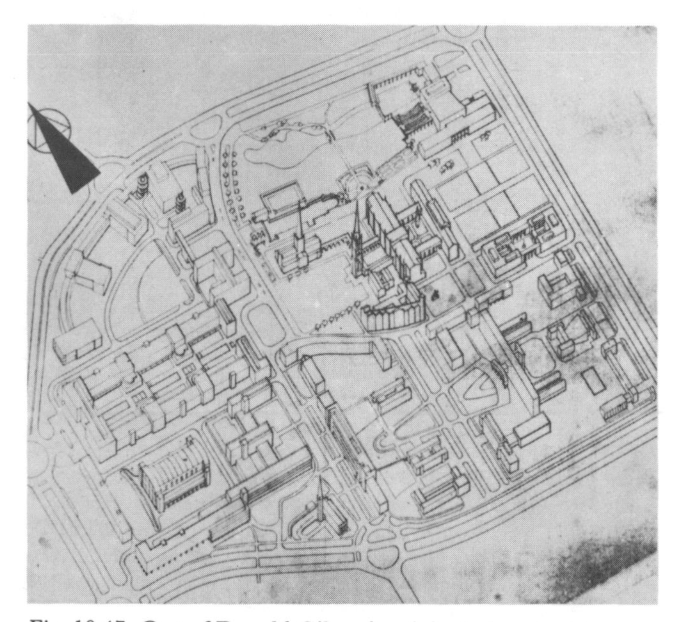

Fig. 10.47. One of Donald Gibson's original plans for Coventry Centre (with P. Johnson-Marshall) with shopping precinct of 1941; compare with final scheme, p. 172.

Fig. 10.48. Segelgatan, Stockholm Centre, first proposals of City Planning Office, 1946.(ii)

(i) Tripp A. *Road Traffic and Its Control*. London, 1938 (1950!).

(ii) *Arkitektur*, II Stockholm. 1962. See p. 176.

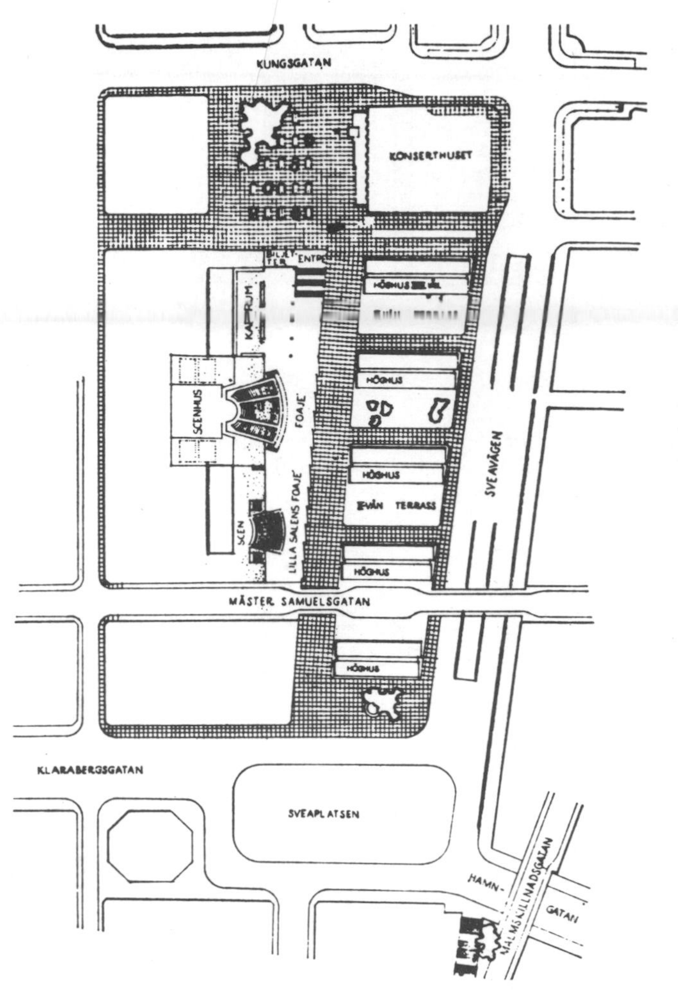

Fig. 10.49. Sergelgatan, plan; compare with final scheme, p.174.

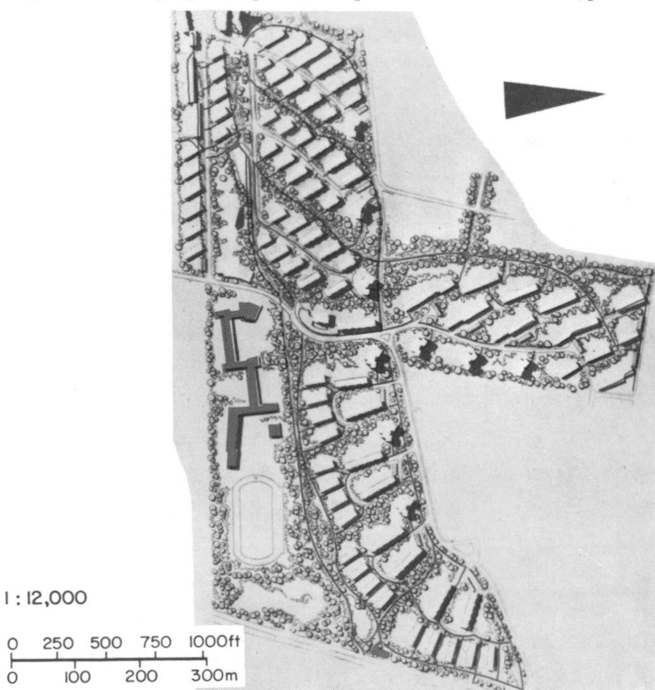

Figs. 10.50. In 1948 Hans Bernard Reichow's Organische Stadtbaukunst(i) was published in Germany and his philosophy of planning included traffic segregation with all circulation taking the branching forms of growth systems in nature. In 1953 he designed Hohnerkamp, Hamburg incorporating the ideas.

(i) by Georg Westermann, Braunschweig 1948.

Fig. 10.51. A focal point at Hohnerkamp.

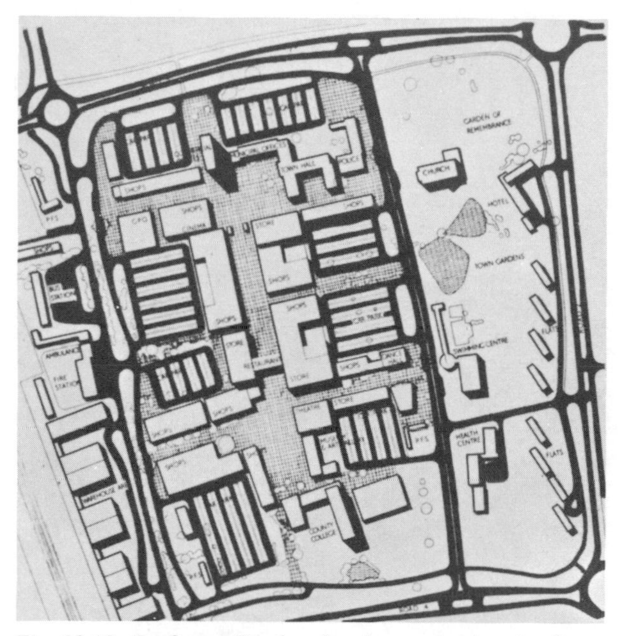

Fig. 10.52. Professor Gordon Stephenson's plan for Stevenage Town Centre (with Clarence Stein and Holliday) in 1950–52 (not built) had a decisive effect on the British New Town Centres which together with the new towns themselves took precious little notice of the motor vehicle and the need for traffic segregation. To what extent the tragic failure of the first fourteen new towns in this respect is due to the professional and/or the lay Development Corporation is not fully established.(i)

(i) From SERT and ROGERS. *Heart of the City*, Ed. Tyrwhit. London 1952. Also *A.J.*, 28 January 1954.

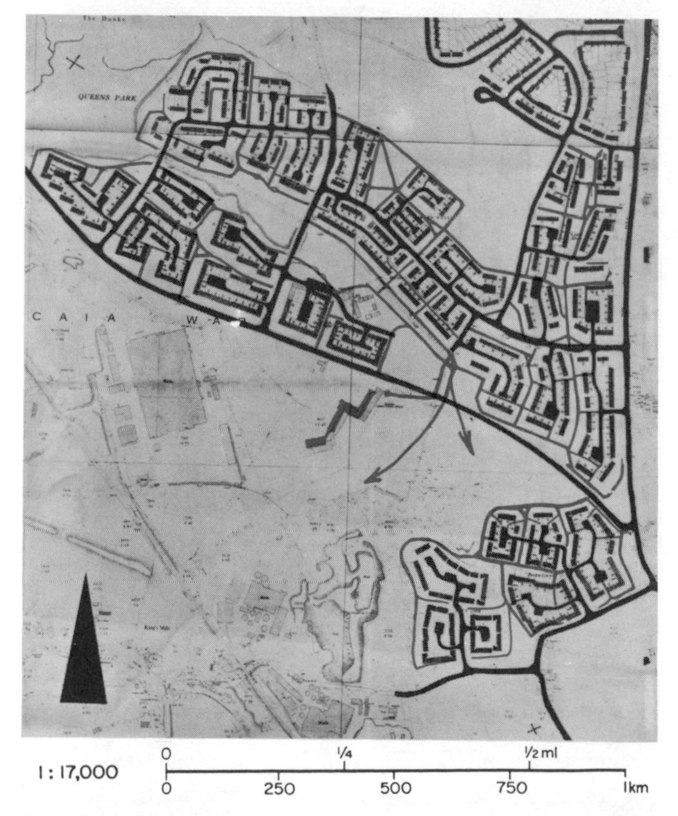

Fig. 10.53. Professor Stephenson's housing scheme in Wrexham, in conjunction with the Borough Engineer, was begun at a time when economy and restrictions were at their height. Criticisms of this scheme have ignored this factor and have often mistakenly blamed the basic Radburn Idea for any shortcomings the scheme may have been deemed to have. Professor Stephenson's influence for traffic segregation was widespread. The whole of this large scheme has not been published before (1950).

Fig. 10.54. A main path, 1954.

I made a return visit to Wrexham in July, 1963, ten years after my first survey. The path system is still bleak. With a little initiative it can, any time, become a welcoming bower of flowers, trees and bushes as intended. The opportunities are built in.

The fruit trees in the cul-de-sac gardens are now growing and making their contribution. The stream and wild grass land on each side along the path system are a wonderful adventure play area.

Fig. 10.55. The little stream in the valley on the path system in Wrexham, left open for recreation is a wilderness now and the stream is polluted. But, when the stream is clean and there is more initiative, there is magnificent scope for pools, bridges, and endless fun and games. The photograph shows that even the very rough terrain of the present is a priceless place for adventure. (How many heads can you find?)

Fig. 10.56. Separate cycle and pedestrian underpass which connect the valley with the other part of the estate and lead to factories nearby.

Fig. 10.57. The contribution of the fruit trees.

Fig. 10.58. In the culs-de-sac at Wrexham, where poor fencing was forced on the designers by shortage of materials, (1951) a fruit tree was planted at the bottom of each garden. Twelve years later this is already making a decisive difference.

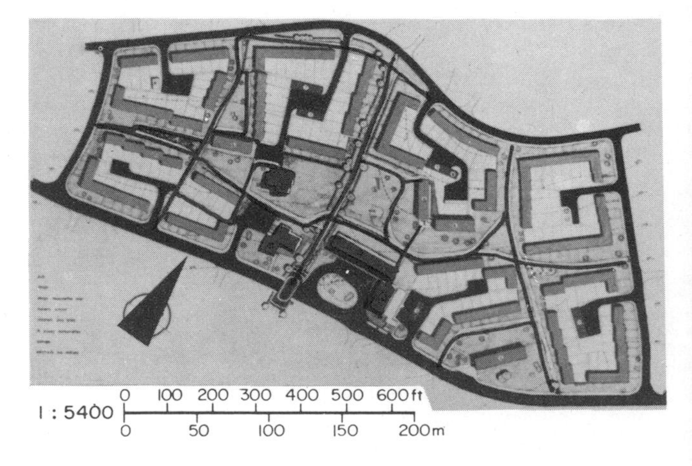

Fig. 10.59 First British plan for super-block with shops and school, Greenhill, Sheffield. Built 1954, the school was not included.

Fig. 10.60. *St. Paul's Area, London, England.*

Design: Sir William Holford.

The area, destroyed by air raids, has been redeveloped in the English tradition of picturesque planning, making a new, quiet, environmental complex. The scheme, first published in 1956, was bound to have a great influence on British Planning and shows a permanent development of the precinctual idea well used in the Festival of Britain by Sir Hugh Casson in 1951.

References

A.R. London, June 1956.
R.I.B.A. Journal, April 1956

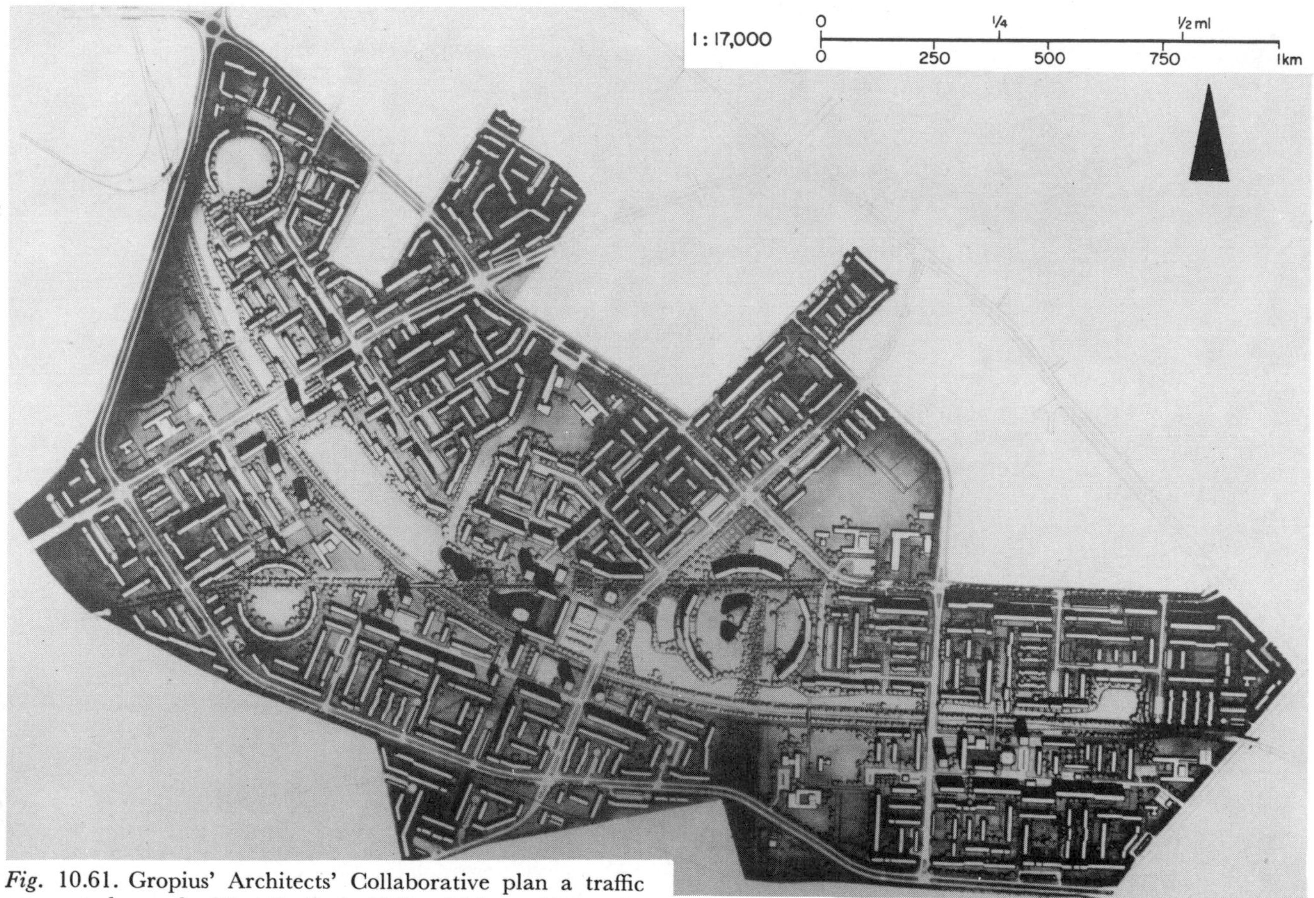

Fig. 10.61. Gropius' Architects' Collaborative plan a traffic segregated area for West Berlin in 1962, which establishes the principle officially for that city.

X.3. Chart of Some Interesting Dates in the History of Planning for Man and Motor Cars

*1922 "Ville Contemporaine". Le Corbusier's exhibition, Paris.

*1923 Regional Planning Association formed at New York.

*1923 *Site Planning in Practice*, by Longstreth Thompson, published London.

*1924 Sunnyside Gardens, built New York. Stein, Wright and Ackerman.

*1925 Immeubles-Villes. Le Corbusier.

*1925 Cité Voisin. Le Corbusier. Paris.

*1928 Radburn, New York, begun. Stein and Wright.

1928 *The New Exploration*, by Benton Mackaye. (Pioneer in Motorways.)

*1930 Algiers Scheme. Le Corbusier.

*1930 Steer Eiler Rasmussen teaches the Radburn Idea at Copenhagen.

*1935 Green Belt, Maryland, begun. Ellington, Wadsworth, Walker, Stein.

1938 *Culture of Cities*, by Lewis Mumford.

*1938 *Road Traffic and Its Control* by Alker Tripp.

1939 Stettin, Estates with Traffic Segregation. City Planner, Reichow.

*1941 Cité Jardin, Lay-man's Radburn. Montreal.

*1941 Baldwin Hills Village built, Los Angeles. 300% car storage. Wilson, Merril, Alexander, Stein, Johnson.

*1941 Coventry, First Precinct in Centre. Donald Gibson, planning officer.

*1944 Hilbersheimer's Scheme for Transformation of Chicago.

*1945 Planning Rotterdam begun, Lijnbaan. Van den Broek and Bakema.

*1946 Coventry begun, Central Precinctual Cross.

*1946 Endsleigh Gardens, Birmingham. Service access, fronts onto green.

* Indicates items illustrated in the text.

*1946 Sergelgatan, first scheme. Stockholm. City Planning Department.

1948 Publication of *Organische Stadtbaukunst*. Reichow.

1949 L'Unité d'habitation, Marseilles, built. Le Corbusier.

1950 *Regeneration der Städte* by Abel. Zürich.

1950 *Towards New Towns for America*, published by the TPR, Stein.

*1951 Queen's Park, Wrexham. G. Stephenson.

*1951 Stevenage New Town Centre, first plan. Stephenson, Stein and Halliday.

*1952 Vällingby, Stockholm. Planned by Markelius.

1952 Smithson's Golden Lane Scheme.

*1953 Hohnerkamp begun, Germany. Planned by Reichow.

*1953 Greenhill, Sheffield begun. City Architect, Womersley.

1953 *Houses 1953*. Ministry of Housing and Local Government, H.M.S.O. Official Radburn recognition.

*1954 Pedestrian Streets in German Towns.

*1954 Sennestadt begun. Competition won by Reichow.

*1956 South Barbican, City of London and L.C.C. pedestrian deck scheme for commercial area.

1957 *The influence of Roads on Social Patterns, Housing Lay-out, Housing Therapy*. Research and Thesis. Paul Ritter.

*1957 Park Hill, Sheffield, begun. City Architect, Womersley.

*1957 Centre and North Barbican Proposals for City of London. Chamberlin, Powell & Bonn.

1958 *Mixed Blessing. The Motor Car in Britain* by C. D. Buchanan.

*1959 Cumbernauld New Town Mark II, Scotland, begun. Chief Architect and Planner, Hugh Wilson.

*1959 First of 34 Man and Motor Exhibitions by Paul Ritter in Europe and North America.

1960 *Radburn Planning: A Reassessment.* Nine Articles, A. J., London, by Paul Ritter.

1960 C. D. Buchanan appointed adviser to the Ministry of Transport.

1960 Housing Lay-out and Children's Play Behaviour by H. Wohlin. Stockholm.

*1960 Willenhall Wood I completed. City Architect, A. Ling, with D. Lyddon.

*1961 Flemingdon Park I built, Toronto, Canada. Planned by Grossman.

1961 *Homes Today and Tomorrow.* Radburn recommended. Ministry of Housing. Great Britain. H.M.S.O.

*1962 *Town Centres, Approach to Renewal.* Joint Bulletin. Ministry of Housing and Ministry of Transport of Britain. Advocates Traffic Free Pedestrian Networks in Renewal Programmes.

1963 *Report by the Minister of Transport.* "There are indications that full motorization cannot be accommodated without circulation and parking arrangements on a scale not so far generally realized. Alternatives may have to be considered to prevent private cars unduly inflating peak hour traffic flows."

1963 The Buchanan Report, H.M.S.O., *Traffic in Towns.*

The Ministry of Transport of Great Britain appointed a Steering Group in 1961 to examine the long-term problems of traffic in urban areas, which published its findings in the Buchanan Report, in November 1963.(i)

Quotations of the most important paragraphs made by the Steering Committee are included in this volume because it is vital to appreciate that the techniques described and illustrated in this book and the basic principles involved have been announced as the basis for government policy in Britain by the Minister of Transport in the House of Commons.(ii)

(i) The MS. of this book was lent to the Ministry Working Group.

(ii) With the sad and important exception of the setting up of the Regional Development Agency.

35. "We draw three main conclusions. The first is that there is no one easy and complete solution to the problem posed by the growth of motor traffic. All the remedies discussed above will have to be used, in one form or another. But secondly, since each line of approach reacts immediately on the others (e.g. any deliberate limitation puts an added burden on public transport, or the building of new urban roads intensifies the parking problem in the centre) it is imperative that they should not be applied haphazard by different authorities reacting to different stimuli and following different time-tables, but in a carefully co-ordinated way after comprehensive analysis and study of the whole complex. Thirdly, and most importantly, any such organised attempt to solve the problem will necessarily involve very large-scale redevelopment of our cities and towns on a significantly different pattern. If we are to have any chance of living at peace with the motor car, we shall need a different sort of city."

37. "The starting point is the principle that traffic and buildings are not two separate things but two facets of the same problem. To the highway engineer, buildings may simply be structures that line, and sometimes obstruct, his roads. But in fact they are the generators of the traffic and the destinations to which it is going. If there were no buildings there would be no traffic—and conversely if there were no traffic there would be very few buildings. Similarly, those who design and locate buildings should not take it for granted that the street system will be able to serve them. The buildings which generate traffic should be integrated with the traffic arrangements in an overall concept of town planning. In so far as this cannot be secured by permissive planning—that is, by regulating the activities of the public—it should be done by positive comprehensive redevelopment. This applies, of course, primarily to the commercial and industrial areas of cities. But it is also applicable to residential areas."

40. "It is clear that any attempt to implement these ideas would result in a gigantic programme of urban reconstruction. We see no reason to be frightened of this. The central sections of most of our cities were very largely built in a few decades of the nineteenth century, and the rebuilding necessary to implement the ideas of the Buchanan Report—which would be very much less than total reconstruction—should not be beyond the powers of a few decades of our century. Moreover, we believe there would be very considerable subsidiary advantages of such a programme. Though the impelling force behind it would be the pressing need to reorganise our cities for the coming volume of motor traffic, it should be possible in many cases to draw an extra dividend in the replacement of slums or unworthy housing. Indeed, it is possible that a vigorous programme of modernising our cities, conceived as a whole and carried on in the public eye, would touch a chord of pride in the British people and help to give them that economic and spiritual lift of which they stand in need."

45. "In any effective programme of urban modernisation, such as we have been outlining, it is possible to distinguish four main stages. First, there must be a clear statement of national objectives. Regional planning cannot work in isolation. Unless there is a policy on a national basis dealing with the location of industry and population, from which would flow policies in respect of roads, ports, air facilities, etc, regional planning cannot be successful. Without such a policy it is impossible to know what populations and kinds of employment must be planned for locally, nor the rate at which development can take place, nor can there be any certainty that some uncontrolled drift of events will not reduce all local plans to futility.

46. "Given such a national policy, the second stage would be to delineate the local areas or 'urban regions' within which the various problems are so interrelated as to demand overall planning. An 'urban region', in nearly every instance, would be much larger than is now embraced within a single city, and not a conurbation, the urban region would have to take in the whole of the surrounding catchment area, going at least as far out as the 'traffic watershed' or limit of car-commuter travel. In a few cases only in this country there would be an overlap, representing an equal pull by two or more large centres; but on the whole the 'spheres of traffic influence' of the big cities and conurbations are clearly ascertainable. The studies of these areas, and the broad regional plans to which they would lead, would require a synthesis of transport with normal town planning considerations on the lines illustrated in the Buchanan Report. Moreover, they would have to escape from the confines of what is at present regarded as financially practicable. But though these regional plans would thus go beyond the existing statutory development plans in various ways, they would clearly be exercises of the same general nature and requiring much the same sort of study. They would be extensions of, rather than substitutes for, existing planning procedures."

47. "Once such a regional plan existed for an area, the third stage would be to get more detailed plans drawn up for the redevelopment of the obsolete and congested parts, for the 'primary road network' found to be necessary and for the considerable displacement and redevelopment that would follow on the construction of the network. Then there would be the fourth stage of execution—of actually bringing about the redevelopment of the obsolete and congested areas, and of driving the roads of the primary network through. These tasks would involve multifarious and very complicated financial and legal problems of land acquisition, multiple ownership and the like."

49. "The new machinery should take the form of a number of Regional Development Agencies, one for each recognisable 'urban region' (and not therefore necessarily covering the whole county). The mandate to the Regional Development Agency should be to oversee the whole programme of urban modernisation in its region, in the sense of seeing that it got done, but not to take over those parts of the whole that are already being effectively done, or that can be effectively done, by the existing authorities. For this task of oversight, the Regional Development Agency should be given far-reaching legal powers and it should be the channel through which all grants for development purposes, from whatever Ministry they originated, were directed. We envisage the Agency being appointed by a Minister (after consultations with the local authorities) and organising itself on business lines, acting through a General Manager rather than a series of committees."

50. "The Regional Development Agency's task should be to act partly through other authorities, partly on its own account."

The working committee illustrate their recommendations with very fully detailed examples of hypothetical solutions to real problems in a small town, a large town, an historic town, a postulated central metropolitan block in England. They use similar techniques, and come to similar conclusions, as those illustrated in this book.

XI. Bibliography

XI. 1. Bibliography: Books

A

ABEL. *Regeneration der Städte*. Zürich, 1950.
Air Pollution. 19 authors. W.H.O., Geneva, 1961.

B

BAKER, FUNERO. *Shopping Centres*. New York, 1958.
BARR C. *Public Authority Housing*. London, 1958.
BARTHOLOMEW H. *Land Uses in American Cities*. Harvard City Planning Series. Vol. XV. Harvard University Press, 1955.
BAUER C. *Modern Housing*. Boston, 1934.
BEAZLEY E. *Design and Detail of the Space between Buildings*. London, 1961.
BENDTSEN P. H. *Town and Traffic*. Copenhagen, 1961.
BRIERLEY J. *Parking of Motor Vehicles*. London, 1963.
BUCHANAN C. D. *Mixed Blessing, The Motor in Britain*. London, 1953.
BURNS W. *British Shopping Centres*. London, 1959.
BURRAGE R. H. and MOGREN E. G. *Parking*. Saugatuck, 1957.

C

CHAPMAN D. *Home and Social Status*. London, 1955.
LE CORBUSIER. *The Home of Man*. London, 1948.
LE CORBUSIER and JEANNERET P. *Ihr Gesamtes Werk von 1910 bis 1929*. Zürich, 1937.
LE CORBUSIER and JEANNERET P. *Ihr Gesamtes Werk von 1929 bis 1934*. Zürich, 1935.
CROWE S. *Tomorrow's Landscape*. London, 1956.
CROWE S. *The Landscape of Roads*. London, 1960.
CULLEN G. *Townscape*. London, 1961.
CULLINGWORTH J. B. *Housing Needs and Planning Policy*. London, 1960.

D

DAHIR J. *Communities for Better Living*. New York, 1950.
DAY A. *Roads*. London, 1963.
DOXIADIS K. A. *Raumordnung im Griechischen Städtebau*. Berlin, 1938.

F

FISKER and MILLECH. *Danske arkitektur Strømninger. 1850–1950*. Copenhagen, 1951.
FOG H. and SAHLIN B. *Liv Mellan Hus*. Stockholm, 1961.
FRISCH M., BURCKHARDT L. and KUTTER M. *Die neue Stadt*. Basel, 1956.

G

GALLION A. B. *The Urban Pattern*. New York, 1960.
GIBBERD F. *Town Design*. London, 1962.
GUTKIND A. *The Expanding Environment*. London, 1953.

H

HAFEN P. *Das Schrifttum über die Deutschen Autobahnen*. Bonn, 1956.
HALMOS P. *Privacy and Solitude*. London, 1948.
Highway Capacity Manual. U.S. Department of Commerce. Washington, 1950.
HILBERSEIMER L. *The New City: Principles of Planning*. Chicago, 1944.

I

INFIELD H. *The American Intentional Communities*. New Jersey, 1955.
INFIELD H. *Sociology of Co-operation*. Loughborough, 1955.

J

JACOBS J. *Death and the Life of Great American Cities*. London and New York, 1961.
JELLICOE G. A. *Motopia*. London, 1961.
JONES J. H. *The Geometric Design of Modern Highways*. London, 1961.

K

KIDDER SMITH G. E. *Sweden Builds*. New York, 1950 and 1957.
KUPER L. *Living in Towns*. London, 1953.

L

LIEPMAN K. *The Journey to Work*. London, 1944.
LOGIE G. *Industry in Towns*. London, 1952.
LOGIE G. *The Urban Scene*. London, 1954.
LORENZ H., KASPER H., and SCHURBA W. *Die Klotoide als Trassierungselement*. Bonn, 1954.

M

MACKAYE B. *The New Exploration*. New York, 1928.
MATSON, SMITH, HURD. *Traffic Engineering*. New York, 1955.
MERKLE, PORTMAN, ARIOLE. *Garten, Menschen, Spiele*. Basel, 1960.
MUMFORD L. *Culture of Cities*. London, 1941.
MUMFORD L. *The City in History*. London, 1961.
MITCHELL R. B. and CHESTER RAPKIN. *Urban Traffic*. New York, 1964.

O

OTTO K. *Die Stadt von Morgen*. Berlin, 1959.

P

PEARSE and CROCKER. *The Peckham Experiment. A Study of the Living Structure of Society*. London, 1943.
PERRY C. The Neighbourhood Unit. A Scheme of Arrangement for the Family Life Community. Vol. VII, *Regional Study of New York and Its Environs*. New York, 1929.
PETTIGREW J. B. *Design in Nature*. Vols. I, II, III. London, 1908.
Planning for a New Town. (Hook). London, 1961.

R

RASMUSSEN S. E. *Towns and Buildings*. Cambridge, Mass., 1951.
REICH W. *Function of the Orgasm*. New York, 1942.
REICH W. *Ether, God and Devil*. New York, 1951.
REICHOW H. B. *Organischer Städtebau*. Braunschweig, 1948.
REICHOW H. B. *Organische Städtbaukunst*. Braunschweig, 1948.
REICHOW H. B. *Die Autogerechte Stadt*. Ravensburg, 1959.
RITTER P. and J. *The Free Family*. London, 1959.
Roads Campaign Council. *New Ways for London, Five Award Winning Schemes*. London, 1960.
RODWIN L. *The Future Metropolis*. New York, 1961.
RODWIN L. *The British New Towns Policy: Problems and Implications*. Harvard, 1956.

S

SCHAEFFER A. A. *Ameboid Movement*. Princetown, 1920.
SCOTT-WILLIAMSON. *Physician Heal Thyself*. London, 1944.
SEGAL W. *Home and Environment*. London, 1953.
SMITH W. & Associates. *Future Highways and Urban Growth*. New Haven, 1961.
SMITH W. & Associates. *Analyses of Central City Access Plans*. Philadelphia, 1959.
STEIN C. *Towards New Towns for America*. Liverpool, 1958.
SUTTIE I. *Origins of Love and Hate*. London, 1935.

T

THOMPSON L. F. *Site Planning in Practice*. London, 1923.
TRIPP A. *Road Traffic and its Control*. London, 1938.
TRIPP A. *Town Planning and Road Traffic*. London, 1942.
TUNNARD C. *The City of Man*. New York, 1953.
TUNNARD C., and PUSHKAREV B. *Man-Made America: Chaos or Control*. Yale, 1963.
TYRWHITT, SERT, ROGERS. *The Heart of the City*. London, 1952.

U

Urban Motorways. Illustrated Report of the London Conference organized by the British Road Federation. London, 1956.

V

VOGLER and KÜHN. *Medizin und Städtebau* I and II. Munich, 1957.

W

Washington Centre for Metropolitan Studies. *Import of Design Transportation and the Metropolis,* Transcript of Conference, Washington, 1963.
WHYTE W. H. *The Organisation Man.* London, 1957.
WIBBERLY G. P. *Agriculture and Urban Growth.* London, 1959.
WIENER N. *The Human Use of Human Beings.* New York, 1954.
WILLIAMS T. E. H. *Urban Survival and Traffic, a Symposium.* London, 1962.
WILLIAMS W. P. *Recreation Places.* New York, 1958.
WILLMOTT P. *The Evolution of a Community.* London, 1963.
WOOD R. C. *Suburbia: Its People and Their Politics.* Boston, 1959.

XI. 2. Bibliography: Articles, Reports and Theses, Reprints of Articles

A

ALKMAAR. B.O. & W. Rotterdam, 18 April 1959.
American Transit Association. Transit Fact Books. Annually.
Andover Town Development. Town Development Division of the Architect's Department of the L.C.C. London, 1962.
Andover: Town Expansion at Andover. *J.T.P.I.* London, December 1962.
Arkitektor II. Stockholm, 1962.

B

BAILEY J. A. The Transportation of People and Goods—Solutions to Area Wide Transportation Problems. October 1962.
BACH I. J. Traffic Planning and Urban Development in Chicago. *T.Q.* Saugatuck, April 1963.
BACON E. N. Philadelphia. *A. Rec.* May 1961.
BACON E. N. *The Space between Buildings.* Harvard, 14 April 1962.
BAGENAL G. H. Planning Against Noise. *A.J.* London, 12 May 1960.
BAILEY J. A. *Transportation in the Philadelphia Metropolitan Region.* Pennsylvania, 19 May 1959.
BAILEY J. A. and BANNELLS J. The Progress of Transportation in the Philadelphia Region. *T.Q.* Saugatuck, April 1958.
BAKEMA and VAN DEN BROEK. Nieuwe Woonformen voor Nord-Kennemorland. *Bow.* Rotterdam, 18 April 1959.
Banbury Central Area. *Civic Trust, Final Report 3 D.* London, 1962.
BANHAM R. Park Hill. *A.R.* London, December 1961.
Barbican Committee. *Residential Development within the Barbican Area.* London, April 1959.
Barbican: Report on the Barbican by Chamberlin, Powell and Bon. London, May 1962.
Baronbackarna. *Plan 1960 and Plan 1962.* Stockholm.
Baronbackarna. University of Manchester. *A.S.J.* No. 8.
BARNES H. A. Traffic Control. *A.J.* London, 26 November 1961.
BARNES H. A. Increasing Street Capacity. 4th World Meeting, International Road Federation. Madrid, October 1962.
BARR C. Housing in the 1960's. *J.R.I.B.A.* April 1962.
Basildon. Planning for the Motor Vehicle at Basildon. *A. & N.B.* London, 11 November 1959.
Basildon New Town. *A.J.* London, June 1956.
Basildon Town Centre. *A.R.* London, January 1956.
Basingstoke Development Group. Basingstoke Town Development. August 1962.
BATE S. M. Express Bus Service on Freeways. *T.Q.* Saugatuck, July 1954.
Bedmont. Housing at Bedmont, Herts. *A.J.* London, 3 January 1962.
Beeston, Ilkeston Road and Stapleford Urban District Council. *Future Housing Estates.* Beeston, 30 October 1959.
Beeston, Housing Architect. *Report on Radburn Principles.* 1959.
BENDIXSON T. M. P. Expansion at Andover. *A.J.* London, 7 November 1962.
BENDIXSON T. M. P. Expanded Towns. *A.J.* London, 14 November 1962.
BENDTSEN S. Traffic Congestion Costs Money. *Plan.* Stockholm, 1953.
BENNETT H. Rebuilding our Cities. The County of London. *J.R.I.B.A.* London, July 1960.
BERBIERS J. L. *St. Peter's Street Improvement Scheme.* City of Canterbury, 1962.
BERRY and ROBINSON. Engineering for Traffic Conference Papers. London, 1963.
Billingham U.D.C. *Development at Low Grange. "Radburn" type layout.* County Planning Office. Durham, 1958.
Bishop's Stortford. *Central Area Draft Plan.* Hertfordshire County Council. Hereford, 1962.

Blackburn Central Development. *A.J.* London, 25 May 1961.

BOILEAU I. Traffic and Land Use. *T.P.R.* Liverpool, April 1958.

BOR W. On the Beeching Report. *A.J.* London, 10 April 1963.

Brandon Estate. Housing at Brandon Estate, Southwark. *A.J.* London, 1 November 1961.

Brandon Wood Estate, Southwark. L.C.C. London, 1960.

BREDSDORFF, RASMUSSEN, SKAARUP. Horsholm. ByPlan 5. 1960.

BRENIKOV P. Planning for the New Mobility. *A.J.* London, 29 December 1955.

Brighton Sea Front Site. *A. & B.N.* London, 28 September 1960.

Bron-y-Mor. *A.J.* London, 26 July 1961 and 30 August 1961.

BROWNE, BLACHNICKI. Over and Under. *A.R.* London, May 1961.

BROWNE K. The Traffic Plan: Enemy or Ally? Townscape. *A.R.* London, 1961.

BROWNE K. Straightjacket. The L.C.C. Daylighting Code *v.* the London Street. *A.R.* London, October 1961 and May 1962.

BROWNE K. South Barbican. *A.R.* London, May 1960.

BRUNNER C. Living with the Motor Vehicle. *A.J.* London, 3 March 1960.

BRUNNER C. Roads and Traffic Economic Aspects. *T.E. & C.* July 1960.

Brunswick Redevelopment Area, Manchester. Manchester, 1962.

BUCHANAN C. D. Reprieve for Piccadilly. *A.J.* London, 16 May 1960.

BUCHANAN C. D. Symposium on Urban Survival and Traffic. Durham University. *A.J.* London, 4 May 1961.

BUCHANAN C. D. Standards and Values in Motor Age Towns. *J.R.I.B.A.* December 1961; *J.T.P.I.* December 1961; *A.J.* 1 November 1961.

BUCHANAN C. D. Transport—the Crux of City Planning. *J.R.I.B.A.* December 1960.

BUCHANAN C. D. Town and Traffic. *J.R.I.B.A.* London, August 1962.

BUCHANAN C. D. Reprieve for Piccadilly. *A.J.* London, 1962.

BUCHANAN C. D. Report of the Public Inquiry concerning development on the Monico site at Piccadilly Circus. *T.P.R.* Liverpool, January 1961. (Full illustrated account.)

BUCK H. Town and Country Planning School. The Redevelopment of Small Towns. Nottingham, 1962.

BURNS W. Problem of the Parked Car. *J.T.P.I.* July/August 1958.

BURNS W. Town Planning Institute—Lessons from the Beeching Report. *Traffic Engineering and Control*. London, August 1963.

BURRELL T. S. Hexham—Plan for a Market Town. *J.T.P.I.* London, December 1961.

C

CALTHROP M. Housing at Bedmont, Herts. Edinburgh, 1962.

Cambridge Village. Covell, Matthews & Partners. March 1963.

Canterbury, St. Peter's Street Improvement Scheme. City of Canterbury, 1962.

Car in Urban Areas. *A.J.* London, 8 October 1959.

Cardiff Central Area. City Architect's Report. Cardiff, January 1962.

CARTER G. E. "Hollow Ring" Town Centres. *A.J.* London, 1 August 1962.

Centre for Safety Education of the Division of General Education, New York University. Personal Characteristics of Traffic-Accidents Repeaters. *T.Q.* Saugatuck, 1948.

CHADWICK G. F. Parking in City Centres. *J.T.P.I.* London, April 1961.

CHADWICK H. and ORAM D. *Beeston Centre Redevelopment*. May 1962. (Private publication.)

CHAMBERLIN, POWELL, BON. *Report on Barbican*. London, 1959.

Chandigarh. *A.Rec.* June 1955.

Chandigarh. *J.R.I.B.A.* London, January 1955.

CHANDLER K. N. Traffic Trends. R.R.L. Slough, 1956.

CHAPMAN D. Convenience, the Achievement of a Desirable Quality in Town Planning. *H. Rel.* February 1950.

CHARLESWORTH G. *Road and Commercial Transport*. Institute of Road Transport Engineers 1954.

Chatham Village. *A.R.* London, May 1960.

CHERMAYEFF S. The New Nomads. *T.Q.* Saugatuck, April 1960.

Chicago Transit Authority. *C.T.A. Transit News*. September 1959.

CHILDS D. *New Towns Reports. 1960–1961.* London, February 1962.

CHILDS D., RIGBY, WHITTLE J. A Study of Environment and Density. *A.J.* London, 8 September 1960.

Child's Eye View Exhibition. Catalogue. Nottingham, 1959.

CHURCHILL J. D. C. Effect of Traffic Engineering Measures on London's Buses. *T.E. & C.* January 1963.

Cités Nouvelles, Centres Urbains. *L'architecture d'aujourdhui.* (Special issue) Paris, April/May 1962.

CLAIRE W. H. Urban Renewal and Transportation. *T.Q.* Saugatuck, July 1959.

CLARK H. F. Housing and the Landscape. Paper to the conference of the Council for the Preservation of Rural England. 1958.

CLAY G. Youngsters Help Build Own Environment. *The Courier Journal—Louisville*, 11 February 1962.

CLAY G. Missed Opportunities and Messed Metropolises. American Society of Planning Officials. California, 20 March 1957.

CLINCH F. H. Urban Highway Requirements. Assessment and Fulfilment. *J.T.P.I.* London, September 1961.

COBURN T. M. Origin and Destination Surveys. Chartered Municipal Engineers, 1962.

COBURN T. M. and GREEN G. R. Speeds on European Motorways. 1959. *International Road Safety Traffic Review*, 1959.
Committee on Uses of Developed Information. Uses of Traffic Accident Records. *T.Q.* Saugatuck, 1947.
Commons Debate on Traffic Chaos. *A.J.* London, 24 December 1959.
CONWAY T., JR. Rapid Transit must be Improved to Alleviate Traffic Congestion. Saugatuck, January 1962.
COOK S. A. G. Kingsway/High Holborn Intersection—Safety for the Pedestrian. Report to Holborn Borough Council Housing and Planning Committee. December 1962.
COPCUTT G. Shopping Facilities in Cumbernauld. *A.J.* London, 1 December 1960.
COPCUTT G. Cumbernauld New Town Centre. *A.J.* London, 5 December 1962.
COVELL, MATTHEWS and PARTNERS. Cambridge Village, March 1963.
COVELL, MATTHEWS and PARTNERS. A Traffic Plan for Central London. London, 1963.
Coventry Development and Redevelopment. Coventry, September 1961.
Coventry Rebuilds. *A.D.* London, December 1958.
Coventry, the City Road Pattern. Draft Joint Report. Coventry, October 1961.
CRAIG J. Town Expansion at Andover. *J.T.P.I.* December 1962.
Cramlington New Town. Challenge and Response. 1961.
CRAWFORD A. Fatigue and Driving. *Ergonomics*. 1961.
Crawley New Town. The Use of Leisure in a New Town. The Report of a Social Survey in Crawley New Town. Workers Educational Association.
The Crisis of Urban England. *A.J.* London, 30 June 1960.
CROCKER P. S. Grangemouth Shopping Centre. *A.J.* London, 26 February 1960.
CROOKE P. Parkhill, Sheffield. *A.D.* London, September 1961.
CROOKE P. Central Development Sheffield. *A.D.* London, September 1961.
CULLEN G. Basildon Town Centre. *A.R.* January 1956.
CULLINGWORTH J. B. *Restraining Urban Growth*. Fabian Society. London, 1960.
Cumbernauld Special Issue. *A. & B.N.* London, 29 March 1961.
Cumbernauld. Car Parking. *A.J.* London, 15 December 1960.
Cumbernauld. Shopping. *A.J.* London, 1 December 1960.
Cumbernauld New Town Centre. *A.J.* London, 5 December 1962.
Cumbernauld Development Corporation. Traffic Engineering Section. Housing Roads and the Housing Unit. March 1963.
Cumbernauld Development Corporation. Traffic Engineering Section. Cul-de-sac Design Procedure. March 1963.
Cumbernauld Central Area Report. Cumbernauld Development Corporation. Chief Architect. Cumbernauld, 1960.
Cumbernauld Development Corporation Preliminary Planning Proposals. Cumbernauld, 1958.

D

DALE R. W. Planning for Traffic in Small Towns. *J.T.P.I.* London, February 1963.
DARK J. W. Why not the Monorail? *J.T.P.I.* London, February 1963.
Darlaston Central Development. *A.J.* London, 2 August 1961.
DAVIES A. B. Planning for the Motor Vehicle at Basildon. *A. & B.N.* London, 11 November 1959 (and correspondence subsequently).
DAVIES E. Traffic Practice. London, 1963.
DAVIS E. Roads and Their Traffic. London, 1960.
DAWSON R. F. F. Survey of Commercial Traffic in London. *Traffic Engineering and Control*. London August, 1963.
Denmark. Housing Layout in Denmark. *J.R.I.B.A.* March 1959.
Deptford. Thamesside Development at Deptford. *A.J.* London, 13 April 1961.
Deptford. *Redevelopment of Royal Victoria Yard, Deptford*. L.C.C. London, April 1961.
Design for Motorised Living. *A.J.* London, 3 November 1960.
Designing for Pleasure and Hardwear in Landscape. *A.J.* London, 2 June 1960.
The Development of a Neighbourhood Unit. *J.R.I.B.A.* London, December 1953.
DOUBLEDAY E. H. Bishop's Stortford. Central Area Draft Plan 1962.
DOUBLEDAY E. H. Traffic In a Changing Environment. *J.T.P.I.* London, September/October 1962.
DOXIADIS C. A. Architecture in Evolution. *J.R.I.B.A.* September and October 1960.
DOXIADIS C. A. Ekistics. Town and Country Planning Summer School 1959.
DOXIADIS C. A. Ekistics. *Apxitektonikh*. January/February 1959.
DOXIADIS C. A. Housing in Baghdad. Government of Iraq, March 1957.
DOXIADIS C. A. The Greek City Plan. *Landscape*, Autumn 1956.
DOXIADIS ASSOCIATES. Eastwick Philadelphia. Philadelphia, 1960.
DOXIADIS ASSOCIATES. Eastwick Philadelphia Project. Philadelphia, 1959.
Dublin. The Future of Dublin. Forgan, April 1962.
Dumbarton, Central Area Redevelopment. *A.J.* London, 14 July 1960.

E

EASTFIELDS. Report on Proposed Layout. County Borough of Northampton, April 1952.
EDBLOM, STROMDAHL, WESTERMANN. Mot en Nymiljo. *Arkitektur 8*. Stockholm, 1962.
EKHOLM and WHITE. Baronbackarna. University of Manchester. *A.S.J.* No. 8.
Elephant and Castle Competition. *A.J.* London, 21 July 1960.
ELKOUBY. Paris and Motopia. *A.J.* London, 23 June 1960.
ERRACO A. B. *Radd Storgatan*. Stockholm, 1958.

F

Factors Influencing the Layout of New Towns. The Association for Planning and Regional Construction. Paris, September 1946.
FALK. Tathe Tstalet. Plan 6. Stockholm, 1960.
Farsta, City of Stockholm. Stockholm, 1960.
Farsta Centrum. *Sartrick ur Arkitektur*. Stockholm, 1961.
FIELD D. *Warrington Caution*. Warrington, 1961.
Flemingdon Park. *The Canadian Architect*, May 1961.
FOREHOE C. and BROWNE K. Shopping Precincts—Townscape. *A.R.* London, 1959.
Fort Worth. *The Greater Fort Worth*. Fort Worth, 1956.
Fresno. Gruen V. Fresno, 1960.
FRY M. Chandigarh. *A. Rec.* June 1955.
FRY M. Chandigarh. *J.R.I.B.A.* London, January 1955.

G

GARDNER MEDWIN R. Rebuilding Rotterdam and Coventry. *The Listener*, 22 September 1955.
GARWOOD F. and DUFF J. T. Changes in Accident Frequency after Changes in Speed Limits in the U.K. Nice, 1960; London, 1960.
GARWOOD F. and JEFFCOATE G. O. Some Facts about Casualties to Pedestrians in Great Britain. *Z. Verkehrssicherheit* 1958.
GEENTY W. A. Billingham Radburn Layout. Durham C.C., 1958.
GENTILI G. Satellite Towns of Stockholm. *Urbanistica*. Milan, September 1958.
GIBBERD F. St. Albans City Centre. *A.J.* London, 22 July 1957.
GILES G. G. and SABY B. E. Skidding as a Factor in Accidents on the Roads of Great Britain. First International Skid Prevention Conference, 1958.
GINSBURG L. B. Swindon, a Breakthrough in Central Area Redevelopment. Official Architecture and Planning. London, November 1962.
GISSANE W. A Report for the Road Injuries Research Group. Birmingham Accident Hospital, England. 6th International Road Safety Congress Report. London, 1962.
GLANVILLE H. and BAKER J. F. A. Urban Motorways in Great Britain. Roads and Road Construction, 1956.
GLECKMAN W. B. Two New Parking Systems. *T.Q.* Saugatuck, July 1961.
GOLD E. The Multiple Family Housing Unit, an Approach to Community Dwelling and Town Planning. October 1961.
GOSS A. Moving Pavements. *A.J.* London, 21 January 1961.
GOSS A. and TETLOW J. D. Homes and Towns in the Motor Age. *H.R.* London, March/April 1963.
GREENWOOD S. Birmingham's New Shopping Centre. *A.J.* London, 18 May 1961.
GREY A. Inchview, Prestonpans, Department of Architecture Research Unit. Edinburgh, February 1963.
Greyfriars Redevelopment. Ipswich. Skipper & Corless. Norwich, 1962.
GRIBBON J. Can Liverpool be Planned from London. *A.J.* 7 February 1962.
GRIME C. Car Design and Operation in Relation to Road Safety. *Practitioner* 1962.
GROES-PETERSEN E. Road Safety Training of Pre-School Children in Copenhagen. 6th International Road Safety Congress Report. London, 1962.
GROSSMAN I. Flemingdon Park. *The Canadian Architect*. May 1961.
GRUEN V. *Fresno*. Fresno, 1960.
GRUEN V. *The Greater Fort Worth*. Fort Worth, 1956.
GRUEN, SMITH. *Shopping Towns U.S.A.* New York, 1960.
GRUEN, ASKWITH. Plan to end our Traffic Jam. *Time Magazine*. New York, 10 January 1960.
GUERIN E. Vällingby. *A. & B.* London, December 1958.
GUNNARSSON O. Optimum Route Location of Main Roads. Göteborg, 1960. (English summary.)

H

Halesowen Town Centre Development. Watson H. M. February 1962.
HALL I. Learning from America's Renewal Schemes for Residential Areas. *H.R.* London, March/April 1963.
HANCOCK T. Oxford Proposals. Wallingford, 1960.
HANCOCK ASSOCIATES. Wallingford, Proposals for the Town. April 1962.
HAND I. and HIXON C. D. Planning, Traffic and Transportation in Metropolitan Areas. *T.Q.* Saugatuck, April 1963.
HARLAKOFF N. Wirtschaftliche Erschliessung von Wohngebieten. Berlin, June 1961.
Harlow Housing Competition. *A.J.* London, 25 May 1961.
HARRIS J. S. The Design of Shopping Centres. *J.T.P.I.* September/October 1961.
Haverhill, Suffolk. Housing at Haverhill. *A.J.* London, 7 November 1962.
Helicopters. *A.J.* London, 25 October 1956.
HENRY and OBERLI. Zukünftiger Stadtverkehr. *B. & W.* Munich, June 1960.
Hereford, County Planning Department, Statement of Principles. Central Area Redevelopment. Hereford, 1962.
HEWITT C. P. How the U.S.A. Solves Traffic Problems. *The Listener*, 10 November 1955.
Hexham. Plan for a Market Town. *J.T.P.I.* London, December 1961.
High Oxford Street. *A.J.* London, 23 March 1961 and 30 March 1961.
High Wycombe. Evidence in Chief Buckinghamshire County Council. 1961.
HILLFIELDS C. D. A. City Architect, Coventry, 1960.
HODGEN R. Prediction of Glasgow's Future Traffic Pattern. *T.E. & C.* March 1963.
HOGAN B. The Future of Dublin. *Forgan*, April 1962.
HOLFORD W. L.C.C. Piccadilly Circus. Future Developments. May 1962.
HOLFORD W. Thoughts of an Architect Planner provoked by Recent Events. *T.E. & C.* May 1963.
HOLLIDAY J. C. Design Policies for the First Review Development Plans. *J.T.P.I.* London, February 1962.
HOLMGREN P. Stockholm Today and Tomorrow. Stockholm, 1 November 1960.
HOMBURGER W. S. Rapid Transit: Present and Future. *T.Q.* Saugatuck, April 1960.
Homes for Today and Tomorrow. H.M.S.O. London, 1961.
The Hook Study. New Town Development. *J.R.I.B.A.* February 1962.
Hook, Planning for a New Town, the Hook Study Summarised. *A.J.* London, 6 December 1961.
Hook Plan. Did the Team Think? *A.J.* London, 7 February 1962.
Horsham Action Group. Town Centre. September 1962.
Horsholm. *ByPlan 5.* 1960.
Housing in Camberwell. *A.J.* London, 25 October 1961.
Housing at Bedmont, Herts. *A.J.* London, 3 January 1962 and 10 January 1962.
HOWGRAVE-GRAHAM H. S. Towards a Car Parking Policy. *J.T.P.I.* London, December 1960.
HOYT H. The Effect of the Automobile on Patterns of Urban Growth. *T.Q.* Saugatuck, April 1963.
HUMPHRIES H. R. Noise and the Motorway. *A.J.* London, 11 July 1962.
HUNTER and LYDDON. Cumbernauld Housing Patterns. *A. & B.N.* 29 March 1961.
Hyde Park's "Cut and Cover" Garage. *A.J.* London, 31 October 1962.

I

Improving the Bye-law Street. *A.J.* London, 10 February 1955.
Inceasing Road Safety by Townplanning. Chalmer's Tech. University. Goteborg, 1960.
Influence of Motor Cars on Towns. *A.J.* London, 5 June 1958.
International Union of Public Transport. List of Publications. Brussels, 1963.
Sixth International Road Safety Congress Report. London, 1962.
Ipswich. Greyfriars Redevelopment. Skipper & Corless. Norwich, 1962.
IVERSEN, LANGE. Vorbildliche Verkaufsstrassen in Deutschen und Ausländischen Städten. *Der Wiederaufbau.* Bremen, December 1958.

J

JEFFERSON J. G. At the Cross Roads in the Motor Age. *A.J.* London, 8 November 1961.
JELLICOE G. A. A Comprehensive Plan for the Central Area of the City of Gloucester. December 1961.
JONES W. W. Car Culture in the U.S.A. Letter, *J.T.P.I.* London, September/October 1960.

K

KEEBLE L. B. Planning at the Cross Roads. *A.J.* London, 22 June 1961.
KENT B. Layout for a Neighbourhood Plan. *A.F.* London, 1938.
KIDD R. Economic Aspects of Housing Layout. Nottingham County Council. June 1955.
KIRBY C. P. Utilization of Parking Meter Space. L.C.C. Architect's Department, 1963.
Kittimat. A New City. *A.F.* London, 1954.
KLEIN A. Solving the Traffic Problem. *T.Q.* Saugatuck, July 1949.
Knightsbridge Green. Morgan & Partners. *A.J.* London, 21 March 1962.
KRIESIS P. Metropolitan Centres. Tractus Politice-Urbanisticus. *J.T.P.I.* London, July/August 1961.

L

LAURIE I. C. Residential Layouts. Thesis, University of Manchester, 1951–1952.
L.C.C. Population, Employment and Transport in the London Region: The Years Ahead. Standing Conference on London Regional Planning. Report by the Technical Panel, 1963.
L.C.C. Redevelopment of the Royal Victoria Yard. Deptford, April 1961.
LEACH A. Multi-Storey Car Parks. *The Builder*, 18 January 1963.
LEDERMANN A. and TRACHSEL A. Playgrounds and Recreation Spaces. *A.D.* London, 1959.
LICHFIELD N. Planning for Urban Renewal. The American Approach. *J.T.P.I.* London, March 1961.
LIEPER J. McM. The Role of the Automobile in Mid-Town Manhattan. *T.Q.* Saugatuck, April 1962.
LIJNBAAN A.D. London, August 1954.
Lillerod. Skaarup, Jespersen. Copenhagen, 1960.
Lincoln Road Mall, Miami Beach. Lipp M. N. *T.Q.* Saugatuck, July 1961.
LING A. Coventry's Pedestrian Shopping Centre. *Keystone*, May/June 1956.
LINN K. Melon Neighbourhood Park. *Landscape Architecture Quarterly*, January 1962.
LINSCOTT P. M. $529,700,000 Rapid Transit System Recommended for Los Angeles. *T.Q.* Saugatuck, April 1961.
LIPP M. N. Lincoln Road Mall, Miami Beach. *T.Q.* Saugatuck. July 1961.
LITTLER F. H. Changing Towns, Planning for Renewal. *J.T.P.I.* July/August 1962.
LODS, HONEGGER. Shopping Centre. Marly-le-Roi. *C du C.T.S.B.* Paris, October 1960.
LODS, HONEGGER. Grand Terres. Marly-le-Roi. *C. du C.T.S.B.* Paris, 1959.
LODS M. Town Planning in an Ancient City. *A.J.* London, 5 April 1951.
LORWENSTEIN L. K. Commuting and the Cost of Housing in Philadelphia. *T.Q.* Saugatuck, April 1963.
LOGIE G. Cumbernauld Road Layout. *A. & B.N.* 29 March 1961.
LOGUE E. J. Can Cities Survive Automobile Age? New Haven Used as a Test Case. *T.Q.* Saugatuck, April 1959.
London, New Ways for London. R.C.C. London, 1960.
London, Oxford Street. Walk Above or Below the Motor Car. *A.J.* London, 12 May 1960.
London Road Competition. *A.J.* London, 1 October 1959.
London Road Competition, Results and Appraisals. *A.J.* London, 24 December 1959.
London Traffic. The "B" ring revived. *A.J.* London, 28 April 1960.
London: A Traffic Plan for Central London. Covell Matthews & Partners. London, 1963.
London, Traffic Plan for London. Smigielski M. K. *J.T.P.I.* London, April 1960.
LORENZ H. *Aesthetische und Praktische Linienführung der Strasse.* XI Internationaler Strassenkongress. Rio de Janeiro, Sektion 1, 1959.
LOVELL S. M. Highway Needs of Britain. Institute of Civil Engineering. London, 1957.
LOWRY J. F. Pilot Project Converts an Eyesore into Community Park in N. Philadelphia. *The Evening Bulletin.* Philadelphia, 27 February 1962.
LU W. Thoroughfare Planning and Goal Definition. *T.Q.* Saugatuck, April 1963.
LUPTON M. Renewal in St. Helens Area, Abingdon. Wallingford, 1961.
LYONS E. Building a Community. *A.D.* London, September 1960.
LYDDON D. and HUNTER. The Cumbernauld Housing Pattern. *A. & B.N.* London, 29 March 1961.
LYNN J. Park Hill Redevelopment, Sheffield. *J.R.I.B.A.* December 1962.

M

MACKAYE B. The New Exploration. A Philosophy of Regional Planning. New York, 1928.
MALFETTI J. L. Scare Techniques and Traffic Safety. *T.Q.* Saugatuck, April 1961.
Malmö Planning Office. *Malmö 1959*, Malmö, 1960.
Malmö 1862–1962. Malmö, September 1962.
MAOO A. F. The Relation of Mass Transport to Total Transportation in Detroit. *T.Q.* Saugatuck, April 1961.
Manchester. Brunswick Redevelopment Area. Manchester, 1962.
Manchester Corn Exchange Redevelopment Project. Hatrell & Partners, 1962.
MANZONI H. J. Public Parking Garages. *J.R.I.B.A.* March 1958.
MARGERISON T. A "Secret Eye" on the Playground. *Sunday Times*, October 1961.
MARKS H. Sub-dividing for Traffic. *T.Q.* Saugatuck, July 1957.

Marly-le-Roi. *Architectural Design*. London, April 1963.
Marly-le-Roi. Grand Terres. *C. du C.T.S.B.* Paris, 1959.
Marly-le-Roi. Shopping Centre. *C. du C.T.S.B.* Paris, October 1960.
Marseilles Experiment. Tomkinson D. *T.P.R.* Liverpool, October 1953.
DE MARSH L. F. John F. Fitzgerald Expressway Boston Central Artery. *T.Q.* Saugatuck, October 1956.
MARSHALL P. J. Rotterdam. *A.J.* London, 27 October 1955.
MARSHALL P. J. I–V Comprehensive Redevelopment. *J.R.I.B.A.* April 1959 and December 1959.
MARSTON W. R. Traffic Plan for Chicago. *T.Q.* Saugatuck, July 1960.
MARTERN B. *Trafiken och det Framtida City*. Stockholm, 1958.
MARTIN, MEMMOTT, BONE. Principles and Techniques of Predicting Future Demands for Urban Area Transportation. Mass., 1963.
MATTHEW R. Building and Planning in the Motor Age. *J.R.I.B.A.* London, September 1962.
MATZUTT DR. M. Road Safety Training of Pre-School Age Children. 6th International Road Safety Congress Report. London, 1962.
MAUCHLEN, WEIGHTMAN and ELPHICK. Cramlington New Town. Challenge and Response. Survey and Plan, 1961.
MAY E. *Das Neue Mainz*. Mainz, 1961.
MAYER A. Complete Communities on the Greenway Principle. *A.F.* London. January and February 1937.
McELROY J. P. Pedestrian Conveyors. *J.T.P.R.* July 1961.
McMILLAN C. P. South Carolina New Speed Law. *T.Q.* Saugatuck, October 1957.
McMILLAN S. C. Recent Trends in the Decentralisation of Retail Trade. *T.Q.* Saugatuck, January 1962.
McMONOGLE. Traffic Accidents and Roadside Features. Washington, 1952.
McGUIRE, WILD. Kitimat. *Canadian Geographical Journal*. November 1959.
McHARG I. C. Can we afford open space. *A.J.* London, 8 March 1956 and 15 March 1956.
MEACOCK H. E. *Off-Street Parking*. Paper to the Institution of Highway Engineers. 1960.
Mechanical Car Park. *A. & B.N.* London, 13 February 1963.
MEDHURST R. The Architecture of New Roads. *J.T.P.I.* November 1960.
Meeting the Transportation Crisis. Passenger Service Improvement Corporation. Philadelphia, 1959.
MEIER R. L. A Communication Theory of Urban Growth. M.I.T. Cambride, U.S.A., 1962.
MERCER J. Modern Tendencies in Estate Layout. Public Works Congress Report, 1954.
Middlesex County Planning Department. Garage and Parking Space in Residential Areas. *J.T.P.I.* March 1960.
MIDDLETON J. T. and CLARKSON D. Motor Vehicle Pollution Control. *T.Q.* Saugatuck, 1961.
MIDWINTER S. W. The Town and the Motor Car. *J.T.P.I.* April 1956. Letter October 1956.
MILLS W. R. Future Trip Length. *T.Q.* Saugatuck, April 1963.
Ministry of Housing and Local Government. Housing 1953. H.M.S.O. London, 1953.
Ministry of Housing and Local Government. Housing 1958. H.M.S.O. London, 1958.
Ministry of Housing and Local Government. Homes for Today and Tomorrow, 1961.
Ministry of Housing and Local Government. Space in the Home. Design Bulletin 6. H.M.S.O. 1963.
Ministry of Housing and Local Government. Town Centres. Approach to Renewal. H.M.S.O. 1962.
Ministry of Housing and Local Government. Residential Areas—Higher Densities. H.M.S.O. 1962.
Ministry of Transport. *Roads in England and Wales. Report of the Year 1961–1962.* H.M.S.O. 1962.
Ministry of Transport. *Parking Survey of Inner London. September 1956.* Interim Report. H.M.S.O. 1957.
Ministry of Transport. *The Transport Needs of Great Britain in the next Twenty Years.* H.M.S.O. 1963.
Minneapolis. City of Minneapolis Planning Commission. Neighbourhood and Community Goals for Minneapolis Living Areas. Housing Series 3. Minneapolis, July 1960.
MOORE R. L. Psychological Factors of Importance in Traffic Engineering. Road Safety in India, 1960.
MORONY L. R. A Realistic Approach to Problems of Motor Vehicle and Highway Use. *T.Q.* Saugatuck, April 1961.
MORRIS R. L. *A.D.* Special Number, The Planning of Philadelphia. London, August 1962
MORRIS R. L. and ZISMAN S. B. The Pedestrian Downtown and the Planner. *J.A.I.P.* August 1962.
The Motor Age I. *Freedom*. London. 22 March 1958.
II. Roads. *Freedom*. London. 29 March 1958.
III. Towns. *Freedom*. London. 5 April 1958.
IV. America. *Freedom*. London. 12 April 1958.
V. Anarchist Attitudes. *Freedom*. London. 19 April 1958.
VI. Solution. *Freedom*. London. 26 April 1958.
Metropolis. *A.J.* Special Issue. London, 1 October 1959.
MUMFORD L. Planning for the Phases of Life. *T.P.R.* Liverpool, April 1949.
MUMFORD L. The Neighbourhood and the Neighbourhood Unit. *T.P.R.* January 1954.
MUMFORD L. A New Approach to Worker's Housing. *International Labour Review*. February 1957.

N

NAIRN I. New Newcastle. *A.J.* London, 20 April 1961.
NAIRN I. L.C.C's plan for New Town, Hook. *Daily Telegraph*, London, 29 September 1961.
Neu-Winsen Reconstruction. Hamburg. *J.T.P.I.* London, September/October 1961.

Neighbourhoods. Three Neighbourhoods Designed to Modern Standards. *H.R.* March/April 1963.
NEWBY R. F. 40 m.p.h. Speed Limit, Effect on Accident Frequency in London Area. *T.E. & C.* London, 1962.
NEWBY R. F. 40 m.p.h. Speed Limit, Effect on Accidents in the London Area. *T.E. & C.* London, 1960.
NEWCOMBE V. Z. Urbanisation in East Pakistan. *J.T.P.I.* London. July/August 1960.
New Newcastle. Traffic Segregation Scheme for City Council. *A.J.* London, 20 April 1961. Letters to the editor, *A.J.* 20 September 1961; *A.J.* 6 September 1961.
New Town Centre Projects. *A.J.* London, 7.8.1963.
NICHOLAS R. The Impact of Motorways on Cities. *A.J.* London, 24 March 1960.
Noise: Final Report. Committee on the Problem of Noise. H.M.S.O. July 1963.
NORRA BISKOPSGADEN. Ekholm & White, Gothenburg, 1959.
North Berwick. Supplement to Report on Redevelopment of High Street Area. County Planning Officer. East Lothian County Council, 1960.
Norwich, Alderson Place. *H.R.* November/December 1961.
Nottingham County Council. Conference of Housing Layout. June 1955.
Nottingham Junior Chamber of Commerce. Traffic Report. Nottingham, 1961.

O

Oakdale Manor, North York, Ontario. *The Canadian Architect*, September 1962.
Orebro. Richardson. Orebro, 1960.
Orreje. *Västerås*. Västerås, 1961.
Oslo, City of Oslo. *Tveita en Forstad til Oslo*. Oslo, 1960.
OU and SHULDINER. Analysis of Urban Transportation Demands. *Transactions North West University Press*, 1963.
Oxford, An Outsider's Plan. *A.J.* London, 4 August 1960.
Oxford Proposals. Hancock T. Wallington, 1960.
Oxford Relieved. *A.R.* London, May 1956.
Oxford Street. Walk Above or Below the Motor Car. *A.J.* London, 12 May 1960.

P

Park Hill. *A.R.* London, December 1961.
Park Hill Redevelopment, Sheffield. *J.R.I.B.A.* December 1962.
Park Hill, an Urban Community. City of Sheffield Housing Development Committee. 1960–1961.
Park Hill, Amenities and Management. Sheffield, 1959.
Park Hill, Sheffield. City Architect. Sheffield, 1961.
Park Hill, Sheffield. *A.D.* London, September 1961.
Park Hill Survey. City of Sheffield Housing Department. September 1962.
Park Hill Development. *A.J.* London, 23 August 1961.
Parking Space. *Times* Correspondence. *A.J.* London, 10 November 1960.
PECKS. Eine Neue Stadt Köln. Cologne, 1958.
Pedestrians Association for Road Safety. A Pedestrian's Bill of Rights. London, March 1961.
Pedestrians on the Ground. *A. & B.N.* London, 26 April 1961.
PENDLETON W. C. Land Use at Freeway Interchanges. *T.Q.* Saugatuck, October 1961.
Personal Characteristics of Traffic Accident Repeaters. *T.Q.* Saugatuck, 1948.
Philadelphia Planning Commission Annual Report, 1957. Philadelphia, 1958.
Philadelphia Planning Commission Annual Report, 1961. Philadelphia, 1962.
Philadelphia City Planning Commission. University City 3. Redevelopment Area Plan. May 1962.
Philadelphia. Planning and Development in Philadelphia. Special Issue. *J.A.I.P.* August 1960.
Philadelphia Comprehensive Plan. The Physical Development Plan for the City of Philadelphia, 1960.
Piccadilly Circus. Ideas for Piccadilly Circus. *A.J.* London, 28 January 1960.
Piccadilly Circus. Reprieve for Piccadilly Circus. *A.J.* London, 16 May 1960.
PSIC Rail Operations. Passenger Volume: Year 1961. April 1962.
PSIC Rail Operations. Passenger Volume: 1962 First Quarter Report. June 1962.
City of Philadelphia. Urban Traffic and Transportation Board. Analysis of Operation Northwest. March 1960.
City of Philadelphia. Urban Traffic and Transportation Board. Market Potential for Pennsylvania Railroad. Chestnut Hill Line. March 1958.
City of Philadelphia. Urban Traffic and Transportation Board. Transportation Requirements—Travel Patterns.

City of Philadelphia. Urban Traffic and Transportation Board. Survey of Railroad Use under Operation North West. Philadelphia, May 1959.
Philadelphia City Planning Commission. Mount Olivet Redevelopment Area Plan. September 1962.
Philadelphia City Planning Commission. Pratt Street Redevelopment Area Plan. June 1962.
Philadelphia City Planning Commission. Haddington Redevelopment Area Plan. March 1962.
Philadelphia City Planning Commission. Center City. Philadelphia, May 1960.
Philadelphia City Planning Commission. Press Comments on Philadelphia's Comprehensive Plan. City Center Plan. 1960.
Philadelphia City Planning Commission. Philadelphia Redevelopment Areas. September 1961.
Philadelphia City Planning Commission. Washington Square Redevelopment Area Plan. August 1961.
City of Philadelphia. Urban Traffic and Transportation Board. Plan and Program 1955. April 1956.
City of Philadelphia. Urban Traffic and Transportation Board. April 1956 May 1960.
Philadelphia City Planning Commission. Capital Program 1962–1967. 30 November 1961.
Downtown Philadelphia: Bacon E. N. A Lesson in Design for Urban Growth. *A.Rec.* May 1961.
Philadelphia's Design Sweepstakes. Thompson S. G. *A.F.* December 1958.
Philadelphia Plan. *A.D.* Special Number. August 1962.
A Key to Open Cities. Miller R.A. *A.F.* February 1958.
City of Philadelphia. Relationship of Assessed Valuation and Sale Price of Real Estate sold in 1958, with Comparisons at Five Year Intervals 1943–1958. Public Information Bulletin. July 1959.
PSIC. Condition of Philadelphia with Respect to a Regional Transportation Plan. Philadelphia, 1960.
Agreement between the City of Philadelphia and the Pennsylvania Railroad Company. 24 April 1962.
Operation North Penn. Exhibit H. Station Parking Facilities. 5 February 1962.
Philadelphia City Planning Commission. Hartranft. Redevelopment Area Plan. 29 August 1961.
Philadelphia Renascence. Study of City Planning and Urban Renewal. Ballam S. R. January 1961.
Office of the City Representative. Bureau of Public Information and Service. Fact Sheet on "Operation Northwest".
Piccadilly. Holford's Plan. *A.J.* London, 16 February 1961.
Piccadilly Circus, Future Development. L.C.C. London, May 1962.
Piccadilly Circus, Report of the Public Inquiry concerning the development of the Monico Site. *T.P.R.* January 1962.
Pierce S. R. Urban Traffic Congestion. Planning for Existing Conditions. *A. & B.N.* London, 11 March 1954.
Pitt G. Neighbourhood Planning in the New Town. Town and Country Planning. July/August 1959.
Planning Department Technical High School, Stockholm. Barn och Billar. Stockholm, 1960.
Planning for the Motor Vehicle and Public Service Vehicle. (i)–(iv) *A. & B.N.* London, 11 November 1959.
Planning for Pedestrians. L.C.C. London, 1962.
Planning for Traffic. I.M.E. Convention Report. October 1961.
Planning: Public Utilities. *A.J.* London, 1 November 1956.
Planning Research Register. *J.T.P.I.* London, 1961.
Pooley F. B. Central Development of High Wycombe. *H.R.* 1961.
Preston B. A Plan for Road Safety. London, February 1961.
Prestonpans. Housing Research Unit. *A.J.* London, 25 February 1960.
Prestonpans, Inchview. Department of Architectural Research Unit. Edinburgh, February 1963.
Price G. L. Experiment in Oxford Street. Letter to the editor. *A.J.* London, 23 June 1960.
Proposal for a Commuter Transportation Demonstration Grant Made to the Administrator Housing and Home Finance Agency. Washington, 4 May 1962.

Q

Quinby H. D. The Rapid Transit District Project of the San Francisco Bay Area. *T.Q.* Saugatuck, April 1961.

R

Rainville W. S. Jr. Transit Faces the Future. *T.Q.* Saugatuck, April 1960.
Ragmaker Dr. J. R. Cyclists Complicate Urban Traffic in the Netherlands. *T.Q.* Saugatuck, April 1948.
Rasmussen E. S. Neighbourhood Planning. *T.P.R.* January 1957.
Rawstorne P. Ideas for Piccadilly Circus. *A.J.* London, 28 January 1960.
Rebuilding City Centres. *J.T.P.I.* London, September/October 1960.
Redditch Town Centre Redevelopment. County Planning Officer. Worcester, 1962.
Reichow H. B. Bau der Sennestadt. *Deutsche Bauzeitschrift.* June 1957.
Reichow H. B. Bau der Sennestadt. *Deutsche Bauzeitschrift.* August 1959.
Reichow H. B. Bau der Sennestadt. *Deutsche Bauzeitschrift.* December 1961.
Reichow H. B. Städtebau und Lärmbekämpfung. *Der Aufbau.* 1962.

REICHOW H. B. Noise Abatement. *A.J.* London, 13 February 1963.
REICHOW H. B. Verkehrs-Chaos. *B.P. Kurier*. November 1962.
Relation of Mass Transportation to Total Transportation in Detroit. *T.Q.* Saugatuck, April, 1961.
REXFORD C. W. Park-ride in St. Louis. *T.Q.* Saugatuck, April 1955.
REYNOLDS P. J. Planning Transport and Economic Forces. *J.T.P.I.* London, November 1961.
REYNOLDS P. J. and WARDROP. Economic Losses due to Traffic Congestion. *T.E. & C.* London, 1960.
RICHARDS J. M. Failure of the New Towns. *A.R.* London, 1960.
RICHARDSON. *Orebro*. Orebro, 1960.
RICKER E. R. Traffic Design of Parking Garages. *T.Q.* Saugatuck, 1957.
Rise Farm. Preview. *A.R.* London, January 1961.
RITTER P. Biofunctional Planning. M.C.D. Thesis submitted Liverpool University. 1952.
RITTER P. Universal Manifestation of Spirals. Nottingham, 1954.
RITTER P. Social Patterns and Housing Lay-out. University of Nottingham, 1957.
RITTER P. Enclosure, O.F. No. 2. Nottingham 1957.
RITTER P. Housing Therapy. O.F. No. 6. Nottingham, 1957.
RITTER P. Man and the Motor Car. O.F. No. 4. Nottingham, 1956.
RITTER P. The Radburn Idea in Great Britain. *H.R.* London, 1959.
RITTER P. Radburn-planering. *Att bo*, Stockholm, April 1961.
RITTER P. Radburn-planering. *Att bo*, Stockholm, June 1961.
RITTER P. Radburn Planning. I. Classic Objections. *A.J.* London, 10 November 1960.
II. A Social Enquiry. *A.J.* London, 17 November 1960.
III. A Social Enquiry, Results. *A.J.* London, 24 November 1960.
IV. Comparative Costs. *A.J.* London, 8 December 1960.
V. Design Principles. *A.J.* London, 12 January 1961.
VI. Design Principles. *A.J.* London, 26 January 1961.
VII. Foreign Examples. *A.J.* London, 2 February 1961.
VIII. Foreign Examples. *A.J.* London, 9 February 1961.
IX. British Examples Under Way. *A.J.* London, 16 February 1961.
RITTER P. Preaching to the Near Converted. *A.J.* London, 20 September 1961.
RITTER P. Radburn Planning. *Ekistics*. Athens, April 1961.
RITTER P. Man and Motor Don't Mix. Nottingham, 1961.
RITTER P. Man and Motor Don't Mix. *Anthos*, I. 1962. Zürich 1962.
RITTER P. Spec-Housing, Special Number. *A.D.* London, May 1962.
RITTER P. Cumbernauld. *Att bo*. Stockholm, 1962.
RITTER P. Rowlett's Hill Housing, Leicester. *A.J.* London, 10 April 1963; also *J.T.P.I.* London April 1963.
Roads Campaign Council. *Divide and Survive*. London, February 1962.
Roads in England and Wales. 1961–1962. H.M.S.O. London, 1962.
Road Research Board of the DSIR. Annual Report (1962). H.M.S.O. 1963.
Road Research Laboratory. Department of Scientific and Industrial Research. List of Publications on Traffic and Safety. Prepared May 1961. (For other R.R.L. publications see authors' names.)
ROBERTSON M. Pilot Project on Willenhall I. Liverpool, 1959.
ROBINSON V. J. Changes and Trends in American Central Areas. *J.T.P.I.* London, June 1962.
ROHE M. Towers and Row Houses, La Fayette, Detroit. *A.F.* New York, May 1960.
ROSNER R. Rotterdam. *A.J.* London, 7 April 1955.
ROSNER R. Homes or Gadgets. *H.R.* London, May/June 1960.
ROSNER R. Hamburg, the Neu-Winsen Reconstruction. *J.T.P.I.* July/August 1960.
ROSNER R. Hamburg, the Neu-Winsen Reconstruction. *J.T.P.I.* June 1960.
ROTH G. J. A Pricing Policy for Road Space in Town Centres. *J.T.P.I.* London, November 1961.
Rotterdam, Lijnbaan, *A.D.* London, August 1954.
RUSCONE, MONTI. Milan Rehabilitation. *A.R.* London, 1960.

S

SCHMIDT R. E. and CAMPBELL M. E. Highway Traffic Estimation. *T.Q.* Saugatuck, 1956.
SCHNEIDER L. M. Impact of Rapid Transit Extensions on Suburban Bus Companies. *T.Q.* Saugatuck, January 1961.
SCHOON E. Hook Report. Letter to the editor. *A.J.* London, 21 March 1962.
Scissors Maisonettes. *A.J.* London, 28 February 1962.
SENIOR D. The Motor Vehicle has come to stay. *J.T.P.I.* London, December 1958.
SENIOR D. Cambridgeshire Development. Village Scheme. *A.J.* London, 24 April 1963.
SERT J. L. Changing Views on the Urban Environment. *R.I.B.A. Journal*. May 1963.

SEYMER N. Apartheid on the Roads. *A.J.* London, 21 May 1960.

SEYMER N. A Survey of Parking Garages. *International Road Safety and Traffic Review.* 1960. World Touring and Automobile Organisation, 1960.

SEYMER N. Regional Planning and Transport. *A.J.* London, 21.8.1963.

SHANKLAND G. City of Liverpool. Planning Consultant's Report 1. Preliminary Proposals for Central Area Redevelopment. Liverpool, 1962

SHANKLAND G. Planning Consultant's Report 2. Provisional Comprehensive Development Area Proposals and Interim Planning Standards. Liverpool, 1962.

SHANKLAND G. Planning Consultant's Report 3. Central Area Redevelopment, Phase one. Liverpool, 1962.

SHANKLAND G. Planning Consultant's Report 4. Central Area Roads New Street System, Roe Street, Hood Street Improvement. Liverpool, 1962.

SHANKLAND G. Planning Consultant's Report 5. Central Area Development. Land Bounded by Old Hall Street, Old Leeds Street, Prussia Street, St. Paul's Square. Liverpool, 1962.

SHANKLAND G. Planning Consultant's Report 6. Central Area Roads: New Street System Mersey Tunnel Entrance Improvement and Flyover. Liverpool, 1962.

SHANKLAND G. Planning Consultant's Report 7. Central Area Roads: Inner Motorway System. Liverpool, 1963.

SHANKLAND G. Planning Consultant's Report 8. Central Area Redevelopment: Two New Centres. Liverpool, 1963.

SHANKLAND G. Planning Consultant's Report 9. Central Area Development: Moorfields Comprehensive Development Area. Liverpool, 1963.

Sheffield, City of Sheffield. History of Corporation Housing Schemes. October 1959.

Sheffield Central Development. *A.D.* London, September 1962.

Sheffield, Ten Years of Housing, 1953–1963. Sheffield, 1962.

SISGARD J. War Between Children and Cars. O.M.E.P. Copenhagen, February 1961.

SKAARUP, JESPERSEN. Lillerod. Copenhagen, 1960.

SKELTON R. R. Crash Barrier Tests on Multiflora Rose Hedges in Public Roads. U.S. Bureau of Public Roads. December 1957.

SKRZYPCZAK-SPAK M. *Der Fussgängerverkehr in den Städten und seine Erschliessungsmöglichkeiten.* Hanover, 1961.

SLAYTON W. L. Urban Renewal and Mass Transportation Planning. *T.Q.* Saugatuck, January 1962.

SMEED R. J. *The Traffic Problem in Towns.* Manchester, 1961.

SMEED R. J. The Influence of Speed and Speed Regulations on Traffic Flow and Accidents. R.R.L. General Report, 5th International Study Week in Traffic Engineering. Nice, 1960.

SMEED R. J. Effect of Some Kinds of Routeing Systems on the Amount of Traffic in the Central Areas of Towns. R.R.L. Department of Scientific and Industrial Research. London, 1962.

SMEED R. J. Theoretical Studies and Operational Research on Traffic and Traffic Congestion. R.R.L. Department of Scientific and Industrial Research, 1957.

SMEED R. J. The Effect of Speed Limits on Speed and Accidents. *Pedestrian,* London, 1961.

SMIGIELSKI M. K. Traffic Plan for London. *J.T.P.I.* London, April 1960.

SMIGIELSKI M. K. London Roads Competition, Results and Appraisals. *A.J.* London, 14 December 1959

SMITHSON A. and P. Golden Lane Competition. *A.R.*, London 1952.

Some Impressions of French Housing. *H.R.* December 1960.

South Bank Development. L.C.C. *A.J.* London, 23 March 1961.

South Bank, Next Development. *A.J.* London, 30 March 1961.

South Carbrain. Proposals for Development. Cumbernauld Development Corporation, 1962.

Specialist Editor No. 7. Planning. Planning: National and Regional Decentralisation and Overspill. *A.J.* London, 28 February 1957.

Spon End C.D.A. City Architect. Coventry, 1961.

STEEL R. and McCULLOCH A. G. Principles and Practice of Town Development. The Basingstoke Scheme. *J.T.P.I.* January 1963.

STEIN C. Kitimat. *A.F.* London, July 1954.

STEIN C. New Towns for America. *T.P.R.* L'pool, No. 3 & 4, 1949.

STEPHENSON G. The Wrexham Experiment. *T.P.R.* Liverpool, 1964.

STEPHENSON G. The Permanent Development of the South Bank. *J.R.I.B.A.* London, 1952.

STEPHENSON and others. The Planning of Residential Areas, Stevenage. *J.R.I.B.A.* London, February 1946.

Stevenage Development Corporation. Revised Proposal, June 1962.

STEWART R. G. Are we Over Emphasising Speed as an Accident Cause? *T.Q.* Saugatuck, October 1957.

STIRLING J. and GOWAN J. Brutalism at Preston: A Housing Scheme. *H.R.* 1961.

Stockholm. *Arkitektor II.* 1962.

Stockholm. City of Stockholm. Stockholm, 1959.

Stockholm. The New City Centre. *A.J.* London, 15 August 1962.

Stockholm. The New City Centre. *A.D.* London, May 1962.

Stockholm. *L'architecture d'aujourdhui,* No. 88. 1960.

Stockholm, A Traffic Route Plan for. English Summary. Stockholm, 1960.

Stockport Corporation. Shopping Promenade. Stockport, 1961.

STONE and YOUNGBERG. Rapid Transit for the Bay Area. San Francisco, September 1961.

St. Paul's Area. *A.R.* London, June 1956. *J.R.I.B.A.* April 1956.

St. Peter's Street Improvement Scheme. City of Canterbury, 1962.

Strangled Cities. *Times Supplement.* London, 4 June 1961.

STREHLENERT. Thors Innerstadt. Stockholm. 1960.

Sutton Coldfield. *A.J.* London, 17 October 1962.

Sutton Coldfield. The Royal Town of Sutton Coldfield. Redevelopment of the Central Area. Consultant's Report to the Borough Council. June 1962.

SWIFT H. L. Merchandising a Bus Service. *T.Q.* Saugatuck, January 1955.

Swindon Central Development. Borough Surveyor. Swindon, 1960.

T

TANNER J. C. Estimation of Vehicle Mileage on the Roads of Great Britain. *T.E. & C.* London, February 1963.

Tapiola: Garden City. Finland, 1959.

TASS L. New Rapid Transit Planning. *T.Q.* Saugatuck, October 1961.

TAYLOR G. BROOKE. Study in Neighbourhood Planning. *Town and Country Planning*, April/May 1958.

TAYLOR H. Houses '53 Review. *A. & B.N.* London, 14 January 1954.

TAYLOR R. Paved with Opportunity. *The Observer.* London, February 1963.

Thames-side Development at Deptford. *A.J.* London, 13 April 1961.

THOMPSON P. Oxford Roads Inquiry. *A.J.* London, 12 January 1961.

TINDALL J. North Berwick Redevelopment of High Street. East Lothian County Council, 1960.

TINDALL J. North Berwick. *T.P.R.* April 1962.

TINDALL, BECKER. Ormiston Road Housing, Tranent. East Lothian, 1957.

TOMKINSON D. The Marseilles Experiment. *T.P.R.* October 1953.

Toulouse-le-Mirail. Cité nouvelles, centre urbaines l'architecture d'aujourdhui. April/May 1962.

Toulouse-le-Mirail. *L'architecture d'aujourdhui.* No. 88, 1960.

Toulouse-le-Mirail. *Architectural Design.* London, April 1963.

T.P.I. Memo. Central Area Development. 12 May 1960.

Tranent. Ormiston Road Housing. Tindall, Becker. East Lothian, 1959.

TRAVIS A. S. The Radburn system, a review of exhibition at Housing centre. *A.J.* London, 20 September 1959.

TURNER R. Garages in Residential Areas. T.P.R. July, 1959.

TWEDELL N. The New Town Village. *A.R.* London, September 1960.

U

Urbanism des Capitales. *L'architecture d'aujourdhui.* Paris, February/March 1960.

Urbanistica 35. Article on New Towns. March 1962.

Urbanistica 36–37. Article on New Towns. November 1962.

Urban Redevelopment. Report of a Committee Appointed by the Civic Trust. London, June 1962.

Urban Shopping Mall. Failure with a Future? *Engineering News Record*, 17 December 1962.

Urban Survival and Traffic. *A. & B.N.* London, 26 April 1961.

Uses of Traffic Accident Records. Committee on Uses of Developed Information. *T.Q.* 1947.

Usworth. Development for a Durham Village. *A.J.* London, 17.7.1963.

V

VAHLEFELD R. and JACQUES. F. *Garages and Service Stations.* London, 1960.

Vällingby. Guerin E. *A. & B.* London, December 1958.

Västeras. Oreje. Västeras, 1961.

Victoria, City of Hong Kong. Central Area Redevelopment. Victoria, August 1961.

Victoria Line. Report by the London Travel Committee to the Minister of Transport and Civil Aviation. H.M.S.O. 1959.

Vocabulary of Traffic Engineering Terms. London, 1963.

W

Walking to School. *Times Editorial Supplement.* London, 7 October 1960.

Wallingford. Proposals for the Town. Hancock Associates. 1962.

WARDROP J. G. and DUFF J. T. Factors Affecting Road Capacity. International Study Week in Traffic Engineering. W.T.A.O. October 1956. Stresa, Italy.

WARDROP J. G. The Capacity of Roads. Reprint *Operational Research Quarterly*, March 1954.

WARDROP J. G. The Traffic Capacity of Weaving Sections of Roundabouts. *Operational Research International Conference Papers. 1957.*

WARDROP J. G. Traffic Capacity of Town Streets. R.R.L. London, 1954.

WARDROP J. G. The Distribution of Traffic on a Road System. Theory of Traffic Flow. Amsterdam, 1961.
Warrington Caution. Warrington, 1961.
Warrington Central Development Scheme, 1958. Analysis. July 1961.
Washington Centre for Metropolitan Studies. The Impact of Design Transportation and the Metropolis. Washington, October 1962.
WATERBURY L. S. Mass Transportation by Monorail. *T.Q.* Saugatuck, January 1961.
WATSON H. M. County Planning Officer. Halesowen Town Centre Development. February 1962.
WEAVER R. C. The Federal Interest in Urban Mass Transportation. *T.Q.* January 1963.
WEDDLE A. E. Housing Layout in Denmark. *J.R.I.B.A.* March 1959.
WEDGEWOOD H. R. East Kilbridc Traffic Survcy. University of Edinburgh, July 1962.
WEHNER B. Improving the Geometric Alignment of Roads. *Traffic Engineering & Control.* May, 1963.
Welfare Island U.S.A. *Interbuild.* London, June 1961.
WILCOXON F. Some Rapid Approximate Statistical Procedures. New York, 1949.
WILLENHALL WOOD I. Report. City Architect. Coventry, 1957.
WILSON H. Design of Shopping Centres. *J.T.P.I.* London, July/August 1958.
WILSON H. Cumbernauld D.C. Preliminary Planning Proposals. Cumbernauld, 1958.
WILSON H. Cumbernauld New Town, Mark II. *A.J.* London, 1 October 1959.
WILSON H. Reports 1–5 Carbrain. Cumbernauld, 1959.
WILSON H. and GIBBS. A. K. Cumbernauld New Town, Mark II. A Plan to Master the Motor Vehicle, the New Road Plan. *A.J.* London, 1 October 1959.
WOHLIN H. Studier au Förskolbarns Lekvanor I Modern Bostadsbelyggelse. Stockholm, 1960.
WOLFE M. R. Shopping Streets and the Pedestrian Rediscovered. *J.A.I.A.* May 1962.
WOMERSLEY J. L. The Architect's Contribution to Housing. *J.R.I.B.A.* London, July 1952.
WOMERSLEY J. L. Coping with the Car Inside Housing Estates. Public Works Congress. November 1960.
WOMERSLEY J. L. Housing the Motor Car. *J.T.P.I.* London, September/October 1961.
WOMERSLEY J. L. Some Housing Experiments on Radburn Principles. *T.P.R.* Liverpool, January 1954.
Worboys Report. H.M.S.O. Manchester, 1963.
Workers Educational Association. The Use of Leisure in a New Town. The Report of a Social Survey in Crawley New Town.
W.H.O. The Poison we Breathe. Geneva, May–June 1961.
WRIGHT J. E. The Houston Transportation Study O–D Survey Procedures. *T.Q.* Saugatuck, April 1961.
WRIGHT M. The Motor Vehicle and Civic Design. *J.R.I.B.A.* London, January 1957.
WRIGHT M. A New Kind of New Town. *J.T.P.I.* London, March 1962.

Z

ZION R. The Landscape Architect and the Shopping Centre. *J.I.L.A.* London, 1962.

XII. Appendices

XII. 1. Conversion Tables[i]

(a) Acres to Hectares

acres		ha
25	10	4
50	20	8
75	30	12
99	40	16
124	50	20
148	60	24
173	70	28
198	80	32
222	90	36

(b) Feet to Metres

ft		m
33	10	3
66	20	6
98	30	9
131	40	12
164	50	15
197	60	18
230	70	21
263	80	24
295	90	27

(i) *Note:* The central column in each table is the key one. Thus, if we take the central column of table (a) to be acres, then the right hand column is the equivalent in hectares; if we take it to be hectares, then the left hand column is the equivalent in acres. For example: 10 acres = 4 hectares; 10 hectares = 25 acres. For full and accurate information see *Conversion Factors and Tables, Part 2. Detailed Conversion Tables.* British Standards Institution. London, January 1962.

(c) Miles to Kilometres

miles		km
6	10	16
12	20	32
19	30	48
25	40	64
31	50	80
37	60	97
44	70	113
50	80	129
56	90	145

(d) Persons per Acre to Persons per Hectare

ppa		ppha
4	10	25
8	20	50
12	30	75
16	40	99
20	50	124
24	60	148
28	70	173
32	80	198
36	90	222

(e) Square Feet to Square Metres

ft^2		m^2
1076	100	9
2153	200	19
3229	300	28
4306	400	37
5381	500	46
6458	600	56
7534	700	65
8611	800	74
9687	900	84

XII. 2. Abbreviations

A. & B. Architecture and Building.
A. & B.N. Architect and Building News.
A.D. Architectural Design.
A.F. Architectural Forum.
A.J. Architect's Journal.
A. Rec. Architectural Record.
A.R. Architectural Review.
B. & W. Bauen und Wohnen.
C.B.D. Central Business District
C.C. County Council.
C. du C.T.S.B. Cahiers du Centre Scientifique et Technique du Batiment.
H.M.S.O. Her Majesty's Stationery Office.
H.R. Housing Review.
H. Rel. Human Relations.
J.A.I.P. Journal of the American Institute of Planners.
J.I.L.A. Journal of the Institute of Landscape Architects.
J.R.I.B.A. Journal of the Royal Institute of British Architects.
J.T.P.I. Journal of the Town Planning Institute.
L.C.C. London County Council.
O.M.E.P. Organisation Mondiale pour l'education préscolaire.
R.C.C. Roads Campaign Council.
R.R.L. Road Research Laboratory.
R.S.H. Royal Society of Health.
T.E. & C. Traffic Engineering and Control.
T.Q. Traffic Quarterly.
T.P.R. Town Planning Review.
W.H.O. World Health Organization.
W.T.A.O. World Touring and Automobile Organization.

XII. 3.

City of Leicester. Planning for Transportation. Programme for determining Communication Pattern.

W. K. SMIGIELSKI, Ing.Arch., M.T.P.I.
CITY PLANNING OFFICER.
November 1962

TRANSPORTATION STUDIES

A. "HOME" INTERVIEW.

PHASE I. GENERATION OF LAND USE—CENTRAL AREA WORK CENTRES.

By interview and records available. Obtain:—
- a) Land use.
- b) Floor space.
- c) No. employees:—
  1. Sex.
  2. Salary range.
- d) No. who own cars—No. used to work.
- e) Age condition of building—Liability for re-development.
- f) Work trip by mode:—
  1. Car—Parking duration.
  2. Train.
  3. Public transport.
- g) Origin of work trip.
- h) Time—Arrival and departure.
- i) If associated with commercial vehicle fleet:—
  1. Number and type.
  2. Typical daily pattern.

By multiple regression analysis relate:
Land use, floor space, generation.

PHASE II. a) DWELLING HOUSE. b) WORK CENTRES.

By interview representative sample of people in homes concerning travel on "control" day. Obtain:—
- a) Destination of each trip.
- b) Mode.
- c) Purpose—Work, business, shopping, education. Social.
- d) Time—Start-Arrival—of each trip.
- e) Type of parking.

For work centres (business, shops, factories, etc.).
Use same method as Phase I.

B. EXTERNAL CORDON TRAFFIC SURVEY.

By direct interview:—
- a) Origin and destination.
- b) Trip purpose.
- c) Vehicle type.
- d) If stopping, parking duration.

C. MASS TRANSPORT SURVEY.

By postcard:—
- a) Origin and destination.
- b) Purpose.

Also:—
- a) Routes.
- b) Schedules.
- c) Passengers—Volumes—Past trends.

TOWN MAP PROPOSALS

FOR EACH TRAFFIC ZONE—TO TERMINATION OF DESIGN PERIOD. OBTAIN:—
1. a) Land use.
   b) Density.
      1. Population trend.
      2. Dwellings.
   c) Probable rateable value.
   d) If work centre:
      1. Type.
      2. No. of employees.
      3. Probable floor space.
   e) Plot ratio.
2. Outline highway system.

GROWTH FACTOR

FOR TOTAL TRIP—ALL MODES.
IN TERMS OF:— FOR EACH ZONE.
- a) Predicted population.
- b) Planned land use.
- c) Predicted motorization.
- d) Predicted trip purpose.
- e) Socio-economic trend.
- f) Overall city growth factor.

PREDICTED TOTAL TRIPS

FOR EACH ZONE. GIVING ALL INTER-ZONAL TRIPS. OBTAINED BY:—
- a) Iteration method—Fratar.
- b) Gravity model—Check.

PREDICTED MODAL SPLIT

GIVING:—
- a) Private car.
- b) Mass transport.
- c) Train.
- d) Walk.

FUTURE TRAFFIC

TOTAL TRAFFIC PER ZONE:—
- a) Total trips.
- b) Purpose.
- c) Mode.
- d) Origin destination.
- e) Times—Start-Arrival.
- f) Type parking required.

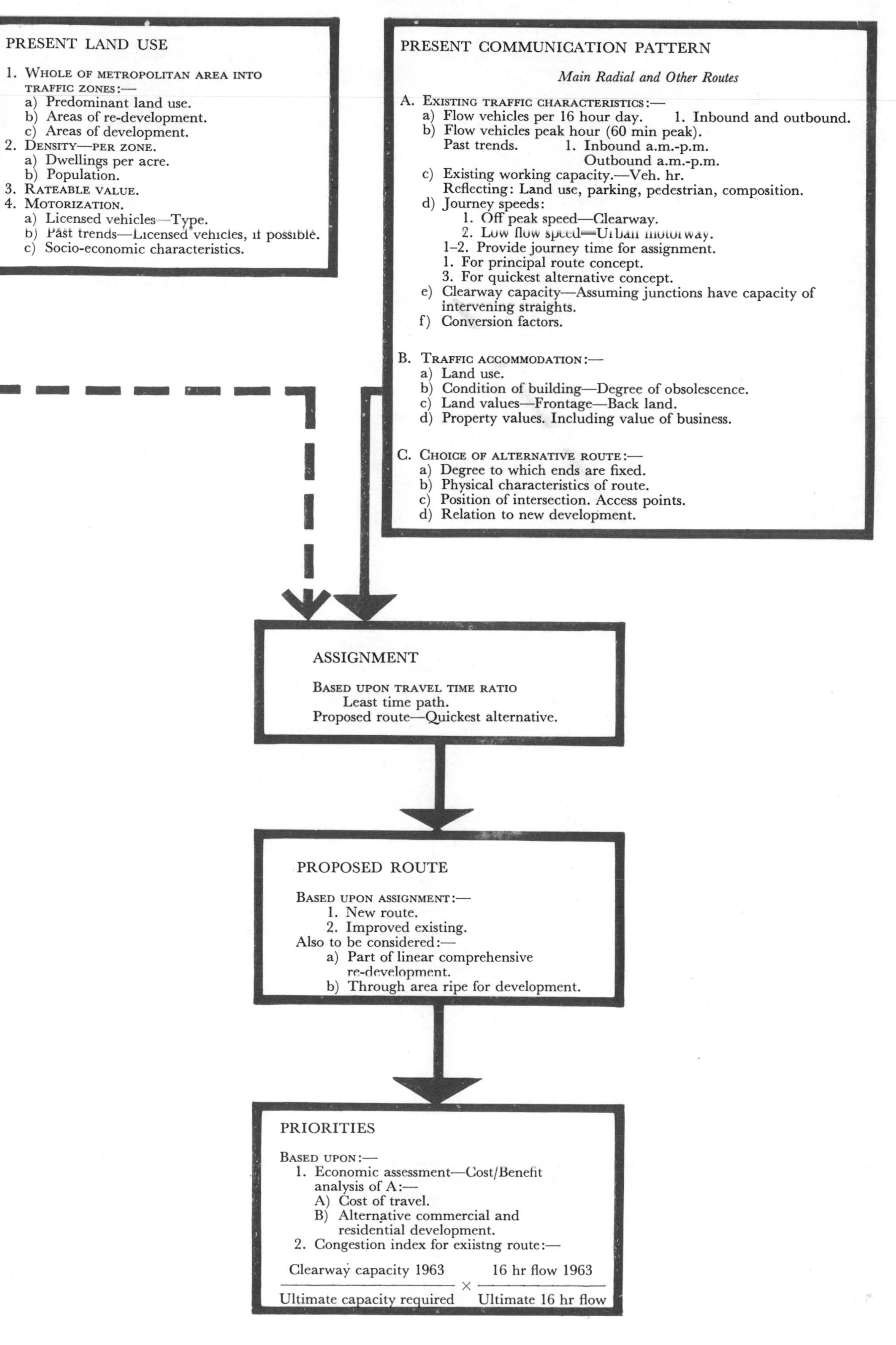
PRESENT LAND USE
1. WHOLE OF METROPOLITAN AREA INTO TRAFFIC ZONES:—
a) Predominant land use.
b) Areas of re-development.
c) Areas of development.
2. DENSITY—PER ZONE.
a) Dwellings per acre.
b) Population.
3. RATEABLE VALUE.
4. MOTORIZATION.
a) Licensed vehicles—Type.
b) Past trends—Licensed vehicles, if possible.
c) Socio-economic characteristics.
PRESENT COMMUNICATION PATTERN
Main Radial and Other Routes
A. EXISTING TRAFFIC CHARACTERISTICS:—
a) Flow vehicles per 16 hour day. 1. Inbound and outbound.
b) Flow vehicles peak hour (60 min peak).
Past trends. 1. Inbound a.m.-p.m.
Outbound a.m.-p.m.
c) Existing working capacity.—Veh. hr.
Reflecting: Land use, parking, pedestrian, composition.
d) Journey speeds:
1. Off peak speed—Clearway.
2. Low flow speed—Urban motorway.
1–2. Provide journey time for assignment.
1. For principal route concept.
3. For quickest alternative concept.
e) Clearway capacity—Assuming junctions have capacity of intervening straights.
f) Conversion factors.
B. TRAFFIC ACCOMMODATION:—
a) Land use.
b) Condition of building—Degree of obsolescence.
c) Land values—Frontage—Back land.
d) Property values. Including value of business.
C. CHOICE OF ALTERNATIVE ROUTE:—
a) Degree to which ends are fixed.
b) Physical characteristics of route.
c) Position of intersection. Access points.
d) Relation to new development.
ASSIGNMENT
BASED UPON TRAVEL TIME RATIO
Least time path.
Proposed route—Quickest alternative.
PROPOSED ROUTE
BASED UPON ASSIGNMENT:—
1. New route.
2. Improved existing.
Also to be considered:—
a) Part of linear comprehensive re-development.
b) Through area ripe for development.
PRIORITIES
BASED UPON:—
1. Economic assessment—Cost/Benefit analysis of A:—
A) Cost of travel.
B) Alternative commercial and residential development.
2. Congestion index for exiistng route:—
Clearway capacity 1963 / Ultimate capacity required × 16 hr flow 1963 / Ultimate 16 hr flow

XII. 4. Index

Access, house, 248.
house, upper-level, 290.
house, service, 232.
house, people, 232.
house, family occasions, 232.
motorway, 78.
pedestrian, 224.
Accessibility, to work, 225.
Accidents, barriers, 70.
to children, 18.
Culs-de-sac, 241.
in housing therapy, 226.
motorways, 75.
single vehicle, 70.
and speed, 73.
stationary objects, 70, 68.
Ackerman, F. L., 319.
Adams brothers, 315.
Administration, problems of, 110.
with Radburn lay-out, 225.
and Strategic Planning, 253.
Air-locks, 289.
Albertslund, Copenhagen, 267.
Alexander, R. E., 260.
Algiers, 323.
Alken, H., 315
Alkmaar, Holland, 139.
Allan, B. J., 118.
Allen, N. P., 121.
Allot, R. A., 301.
Alm, B., 234, 274, 277.
American Automobile Manufacturers Assoc., 88.
Amsterdam, Holland, 168.
Andover, 156, 210, 234.
Anthropology, 4.
Anti-fume devices, 23.
Area Ratio Method, 109.
Areas, town, 113.
Arkhammar, 296.
Arterial highway, 105.
Ash, R. J., 300.
Ashworth, G. W., 121, 180.
Aspect, double, 291.
Aspley Estate, Nottingham, 42.
Attenborough, M. J., 213.
d'Autenil Richard, R. P. J., 271.

Bacon, E. N., 14, 176, 177, 268.
Bagdad, 42.
Baggesen, T., 323.
Baines, G. G., 208, 232.
Bakema, J. B., 139, 212, 314, 328.
Baldwin Hills Village, 243, 260, 328.
Baldwin, K., 298.
Baltimore, 95.
Banbury, 197.
Baniewicz, T., 112.
Bannerman, A. H., 118.
Barbican, London, 288.
Barlow, R. A., 290.
Barnes, H. A., 160.
Barnes, W., 266.
Barnett, W., 301.
Baronbackarna, 235, 274.
Barriers, between path and road, 237.
Basildon, dimensions of service culs-de-sac, 251.
main centre, 153, 221.
parking provision, 109.
path system, 33.
Lee Chapel, North, 246.
Basingstoke, 156, 233.
Beazley, E., 41.
Bedmont, 282.
Beeching Report, 64, 65.
Beeston, 206.
Fishpond, 247, 251.
Ilkeston Road, 266.
Bell Green, Coventry, Economic Analysis, 230.
Bellis, W. R., 70.
Belonging, 115, 225, 234.
Bendixson, T. M., 211.
Bendtsen, P. H., 62, 110, 153, 159, 162.
Bennett, H., 121, 123, 213, 265, 269, 299.
Bent, A., 300.
Berbiers, J. L., 212.
Bermondsey, London, 235.
Bigger, F., 319.
Bigwood, R. T., 218.
Billingham, 228.
Bing, A. M., 319.
Biological function of walking, 38.
Birmingham, Primrose Hill, 300.
Birmingham University School of Traffic Engineering, 64.
Birmingham School of Planning, 64.
Bishop's Stortford, 201.
Biskopsgaden, 277.
Blachnicki, H., 298, 304a.
Blackeberg, 43.
Bonneson, 111.
Bor, W., 174.
Boston, 103, 35, 163.
Bracknell, 109.
Branching systems, 80, 133, 134.
Brandon Estate, London, 299.
Brecher, 148.
Bredsdorff, P., 208, 267.
Bremen, 157.
Bridges, foot, 202, 216.
convenience, 51.
Manahawkin Bay Bridge, 59.
linking parking and precinct, 210.
position and importance, 253.
Briggs, F., 262.
Bright, J., 319.
Brighton, Sea Front, 216.
Brindell, D., 194.
British Retailers Assoc., 167.
Broek van den, 139, 212, 314, 328.
Brolin, 296.
Brooklyn, 78.
Brown, L., 266.
Browne, K., 289.
Browning, E., 118.
Brunner, C., 165.
Brunswick, Manchester, 300.
Bruzeluis, N. G., 70.
Buchanan, C. D., 33, 166, 329, 330.
Buckthorpe, L. W., 118, 280.
Buffets, Les, Fontenay aux Roses, 303.
Burrage & Mogren, 84, 109.
Buczynski, J. K., 218.
Bye-law Housing, regeneration, 309.

Cable Cars, 54.
Caesar, 314.
Cairns, J. G., 213.
California, 33, 54.
Callaghan, G., 118.
Calthrop, M., 283.
Cambridge, England, 52.
Cambridge, Mass., 52.
Cambridge Village, 147.
comparison of path systems, 55.
Campbell, K. J., 123.
Canaletto, 26.
Candilis, G., 126.
Canopies, 212.
Capacity, road, 59, 67, 70.
and commercial vehicles, 93.
factors affecting, 71, 160.
factors limiting, 70.
of junction, 240.
peak period increase, 160
possible, 70, 105.
practical, 105.
of uninterrupted streets, 72.
at various speeds, 71.
Car ownership, 60, 108.
and income, 62.
increase, 160, 249.
occupancy, 67.
prediction, 110.
Car parks, connected, 170.
Car storage, 235.
alternatives, 235.
collective, 238.
underground, 289.
underground and multilevel, 235.
Cars, collection of derelict, 92.
Carbrain, 41, 43, 258.
cul-de-sac, 246.
dwelling plan, 232.
superblock, 252.
Carss, J. M., 213.
Carveyors, 55.
Casson, Sir Hugh, 39, 327.
Catcott, E. J., 23.
Caversham, 39, 262.
Centres, car-free, care-free, 152.
high density, 212.
linear, 191.
multi-level, 119, 129, 186, 192.
multi-level, supplied underground, 131.
service underground, 186.
superblock, 252, 253.
Chadwick. H., 206, 266, 247.
Chalk, J. W., 213.
Chamberlin, Powell, Bon, 289, 329.
Chandler, K. N., 60.
Chandigarh, 80, 140.
Chapman, D., 51.
Character, roads and path, 35.
Chariots, 314.
Charles II, 314.
Chase, S., 319.
Chateau-neuf-sur-Loire, 96.
Chatham Village, 319.
Chermayeff, S., 4, 5, 6,
Chester, 45, 208.
Chicago, car use and trips taken, 88.
expressway, 96.
motorization forecast, 60.
public transport, use, 89.
trips to C.B.D., 91.
World Fair, 53.
Chicago Transit, Authority, 99, 100.
Children, accidents to, 18.
pre-school age, 18.
play of, 202.
safety in culs-de-sac, 251.
and shopping, 90.
and school, 89.
Child's Eye View Exhibition, 11.
Christian ethics, 16.
Churchill, J. D. C., 160.
Circulation, 154.
pedestrian and vehicular, 155.
traffic, 156.
Circular town, 111, 112.
Cité Industrielle, 314.
Cité Jardin, 33, 271, 328.
Citizen participation, 5, 225.
Citizens Traffic Safety Board, 167.
City Planning Commission, Philadelphia, 101.
Clark, F., 218.
Clements Hill, 234, 269.
Cleveland, 95.
Climate, 51.
effect on capacity, 71.
Clunie, W. L., 286.

Coal Exchange, Barbican, 289.
Coburn, T. M., 73.
Collector roads, 224, 241.
Cologne, 113, 133.
  comparison of path systems, 56.
Collerton, A., 301.
Commercial vehicles, vans, 93.
  lorries, 93.
  growth, 93.
  numbers, 93.
  and road capacity, 93.
  size, 93.
Common Functional Principles, 4.
Communal, orchards, 146.
  openspace, 228.
Communication, 59, 165, 168.
  radio, 162.
Community trust, 147.
  feeling, 276.
Concentration, and slow driving, 73.
Congestion, cost of, 167, 67.
Contact, sociological, 226.
Contini, E., 192.
Contours, 256.
  terrace housing, 256.
Contrasts, 39.
Convenience, 51, 52, 60, 62, 111, 112.
Conversion, new town to traffic segregation, 139.
Cook, P., 194, 297, 298.
Cooper, Miss S., 195.
Co-operation, 5, 6.
  functional, 225, 271.
Copcutt, G., 60, 110, 118.
Copenhagen, and cyclist, 52.
  use of public transport, 89.
  patio housing, 42.
  full motorization, 165.
  renewal, 310.
  Medical Association Housing, 316.
  Rasmussen's pupils, 323.
le Corbusier, 60, 140, 314, 322, 323, 328.
Corner shop, 308, 310.
Cost comparison, Radburn, 228.
  comparison of street and sewer works, 228, 229.
  comparison single/double footpath, 230.
  comparison of lay-out economics, 231.
Cost differences, ring road, motorway, 161.
Courtyard housing, climate, 284.
  and rural community, 284.
Covell, Matthews and Partners, 147.
Cowley, G. W., 201.
Cox, O. J., 121.
Coventry, Hillfields, 45, 303.
  house numbering, 257.
  original precinct plan, 324, 328.
  Spon End, 297.
  Town Centre, 70.
  Willenhall I, 251, 278.
  Willenhall II, 251.
Craig, J. C., 121, 123, 211.
Crawford, N., 192.
Crawley, 109.
Creation, 4.
Creative experiment, 5, 226.
Criminals, 33.
Cromptom, D., 213.
Cropper, R., 298, 304a.
Crowds, people and cars, 33.
Crowe, S., 75.
Cul-de-sac, design, 237, 239, 240, 241.
  multiple, 242.
  path, link, 250.
  Pioneer projects, early 50's, 243.
    mid 50's, 244.
    late 50's, early 60's, 246.
  without path connection, 251.
Cullen, G., 41.
Culpin, C., 248.
Cumbernauld Development Corporation, 109.
Cumbernauld, 117, 329.
  area and walking distances, 113.
  Carbrain, 247, 231, 235, 236, 237, 251.
  car usage, 61.
  central area, retailing and parking, 110.
  and cyclists, 51.
  junctions, 80.
  Kildrum, 280.
  open space, 114.
  park, 246, 251.
  parking, 235.
  parking needs, 60, 109.
  path systems, comparison, 56.
  peak hour flow, 108.
  pedestrian environment, 39.
  play spaces, 43.
  Seafar, 247.
  sheep, 227.
  South Carbrain, 290.
    proposed lay-out, 290.
    building form, 291.
    car provision, 291.
    open space, 292.
  split level housing, 41.
  underpasses, design cost alternatives, 254.
Curry, D. J., 213.
Curvature of path, 21.
Cuttings, 68.
Cycle tracks, 170, 199.

Dallas, 163.
Dallington Fields, 229.
Dalton, C., 195.
Damage, 10.
Danger, 249.
  diversion of attention, 58.
  grid iron system, 306.
  at junctions, 240.
  and path system, 236.
  and speed, 73.
Danish Institute for Shopping Centre Research, 167.
Danish Medical Association, 316.
Dartford-Purfleet, 96.
Davies, A. B., 22, 246.
Davidson, C. H., 269.
Davighi, J. A., 118.
Day, A., 64, 16, 164.
Deck Housing, 239.
  Cumbernauld, 117.
  Royal Victoria Yard, Deptford, 299.
  Park Hill, 286.
  Toulouse le Mirail, 126.
  and traffic segregation, 228.
  Washington, 307.
  Highfields, Leicester, 304a.
Deck system, 256.
Delivery, 90, 158, 233.
Demers, Mrs., 6.
Density, 105, 306.
  Baldwin Hill, 260.
  and car space, 235.
  critical, 105.
  Cumbernauld, 234.
  neighbourhoods, 114.
  and population, 111.
  residential increase in, 112.
  of towns and population, 112.
  and town size, 111.
  and walking, 61.
  of workers, 67.
Denton, J. A., 118.
Deptford, 299.
Design, criteria of, 7.
  of pedestrian spaces, 227.
  of speed, 105
  three- and four-dimensional, 108.
  town, walking distances, 111.
  of vehicles, 18.
Desire lines, 22, 39, 51, 64, 79, 105, 114, 236, 297.
Detail, 43.
Detroit, 89, 154.
Detroit, Lafayette, 276.
Diamond, Redfern and Partners, 262.
Dimitrijevic, 134, 301.
Diplock, R., and Associates, 216.
Disneyland, 90, 96, 43.
Distance, focal, 58.
  passing sight, 105.
  walking, 112.
Distributor roads, 79.
District heating, 146.
Dony, P., 126.
Downs, A., 161.
Doubleday, E. H., 201.
Doxiadis, 16, 42, 115, 116.
Doxiadis Associates, 268.
Driver, emotional state, 59,
  Metropolitan, 72.
  training, 59.
Dual carriageway, 105.
Duff, J. T., 15, 167.
Duncan, G., 218.
Dunn T. & Co., 315.
Düsseldorf, 306.
Dwelling plans, 232.
  Baldwin Hills, 260,
  Bedmont, 282.
  Carbrain, 258.
  Clements Area, Haverhill, 269.
  Coventry, 233.
  examples, 233.
  Inchview, 283.
  Park Hill, 286.
  relation of dwelling to access, 232.
  relation of dwelling to lay-out, 233.
  Washington, 301.
  workable, 225.

Earle, J. H., 218.
Eastburn, Radburn, economic analysis, 229.
East Kilbride, 109.
East Lothian County Council, 283.
Eastwick, Philadelphia, 267.
East York, Canada, 289.
Ecochard, M., 134.
Ecological considerations, 23.
Ecological criteria, 227.
Economic considerations, 67.
Economics of growth, 115.
  of shopping streets, 167.
  and administrative criteria, 227.
Education, 4.
  of children with traffic, 34.
Efficiency, road, 70, 71.
Eggeling, F., 136.
Ekholm, A. A., 294, 277.
Ekstrand, 300.
Elizabeth, Australia, 222.
Ellison, M. J., 121, 123.
Elm Green, I and II, 230, 281.
Elongated centre, 114.
Elsinore, Denmark, 247.
Elysian Park, Los Angeles, 169.
Emancipation, therapeutic, 227.
Embankments, 68, 181.
Emotional limp, 7.
Endsleigh Gardens, Birmingham, 328.
Engleback, N., 213.
Environment, 40.
  creation of, 144, 152, 224.
  existing, 253.
  pedestrian, 115.

Erith, 123.
comparison of path systems, 55.
Ervi, A., 144.
Escalators, 54.
Essen, 215.
Exercise, 224.
Expansion towns, 156, 210, 269.
Experiment, creative, 5, 226.

Falk, K., 274, 277.
Family isolation, 225.
Farchoc & Brown, 38.
Fares, 64.
Farringdon, 315.
Farsta, 130.
comparison of towns and walking distance, 113.
multi-level centre, 110.
shopping centre, 42.
Fashion, 18, 39.
Faulder, E. D., 123.
Federal Highways Department, Philadelphia, 101.
Fencing, slatted, 280.
Fenter, D. G., 300.
Fidler, S., 300.
Fishpond Radburn, economic analysis, 229.
Fisker & Millech, 316.
Fitness, 38.
Flats, 239.
Flemingdon Park, East York, 289.
garaging, 235.
Flexibility, 35.
Flow increase, factors affecting, 159.
Flow line, 105.
Flyovers, 219.
Focus, distance, 58.
Fontenay-aux-Roses, 44, 303.
Forsman, 296.
Fort Worth, 192.
Foster, Miss I., 121.
Fountain, E. W., 298.
Fox, C., 194.
Freeway, rural, 75, 105, 110.
Francois, J., 126.
Fresh air, 224.
Friendliness, path system, 225.
Frustration, 22.
Fry, M., 140.
Fulton, Minneapolis, 308.
Fumes and poison, 23, 159, 163, 235.
Functional principles 4.
Function of roads, 146.

Garage, blocks, 310.
circular, 280.
communal, 134, 162.
Court, 240.
roofs, as walkways, play spaces, 296.
provision, 138, 240.
split level, 300.
surface and underground mixed, 295.
underground, 289, 290, 294, 296.
underground, underdeck, 298.
Garden court blocks, 319.
Gardening, 123.
Garnier, T., 314, 322.
Garside, D. H., 118, 258.
Geddes, P., 148.
Gentili, G., 129.
Gibberd, F., 246.
Gibbs, A. K., 118.
Gibson, D., 170, 172, 314, 324, 328.
Gibson, R., 39, 262.
Gillespie, W., 118.
Gilmour, A., 299.
Ginsburg, L. B., 199.
Gissane, W., 18.
Glare, 58.
Gleckman, W. B., 87.
Gollins, Melvin, Ward, 219.
Goodman, C. H. Associates, 148.
Goss, A., 53.
Gosselin, J. A., 271.
Gothenburg, footpath, 39, 40.
traffic segregation, 158.
urban path, 12.
Gothenberg, Biskopsgaden, 277.
Graham, Miss S. D., 123.
Green, D., 194.
Greenbelt, 35, 320, 321, 320.
Greenhill, Sheffield, 231, 326, 328.
Greenway system, 218.
Gregory-Jones, D. M., 180.
Greub, W., 192.
Greyfriars, Ipswich, 216.
Grid-iron, 21, 158, 322.
restructuring, 306.
accident pattern, L. A., 307.
Grime, G., 93.
Griffith, C., 300.
Groes-Petersen, E., 18.
Grossman, I., 289, 329.
Grove, D. T., 299.
Growth factors, town, 62.
principles of, 115, 116.
pattern, 115.
principles, 115.
Gruen, V., 54, 55, 192.
Guard rails, 237.
Guerin, E., 129.
Guldheden, 19.
Gwilliam, T. S., 194.

Haase, D., 138.
Halmos, P., 224.
Hamburg, Hohner Kamp, 325.
Hamilton, 58.
Hammersmith, London, 219.
Hampson, I. G., 299.
Hardy, D. C., 121.
Harlakoff, N., 240.
Harlow, 109, 113, 231, 246, 251.
Harlow Competition, 248.
Harrington, R., 180.
Hässelby Strand, 129.
Hatfield, 109.
Hausman, 33.
Haverhill, 233, 234, 269.
Hayes, D., 298.
Hayes, F., 266.
Health, 38, 55.
social, 226.
and safety, 237.
Heating points, 235.
Heineman, H., 23.
Helicopter, 7.
ports, 192.
Hemel Hempstead, 42, 109.
Henry, A., 273.
Herron, R. J., 213.
Highland Branch Rapid Rail Transit, 103.
Highway Research Board, 75.
High Wycombe, 205.
Hilberseimer L. 312.
Hillfields, Coventry, 303.
Hillier, Parker, May & Bowden, 197.
Hobbies, provision for, 225.
Hodgem, R., 93.
Holidays, week-end, 90; long, 91.
Holford, Sir W., 220, 327.
Holgren, P., 172.
Hollamby, E., 123, 299.
Holland, 53.
Holliday, 214.
Honneger, J. J., 273.
Hook, 116.
car provision, 108, 247.
and cycling, 52.
housing, high density, 121.
parking, estimates, 109.
multi-level, 110.
path systems, comparison, 56.
walking distances, comparison, 113.
Horizontal segregation, 34, 157.
Horseshoe, 189.
Horsfall, G., 213.
Horsholm, 208.
House performance, 239.
Housing, split level, 41.
therapy, 225.
Housing Research Unit, 283.
Hovercraft, 7, 70.
Howitt, C., 42.
Hoyt, H., 7.
Humour, 41.
Humphreys, H. R., 24.
Hunter, J. R., 118.
Huntingdon, L.C.C., 231, 247, 251, 265.
Hyde Park, Sheffield, 286.
Hydrants, 235.

Idealism, 7.
Ilkeston Road, economic analysis, 229.
Immeubles-Villas, 322, 328.
Impact, 18.
Inchview, Prestonpans, 283.
Infield, Prof. H., 5, 224.
Inertia of society, 7.
Information, Public Relations, 55.
Ingham & Boyd, 319.
Injury, 17, 18.
Institute of Community Studies, 4.
Integration, of knowledge, 4.
environmental, 108.
road, 205.
Interrelatedness, of factors, 7.
Interprofessional integration of knowledge, 4.
Intersections, 3 or 4 leg, 306.
Interwar housing, 310.
examples, 311.
Intimacy, provision, 283.

Jacobs, Jane, 22, 319.
J.A.K., 62.
Jeanneret, P., 140, 322.
Jellicoe, G. A., 52, 324.
Jenkin, D. C. H., 269.
Jiggen, A. H. F., 209.
Johnson, R. D., 260.
Johnson-Marshall, P., 218, 314, 324.
Jones, A. C. C., 123, 299.
Jones, D. G., 299.
Jones, P. E., 269.
Josic, A., 126.
Journeys, 108.
commuter, comparative costs, 166.
convenient, 236.
road capacity, 89.
to school, 225.
to work, 89.
Junctions, 72, 79.
capacity, 240
flow, 81.
frequency, 240.
grade separated, 322.
2-way streams, 79.
4-way streams, 79.

Keilty, J. W., 154.
Kennedy, T. E., 213.

Kentucky Road, Ontario, 295.
Kerbside parking, abolition, 160, 164.
and sales, 167.
Kick about pitches, 293.
Kiel, 52.
Kildrum, Cumbernauld, 42, 280.
Kings Heath economic analysis, 229.
Kirby, C. P., 163, 164.
Kirkcaldy, 244, 251.
Kitimat, 143.
Klein, A., 114.
Klein, J., 294, 295.
Knight, H. R. E., 299.
Knightsbridge Green, 220.
Knowles, J., 194.
Kohn, R. K., 319.
Kortedela, 19.
Kristensen, E., 42.
Kuper, L., 224.

Lacey, G. I., 121.
Lafayette Park, Detroit, 235, 276.
Land use, multiple, 65, 114, 165, 290.
developed, 193.
Landscaping, 218.
capital cost, 228.
running cost, 228.
Lagneau, 134.
Latimer, J., 290.
Lay-bys, 237.
Leaker, D. R., 118.
Leasehold, 285.
Lee, E., 194, 311.
Legal considerations, path system, 33.
traffic segregation, 33.
Leibrand, K., 14.
Leicester, transportation study, 62, 63.
short-term parking, 164.
Rowlatts Hill, 298.
Highfields, 304a.
Leicester Square, 191, 247
Leiper, J. Mc. M., 153.
Lenderyou, E. G., 121.
Letchworth, Jackson Estate, 266.
Liepman, K., 32.
Lifts, 54.
Lighting, 41.
coloured, 235.
contrast, 278.
economy, 228.
low-level, Park Hill, 287.
road, 59.
road, indicating grade, 136.
space for, 228.
Lijnbaan, Rotterdam, 212, 328.
Limited access, 306.
accident history, Los Angeles, 307.
Lincoln Road, Miami, 68, 157.
Lindman, G., 300.
Lindström, Prof. S., 39.
Linear, centre, 191.
city concept, 139.
playground, 119.
Ling, A., 172, 247, 278, 297, 329.
Lipp, M. N., 95.
Livability, 224.
Liverpool, 61, 180.
Lloyd, F. J., 95.
Lobb, H., & Partners, 88.
Lock, M., 202.
Lods, M., Honneger, 273.
Logie, G. C., 108, 123.
London, 98, 189.
London, Brandon Estate, 299.
London Traffic Management Unit, 62.
Loneliness, 226.
Lookout, vehicle, 237.
Loops, 241, 242, 308.
Loop roads, 213.
Lorenz, H., 75.
Los Angeles, grid iron planning, 306.
peak hour travel, 159.
public transport use, 89, 94.
townscape, 4.
urban motorway, 12.
Lovell, S. M., 68, 69.
Lucerne, 45.
Ludgate Circus, 315.
Lyddon, D., 118, 236, 258, 278, 329.
Lynn, J., 286.
Lyons, E., 285.

MacKaye, B., 328.
MacLaughlin, Miss A. F., 123.
Macrocosm, 39.
Madin, J., 264.
Madje, J., 5, 6.
Major road, 105.
Malfetti, J. L., 55.
Malmö, Sweden, 300.
Man, as driver, 58.
Manahawkin Bay Bridge, 59.
Manchester, Brunswick development, 300.
Mansize space, 39, 41, 258.
Markelius, S., 328.
Market place, 157.
Markham, J., 123.
Marshall, J. P., 212.
Martern, B., 173
Martin, R. L., 213.
Marly-le-roi, 235, 273.
Marymead, Stevenage, 230.
Mathew, Sir R., 218, 283.
Matson, Smith and Hurd, 83.
Matzutt, Dr. M., 18.
Masser, I., 180.
Massey, C., 195.
Mayer, A., 140, 144.
Mayne, R., 44.
Maximum walking distance, 112.
McCarter, W. J., 99.
McElroy, J. P., 53.
McMonogle, 58.
McCulloch, A. G., 118.
McWilliam, G. A., 218.
Meander, horizontal, 68.
vertical, 68.
Mears, C. B., 123.
Meats, A., 180.
Mechanical aids, 52, 158.
Mellanheden, Malmö, 300.
Merril, E., 260.
Miami, Lincoln Road Mall, 157, 168.
Micro-climatic factors, 23, 35.
Microcosm, 39.
Middlesborough, 51.
Middleton-Harwood, J., 15, 62.
Middleton, J. T., and Diana Clarkson, 23.
Millard, C., 45, 51, 194, 195.
Mill Creek Development Area, 308.
Ministry of Housing, 233, 248, 249.
Ministry of Transport, 60, 61, 68.
Minneapolis, 44, 163, 237, 306.
Minneapolis, Fulton neighbourhood, 308.
Momentum, 10, 16.
Monorail, 53.
Monotony, 58.
Montreal, 33, 54.
Cité Jardin, 271.
Moore, R. L., 51, 70.
Moral considerations, path system, 33.
Morgen and Partners, 220.
Morris, H. C., 121.
Moscardini, 118.
Moscow, 98.
Motopia, 52.
Motorization, assumptions, 110.
forecast, 60.
full, 110.
cost of, 165.
and town size, 111.
Motorway (*see also* Freeway), 105.
cost, 77.
design, 75.
disapproved, 154.
effects on other streets, 161.
conclusions, 161.
introduction of, 160.
land use, 77.
planning, 161.
planning implications, 160.
rural, 75.
and safety, 75.
sunken or raised, 161.
tangential, 195.
in townscape, 161.
in tunnel, 189.
unfinished, 154.
urban, 12, 78.
views from, 159.
Movement, 10, 14.
mechanical, 24.
Moving pavements, 191, 192.
Multi-flora rose, 70.
Multiple level connections, 159.
Multiple Shops Federation, 167.
Mumford, L., 24, 33, 90, 314, 319, 328.

Napper, J. H., & Partners, 301.
Neal, I. A., 298.
Needs and nature of vehicles, 57.
Neighbourhood, concept, 115.
Cologne, 133.
definition, 115.
density, 115.
growth, 115.
identification, 286.
Kitimat, 144.
Neu-Winsen, 138.
path system, 115.
self-sufficient, 134.
sociological idea, 115.
Vällingby, 129.
Neu-Winsen, 138.
New towns, centre, linear, 126.
Basildon, 221.
centre in valley, 122.
functional patterns, 139.
linear concept, 139.
parking, in centres, 109.
needs, 60.
privately built, 134, 144, 147.
redevelopment, 68.
social function, 115.
speculative built, 148.
with traffic segregation, 108, 114.
New York, 33, 153.
New York City Planning Commission, 153.
Noise, 23, 38.
Noise disturbance, 292.
North Berwick, 168.
Northampton, 227, 229, 251.
North York, Ontario, 294.
Norwood, K., 192.
Nottingham, 155, 194, 310.
Aboretum, 316.
Aspley Estate, 42.

Oakdale Manor, Ontario, 294.
Olmstead, New York, 316.
Openings, onto roads, 68.
Open space, 292, 256.
incidental, public, 293.
Opportunism, 7.
Optimum town size, 111.

Oram, D., 206.
Orebro, 274.
Organic rhythm, 68.
  flow, 80.
Origin Destination Study, 105.
Ormiston Road, Tranent, 231.
Orreje & Co., 199.
Otis Elevator Co., 53.
Overpass, 201.
  experiment, 166.
Oxford, 45.
Oxford Street, 315.

Paddington, 315.
Palairet, R. C. W., 121.
Palmer, J. R. B., 118.
Paris, 33, 98.
Park and ride, 155, 173, 195.
Park Hill, Sheffield, 44, 115, 286, 323.
Parking, areas, 193.
  estimates, 162.
  kerbside, abolition, 160.
  linked, 189.
  mechanical, 88.
    cost, 88.
  multi-storey, 67, 152, 158, 189.
  multiple use, 84.
  municipal, 153.
  New Towns, space, 189.
  offences, 162.
  off-street, 153.
  possibilities, 84, 86.
    2-way straight ramp, 85.
    parallel straight ramp, 85.
    opposed straight ramp, adjacent parking, 85.
    opposed straight ramp, clearway type, 86.
    silo garage, 87.
    park-a-back system, 87.
  position, 84.
  requirements, general and basic, 84.
  restriction, 68.
  single landuse, 84.
  single or multi-storey, 157.
  short-term, 153.
  and trade, 163.
Parking meters, 164.
  cost, 164.
Parkleys Ham Common, 235, 285.
Participation, citizen, 5.
Passenger Service Improvement Corporation, 101.
Path Analysis, 236.
  character, 29, 31, 33, 34, 35, 41, 51.
  contact with ring road, 237.
  –cul-de-sac link, 250.
  curving, 273.
  environment, 236.
  furniture, 236.
  survey, 156.
  system, 114.
    comparison of, 55, 56.
    growth of, 115.
    of Routes, 236.
Patio housing, 41, 42, 228, 283.
Pattern, growth and development, 115.
Pavement, 75, 240.
Peak period, 67, 68.
  capacity increase, 160.
Pears, Mrs. W. A., 123.
Pearse & Crocker, 225.
Pecks, 133.
Pedal cycle, 52.
Pedestrains, 54, 90, 192.
Pedestrian, access, 224.
  conveyors, 53.
  deck, 210.
  effect on capacity, 71.
  environment, 39, 237.
  islands, 148.
  multi-level environment, 213.
  multi-level terraces, 134.
  needs along path system, 36.
  network, 199.
  podium, 138, 219.
  routes, 114.
  ways, 156.
Pedestrian street, 309.
  link between centre and public transport, 215.
Percival, A. A., 121.
Peripheral motorways, 161.
Perkins, A. C., 121.
Perrotet, 301.
Perry, C., 115.
Peterlee, 109.
Pettigrew, J. B., 21.
Piazza del Palio, 46, 47, 48, 49.
Piazza, study of urban, 46–51, 220.
Piccadilly Circus, 191, 220.
Pickering, M. N., 121, 269.
Piot, H., 126.
Philadelphia, 96, 101, 102, 308.
Philadelphia, Eastwick, 268.
Philadelphia Planning Commission, 14.
Planning, pattern, 6.
  scientific method, 6.
  solution, 34.
Play, activity, 38.
  equipment, 310.
  spaces, 248.
    covered, 253.
  streets, 35, 309.
Playing fields, 114.
Plot shapes, 238.
  alternatives, 238, 239.
Pocock, R. C. F., 213.
Point blocks, 234.
  research, 286.
Poison, 4, 23, 38.
  social, 226.
Pollution, 23, 24.
Pompeii, 314.
Pooley, F. B., 205.
Precincts, creation of, 155.
  linked, 189.
Prefabrication, 123, 212, 277.
Prestonpans, Inchview, 283.
Primrose Hill, Birmingham, 300.
Principles of growth, 115.
Privacy, 224, 232, 234, 280, 291.
Private cars, anthropometrics, 92.
  design and accidents, 92.
  form, factors affecting, 91.
    fashion, 91.
    prestige, 91.
    lack of road and parking space, 91.
    convenience, 91.
    running cost, 92.
    wind resistance, 92.
    wheels, 92.
    manufacture, 92.
    safety, 92.
    material damage and disposal, 92.
    engine, 92.
  functions, 88.
  protection, 92.
  use, 88, 91.
Probert, G. S., 201.
Processions, 217.
Propaganda, 55, 73.
Proudlove, A., 16, 180.
Psychosocial criteria, 225.
Psychosomatic criteria,
  Radburn possibilities, 224.
Public relations, 176.
Public transport, 32, 51, 54.
  in Centre, 131.
  under Centre, 129.
  covered links between, 202.
  economics, 94.
  general attitude, 94.
  integrated, 192.
  lack of, 215.
  linked with shopping, 205.
  need for, 154, 158.
  principles, 95.
  social aspects of, 32.
  systems, 112.
  types:
    buses, 95.
    pedestrains, 54.
    monorail, 96.
    rail, 96.
    self-guided bus trains, 98.
    trams, 189.
    underground, 98.
Pushkarev, 59, 75, 76.
Priestman, M., 266.

Qualls & Cunningham, 148.

Racksta, 249, 256.
Radburn, 113, 224, 320, 321, 243, 328.
  administrative criteria, 227.
  capital expenditure, 227.
  criterion of creative experiment, 228.
  development, 228.
  economic criteria, 227.
  general maintenance, 227.
  home maintenance, 227.
  investment, 228.
  lay-out, 248.
  principles, 224.
  rents, 227.
  selling price, 227.
  unity with, agriculture, 227.
    industry, 227.
Radial roads, 67.
Raffloer, Miss M. E., 121.
Rail transport, 108.
Rainer, Professor, 41, 297, 298.
Ramp, 105, 237, 253.
  approach, 87.
Ramped tracks, Italy, 35.
Rapid Transit, *see* Public transport.
Rasmussen, S. E., 208, 314, 323, 328.
Rawdon, B., 192.
Rawlinson, J., 68.
Reaction time, 70.
Reay, D., 214.
Rectangles, 112.
Redfern, A., 262.
Regional Planning Association, 328.
Reich, W., 7, 224.
Reichow, H., 24, 59, 68, 74, 80, 97, 134, 136, 306, 314, 325, 328.
Reichow's principles, 133, 136.
Relaxation, 224.
Reno, 88.
Research Project, 27.
  design, 27.
  interview, 27.
  observations, 27, 29, 31.
  results, 27, 28.
  opinions, impressions, observer, 28, 29.
    inhabitants, 30.
  summary, 31.
Residential areas, examples, 258.
  traffic segregated, 258.
  privately developed, 262, 268, 285, 289.
  speculative development, 285.
Residents Committee, 147.
Residential renewal, 305.
  examples, 308.
  general principles, 306.

Reston, Washington, 5, 38, 88, 148, 149.
Reynolds, D. J., 167.
Rexford, C. W., 102.
Rhythm, 10.
organic, 68.
Rhythmic stimuli, 58.
Ribet, E., 269.
Richardson, 274.
Richmond, G. I., 286.
Ricker, R. E., 85.
Ring route, 79, 105, 224.
access, 237.
Ring and spoke, 152.
Ritter, P., 16, 165, 194, 195, 227, 262, 266, 329.
Roads, arterial, 105.
basic alternatives, 241.
building administration, 68.
capacity, 59, 70.
factors affecting, 160.
and parking, 163.
and commercial vehicles, 93.
character, 35.
crossroads, blocked, 308.
design, 240, 75.
distributor, 79.
dual carriageway, 105.
efficiency, 67, 75.
examples, 243.
functions, 140.
high and low level, 110.
main, 79.
radial, 67.
residential feeder, 240.
shape, space, noise and danger, 68.
signs, 257.
surface, 35.
types, 75.
Road Research Laboratory, 60, 61, 68, 71, 72, 94, 166.
Roberts, J. A., 213.
Robertson, L. R., 198.
Robertson, M., 257, 290.
Robinson, D. C., 180.
Rogine, I., 146.
Rohe, M. van de, 54, 276.
Roos, T., 300.
Rose, V., 197.
Rosner, R., 138, 212.
Routes, capacity, 10.
pedestrian, 114.
selection of, 236.
shortest, 236.
of vehicles, 68.
Rowlatts Hill, Leicester, 247, 298.
Rows, Chester, 208.
Royal Victoria Dock Yard, Deptford, 314.

Sabende, 135.
comparison of path systems, 55.
Safety Congress, 6th International, 70.
"Safety Walks", 308.
Sales, effect of parking on, 153.
Salford, 218.
San Francisco, 103, 104, 154.
rapid transit proposals, 104.
Satellite towns, 129, 131.
Satterlee & Smith, 148.
Sauffege, 96.
Scadgell, W. J., 121.
Scale, 11, 38, 108.
space-time, 14.
pedestrian, intimate, 225.
Scare techniques, 55.
Schaeffer, A. A., 21.
Schools and universities and car use, 89.
Schoon, G. E., 121, 180.
Scientific method of planning, 6.
Scissor housing, 123, 299.
Scoffham, E., 301.
Scott, A. S., 118.
Scott-Williamson, 224.
Sculptured outlines, 213.
Searle, D. A., 180.
Sears, H., 294, 295.
Seattle, 53, 96, 97, 316.
Segregation, 68
history, 313.
horizontal, 157, 289, 315, 316.
vertical, 158, 315.
time, 158, 314.
pedestrian–vehicular, 180.
vehicular in residential-areas, 249.
see also separation.
Senior, D., 147.
Sennestadt, 59, 136, 328.
comparison of path systems, 55.
Separation, traffic, 34, 156.
see also segregation.
Sergelgaten, Stockholm, 174, 324, 328.
Sert & Rogers, 324.
Services, installation, 227.
economics, 227, 228.
Services, and storage, 159.
low level, 202.
Service, roads, 79.
lane, 234.
Shankland, C. G. L., 61, 121, 180.
Shape, town, 114.
Sheach, L., 290.
Sheffield, 158.
Greenhill, 243, 251.
Park Hill, 44, 115, 286, 323.
Sheep, 118.
Shephall Centre, Stevenage, 215.
Shopping, 90.
street, 216, 217.
economics, 167.
two-level, 172.
"way", 324.
Short, R., 195.
Sidewalks, 35, 308.
Sienna, 46.
Signs, 59.
sky, 220.
Simmonds, E., 180.
Simpson, R., 192.
Size, 10, 12, 13, 39.
Skaarup, H. H., 208.
Skipper & Corless, 216.
Skrzypcak-Spak, M., 111, 112, 113.
Smeed, R. J., 15, 16, 65, 67, 73, 80, 93, 110, 160.
Smigielski, K., 4, 62, 189, 247, 298.
Smog, 23.
Smith, A. V., 286.
Smith, C. W., 201.
Smith, I. S., 286.
Smith, W., and Associates, 61, 89, 95, 160.
Smithson, A. and P., 126.
Snellman, 296.
Snow, J., 286.
Societies attitude to accidents, 16.
Social contact, 225.
Social crisis, 226.
Social health, 226.
Social function of new towns, 115.
Social therapist, 5.
therapy, 7.
patterns, 27.
Sociological considerations, 27.
Sociology, 7.
Socio-psychological factors, 34.
Solna, Stockholm, 235, 296.
Solon, D., 192.
Sound, 24.
Sound barrier, 68.
South Bank, London, 213.
South Carolina, 73.
Southland, B. H., 192.
Space, need for, 225.
Spacing, between vehicles, 71.
Speed, 10, 12, 13, 21, 38, 67, 70, 73.
and accidents, 73.
average peak hour, 159.
and capacity, 73.
and character, 73.
and climate, 74.
and danger, 73.
design, 58.
flow relation, 167.
minimum, 167.
optimum, 105.
permissible, 74.
regulations, effect of, 73.
and space, 74.
on urban roads, 159.
vision, 58.
Spirals, 19.
Spiral planning, 75.
Split-level housing, 41.
Spon End, Coventry, 297.
Sport and Cultural Events, 90.
Spring, P., 298.
Sputnik, Russia, 146.
Comparison of path systems, 56.
Standardization of house construction, 262.
Station, underground, 174.
Statistics, interpretation of, 91.
Steam engine, 314.
Stedman, J., 269.
Stein, C., 129, 144, 214, 224, 260, 309, 310, 314, 319, 320, 325, 328.
Stephenson, Professor G., 214, 243, 314, 319, 325, 328.
Steps, 253.
Stevenage, 32, 110, 164, 214.
town centre, 325, 328.
Elm Green, 246, 251, 257, 281.
Pin Green, 281.
Shephall Centre, 215.
Stewart, A., 218.
Stewart, R. G., 73, 218.
St. Louis, 102.
St. Louis Public Service Company, 102.
Stockholm,
Farsta, 42, 110, 113, 130.
public transport, 95.
escalators and lifts, 54.
underground railway, 98.
Solna, 296.
traffic accidents, 80.
traffic route plan, 173.
underpasses, 39, 254.
Vällingby, 113, 128, 129.
Stone & Youngberg, 103.
Stourport, 243, 251.
St. Paul's, London, 327.
Strain, release from, 227.
Streatfield, D., 299.
Street, survey of, 29, 31.
Street names and numbers, 257.
possibilities, 257.
Strøget, Copenhagen, 158, 216.
Stub blocks, 290.
Sturni, G. P., 70.
Suburban shopping centres, 90, 152.
Sunnyside, 310, 319, 328, 309.
Superblocks, 115, 314.
application in comprehensive redevelopment, 253.
basic factors, 252.
basic possibilities, 252, 253.
Carbrain, 252.
Cumbernauld, 258.
design, 252.
Radburn, 283.
unrelated, 281.
Superelevation, 74.

Survey, 27 (*see also* Research).
Suspicions, 226.
Suttie, I., 224.
Sutton Coldfield, 202.
Svenson, K., 267.
Swindon, 198.
Symbols, road, 59.
Symbolism, architectural, 140.

Tactility, 10, 39.
Tangential, road system, 67, 75, 189.
Tapiola, Finland, 144.
Tass, L., 98.
Taylor, W. C., 118.
Terraces, 234.
Terrace housing, 256.
Texture, 39.
Thivierge, P., 271.
Thomas, M. D., 23.
Thompson, L., 322, 328.
Thomson, W. C., 118, 280.
Thornberg, 300.
Thurston, 58.
Ticker, D., 262.
Time contour map, 105.
Time segregation, 34.
Tindall, J., 168.
Tivoli Gardens, 54.
Toddlers, play, 225.
Toronto, Ontario, 98, 295.
Toulouse le Mirail, 113, 116, 126.
comparison of path systems, 56.
Town, circular, 112.
circular, town pattern, 111.
design, 111.
planning, 38.
size, 111.
shape, 114.
and walking distance, 113.
Trafalgar Square, 191.
Traffic, accidents, 16, 17.
arrangements, 34.
commuter, 195.
engineering, 16.
for pedestrians, 236.
flow, 53, 106.
generation and movement, 68.
interchange, 105.
lane, 106.
loss of, 216.
improvement, cost, 166.
multiple-level, 184.
multiple streams, 158.
in New Towns, 108.
pattern change, 216.
tidal, 105.
volume, 106.
segregation, history, 313.
summary, 314.
Transportation, studies, 62.
surveys, 62.
Travelling time, 112.
Travolators, 53.
Trend planning, 52, 64, 65.
Trend, traffic, 60.
Treppenstrasse, Kassel, 215.
Tripp, A., 189, 306, 324, 328.
Trip percentages, 88.
Trolleys, 54, 90.
Trout, F., 194.
Tunnard & Puskarev, 59, 75.
Turin, 97.
Turner, R., 21.
Turning circles, 250, 242.
Twedell, N., 197, 278.

Underground railway, 322, 315, 173.
Underpasses, 51, 237, 253, 316, 321.
alternatives, design and cost, 255.
cost, 253.
position and importance, 253.
shapes, 253.
on shopping arcade, 189.
Unit Sales Method, 109.
Universal ownership, 74.
Urban renewal, basic principles, 155.
comparative values, 168.
design for, 155.
economic considerations, 165.
general policy, 152.
private development, 199, 208, 220.
Urban Traffic and Transportation Board, 101.
Urban transportation facilities, 100.
Usworth development, Co. Durham, 301.
Utzon, J., 247.

Vällingby, Stockholm, 113, 128, 129, 256, 328.
Hasselby Strand, 250.
Vallis, 283.
Values, 4.
Value judgement, 4.
Variety, 43.
Västerås, 199.
Vehicle, design, 18.
growth and population, 60.
routes, 68.
and town space, 65.
Venice, 24, 41.
Vertical segregation, 34, 158, 159.
Vibration, 24.
Victoria, Hong Kong, 188.
Victorian housing, access, 319.
Village, Bedmont, 282.
Cambridge, 147.
Usworth, Co. Durham, 301.
Ville Contemporaine, 60, 323, 328.
Vincent, L., 215, 281.
Vinci, Leonardo da, 315.
Violence and the path system, 33.
Visiting, 90.
Visitors, 235.
Visual impact, 248.
Visual qualities, 39.
Voisin, Paris, 322, 328.
Volrat, Tham, 39.
Volumes, 39.

Wadsworth, 321.
Walker, 321.
Walking, 38, 60.
comparison, distance and towns, 113.
convenient time, maximum, 111.
distance, maximum, 112.
distance, increase in, 112.
towns designed for, 114.
speeds, 111.
Walk ups, 291.
Wall walk, 208.
Wardrop, J. G., 160, 166.
Wardrop & Duff, 15, 71.
War Office, 297, 298.
Warren, B. F., 286.
Washington, 59.
Reston, 148.
Co. Durham, 301.
Waterhouse, A., 213.
Watford, R. D. C., 283.
Weather-resistant finishes, 235.
Weaver, R. C., 95.
Webin Communities, 289.
Weill, 134, 301.
Welsh's Farm, Stourport, 231.
Welwyn, parking needs, 109.
West, F. G., 123, 213, 269.
Westerman, G., 324.
Westwood, P. A., 299.
White, 140, 144.
White, S., 274, 277.
Whittle, J., 299.
Whittle, Mrs. P., 218.
Whittlesey & Conklin, 148.
Whyte, W. H., 5.
Wibing, N., 39.
Wilcoxon, F., 31.
Willenhall Wood, Coventry:
cars, and social pattern, 27.
storage, 235.
impression given, 225.
path system, 236.
pedestrian environment, 43.
Willenhall I, 244, 257, 278.
economic analysis, 230.
Willenhall II, 244, 257, 279.
Williams, M., 266.
Willmot, P., 115.
Wilshaw, V. E, 180.
Wilson, H., 118, 246, 247, 258, 280, 290, 329.
Wilson, L., 260.
Winder, W. J., 286.
Windinge, B., 323.
Windows, 234.
Window shopping, 53.
Winter, J., 298.
Wohlin, H., 329.
Woods, S., 126.
Womersley, J. L., 243, 286, 328.
Work democracy, 5.
Worskett, R., 197.
Wortman, Professor, 155.
Wrexham, culs-de-sac, 326.
lay-out, 243.
dimensions, 251.
housing scheme, 325.
Radburn, economic analysis, 231.
street numbering, 257.
Wrexham, revisited, 326.
Wright, H., 309, 314, 319.
Wupperthal, 97.

York, wall walk, 155, 156.
Youngman, G. P., 118.

Zwicker, B., 192.

XII.5. French Translation

XII.5.1. French Translation of Contents List

I. Témoignages de Reconnaissance

I.1. Inspiration 2

I.2. Aide personnelle 2

I.3. Renseignements et autorisations 2

I.4. Remerciements officiels 2

I.5. Présentation 2

II. Analyse et Introduction

II.1. Fonction du présent livre 4
II.1.1. Coordination des connaissances 4
II.1.2. L'homme et l'automobile 4
II.1.3. Préjudices et conflits 4

II.2. L'environnement et la méthode scientifique 4
II.2.1. Les limitations actuelles de la science 4
II.2.2. Criteriums 5
II.2.3. L'expérimentation créatrice 6
II.2.4. Les responsabilités 6

II.3. Possibilités et systèmes du planning 6
II.3.1. Les possibilités du planning 6
II.3.2. Profusion d'alternatives 7
II.3.3. L'idéaliste opportuniste 7
II.3.4. Fondement d'une critique rationnelle 7

III. La Relation Homme–Véhicule

III.1. Résumé des caractéristiques de la notion homme–véhicule 10

III.2. Caractères physiques de l'homme et des véhicules 11
III.2.1. Échelle 11
III.2.2. Grandeur relative et cumulative 13
III.2.3. Associations tactiles 13

III.3. Le mouvement 14
III.3.1. Vitesse et portée 14
III.3.2. Capacité routière 14
III.3.3. Vitesse acquise et danger 16
III.3.3.1. Les accidents dans les secteurs à circulation dissociée 19
III.3.4. Courbure des routes 21
III.3.5. Frustration mutuelle 22

III.4. Considérations écologiques 23
III.4.1. Émanations et poison 23
III.4.2. Bruit 24
III.4.3. Vibrations 26

III.5. Considérations sociologiques 27
III.5.1. Influence de la voiture particulière sur les systèmes sociaux 27
III.5.1.1. Effets constructifs 27
III.5.1.2. Effets destructifs 27
III.5.2. Projet de recherches 27
III.5.2.1. But du project de recherches 27
III.5.2.2. Présentation des résultats 27
III.5.2.3. Résumé de l'ensemble de l'étude 31
III.5.2.4. Étude entreprise à Stevenage 32
III.5.3. Aspects sociaux du problème des transports publics 32
III.5.4. Considérations morales et légales 32
III.5.5. Masses humaines et masses automobiles 33
III.5.6. Facteurs psychosociaux 34

III.6. Solutions apportées par le planning 34
III.6.1. Organisation fonctionnelle de la circulation 34
III.6.2. Caractères des routes et des chemins 35
III.6.3. Analyse graphique des besoins de mouvement et de repos du piéton 36

IV. Les Besoins de l'Homme en tant qu'Organisme

IV.1. Fonctions biologiques de la marche 38

IV.2. Qualités requises pour assurer un bon environnement aux piétons 39
IV.2.1. Espace à la dimension de l'homme 41
IV.2.2. Questions de détails 43
IV.2.3. Variété infinie 44

IV.3. Étude d'une Piazza urbaine pour piétons 46
IV.3.1. Relevé des entrées et des sorties du public, fait de minute en minute 46
IV.3.1. Analyse visuelle 47
IV.3.3. Tentative d'explication du bon fonctionnement du système-piazza 50

IV.4. Agréments et facilités 51

IV.5. Moyens mécaniques 52
IV.5.1. Cycles à pédales 52
IV.5.2. Transporteuses pour piétons 53
IV.5.3. Escaliers roulants et ascenseurs 54
IV.5.4. Chariots 54
IV.5.5. "Pédestrains" 54
IV.5.6. Conveyor pour automobiles 55

IV.6. Renseignements 55

IV.7. Comparaison des systèmes de voies 55

V. Besoins et Nature des Véhicules

V.1. L'homme en tant que conducteur **58**
V.1.1. Perception 58
V.1.2. L'équipement routier 59
V.1.3. Formation 59
V.1.4. Etat émotif 59

V.2. Généralities sur le développement de la Motorisation **60**
V.2.1. Etude et examen du problème des transports 62
V.2.2. Le véhicule dans l'agglomération urbaine 65
V.2.3. Examen général du problème économique 67

V.3. Itinéraire des véhicules **68**
V.3.1. Aministration de l'aménagement routier 68
V.3.2. Espace, état, bruit et dangers routiers 68
V.3.3. Capacité 70
V.3.4. Vitesse 73
V.3.5. Types de routes 75
V.3.5.1. Généralités 75
V.3.5.2. Autoroutes et routes à circulation protégée 75
V.3.5.3. Routes nationales 79
V.3.5.4. Routes de service 79
V.3.5.5. Résumé 79
V.3.6. Généralités sur les carrefours 79
V.3.6.1. Délit des voitures 81
V.3.6.2. Carrefours de 3 routes 81
V.3.6.3. Carrefours de 4 routes 82

V.4. Le Parking **84**
V.4.1. Nécessités générales et fondamentales 84
V.4.2. Parkings de niveau avec le sol 84
V.4.3. Garages-parking à niveaux multiples 85
V.4.4. Parkings mécaniques 88

V.5. Fonctions des véhicules particuliers **88**
V.5.1. Trajet jusqu'au lieu du travail 89
V.5.2. Ecoles et universités 89
V.5.3. Courses 89
V.5.4. Visites 90
V.5.5. Evènements sportifs et culturels 90
V.5.6. Week-ends et demi-congés 90
V.5.7. Grandes vacances 90
V.5.8. Interprétation des statistiques 91

V.6. Types de véhicules particuliers **91**
V.6.1. Facteurs modifiant le type de véhicules 91

V.7. Véhicules de commerce, camionnettes et camions **93**
V.7.1. Nombre et accroissement 93
V.7.2. Caractéristiques de la dimension, de la vitesse et du débit 93
V.7.3. Influence sur la circulation 93

V.8. Transports publics **94**
V.8.1. Attitude générale 94
V.8.2. Principes des systèmes de transports publics 95
V.8.3. Types de transports rapides 95
V.8.3.1. Autobus 95
V.8.3.2. Systèmes à deux rails 96
V.8.3.3. Monorail 96
V.8.3.4. Chemins de fer souterrains 98
V.8.3.5. Trains-bus autoguidés 99
V.8.4. Comparaison de traits caractéristiques 100
V.8.5. Exemples 101
V.8.5.1. Philadelphie 101
V.8.5.2. St. Louis 102
V.8.5.3. Boston 103
V.8.5.4. San Francisco 103

V.9. Lexique **105**

VI. Villes Nouvelles á Circulation Dissossiée

VI.1. Nécessités fondamentales **108**

VI.2. Circulation automobile dans les villes nouvelles, généralités **108**
VI.2.1. Parking dans le centre des villes nouvelles 109
VI.2.2. Cumbernauld, recherches sur C.A./parking autour des grands magasins 110

VI.3. Centre entièrement inaccessible aux véhicules dans une ville de 250.000 habitants **111**

VI.4. Planification basée sur le parcours à pied **111**
VI.4.1. Densité et importance de la ville 111
VI.4.1.1. Etude sur l'importance de la ville, sa population et sa densité 111
VI.4.1.2. Schémas illustrant la densité d'une ville 112
VI.4.2. Comparaison entre les parcours à effectuer à pied dans différentes villes 113
VI.4.3. Configuration de la ville 114

VI.5. Utilisation de la terre **114**

VI.6. Espaces verts **114**

VI.7. Le concept du voisinage **115**

VI.8. Principes de l'accroissement **115**

VI.9. Exemples **117**
VI.9.1. Cumbernauld, Ecosse 117
VI.9.2. Hook, Angleterre 121
VI.9.3. Erith, Angleterre 123
VI.9.4. Toulouse-le-Mirail, France 126
VI.9.5. Vällingby, Suède 128
VI.9.6. Farsta, Suède 130
VI.9.7. Cologne, Allemagne 133
VI.9.8. Sabende, Guinée 134
VI.9.9. Sennestadt, Allemagne 136
VI.9.10. Neu-Winsen, Allemagne 138
VI.9.11. Alkmaar, Hollande 139
VI.9.12. Chandigarh, Indes 140
VI.9.13. Kitimat, Canada 143
VI.9.14. Tapiola, Finlande 144
VI.9.15. Sputnik, Russie 146
VI.9.16. Cambridge Village, Cambridgeshire, Angleterre 147
VI.9.17. Reston, près de Washington, U.S.A. 148

VII. Aménagement Urbain

VII.1. Politique général 152
VII.1.1. Le cas de New-York 153
VII.1.2. Le cas de San Francisco 154
VII.1.3. Le cas de Détrot 154

VII.2. Projet d'aménagement urbain 155
VII.2.1. Principes fondamentaux 155
VII.2.2. Dissociation horizontale de la circulation 157
VII.2.2.1. Dissociation dans le temps 158
VII.2.3. Dissociation verticale, les piétons étant au-dessus 158
VII.2.4. Dissociation verticale, les véhicules étant au-dessus du niveau du sol 159
VII.2.5. Dissociation verticale, les véhicules étant au niveau du sol, les piétons au-dessous 159

VII.3. Voies urbaines 159
VII.3.1. Amélioration du système routier existant 159
VII.3.2. Création d'autoroutes ou de routes à circulation protégée 160
VII.3.3. Aspect d'une ville à autoroutes 161

VII.4. Le parking 162
VII.4.1. Evolutions du parking 162
VII.4.2. Autres facteurs affectant le parking 163
VII.4.3. Parking côté trottoir 164
VII.4.4. Parkings à compteurs 164

VII.5. Considérations économiques, Généralités 165
VII.5.1. Prix de revient d'une motorisation totale 165
VII.5.2. Prix de revient comparés des trajets d'abonnés 166
VII.5.3. Prix de revient des améliorations de la circulation 166
VII.5.4. Frais entraînés par l'encombrement 167
VII.5.5. Régime économique des rues commercielles réservées aux piétons 167
VII.5.6. Valeurs comparées 168

VII.6. Exemples 170
VII.6.1. Coventry, aménagement et plan de la circulation 170
VII.6.1.1. Coventry, centre commercial 172
VII.6.2. Stockholm, aménagement et plan de circulation 173
VII.6.2.1. Stockholm, Sergelgaten 174
VII.6.3. Philadelphie, aménagement et plan de la circulation 176
VII.6.3.1. Section est du marché 177
VII.6.4. Liverpool, plan de la circulation 180
VII.6.4.1. Quartier de Ravenseft 182
VII.6.4.2. Centre civique et social 184
VII.6.4.3. Paradise Street, Quartier du Strand 184
VII.6.5. Newcastle, aménagement et plan de la circulation 186
VII.6.5.1. Centre, Ensemble de l'aménagement 187
VII.6.6. Cité de Victoria, Hong Kong, Centre 188
VII.6.7. Londres, Plan de la circulation 189
VII.6.7.1. Quartier de Trafalgar Square 191
VII.6.8. Fort Worth, Texas, Reconstruction du la centre 192
VII.6.9. Nottingham, Centre et plan de la circulation 194
VII.6.10. Banbury, réaménagement du centre 197
VII.6.11. Swindon, réaménagement du centre 198
VII.6.12. Västerås, Suède, Aménagement et plan de la circulation 199
VII.6.13. Bishop's Stortford, réaménagement du centre 201
VII.6.14. Sutton Coldfield, réaménagement du centre 202
VII.6.15. High Wycombe, réaménagement du centre 205
VII.6.16. Beeston, réaménagement du centre 206
VII.6.17. Horsholm, Danemark 208
VII.6.18. Cité de Chester, aménagement du centre 208
VII.6.19. Andover, aménagement de la ville 210
VII.6.20. Lijnbaan, Rotterdam, Hollande 212
VII.6.21. Aménagement de South Bank, Londres 213
VII.6.22. Stevenage, centre 214
VII.6.23. Stevenage, centre de Shephall 215
VII.6.24. Treppenstrasse, Kassel, Allemagne 215
VII.6.25. Rue commerciale pour piétons à Essen, Allemagne 215
VII.6.26. Réaménagement de Greyfriars, Ipswich 216
VII.6.27. Fronts de mer, Brighton 216
VII.6.28. Strøget, rue commerciale à Copenhagen, Danemark 216
VII.6.29. Zone de réaménagement d'ensemble de Ellor Street, Salford 218
VII.6.30. Aménagement du centre de Hammersmith, Londres 219
VII.6.31. Piccadilly Circus, Londres 220
VII.6.32. Knightsbridge Green, Londres 220
VII 6.33. Ville nouvelle de Basildon, centre principal 221
VII.6.34. Elizabeth, l'Australie, centre principal 222

VIII. Circulation Dissociée dans les Quartiers Résidentiels

VIII.1. Echelle de l'habitabilité appliquée aux principes de Radburn 224
- VIII.2.1. Critériums psychosomatiques: possibilités de Radburn 224
- VIII.1.2. Critériums psychosociaux 225
- VIII.1.3. Critériums écologiques 227
- VIII.1.4. Critériums économiques et administratifs 227

VIII.2. Prix de revient 228
- VIII.2.1. Comparaison des prix de revient des rues et des travaux d'égouts 228
- VIII.2.2. Analyse économique 229

VIII.3. Plans d'habitations 232
- VIII.3.1. Analyse de la relation des plans d'habitations avec l'accès 232
- VIII.3.2. Entreposage des voitures 235
- VIII.3.3. La dépense et la densité 235

VIII.4. Voies, système d'itinéraires 236
- VIII.4.1. Analyse 236

VIII.5. Configuration des terrains 238
- VIII.5.1. Maisons 238
- VIII.5.2. Appartements et deck-housing 239

VIII.6. Routes secondaires résidentielles 240
- VIII.6.1. Dimensions et fréquence des carrefours 240
- VIII.6.2. Plan des routes secondaires—généralités 240
  - VIII.6.2.1. Culs-de-sac 240
  - VIII.6.2.2. Voies de raccordement 240
- VIII.6.3. Les possibilités fundamentales 241
- VIII.6.3.1. Les impasses 241
- VIII.6.3.2. Les routes de contournement 242
- VIII.6.4. Exemples 243
  - VIII.6.4.1. Projets de pionnier autour des années 1950 243
  - VIII.6.4.2. Projets autour des années 1955 244
  - VIII.6.4.3. Projets autour des années 1955–60 246
  - VIII.6.4.4. Dimensions de culs-de-sac secondaires 248
  - VIII.6.4.5. Comment lier les impasses et les sentiers 249
- VIII.6.5. Mélange des piétons et des véhicules, dans les culs-de-sac 250

VIII.7. Plan de "supercité" 252
- VIII.7.1. Facteurs fondamentaux 252
- VIII.7.2. Applications dans le réaménagement 253

VIII.8. Passages souterrains et ponts 253
- VIII.8.1. Position et importance 253
- VIII.8.2. Alternatives de plans et de prix 255

VIII.9. Espaces verts 256

VIII.10. Contours 256

VIII.11. Noms et numéros des rues 257

VIII.12. Exemples de zones résidentielles à circulation dissociée 258
- VIII.12.1. Cumbernauld, près Glasgow 258
- VIII.12.2. Village de Baldwin Hills, Los Angeles 260
- VIII.12.3. Caversham, Berks. 262
- VIII.12.4. Bron-y-Mor, Pays de Galles 263
- VIII.12.5. Huntingdon, L.C.C. 265
- VIII.12.6. Lotissement Jachson, Letchworth 266
- VIII.12.7. Ilkeston Road, Beeston, Nottinghamshire 266
- VIII.12.8. Albertslund, Copenhague 267
- VIII.12.9. Eastwick, Philadelphie 268
- VIII.12.10. Quartier Clements, Haverhill, Suffolk 269
- VIII.12.11. Cité-jardin, Montréal 271
- VIII.12.12. Marly-le-Roi, Paris 273
- VIII.12.13. Baronbackarna, Orebro 274
- VIII.12.14. Parc Lafayette, Titre I, Détrot 276
- VIII.12.15. Biskopsgaden, Göteborg 277
- VIII.12.16. Bois de Willenhall I, Coventry 278
- VIII.12.17. Kildrum, Cumbernauld 280
- VIII.12.18. Elm Green I and II, Stevenage 281
- VIII.12.19. Bedmont, Hertfordshire 282
- VIII.12.20. Inchview, Prestonpans, Edinbourg 283
- VIII.12.21. Parkleys, Ham Common, Londres 285
- VIII.12.22. Park Hill, Sheffield 286
- VIII.12.23. Barbican, Londres 288
- VIII.12.24. Parc de Flemingdon, East York, Toronto 289
- VIII.12.25. South Carbrain, Cumbernauld 290
- VIII.12.26. Oakdale Manor, North York, Toronto 294
- VIII.12.27. Kentucky Road, Toronto 295
- VIII.12.28. Solna, Suède 296
- VIII.12.29. Spon End, Coventry 297
- VIII.12.30. Logements réservés aux familles. Ministère de la Guerre Britannique (A) 297
- VIII.12.31. Logements réservés aux familles. Ministère de la Guerre Britannique (B) 298
- VIII.12.32. Rowlatts Hill, Leicester 298
- VIII.12.33. Lotissement Brandon, Londres 299
- VIII.12.34. Royal Victoria Yard, Deptford, Londres 299
- VIII.12.35. Zone de réaménagement de Brunswick, Manchester 300
- VIII.12.36. Primrose Hill, Birmingham 300
- VIII.12.37. Almhog, Nydala, Hermodsdal, Malmö 300
- VIII.12.38. Mellanhaden, Malmö 300
- VIII.12.39. Washington, Durham 301
- VIII.12.40. Fontenay-aux-Roses, Paris 304
- VIII.12.41. Hillfields, Coventry 304
- VIII.12.42. Highfields, Leicester 304a

IX. Aménagement Résidentiel

IX.1. Principes généraux **306**
IX.1.1. Recherches sur les accidents dûs au système du "gril" 306

IX.2. Exemples **308**
IX.2.1. Plan d'amélioration des rues, Quartier de Pershing Field, environs de Fulton, Minneapolis, U.S.A. 308
IX.2.2. Zone de réaménagement de Mill Creek, Philadelphie, Commission de planification urbaine, 1954 308
IX.2.3. Arrêté britannique sur le logement, Nottingham 309
IX.2.4. Dégagement sur l'arrière des immeubles 310
IX.2.5. Le logement entre les deux guerres, Nottingham 310
IX.2.6. Chicago, la transformation d'un cité 312

X. Historique de la Circulation Dissociée

X.1. Résumé **314**

X.2. Historique illustré de la circulation dissociée **314**
X.2.1. Exemples d'avant l'ère du moteur à essence 314
X.2.2. Planning pour "Vivre malgré l'automobile" 319
X.2.3. Planning pour la dissociation de la circulation en Europe 322

X.3. Graphique des dates de l'histoire du planning pour l'homme et l'automobile **328**

XI. Bibliographie

XI.1. Livres 331

XI.2. Rapports, articles, thèses 333

XII. Appendices

XII.1. Tables de conversion 347

XII.2. Abréviations 347

XII.3. Schéma d'étude sur les transports 348

XII.4. Index 350

XII.5. Traduction française 357
XII.5.1. Tables des matières 357
XII.5.2. Analyse et introduction 366

XII.6. Traduction allemande 371
XII.6.1. Table des matières 371
XII.6.2. Analyse et introduction 380

XII.5.2. French Translation of Section II.

II. Analyse et Introduction

II. 1. Fonction du livre. 366
- II. 1.1. Coordination des connaissances. 366
- II. 1.2. L'homme et l'automobile. 366
- II. 1.3. Préjudices et conflits. 366

II. 2. Environnement et méthode scientifique. 367
- II. 2.1. Limitations actuelles de la science. 367
- II. 2.2. Critériums. 367
- II. 2.3. L'expérimentation créatrice. 368
- II. 2.4. Responsabilités. 368

II. 3. Possibilités et systèmes du planning. 369
- II. 3.1. Les possibilités du planning. 369
- II. 3.2. Alternatives multiples. 369
- II. 3.3. L'idéaliste opportuniste. 369
- II. 3.4. Fondement d'une critique rationnelle. 370

II. Analyse et Introduction

II. 1. Fonction du livre.

II. 1.1. *Coordination des connaissances*

Architectes, urbanistes, ingénieurs, administrateurs et législateurs, ainsi que ceux qui sont chargés d'aménager des secteurs, tous font des plans. Les principes fondamentaux et fonctionnels communs ne sont ni décrits, ni mis en évidence, ni rappellés à l'attention, ni même définis, dans l'enseignement ou dans les manuels. Ainsi nous manque un fondement rationnel pour décider et juger; le point de départ, la matrice, le terrain de base des activités interprofessionnelles font défaut. Rien d'étonnant que, dans la pratique, il n'y ait guère d'actions concentrées collectivement.

Ce livre se propose, non seulement de dispenser, mais encore de coordonner les connaissances et les méthodes d'approche, et de mettre en lumière les éléments fondamentaux communs aux différentes professions. Prenant comme mobile primordial, pour l'ensemble des professions, la création d'un bon environnement, le lecteur trouvera ce qui est directement applicable à son cas, et dans le contexte du livre, comment cela s'intègre dans l'ensemble du plan.

II. 1.2. *L'homme et l'automobile*

L'automobile est le véhicule le plus utile et aux usages les plus variés que l'homme ait inventé.

Dans l'état primitif où elle est actuellement, des gaz malsains émanent encore des moteurs et des carburants mais des modèles plus poussés apparaîtront et seule est rationnelle une attitude positive, faisant des plans pour la meilleure utilisation possible des véhicules.

Il semble peut probable qu'aucune invention ne supplante le véhicule à roues si ce n'est l'automobile volante qu'est l'Hovercraft. Bien que la qualité de la surface perdrait alors de son importance, il faudrait prévoir la place et les itinéraires comme pour une voiture, étant donné le coussin d'air qui la relie au sol. Les transports aériens ne prendront d'importance que lorsque des dispositifs électroniques pourront guider suivant des routes individuelles et à des hauteurs variées, une sorte d'appareil aérodynamique, satisfaisant du point de vue pratique alors que l'hélicoptère d'aujourd'hui ne l'est pas.

Les chapitres suivants analysent les différentes natures d'hommes et de véhicules, afin que leurs besoins respectifs puissent être évalués le plus précisement possible.

Un environnement écologique harmonieux pour l'homme, dans lequel un usage du véhicule jouera le rôle décisif, tel est le but.

Cette publication est opportune, non seulement parce que une telle information est d'un pressant besoin, mais aussi, parce que, à l'heure actuelle, en 1963, toutes les solutions de base qui semblaient théoriquement possible, ont été mises en pratique et par conséquent peuvent être discutées et illustrées par des exemples. Qu'elle soit regardée comme un "animal domestique" (Smigielsky) ou comme un "tyrannosaure deux tons" (Chermayeff) *la possession et l'utilisation* universelles de la voiture exigent une nouvelle façon d'aborder presque tous les problèmes d'environnement, depuis le plan d'une maison, jusqu'à l'organisation de chaque région par zone.

II. 1.3. *Préjudices et Conflits*

Sans organisation du champ des connaissances, sans système de coordonnées, le tableau est rendu encore plus confus par les troubles suscités par une friction continuelle, dans les secteurs encombrés d'hommes et de véhicules qui n'ont pas été prévus pour cet usage mixte.

Le ressentiment du piéton et du cycliste, qui ne peut plus respirer un air pur et se coit chassé de la chaussée, à sa raison d'être. Il est normal que le citadin endurci soit horrifié à la vue du véritable tapis de voitures qui recouvre jusqu'au moindre recoin de sa ville. De même l'irritation de conducteur est justifiée, bien que piéton lui-même auparavant, peut-être. Rendu furieux d'abord par les embouteillages, alors qu'il voudrait pouvoir profiter de la grande rapidité que lui permettrait son moyen de locomotion, il le sera plus tard, en arrivant à destination, du fait qu'il devra continuer à rouler, simplement faute de trouver une place où parquer même si la permission lui en était accordée.

Menacer l'un des partis d'un type d'urbanisme comme celui de Los Angeles, et pronostiquer à l'autre de longues marches fatigantes, n'aide en rien à clarifier la situation. La peur et la haine n'incitent pas à la raison. Seul reconnaître les défauts du système qui est à la base du dilemme que présente notre position actuelle et comprendre les besoins et les fonctions propres de l'homme et de son aide merveilleusement utile qu'est l'automobile, nous ménera aux solutions admissibles, hardies, bien équilibrées et originales qui sont requises.

Ceux qui sont incapables de voir qu'il y a là un problème grandissant et véritablement effrayant, ceux qui s'habituent à la médiocrité et au massacre de vies humaines, ou considèrent qu'il suffit de copier les solutions déjà adoptées: ceux-là ont besoin d'être secoués pour sortir de leur contentement facile.

II. 2. Environnement et méthode scientifique

II. 2.1. *Limitations actuelles de l'étude scientifique*

Qu'est-ce qu'un bon environnement? Quelles sont les valeurs fondamentales qui doivent nous guider dans le planning? Nous ne le savons pas, parce que l'outillage nécessaire n'existe pas.

D'après une opinion erronée largement répandue, la sociologie doit nous donner les réponses à ces questions. Voilà une faute grave: c'est mettre toute sa confiance dans une discipline qui exclue spécifiquement les jugements de valeur et par conséquent ne recommande aucun plan. Même si on inclut l'approche plus fructeuse de l'anthropologie, comme le fait l'"Institute of Community Studies", ce n'est encore qu'étudier ce qui existe, ce que par expérience nous savons souvent être mal conçu d'avance. Toute idée originale est exclue. La sociologie sous sa forme actuelle n'est pas destinée à donner des informations propres à servir de guide aux urbanistes dans les subtils problèmes qualitatifs pour lesquels il est urgent qu'ils reçoivent information et inspiration.

L'attitude de l'étude scientifique est fréquemment critiquée du dedans. John Madge dit dans son livre: "L'outillage de l'économie sociale",(i) étude approfondie, "il n'y a pas de science dans laquelle la recherche de la connaissance objective soit plus vaine que dans l'économie sociale". Et plus loin il suggère que "... l'étude de la sociologie est un mouvement aussi bien qu'une recherche des faits".

Pourtant il sait que ses idées ne trouveront aucun écho chez les sociologues académiques britanniques. Une telle oeuvre est considérée comme étant "... d'un niveau plutôt bas" et comme ayant utilisé "... des techniques périmées".

L'ingénieur des transports se trouve dans la même position difficile. Bien que ses connaissances ne soient pas étayées par un système éprouvé et sur lequel il puisse compter, il est appelé à prendre des décisions qui impliquent un nombre immense de variables, dont les effets sont, pour l'instant, incalculables, qui souvent échappent à son autorité et dont la communauté dont il fait partie ne peut assumer les frais. La rationalisation du transport de tout ce qui est nécessaire pour subvenir aux besoins d'un secteur donné, est une tâche complexe, qui demande beaucoup d'idées originales. Nous devons conclure que, avec leurs méthodes actuelles, les sciences et technologies étant incapables d'étudier l'homme et son environnement, elles ne peuvent nous apporter une aide efficace.

II. 2.2. *Criteriums*

Les médecins savent qu'une bonne nourriture, de l'air pur et de l'exercice sont choses désirables du point de vue biologique. Rien d'équivalent ne nous guide, quant à savoir ce qui est sain sociologiquement, même du point de vue conformiste. L'urbaniste, d'après son bon sens et les connaissances disponibles en dehors des domaines dévastés par la stérilité académique, à la fois philosophe, prophète et thérapeute social, doit se faire une opinion par lui-même. A tout âge, l'homme éprouve le besoin d'avoir des occasions de coopérer, dans ses travaux, ses cérémonies, ses jeux, ses amitiés, ses amours. Chacun en a conscience, et les travaux de gens comme le professeur H. Infield le mettent en évidence.

On insiste, ad nauseam, sur le besoin d'intimité. Il est certain que c'est un besoin fondamental. On peut trouver d'habiles solutions, mais, pour l'essentiel, tout ce qu'il faut, c'est un rideau et un gros mur. Souvent la question est de trouver une certaine intimité dans sa propre maison, au sein de sa propre famille, ce qui est plus difficile. Et souvent aussi, la meilleure solution, c'est de quitter les lieux. Ceci nous amène à nous occuper de groupes sociaux autres que la famille.

Il est de beaucoup plus important d'organiser la coopération. C'est un problème complexe, mais qui renferme d'immenses possibilités à peine effleurées jusqu'à présent. Ce domaine, plus que tout autre, va influé sur le futur développement de l'humanité, du point de vue sociologique de l'organisation en groupes, de la participation du citoyen et de la démocratie du travail.

Chermayeff, dans un discours pénétrant, nous offre sa vision de l'avenir où mouvement et mobilité sont bien en perspective.

"Tout ceci présuppose un examen très attentif de tout le spectre des expériences relatives à la mobilité, à la vitesse et à la dimension de l'espace habité. Ce n'est qu'alors que nous pourrons, à l'une des extrémités du spectre, réorganiser le monde des piétons: plus que le monde des piétons—le monde de l'homme au repos dans le plein sens du mot. Si, couchés dans l'herbe, nous regardons un crocus près de nous, nous avons du monde une vision de ver de terre, et ce monde est beau. Si, à une autre échelle, nous montons dans un avion à réaction, nous ne souhaiterons priver personne du plaisir que procure ce surprenant miracle, ni de cette vision entièrement nouvelle de l'univers et des charmes du paysage. Je veux jouir de tout, depuis le crocus jusqu'à la vue du ciel, et je préfère toutes ces joies à deux voitures par garage ... .

Je suis absolument convaincu que ce qui est fait pour rester, ce n'est pas la "cité", mais la conglomération de gens qui se rassemblent pour des

(i) Madge, J. *The Tools of Social Science*. London, 1953.

raisons valables. J'espère que nous allons cultiver cette tendance au rassemblement à des fins nouvelles aussi bien qu'anciennes, mais non pas à cet horrible niveau où les gens finissent vraiment par ne plus pouvoir se supporter. Il n'est plus question de se réunir pour la simple raison qu'on est voisins; on se réunit parce qu'on a quelque chose en commun ou une tâche à accomplir. Je suis convaincu qu'il va y avoir un regroupement, en grande partie dû aux contraintes exercées par l'accroissement de la population, le progrès de la technologie et la complexité infinie de nos systèmes. Et je ne veux pas instituer un système de transports aux dépens d'autres valeurs.(i)

L'image qui s'impose, pour l'avenir, comprend un usage sélectif, pratique et rationnel de la voiture particulière, de même qu'une participation plus active des citoyens aux développements des plans et à l'utilisation de l'environnement prévu pour accroître les groupes sociaux d'aujourd'hui.

La société fait preuve d'une mobilité grandissante qui réclame des aménagements propres à faciliter l'intégration et à hâter l'enracinement.(ii)

II. 2.3. *Expérimentation créatrice*

C'est l'homme tout entier, le philosophe sensible et non pas simplement le savant objectif qui est en chacun de nous, qui seul saura tenir compte des implications fonctionnelles et morales, de l'imagination, des aspirations. L'urbaniste doit être un homme capable d'imaginer, d'évoquer la vision d'un autre, de s'y soumettre et puis de travailler à sa réalisation.

Comme l'a dit Madge, la méthode scientifique se limite, sans raison, à l'isolation des facteurs et à leur analyse. Exclure l'élément subjectif qu'apporte l'expérimentateur, tel est invariablement le but. Ce qui est particulièrement nécessaire, dans le "planning", c'est une attitude opposée, bien que tout aussi efficace et tout aussi valable comme méthode scientifique hypothétique et expérimentale.

Normalement l'hypothèse est prouvée numériquement mais là encore ce n'est pas valable pour le "planning" où les résultats de l'expérience peuvent apporter des advantages originaux.

Examinons cette assertion "je peux atteindre l'autre berge, d'un jet de pierre", en tant qu'hypothèse. L'expérience comprend l'ingéniosité déployée à trouver l'endroit le plus étroit de la rivière, à utiliser le vent, à découvrir la pierre qu'il faut, etc..... En d'autres mots, les différents facteurs ne sont pas isolés. On en recherche ardemment le plus grand nombre qui puissent contribuer à prouver l'exactitude de l'hypothèse, à n'importe quel moment, et contribuer au succès de l'expérience faite en atteignant l'autre rive avec la pierre.

(i) Chermayeff, S. *The New Nomads*. T. Q. Saugatuck. April, 1960.

(ii) Whyte, W. H. *The Organisation-Man*. London, 1957.

Tel est le concept de base de l'Expérimentation Créatrice que j'ai mentionnée comme un moyen d'introduire et d'essayer fructueusement des idées nouvelles dans le "planning". Les résultats de telles expériences seront substantiels et significatifs. Les participants sont entraînés d'une façon dynamique dans les processus qui améliore et développe constamment les méthodes et l'environnement. Les rapports et l'action réciproque de tous les facteurs étant d'importance primordiale, ils sont étudiés et modifiés si cela peut être d'une aide quelconque. La théorie, parfois admise, qu'il est préférable que les gens ne sachent pas qu'ils sont les sujets d'une expérience, est tristement inefficace et dangereuse, et est déplacée dans l'expérimentation créatrice, où cette notion même aide puissamment au succès de l'entreprise.(i)

Dans de telles expériences, la réussite implique la possibilité d'une vie agréable pour gens de tous âges, et progressivement de moins en moins pleine de déceptions et de misère. On ne peut pas voir cela d'une façon objective. L'urbaniste doit le vouloir de toute sa force et tâcher d'atteindre son but par tous les moyens, comme dans le cas de la pierre et de la rivière.

Les planificateurs devront analyser les besoins et les ressources, créer d'une façon imaginative en rapport avec l'analyse, organiser en temps opportun l'exécution du plan, et instiguer une Communication que permette d'initier les gens à en faire un usage approprié. La conclusion concernant les principes fondamentaux qui gouvernent un projet étant atteinte, il faut rejeter l'emphase sur la meilleure façon de les mettre en oeuvre. Les principes sont l'hypothèse: l'expérimentation créatrice réside dans la réussite de la réalisation.

Il est important, dans de telles expériences, de tenir compte du plus grand nombre de projets possible et de rassembler soigneusement les résultats pour grossir le petit volume de connaissances utiles.

II. 2.4. *Responsabilités*

Nous vivons dans un siècle où l'homme du commun tient la première place. Le planning est un luxe destiné à tout le monde. Il doit y avoire une voiture automobile au service de chaque famille. Nous sommes en présence d'un défi de proportions énormes. Ce n'est pas simplement une question de logements ou d'esthétique, ni de circulation bien

(i) Le rapport de Mrs. Demers sur Park Hill à Sheffield, montre que la crainte d'importuner les locataires en les faisant participer aux recherches, n'était pas fondée. La pensée d'être les sujets d'une expérience importante plaît aux gens. Ville de Sheffield.

règlementée. La tâche consiste à procurer un environnement qui permette de mieux satisfaire les besoins primordiaux. Etant donné un service véhiculaire, efficace, les attributs positifs d'une merveilleuse mobilité et la planification, par ceux qui désirent coopérer, des collectivités qui se rendent compte des avantages de la coopération, augmentent les alternatives offertes à chaque individu, accroîssant ses capacités et multipliant les occasions favorables. Raisonnablement, nous devrions prévoir une plus large somme à dépenser par individu pour l'aménagement de l'environnement, que nous ne l'avons fait jusqu'ici. Il est, pour le moment, aussi répandu que peu perspicace d'avoir comme unique but la réduction des impôts. Il s'en suit une lourde responsabilité. Aux U.S.A., il est résulté, d'un certain nombre de facteurs, des solutions qui, dans l'ensemble, ne sont pas applicables ni acceptables en Europe, ni d'ailleurs dans nombre d'autre pays sous-développés. Il nous faut réévaluer les besoins et les fonctions de l'homme et de l'automobile à propos de chaque environnement, pris séparément. "Les pays sous-développés" (ainsi que les villes nouvelles!) "peuvent parvenir à réglementer la circulation sans grand mal pour le pionnier."(i)

L'un après l'autre, tous les pays sont de plus en plus enrichis et encombrés de gens et de voitures. Le nombre des habitants de la terre augmente de plus de 100.000 par jour % l'équivalent d'une ville. Et pourtant, dans le monde entier, les recherches faites sur tous les aspects du problème de l'environnement sont incroyablement insignifiantes et inefficaces.

II. 3. Possibilités et systèmes du planning

II. 3.1. *Les possibilités du planning*

Le planning est un instrument puissant. Le besoin et l'usage des moyens à transport privés dans le centre des villes peut être réduit grâce à l'aménagement. Le nombre des piétons et des cyclistes peut être considérablement accru par un environnement engageant où ils se sentiront protégés, ou réduit à l'extrême par des conditions, défavorables. L'aménagement peut détruire la vie d'un quartier entier, dans une ville, en le rendant inaccessible aux transports privés sans toutefois assurer des transports publics. Le planning à offrir des possibilités passionnantes par leur étendue, nous, une insuffisance monstrueuse. La masse énorme des conséquences s'en fait encore sentir. Nous y sacrifions nos plaisirs, nous nous sacrifions corps et âme. L'iniquité des pères (de la cité) retombe encore sur la troisième et la quatrième génération.

(i) Chermayeff, S. *The New Nomads.* T. Q. Saugtauck, April, 1960.

II. 3.2. *Alternatives multiples*

Le fait que toute action affecte étroitement plus d'un domaine, complique le travail du planificateur. Cette question est particulièrement grave à l'heure actuelle, étant donné que la corrélation des factums n'est pas même reconnue, ni tenue en ligne de compte dans les calculs et les précisions, les orientations et les études. Résaux ferroviaires et routes, développement économique et problèmes d'expansion, fabrication automobile et mesures routières, et nombre d'autres problèmes étroitement liés, sont traités séparément, entraînant des résultats d'une bêtise désastreuse.

Les facteurs étant de nature étroitement liées, on peut aussi attaquer tous les problèmes sous un certain nombre d'angles. Prenons par exemple une route extrêmement congestionnée:

On pourrait l'élargir,

ou construire une autoroute;

on pourrait supprimer les obstructions telles que piétons, feux et autorisations de stationner;

on pourrait dévier la circulation;

on pourrait rendre plus attrayants les accotements et les transports publics;

ou déplacer les lieux de destination vers lesquels se dirige le flot de la circulation;

ou encore échelonner la réparation des heures de travail, cause de la congestion;

le parking, au lieu de destination, pourrait être d'un tarif plus élevé, décourageant ainsi les trajets inutiles;

ou, avec le même résultat, on pourrait instituer une route à péage;

on pourrait utiliser des véhicules plus petits;

ou avoir un plus grand nombre de passagers par voiture;(i) toutes les combinaisons ci-dessus pourraient être essayées.(ii)

Il est ironique de penser que la solution la plus évidente, celle qui consisterait à élargir la route, serait sans doute la moins efficace.

En fait, une connaissance approfondie d'un problème en particulier et des solutions possibles, ainsi que de la série d'effets qu'aura toute action, a plus de chance d'apporter une réponse satisfaisante. Surtout si cela fait partie d'une vision plus ample où d'un plan plus d'ensemble soigneusement étudié.

II. 3.3. *L'idéaliste opportuniste*

Le planificateur doit se garder de l'idéalisme. A moins de reconnaître clairement que l'émotivité de l'homme en fait un être boiteux, à moins de

(i) Hoyt, H. The Effect of the Automobile on Patterns of Urban Growth. *Traffic Quarterly*, Avril, 1963.

(ii) Le nombre total de combinaisans possibles et de 1.024.

comprendre ses motifs cachés, nombreux et variés, d'en tenir compte et de les déjouer aussi efficacement que possible, il est inévitable de devenir cynique. Il se pourrait que tout le monde connaisse la cause d'une blessure, sache qu'il serait aisé de supprimer cette cause et pourtant que nul ne tente rien d'efficace dans ce but. L'inertie de la société, si exaspérante pour un esprit vif et jeune et qui décroule de cette infirmité émotive, doit être considérée comme une maladie. Alors, étant diagnostiquée dans ses diversas formes, elle sera soignée selon les possibilités dont on dispose: depuis l'excision de la partie affectée, jusqu'à une plus grande manifestation d'amour, toute la gamme de la thérapie sociale est applicable. Mais il restera tout un côtê important de cette inertie de la société, de cette inaptitude à l'action. Il faudra manier la compromis intelligemment afin de ne changer le thème principal du projet que le moins possible.

L'urbanisme devra être opportuniste de deux manières. L'opportunisme est un principe naturel de la croissance,—le haricot grimpera le long du tuteur disponible, tout en tâtonnant pour tâcher de trouver mieux,—et non simplement le fait d'emprunter dix shillings à quelqu'un, pour, le lendemain, s'embarquer pour l'Extrême Orient sur un bateau omnibus.

Les premières occasions se présentant à l'urbaniste surgissent des possibilités qui se trouvent à l'état latent dans toute situation, et il doit en tirer le meilleur parti possible. La deuxième façon, pour lui, d'être opportuniste, concerne la réalisation du plan, quel qu'il soit. Faire ce qui est possible, quand, où et avec qui cela est possible, devrait activer le cours l'exécution du plan. Et, à l'exemple du haricot, il faut veiller attentivement à profiter des possibilités, dans toute l'acception du terme: c'est là que réside toute une partie de l'art. Plus le plan d'ensemble sera vaste, souple, hardi, fouillé, plus grand sera le nombre des possibilités utilisées qui correspondent au système dynamique du plan. Le mieux est de considérer un idéal comme étant un but dynamique pour lequel nous travaillons en temps opportun.

II. 3.4. *Fondement d'une critique rationnelle*

Les exemples examinés dans ce livre embrassent un nombre d'idées très variées et il est évident que l'auteur a ses préférences. Cependant il les considère comme étant toutes dignes d'être essayées et évaluées quand elles se font jour.

Toutefois il faut d'abord comprendre les critères sur lesquels se basent les auteurs de chacundes projets: On peut les considérer comme non-valables. Mais quoi qu'il en soit, le projet lui-même doit être jugé d'après les criterès de son auteur et non d'après ceux du critique. On ne peut donner de conseil plus utile sur la leçon à tirer du travail d'autrui. Voilà ce qu'il nous faut absolument savoir: est-ce que nous rejetons les critères de l'auteur du plan, ou bien ceux de la société? ou encore, est-ce que nous reprochons au projet son incapacité à y répondre?

(Translation by Prof. Lewis Thorpe and Mme. Nicole Jolivet.)

XII.6. German Translation

XII.6.1. German Translation of Contents List.

I. Anerkennung

I.1. Inspiration 2

I.2. Persönliche Hilfe 2

I.3. Information und Erlaubnis 2

I.4. Offizielle Anerkennung 2

I.5. Darstellung 2

II. Zusammenhang und Einleitung

II.1. Funktion des Buches 4
- II.1.1. Integrierung von Kenntniss 4
- II.1.2. Mensch und Motor 4
- II.1.3. Vorurteil und Konflikt 4

II.2. Umwelt und Wissenschaftliche Methode 4
- II.2.1. Begrenzung der Wissenschaft in der Gegenwart 4
- II.2.2. Prinzipien und Prüfsteine 5
- II.2.3. Das Schöpferische Experiment 6
- II.2.4. Verantwortung 6

II.3. Macht und Muster der Pläne 6
- II.3.1. Macht der Pläne 6
- II.3.2. Überfluss an Auswahl 7
- II.3.3. Der Opportunistische Idealist 7
- II.3.4. Basis fur Rationale Kritik 7

III. Das Verhältnis Zwischen Mensch und Fahrzeug

III.1. Zusammenfassung der Haupt-Merkmale von Mensch und Fahrzeug 10

III.2. Äusserliche, Merkmale von Mensch und Fahrzeug 11
III.2.1. Massstab 11
III.2.2. Relative und Zusetzliche Grösse 13
III.2.3. Tastgefühlverbindungen 13

III.3. Bewegungsvorgänge 14
III.3.1. Geschwindigkeit und Reichweite 14
III.3.2. Wegbelastung 14
III.3.3. Triebkraft und Gefahr 16
III.3.3.1. Unfälle in Verkehrsgetrenten Gegenden 19
III.3.4. Bewegungskurven 21
III.3.5. Gegenseitige Vereitelung 22

III.4. Ekologische Betrachtungen 23
III.4.1. Gase und Gift 23
III.4.2. Lärm 24
III.4.3. Vibrationen 26

III.5. Soziologische Betrachtungen 27
III.5.1. Einfluss des Privat-Autos auf Soziale Verbindungen 27
III.5.1.1. Positive Folgen 27
III.5.1.2. Schadende Folgen 27
III.5.2. Ein Forschungs-Projekt 27
III.5.2.1. Entwurf des Forschungs-projekts 27
III.5.2.2. Darstellung der Resultate 27
III.5.2.3. Zusammenfassung der ganzen Untersuchung 31
III.5.2.4. Untersuchung in Stevenage 32
III.5.3. Soziale Gesichtspunkte des öffentlichen Verkehrs 32
III.5.4. Moralische und Gesetzliche Betrachtungen 32
III.5.5. Menschen Gedränge und Auto Ansammlungen 33
III.5.6. Sozialpsychologische Faktoren 34

III.6. Stadtplanungslösungen 34
III.6.1. Funktionelle Verkehrsanordnungen 34
III.6.2. Eigenschaften von Strassen und Fusswegen 35
III.6.3. Graphische Analyse der Bewegungsnotwendigkeiten für Fussgänger während Bewegung und Ruhe 36

IV. Bedürfnisse des Menschen als Organismus

IV.1. Biologische Funktion des Gehens 38

IV.2. Notwendige Eigenschaften des Fussgängerbereiches 39
IV.2.1. Räume der Menschlichen Grösse Angepasst 41
IV.2.2. Wichtigkeit der Einzelheiten 43
IV.2.3. Endlose Abwechslung 44

IV.3. Studie eines Städtischen Piazzas für Fussgänger 46
IV.3.1. Konstante Untersuchung des Kommens und Gehens 46
IV.3.2. Visuelle Analyse 47
IV.3.3. Versuch zu Erklären wieso der Piazza gut Funktioniert 50

IV.4. Bequemlichkeit 51

IV.5. Mechanische Hilfsmittel 52
IV.5.1. Fahrräder 52
IV.5.2. Laufende Bänder für Fussgänger 53
IV.5.3. Rolltreppen und Aufzüge 54
IV.5.4. Karren 54
IV.5.5. "Pedestrains" 54
IV.5.6. "Carveyors" 55

IV.6. Auskunft 55

IV.7. Fussgangssystemvergleich 55

V. Notwendigkeiten und Wesen des Fahrzeugs

V.1. Der Mensch als Fahrer 58
V.1.1. Wahrnehmungsvermögen 58
V.1.2. Strasseneinrichtungen 59
V.1.3. Ausbildung 59
V.1.4. Gemützustand 59

V.2. Motorisierung, Vorhersage und Allgemeines 60
V.2.1. Transport Studien und Untersuchungen 62
V.2.2. Fahrzeuge und Städteraum 65
V.2.3. Allgemeine Wirtschaftliche Betrachtungen 67

V.3. Fahrzeugverkehrsbahnen 68
V.3.1. Verwaltung der Strassenplanung 68
V.3.2. Strassenraum, Form, Lärm und Gefahr 68
V.3.3. Leistungsfähikeit 70
V.3.4. Geschwindigkeit 73
V.3.5. Strassenarten 75
V.3.5.1. Allgemeines 75
V.3.5.2. Autobahnen 75
V.3.5.3. Hauptstrassen 79
V.3.5.4. Zufahrtstrassen 79
V.3.5.5. Zusammenfassung 79
V.3.6. Knoten im Allgemeinen 79
V.3.6.1. Fahrstrom an Knotenpunkte 81
V.3.6.2. Knotenpunkte von Drei Strassen 81
V.3.6.3. Knotenpunkte von Vier Strassen 82

V.4. Parken 84
V.4.1. Allgemeine und Grundsätzliche Erfordernisse 84
V.4.2. Parkplätze, auf ebener Erde 84
V.4.3. Mehrstöckige Parkenanlagen 85
V.4.4. Mechanische Parkenanlagen 88

V.5. Funktion des Privat-Autos 88
V.5.1. Berufsverkehr 89
V.5.2. Schulen und Universitäten 89
V.5.3. Einkaufen 89
V.5.4. Besuche 90
V.5.5. Sport und Kulturelle Ereignisse 90
V.5.6. Wochenend und Halbtägige Ferien 90
V.5.7. Lange Ferien 90
V.5.8. Deutung der Statistik 91

V.6. Personenkraftwagenform 91
V.6.1. Faktoren Welche Form Beeinflussen 91

V.7. Handelswagen und Lastwagen 93
V.7.1. Anzahl und Wachstum 93
V.7.2. Eigenschaften der Grösse, Geschwindigkeit und des Verkehrsflusses 93

V.8. Öffentlicher Transport 94
V.8.1. Allgemeine Stellung 94
V.8.2. Prinzipien Öffentlicher Transport Systeme 95
V.8.3. Typen von Schnelltransport 95
V.8.3.1. Autobus 95
V.8.3.2. Zweischienige Bahnen 96
V.8.3.3. Einschienenbahnen 96
V.8.3.4. Untergrundbahnen 98
V.8.3.5. Selbstgelenkte Autobuszüge 99
V.8.4. Vergleich Typischer Züge 100
V.8.5. Beispiele 101
V.8.5.1. Philadelphia 101
V.8.5.2. St. Louis 102
V.8.5.3. Boston 103
V.8.5.4. San Francisco 103

V.9. Glossary of terms 105

VI. Neue Städte mit Verkehrstrennung

VI.1. Grundsätzliche Bedürfnisse **108**

VI.2. Fahrzeugverkehr in neuen Städten, Allgemeines **108**
VI.2.1. Parken in Zentrum von neuen Städten 109
VI.2.2. Cumbernauld, Untersuchung des Zentralen Handels und Parken 110

VI.3. Vollständig Motorisiertes Stadt-Zentrum für 250,000 **111**

VI.4. Stadtentwurf auf Grund von Gehdistanzen **111**
VI.4.1. Siedlungsdichte und Stadtgrösse 111
VI.4.1.1. Stadtgrösse, Einwohnerzahl und Siedlungsdichte 111
VI.4.1.2. Stadt-Siedlungsdichte-diagramme 112
VI.4.2. Vergleich von Städten und Ihrer Gehdistanzen 113
VI.4.3. Stadtform 114

VI.5. Flächennützung **114**

VI.6. Grundflächen **114**

VI.7. Die Nachbarschaftsidee **115**

VI.8. Wachstumsprinzipien **115**

VI. 9. Beispiele **117**
VI.9.1. Cumbernauld, Schottland 117
VI.9.2. Hook, England 121
VI.9.3. Erith, England 123
VI.9.4. Toulouse-le-Mirail, Frankreich 126
VI.9.5. Vällingby, Schweden 128
VI.9.6. Farsta, Schweden 130
VI.9.7. Köln, Deutschland 133
VI.9.8. Sabende, West Guinea 134
VI.9.9. Sennestadt, Deutschland 136
VI.9.10. Neu-Winsen, Deutschland 138
VI.9.11. Alkmaar, Holland 139
VI.9.12. Chandigarh, Indien 140
VI.9.13. Kitimat, Kanada 143
VI.9.14. Tapiola, Finnland 144
VI.9.15. Sputnik, Russland 146
VI.9.16. Cambridge Village, Cambs., England 147
VI.9.17. Reston, Vereinigte Staaten 148

VII. Städtischer Wiederaufbau

VII.1. Allgemeines Prinziep **152**
VII.1.1. Der Fall New York 153
VII.1.2. Der Fall San Francisco 154
VII.1.3. Der Fall Detroit 154

VII.2. Entwurf für Städtischen Wiederaufbau **155**
VII.2.1. Grundsätzliche Prinzipien 155
VII.2.2. Horizontale Verkehrstrennung 157
VII.2.2.1. Verkehrstrennung durch Zeiteinteilung 158
VII.2.3. Vertikale Verkehrstrennung, Fussgänger oben 158
VII.2.4. Vertikale Verkehrstrennung, Fahrzeuge über dem Erdboden 159
VII.2.5. Vertikale Verkehrstrennung, Fahrzeuge am Erdboden, Fussgänger unterirdisch 159

VII.3. Stadtstrassen **159**
VII.3.1. Verbesserung vorhandener Strassensysteme 159
VII.3.2. Einführung von Autobahnen 160
VII.3.3. Autobahnen im Stadtbild 161

VII.4. Parken **162**
VII.4.1. Parkenvoranschlag 162
VII.4.2. Andere Faktoren die Parken beeinflussen 163
VII.4.3. Bordschwellenparken 164
VII.4.4. Parken Automaten 164

VII.5. Allgemeine Wirtschaftliche Betrachtungen **165**
VII.5.1. Kosten volständiger Motorisierung 165
VII.5.2. Vergleichende Kosten im Pendelverkehr 166
VII.5.3. Kosten Von Verkehrsverbesserungen 166
VII.5.4. Kosten der Strassenverstopfung 167
VII.5.5. Wirtschftslehre der Verkehrsfreien Einkaufstrassen 167
VII.5.6. Wertevergleich 168

VII.6. Beispiele **170**
VII.6.1. Coventry, Verkehrs und Entwicklungsplan 170
VII.6.1.1. Coventry Einkaufsbezirk 172
VII.6.2. Stockholm, Verkehrs und Entwicklungsplan 173
VII.6.2.2. Stockholm, Sergelgaten 174
VII.6.3. Philadelphia, Verkehrs und Entwicklungsplan 176
VII.6.3.1. Market East Section 177
VII.6.4. Liverpool, Verkehrsplan 180

VII.6.4.1. Ravenseftgegend 182
VII.6.4.2. Stadtkern und Sozialer Mittel punkt 184
VII.6.4.3. Paradise Street—Strand Gegend 184
VII.6.5. Newcastle, Entwicklungs und Verkehrsplan 186
VII.6.5.1. Stadtzentrum, Gesamptplan 187
VII.6.6. City of Victoria, Hong-Kong, Stadtzentrum 188
VII.6.7. London, Verkehrsplan 189
VII.6.7.1. Trafalgar Square Gegend 191
VII.6.8. Fort Worth, Texas, Stadtzentrum Wiederaufbau 192
VII.6.9. Nottingham, Stadtzentrum und Verkehrsplan 194
VII.6.10. Banbury, Stadtzentrum, Wiederaufbau 197
VII.6.11. Swindon, Stadtzentrum Wiederaufbau 198
VII.6.12. Västerås, Schweden, Entwicklungs und Verkehrsplan 199
VII.6.13. Bishop's Stortford, Stadtzentrum Wiederaufbau 201
VII.6.14. Sutton Coldfield, Stadtzentrum Wiederaufbau 202
VII.6.15. High Wycombe, Stadtzentrum Wiederaufbau 205
VII.6.16. Beeston, Stadtzentrum, Wiederaufbau 206
VII.6.17. Horsholm, Dänenmark 208
VII.6.18. Chester City, Stadtzentrum Wiederaufbau 208
VII.6.19. Andover, Wiederaufbau der Stadt 210
VII.6.20. Lijnbaan, Rotterdam, Holland 212
VII.6.21. South Bank, Wiederaufbau 213
VII.6.22. Stevenage, Stadtzentrum 214
VII.6.23. Stevenage, Shephall Zentrum 215
VII.6.24. Treppenstrasse, Kassel, Deutschland 215
VII.6.25. Einkaufstrassen für Fussgänger 215
VII.6.26. Greyfriars, Ipswich, Wiederaufbau 216
VII.6.27. See-front, Brighton, England 216
VII.6.28. Strøget, Einkaufstrassen, Copenhagen, Dänenmark 216
VII.6.29. Ellor Street Stadt-Umbau, Salford 218
VII.6.30. Hammersmith Zentrum, London 219
VII.6.31. Piccadilly Circus, London 220
VII.6.32. Knightsbridge Green, London 220
VII.6.33. Basildon, Neue-Stadt, Hauptzentrum 221
VII.6.34. Elizabeth, Australien, Hauptzentrum 222

VIII. Verkehrstrennung in Wohnvierteln

VIII.1. Lebensmassstab an Radburn Prinzipien angewandt 224
VIII.1.1. Psychosomatische Prüfsteine: Radburn Möglichkeiten 224
VIII.1.2. Psychosoziologische Prüfsteine 225
VIII.1.3. Ekologische Prüfsteine 227
VIII.1.4. Wirtschaftliche und Administrative Prüfsteine 227

VIII.2. Kosten 228
VIII.2.1. Vergleich der Strassen und Kanalisationskosten 228
VIII.2.2. Wirtschaftliche Analyse 229

VIII.3. Wohnpläne 232
VIII.3.1. Analyse der Beziehungen von Wohnplänen und Zugang 232
VIII.3.2. Automobil-Einlagerung 235

VIII.4. Fusswegsysteme 236
VIII.4.1. Analyse 236

VIII.5. Bauplatzformen 238
VIII.5.1. Häuser 238
VIII.5.2. Wohnungen und "Deckhousing" 239

VIII.6. Zufahrtstrassen in Wohnvierteln 240
VIII.6.1. Grösse und Anzahl von Knoten 240
VIII.6.2. Zufahrstrassenentwurf, Allgemeines 240
VIII.6.2.1 Sackgassen 240
VIII.6.2.2. Schlingen 240
VIII.6.3. Grundlegende Alternativen 241
VIII.6.3.1. Sackgassen 241
VIII.6.3.2. Schlingen 242
VIII.6.4. Beispiele 243
VIII.6.4.1. Pionier Entwürfe im Beginn der Fünfzigerjahre 243
VIII.6.4.2. Entwürfe dere mittleren Fünfzigerjahre 244
VIII.6.4.3. Entwürfe der letzten Fünfzigerjahre und des Beginns der Sechziger-jahre 246
VIII.6.4.4. Ausmasse von Sackgassen 248
VIII.6.4.5. Verbindungen zwischen Sackgassen und Fusswegen 249
VIII.6.5. Vermischung von Fussgängern und Fahrzeugen in Sackgassen 250

VIII.7. "Superblock" Entwurf 252
VIII.7.1. Grundliegende Faktoren 252
VIII.7.2. Anwendung im Wiederaufbau 253

VIII.8. Unterführungen und Brücken **253**
VIII.8.1. Lage und Wichtigkeit 253
VIII.8.2. Entwurf und Kostenauswahl 255

VIII.9. Grünflächen **256**

VIII.10. Höhenlinien **256**

VIII.11. Strassennamen und Nummern **257**

VIII.12. Beispiele von verkehrsgetrennten Wohnvierteln **258**
VIII.12.1. Carbrain, Cumbernauld, nr. Glasgow 258
VIII.12.2. Baldwin Hills Village, Los Angeles 260
VIII.12.3. Caversham, Berks. 262
VIII.12.4. Bron-y-Mor, Wales 263
VIII.12.5. Huntingdon, Hunts. 265
VIII.12.6. Jackson Estate, Letchworth 266
VIII.12.7. Ilkeston Road, Notts. 266
VIII.12.8. Albertslund, Kopenhagen 267
VIII.12.9. Eastwick, Philadelphia 268
VIII.12.10. Clements Area, Haverhill, Suffolk 269
VIII.12.11. Cité Jardin, Montreal 271
VIII.12.12. Marly-le-Roi, Paris 273
VIII.12.13. Baronbackarna, Orebro, Schweden 274
VIII.12.14. Lafayette Park Title, I, Detroit 276
VIII.12.15. Biskopsgaden, Göteborg 277
VIII.12.16. Willenhall Wood I, Coventry 278
VIII.12.17. Kildrum, Cumbernauld 280
VIII.12.18. Elm Green I and II, Stevenage 281
VIII.12.19. Bedmont, Hertfordshire 282
VIII.12.20. Inchview, Prestonpans, Edinburgh 283
VIII.12.21. Parkleys, Ham Common, London 285
VIII.12.22. Park Hill, Sheffield 286
VIII.12.23. Barbican, London 288
VIII.12.24. Flemingdon Park, East York, Toronto 289
VIII.12.25. South Carbrain, Cumbernauld 290
VIII.12.26. Oakdale Manor, North York, Toronto 294
VIII.12.27. Kentucky Road, Toronto 295
VIII.12.28. Solna, Schweden 296
VIII.12.29. Spon End, Coventry 297
VIII.12.30. Ehepaar Quartiere, British War Office (A) 297
VIII.12.31. Ehepaar Quartiere, British War Office (B) 298
VIII.12.32. Rowlatts Hill, Leicester 298
VIII.12.33. Brandon Estate, London 299
VIII.12.34. Royal Victoria Yard, Deptford, London 299
VIII.12.35. Brunswick Wiederaufbaugegend, Manchester 300
VIII.12.36. Primrose Hill, Birmingham 300
VIII.12.37. Almhog, Nydala, Hermodsdal, Malmö 300
VIII.12.38. Mellanhaden, Malmö 300
VIII.12.39. Washington, Durham 301
VIII.12.40. Fontenay-aux-Roses, Paris 304
VIII.12.41. Hillfields, Coventry 304
VIII.12.42. Highfields, Leicester 304a

IX. Wiederaufbau und Verbesserung von Wohnvierteln

IX.1. Allgemeine Prinzipien **306**
IX.1.1. Untersuchung von Unfällen in Rasterplanung 306

IX.2. Beispiele **308**
IX.2.1. Strassenvebesserung, Pershing Field gegend, Fulton Nachbarschaft Minneapolis, U.S.A. 308
IX.2.2. Mill Creek Wiederaufbaugegend, Philadelphia City Planning Commission, 1954 308
IX.2.3. Brittisches Lokal-Gesetz Wohnviertel, Nottingham 309
IX.2.4. Ausräumung hinter Wohnungsbauten 310
IX.2.5. Verbesserung von Wohnvierteln Zwischen den Weltkriegen Erbaut, Nottingham 310
IX.2.6. Grossstadt Umwandlung, Chicago 312

X. Geschichte der Verkehrstrennung

X.1. Zusammenfassung **314**

X.2. Illustrierte Geschichte der Verkehrstrennung **314**
X.2.1. Beispiele vor Benzin Motoren 314
X.2.2. Planung für "Das Leben trotz des Automobiles" 319
X.2.3. Planung der Verkehrstrennung in Europa 322

X.3. Tabelle von Daten in der Geschichte der Planung für Mensch und Motor **328**

XI. Literaturverzeichnis

XI.1. Bücher 331

XI.2. Berichte, Artikel, Dissertationen 333

XII. Anhänge

XII.1. Umwandlungs Tabellen 347

XII.2. Abkürzungen 347

XII.3. Schema einer Verkehrs Untersuchung 348

XII.4. Sachregister 350

XII.5. Französische Übersetzung 357
XII.5.1. Inhaltsverzeichnis 357
XII.5.2. Zusammenhang und Einleitung 366

XII.6. Deutsche Übersetzung 371
XII.6.1. Inhaltsverzeichnis 371
XII.6.2. Zusammenhang und Einleitung 380

XII.6.2. German Translation of Section II

II. Zusammenhang und Einleitung

II. 1. Funktion des Buches. 380

II. 1.1. Integrierung von Kenntniss. 380
II. 1.2. Mensch und Motor. 380
II. 1.3. Vorurteil und Konflikt. 380

II. 2. Umwelt und Wissenschaftliche Methode. 380

II. 2.1. Begrenzung der Wissenschaft in der Gegenwart. 380
II. 2.2. Prinzipien und Prüfsteine. 381
II. 2.3. Das Schöpferische Experiment. 382
II. 2.4. Verantwortung. 382

II. 3. Macht und Muster der Pläne. 383

II. 3.1. Macht der Pläne. 383
II. 3.2. Überfluss an Auswahl. 383
II. 3.3. Der Opportunistische Idealist. 383
II. 3.4. Basis für Rationale Kritik. 384

II. Zusammenhang und Einleitung

II. 1. Funktion des Buches

II. 1.1. *Kenntnissintegrierung*

Architekten, Städtebauer, Ingenieure, Spekulatoren, Verwalter und Gesetzgeber, sie alle entwickeln Pläne. Allgemein gültige, funktionelle Prinzipien werden in der Erziehung oder in Textbüchern nicht beschrieben, betont, oder empfhohlen, nicht einmal definiert. Es gibt keine ordentliche Basis für Entscheidungen oder Urteile. Ein Anfangspunkt und ein Zusammenhang der interprofessionalen Aktivitäten existieren nicht. Es ist kein Wunder, dass in der Praxis wenig Zusammenarbeit zu finden ist.

Die Absicht dieses Buches ist nicht nur Kenntnisse zu vermitteln, sondern auch dieselben und die Methoden der verschiedenen Professionen zu vereiningen und die gemeinsamen Elemente zu illustrieren. Nimmt man als treibende Kraft für alle den Aufbau einer guten Umwelt, dann wird der Student erkennen, was sich direkt auf ihn bezieht und wie, im Zusammenhang mit dem Buch als Ganzes, dies ein Teil des Bildes ist.

II. 1.2. *Mensch und Motor*

Das Automobil ist das vielseitigste und nützlichste Fahrzeug das von Menschen erfunden wurde. Im gegenwärtigen, primitiven Zustande erzeugen Motor und Brennstoffe noch giftige Gase, aber besser entwickelte Typen werden sicher folgen. Nur eine positive Einstellung und Planung für die günstige Verwendung von Fahrzeugen ist rationell.

Es ist wahrscheinlich, dass keine Entwicklung, ausgenommen die des "Hovercraft", das Fahrzeug mit Rädern ausser Gebrauch setzen wird. In diesem Fall würden Oberflächen unwichtig werden, und die Raumbedürfnisse und Bahnen wären dem des Automobiles ähnlich, da es mit dem Erdboden durch eine Luftschicht verbunden ist. Flugzeuge werden unwichtig bleiben, bis elektronische Hilfsmittel erfundenwerden, die neue, aerodynamische Apparate, entlang individueller Wege in verschiedenen Höhen lenken können, in einer Weise, die mit dem Hubschrauber nicht möglich ist.

II. 1.3. *Vorurteil und Konflikt*

Ohne ein geordnetes wissenschaftliches Bereich, ohne Nachweissystem, wird das Bild weiterhin durch Gefühle verwirrt, die durch unmittelbaren und unablässigen Einfluss auf das Leben in Gebieten erregt werden, die nicht für solchen Verkehr geplant wurden, jedoch von Menschen und Fahrzeugen wimmeln.

Der Unwille des Fussgängers und Radfahrers, die keine frische Luft atmen können und von der Strasse verdrängt werden, ist ganz logisch. Der Verdruss des Chauffeurs ist auch gerechtfertigt, obwhol er vielleicht der vorher Erwähnte ist, jedoch jetzt am Lenkrad sitzt. In diesem Fall ist er wütend über die Überfüllung, weil er von der grossen Geschwindigkeit seines Fahrzeuges Gebrauch machen will; und später wieder, wenn er am Ziel ankommt und weiterfahren muss, weil kein Parkplatz und keine Parkenerlaubnis vorhanden sind.

Man kann auch kein klares Bild schaffen, indem man einer Partei mit einem Stadtbild von Los Angeles droht und einer anderen Partei mit einer Vorhersage von langen, ermüdenden Gehdistanzen. Angst und Hass fördern keine Vernunft. Nur die Anerkennung dessen, was unserer heutigen Lage zu Grunde liegt und woraus unser Dilemma entspringt und das Verständnis für die Bedürfnisse und Funktionen des Menschen und seines hervorragenden Hilfsmittels, des Automobils, wird zu den kühnen, ausgeglichenen Lösungen führen, die erforderlich sind. Diejenigen, die nicht begreifen können, dass ein wachsendes und wirklich beängstigendes Schauspiel existiert oder die jenigen die sich an Gemetzel und Mittelmässigkeit im Menschenleben gewöhnt haben, oder angenommene Lösungen als nachzuahmende Mode betrachten, solche Leute müssen aus ihrer Gleichgültikeit aufgerüttelt werden.

II. 2. Umwelt und Wissenschaftliche Methode

II. 2.1. *Begrenzung der Gegenwärtigen Wissenschaft*

Was ist eine günstige Umwelt? Welches sind die grundsätzlichen Werte, welche uns bei der Städteplanung leiten sollten? Wir wissen es nicht, da die Werkzeuge zur Entdeckung noch nicht entwickelt sind. Es ist ein weitverbreiteter Irrtum, dass die Soziologie eine Antwort auf diese Fragen finden sollte. Das ist ein gefährlicher Fehler. Man würde einer Wissenschaft vertrauen, die speziell Werturteile und selbst Ratschläge ausschliesst. Auch wenn die nützlichere Annäherung der Anthropologie mit inbegriffen ist, wie bei dem "Institute of Community Studies," ist es noch immer ein Studium des Existierenden und wir wissen schon durch Erfahrung, dass dieses Existierende fehlerhaft erdacht wurde. Originelle Ideen sind ausgeschlossen. Soziology ist im heutigen Zustand nicht dazu bestimmt, Architekten Informationen zu geben, um sie bei den qualitativen, subtilen Problemen zu leiten, für welche Auskunft und Inspiration dringend erforderlich sind.

Der Stand der Wissenschaft wird häufig innerhalb soziologischer Kreise kritisiert. Das Buch *Tools of Social Science*, von John Madge sagt: "... in keiner

Wissenschaft ist die Verfolgung objektiver Kenntnisse nutzloser, als in den Sozialwissenschaften". Später schlägt er vor: "... Die soziale Untersuchung ist eine Bewegung, sowie auch eine Tatsachensuchende Forschungsreise." Doch er weiss, dass seine Ansichten in Britisch-akademischer Soziologie keinen Wiederhall finden. Solche Arbeit hat "... einen ziemlich erniedrigten Stand", und hat "... veraltete Technik" benützt.

Der Verkehrsingenieur ist in einer ähnlichen Verlegenheit. Ohne einen verlässlichen oder geprüften Arbeits-Umriss der Kenntnisse wird er aufgefordert, Entscheidungen zu fällen, welche eine ungeheure Zahl von Variationen einschliessen, deren Folgen gagenwärtig gänzlich unberechenbar sind, häufig ausserhalb seiner Kontrolle, liegen und die Kosten übersteigen, die sich seine Gemeinde erlauben kann. Rationeller Transport aller verlangten Artikel in einem gegebenen Distrikt ist eine komplizierte Aufgabe, die viel originelles Denken erfordert. Wir müssen daher schliessen, dass Wissenschaft und Technologie mit ihren heutigen, wirkungslosen Methoden dem Studium des Menschen und seiner Umwelt nicht in angemessener Weise helfen können.

II. 2.2. *Prüfsteine*

Aertzte wissen, dass gutes Essen, frische Luft and körperliche Übung biologisch wünschenswert sind. In der Soziologie haben wir keine Vervollständigung dieses Grundsatzes, die uns in der Frage was gesund ist helfen kann.

Der Planende, der mit gesundem Menschenverstand beginnt und sich der vorhandenen Kentnisse bedient, die von akademischer Sterilität nicht verkümmert wurden, muss sich sein eigenes Bild formen und gleichzeitig als Philosoph, Prophet und Sozialtherapeut handeln.

Die Notwendikeit für Gelegenheiten zur Zussammenarbeit, für alle Alterstufen, in Arbeit, Spiel oder Feier, Freundschaft oder Liebe, kann nur von allgemeiner Erkenntnis herrühren und aus dem Inhalt der Werke von Menschen wie Professor H. Infield stammen.

Das Bedürfniss für ein Privatleben wird zum Überdruss betont. Es ist sicher ein grundsätzliches Bedürfniss. Geschickte Lösungen sind möglich, aber eigentlich sind nur eine dicke Wand und Vorhänge nötig. Oft ist die Frage, sich innerhalb des Hauses von seiner Familie zurückziehen zu können schwieriger zu lösen und häufig ist es der beste Ausweg, weg zugehen. Dies führt zur Versorgung von anderen sozialen Gruppen als die Familie.

Für Zusammenarbeit zu planen ist viel wichtiger. Es ist komplizierter die Möglichkeiten aber sind enorm und die Sache ist kaum in Angriff genommen worden. Dies vielleicht mehr als irgend etwas anderes wird die nächste Entwicklung des Menschen vom soziologischen Standpunkt beeinflussen, die Organisation in Gruppen, die Teilnahme der Bürger und die Arbeitsdemokratie.

Chermayeff zeigt in einem durchdringenden Vortrag was für eine Rolle Bewegung und Beweglichkeit in seiner Vision der Zukunft spielen:

"All dies setzt eine sehr vorsichtige Prüfung des ganzen Spektrums von Erfahrungen voraus, was Beweglichkeit, Geschwindikeit, und Dimensionen bewohnten Raumes betrifft; nur dann werden wir, an einem Ende des Spektrums, die Fussgängerwelt zurückgewinnen können; und, mehr als die Welt des Fussgängers, die Welt des ruhenden Menschen im vollsten Sinne des Wortes. Wenn wir in der Nähe eines Krokus im Grass liegen, haben wir die Aussicht eines Wurms auf die Welt, und es ist eine sehr schöne Welt. Steigen wir in einem Düsenflugzeug auf, dann wollen wir niemanden die Freude an diesem erstaunenden Wunder nehmen, den ganz neuen Blick auf das Universum und die Schönheit der Landschaft. Die ganze Region vom Krokus bis zum Wolkenblick möchte ich haben, und ich möchte lieber alle diese Freuden geniessen, als zwei Wagen in jeder Garage stehenhaben. ..."

"... Ich bin volkommen überzeugt, dess es nicht 'die Stadt' ist, die dauernd bei uns bleibt sondern Ansammlungen von Menschen, die aus guten Gründen zusammengekommen sind. Ich hoffe, dass wir diese Art von Zusammenkommen fur alte und neue Zwecke kultivieren werden aber nicht das greuliche Zusammensein, wo in Wirklichkeit jeder den andern zu hassen anfängt. Wir treffen einander nicht länger nur weil wir Nachbarn sind, wir kommen zusammen, weil wir Interessen teilen oder eine gemeinsame Arbeit ausführen wollen. Ich bin überzeugt, dass wir unsere Gruppenordnung wegen des Zwangs der Bevölkerungszunahme, der Entwicklung der Technik und der unbegrenzten Verwicklung unserer Systeme ändern werden. Und ich möchte kein Transportsystem auf Kosten anderer Werte billigen."

Das Zukunftsbild, das hervorgeht, schliesst das ausgewählte, praktische und rationelle Benutzen von Privatwagen ein, ebenso grössere Mitarbeit der Bürger im Entwurf wie auch im Ausnutz seiner Umwelt, die für eine Erweiterung der sozialen Gruppen von heute geplant ist. Die Gesellschaft zeigt eine wachsende Beweglichkeit, welche Einrichtungen verlangt, die ein Zugehörigkeitsgefühl fordern und die Einwurzelung beschleunigen. Doch veilleicht ist das Werk beinahe vollendet: Reston, die erste ausgeglichene Neue Stadt in den Vereinigten Staaten wurde in 1963 mit der Überzeugung entworfen, dass "Amerikaner die Stabilität einer lebenslänglichen Zugehörigkeit zu einer Gemeinschaft wünschen." Sie sind "der Heimatlosigkeit müde." Die grosse Beweglichkeit, welche die Bewohner des Westens zeigen, kann ein Zeichen ihres Wunsches sein, schnell materielle Vorteile zu erringen. Dies ist kein Endzweck. Sesshaft zu bleiben und reich zu werden, könnte im Zeitalter des Überflusses bei vorausgesetzten Reise und Verbindungsmöglichkeiten der wirkliche Luxus sein.

II. 2.3. *Das Schöpferische Experiment*

Es braucht den ganzen Mann, den fühlenden Philosophen, nicht nur den objektiven Wissenschaftler, in jedem von uns, um die funktionellen, ethischen Begriffe, Vorstellungen und Bestrebungen gründlich in Betracht zu ziehen. Der Architekt muss ein Mensch sein, der die Vorstellungskraft besitzt, die Vision eines andern gutzuheissen und sodann an der Verwirklichung zu arbeiten.

Wie Madge gesagt hat, die Wissenschaftliche Methode setzt sich selbst unnötige Grenzen mit der Vereinzelung von Faktoren und deren Analyse. Das subjektive Element des Forschers auszuschalten, scheint das allgemeine Ziel zu sein. Was bei der Planung besonders gebraucht wird, ist eine wirkungsvolle, jedoch gegensätzliche Annäherung, mit der jedoch die Wissenschaftliche Methode mit Hypothese und Experiment gültig verbleibt. Gewöhnlich wird die Hypothese zahlenmässig geprüft, aber dies hat keine Gültikeit für Städtebau, wo originelle Vorteile als Resultat eines Experimentes erscheinen können.

Betrachten wir die Aussage "Ich kann einen Stein über den Fluss werfen" als Hypothese. Das Experiment schliesst dann die notwendige Intelligenz ein, den engsten Platz zu finden, den Wind zu benützen, den richtigen Stein zu finden usw, usw. Mit anderen Worten, die vielen Faktoren stehen nicht einzeln. So viele, wie zu einer gegebenen Zeit zur Richtigkeit der Hypothese und zum Erfolg des Experimentes führen können den Stein hinüberzubringen, werden eifrig gesucht.

Dies ist der Grundbegriff des schaffenden Experiments, welches ich als Mittel zur wirksamen, fruchtbarer Einführung neuer Ideen formuliert habe. Die Resultate eines solchen Experimentes werden wesentlich und von Bedeutung sein. Sein Vorgang verbindet Teilnehmer in dynamisher Weise, indem sie die ganze Zeit Methoden und ihre Umgebung verbessern und entwickeln. Die Beziehung und gegenseitige Beeinflussung aller Faktoren sind von grosser Wichtigkeit; sie werden beachtet und geändert wo dies helfen kann. Die angenommene Theory, dass Leute nicht wissen sollten, dass sie an einem Experiment teilnehmen, ist falsch und gefährlich und hat keinen Platz im Schaffenden Experiment, da gerade das Bewustsein, ein Teil des Experimentes zu bilden eine starke Hilfe zum Erfolg bedeutet.(i) Erfolg in solchen Experimenten schafft eine gute Möglichkeit für ein Leben für alle Altersgruppen und allmählich weniger Elend und Vereitelung von Plänen und Wünschen.

Man kann darüber nicht objektive sein. Der planende Mensch muss einen starken Wunsch haben und mit allen Mitteln danach streben, genau so, wie er den Stein über den Fluss werfen wollte.

Diejenigen die planen, sollten Bedürfnisse, Mittel und Rohstoffe analysieren, gemäss der Analyse mit Einbildungskraft schaffen, die Ausführung des Entwurfs passend organisieren und Information zur richtigen Benützung anzuordnen. Wenn man zu dem Schluss gekommen ist, welches die grundsätzlichen Prinzipien sind, die ein Projekt beherrschen sollen, verschiebt sich die Wichtigkeit auf die besten Wege diese Prinzipien auszuführen. Die Prinzipien sind die Hypothese; das Schaffende Experiment liegt in der erfolgreicnen Verwirklichung.

Es ist wichtig, dass so viele Entwürfe wie möglich solchen Experimenten unterworfen und die Resultate sorgfältig Gesammelt werden sollten um das kleine Volumen nützlicher Kenntnisse zu vergrössern.

II. 2.4. *Verantwortung*

Dies ist das Jahrhundert des einfachen Menschen. Der Luxus der Stadtplanung ist für alle. Kraftfahrzeuge sollten jeder Familie dienen. Das ist alles eine Herausforderung von riesiegen Proportionen. Es ist nicht nur eine Sache des Raumbedarfs, des schönen Anblicks oder des geregelten Verkehrsstromes. Es ist die Aufgabe, eine Umwelt zu schaffen, die der Befriedigung primärer Bedürfnisse dient. Die positiven Eigenschaften der erstaunenden Beweglichkeit eines wirksamen Fahrzeugverkehrs, und die Planung von Gemeinschaften, die den Vorteilen der Zusammenarbeit Rechnung tragen, ergeben eine grössere Auswahl für jedermann, was seine Kapazität and Möglichkeiten erweitert. Es ist vernünftig, für unsere Umgebung grössere Ausgaben pro Person zu erwarten als bisher. Die Verringerung der Steuern als einziges Ziel zu erstreben, ist heutzutage ebenso weit verbreitet wie kurzsichtig.

Diese Angelegenheit bringt schwere Verantwortungen mit sich. In den Vereinigten Staaten haben eine Anzahl Faktoren Lösungen ergeben, die als

(i) Frau Demer's Bericht von Park Hill, Sheffield, (1962) und die Arbeit des Verfassers zeigen, dass die Angst den Mieter durch seinen Beitrag an der Untersuchung verärgert zu haben, gundlos waren. Leute nehmen mit Vergnügen an ainem wichtigen Experiment teil.

Ganzes gesehen zu Europa keine Beziehungen haben, und von diesem Staat, so wie weniger entwickelten Ländern, nicht annehmbar sind. Wir müssen vom neuem die Funktionen und Bedürfnisse des Menschen und des Motors in jeder Umgebung abschätzen. "Minder entwickelte Länder" (und Neue Städte!) "können eine Verkehrsordnung ohne Pionier Schwierigkeiten erzielen" (Chermayeff). Ein Land nach dem andern wird reicher, dichter bevölkert und mit Kraftwagen überfüllt. Die Bevölkerung der Erde vermehrt sich täglich um mehr als 100,000, jeden Tag entsteht eine neue Stadt, und dennoch, ist Forschung in allen Fragen, die unsere Umwelt betreffen, ungeheuer klein und wirkungslos.

II. 3. Macht und Muster der Pläne

II. 3.1. *Die Macht der Pläne*

Ein Plan ist ein mächtiges Instrument. Bequemlichkeit kann die Notwendigkeit und den Gebrauch von Privatautos im Stadt-Zentrum reduzieren, geschützte, einladende Einrichtungen können Fussgänger und Radfahrer anziehen, oder sie bei ungünstigen Zuständen ausschliessen. Ein Plan kann einen ganzen Stadtteil zerstören, wenn er für Privatwagen nicht zugänglich gemacht wird, ohne für öffentlichen Transport zu sorgen. Die Möglichkeiten im Städtebau sind ausserordentlich gross, unsere Mängel ungeheuer. Eine enorme Anzahl der Übel wird fortgesetzt. Wir opfern Freude, wir opfern Leib und Leben. "Die Missethaten der (Stadt)-Väter werden, bis ins dritte und vierte Glied an den Kindern heimgesucht".

II. 3.2. *Überfluss und Alternative*

Die Arbeit des Entwerfers wird durch die vielfältige Beziehung zwischen vielen Sphären und deren Wirkungen kompliziert. In der Gegenwart ist dies besonders ernst, da diese Beziehumgen weder annerkannt, noch in Rechnungen, Vorraussagen und Untersuchungen, eingeschlossen werden. Eisenbahnen und Strassen, wirtschaftliche Entwicklung und Bevölkerungsentlastung der Gross-Städte, Autoerzeugung und Strassenbau, und eine Reihe eng verbundener Faktoren werden seperat behandelt, woraus sich unheilvolle, unsinnige Resultate ergeben.

Die Beziehung der Umstände aufeinander erlaubt es, Probleme von verschiedenen Gesichtspunkten aus angreifen zu können. Nehmen wir zum Beispiel eine sehr überfullte Strasse:

Sie könnte verbreitert werden;

eine Autobahn könnte gebaut werden;

Hindernisse könnten von der Strasse entfernt werden; zum Beispiel Fussgänger, Roboten und geparkte Wagen;

Fusswege und öffentlicher Transport könnten anziehender gestaltet werden;

man könnte die Lage der Verkehrsziele ändern;

die Arbeitstunden, welche die Stauung des Verkehrs verursachen, könnten verlegt werden;

Parkenkosten am Ziel könnten erhöht werden, um dadurch Leute vom Fahren abzuhalten;

für das Benützen der Strasse könnte gezahlt werden, um ähnliche Resultate zu erziehlen;

kleinere Fahrzeuge könnten benützt werden;

meherer Leute könnten in jedem Wagen fahren;

jede Kombination der obigen Ideen könnte versucht werden.

(Die Zahl aller möglichen Kombinationen ist 1024.)

Die augenscheinlichste, erste Lösung, die Strasse breiter zu machen, ist ironischerweise wahrscheinlich die Wirkungsloseste.

Es ist klar, dass die gründliche Kenntniss eines bestimmten Problems, der möglichen Lösungen, und des "chain-effect" jeder Handlung, eine wirkungsvolle Lösung wahrscheinlicher machen. Besonders wenn es der Teil einer grösseren Vision oder eines sorgfältig erdachten Gesamtplanes ist.

II. 3.3. *Der Opportunistische Idealist*

Beim Städtebau muss man sich vor dem Idealismus hüten. Man muss die chronische Stauung der menschlichen Gefühle (emotional limp) klar erkennen die vielen versteckten Gründe gut verstehen, diesen Rechnung tragen und streben diesen, in jeder möglichen Weise entgegenzuhandeln, sonst ist Zynismus unvermeidlich. Es gibt Dinge, die, wie wir wissen, für Unfälle verantwortlich und leicht zu entfernen sind, doch handelt niemand in einer wirksamen Weise dagegegn. Die Trägheit der Gesellschaft, welche alle diejenigen, die geistig jung und beweglich sind, aufreizt, und welche von der Stauung der Gefühle herrührt, muss als eine Krankheit betrachtet, in ihren verschiedenen Formen erkannt und mit allen vorhandelnden Mitteln kuriert werden: vom Herausschneiden des kranken Teiles durch Chirurgie bis zur Behandlung mit Liebe, der ganze Umfang sozialer Therapien ist wesentlich. Aber ein beträchtlicher Teil der Trägheit, des "Stillsitzens", wird weiter gehn. Kompromisse sollten intelligent durchgeführt werden, so dass der Kern einer Vision so wenig wie möglich geändert wird.

Opportunismus ist ein Naturgesetz des Wachstums: Die Bohne klettert am vorhandenen Stock hinauf und sucht zur selben Zeit rings herum, um etwas besseres zu finden. Es ist nicht nur das Vorhaben zehn Mark von jemandem zu leihen, der morgen in einem langsamen Dampfer in den Fernen Osten fährt.

Der Städtebau sollte in zwei Weisen opportunistisch sein.

Die erste Gelegenheit zum Opportunismus im Städtebau ergibt sich aus den latenten Möglichkeiten jeder Situation, die voll ausgenütz werden sollte. Die zweite Gelegenheit betrifft die Verwirklichung irgendeines Entwurfes. Dasjenige auszuführen, was möglich ist, wenn es möglich ist, wie und mit wem es möglich ist, sollte den Fortschritt jedes Planes beschleunigen. Wie bei der Bohne ist eine scharfe Ausschau nach allen Möglichkeiten jeder Art ein Teil der Kunst. Je biegsamer, umfassender, und kühner der Gesamtplan ist, desto mehr werden verwirklichte Möglichkeiten, nach dem dynamichen Muster des Entwurfes, sinn und plangemäss in Erscheinung treten. Ideale werden am besten als dynamische Ziele angesehen, welchen wir mit Opportunismus zustreben.

II. 3.4. *Basis für Rationale Kritik*

Die Beispiele die in diesem Buch illustriert sind, schliessen eine grosse Variation von Ideen ein und der Verfasser bevorzugt offenbar die einen vor den andern. Jede Idee ist jedoch wert befunden worden, versucht und geprüft zu werden. Doch zuerst müssen wir das Kriterium des Entwerfers verstehen. Wir können es als wertlos ansehen; aber ob es das ist oder nicht der Entwurf selbst muss nach der Erfüllung der gestelleten Aufgabe und keinesfalls nach den Begriffen des Kritikers beurteilt werden. Man kann für das Lernen vom Werk anderer kaum einen wichtigeren Rat erteilen. Es ist entscheidend festzustellen, ob wir die Prinzipien des Künstlers oder unserer Gesellschaft abweisen, oder ob die Pläne mit Fehlern behaftet sind, die eine Ausführung der Prinzipien unmöglich machen.

(Translation by Else Ritter and Paul Ritter.)